DISCARD
Otterbein University
CourtrightMemorialLib

Courtright Memorial Library
Otterbein College
Westerville, Ohio 43081

# Micelles, Microemulsions, *and* Monolayers

# Micelles, Microemulsions, and Monolayers

## Science and Technology

edited by

## Dinesh O. Shah
*Charles A. Stokes Professor of Chemical Engineering and Anesthesiology*
*Director, Center for Surface Science and Engineering*
*University of Florida*
*Gainesville, Florida*

MARCEL DEKKER, INC.　　NEW YORK • BASEL • HONG KONG

**Library of Congress Cataloging-in-Publication Data**

Micelles, microemulsions, and monolayers : science and technology /
  edited by Dinesh O. Shah.
     p.   cm.
   Includes bibliographical references (p.   –   ) and index.
   ISBN 0-8247-9990-9 (acid-free paper)
     1. Surface active agents--Congresses.   2. Micelles--Congresses.
   3. Emulsions--Congresses.   I. Shah, D. O. (Dinesh Ochhavlal).
   TP994.M54   1998
   541.3'3--dc21                                                            97-52810
                                                                                CIP

The publisher offers discounts on this book when ordered in bulk quantities. For more information, write to Special Sales/Professional Marketing at the address below.

This book is printed on acid-free paper.

**Copyright © 1998 by MARCEL DEKKER, INC. All Rights Reserved.**

Neither this book nor any part may be reproduced or transmitted in any form or by any means, electronic or mechanical, including photocopying, microfilming, and recording, or by any information storage and retrieval system, without permission in writing from the publisher.

MARCEL DEKKER, INC.
270 Madison Avenue, New York, New York  10016
*http://www.dekker.com*

Current printing (last digit):
10  9  8  7  6  5  4  3  2  1

**PRINTED IN THE UNITED STATES OF AMERICA**

# In Memoriam

Vinod Kumar Pillai
(1968–1997)

Dr. Vinod Kumar Pillai was born in Tiruvalla, India on January 13, 1968. He attended St. Xavier School in Delhi, India. In 1990 he received his B.S. degree in chemical engineering at the Indian Institute of Technology, Delhi, India. Subsequently, he came to the University of Florida as a graduate student and worked in Professor Dinesh O. Shah's research group from 1990 to 1995. During the course of his Ph.D. thesis research, he worked on nanoparticles of magnetic materials and superconductors, and investigated the role of interfacial rigidity and chain-length compatibility on the reaction kinetics in microemulsions. After receiving his Ph.D. in 1995, Vinod completed one year of postdoctoral work at McMaster University in Canada. He then returned to Professor Shah's laboratory as a postdoctoral associate in 1996.

He was the coordinator of the International Symposium on Micelles, Microemulsions, and Monolayers: Quarter Century Progress and New Horizons hosted by the Center for Surface Science and Engineering at the University of Florida August 28–30, 1995. Unfortunately, he passed away on March 10, 1997. This book is dedicated to the memory and invaluable scientific contributions of Dr. Vinod Pillai, who was a dear colleague, ideal student, enthusiastic postdoctoral associate, and really good friend to everyone who knew him.

# Preface

This book is based on the plenary and invited lectures presented at the International Symposium on Micelles, Microemulsions, and Monolayers: Quarter Century Progress and New Horizons, which was hosted by the Center for Surface Science and Engineering at the University of Florida August 28-30, 1995. The papers presented at this symposium reviewed the progress achieved in the past 25 years (1970-1995) and described the new horizons for future research on micelles, microemulsions, and monolayers, as well as their technological significance.

The chapters cover the new insights into the dynamic properties of interfaces and the kinetics of micellization in relation to foaming, emulsification, wettability, detergency, and solubilization phenomena. Several chapters describe the key advances in the use of microemulsions in technological areas such as preparation of nanoparticles, enzymatic reactions, and pharmaceutical and drug delivery systems, as well as polymerization of microemulsions to produce microlatexes. The symposium also covered the major developments in self-assembled and insoluble monolayers at the gas/liquid and liquid/liquid interfaces as well as Langmuir-Blodgett films.

The Center for Surface Science and Engineering is most grateful to the National Science Foundation, the University of Florida, and the following companies for their generous support of this symposium: ALCOA, Inc.; Alcon Laboratories; ICI Surfactants; Kimberly-Clark Corporation; Kraft General Foods; LG Chem Ltd.; LG Cosmetics & Household Products; Lonza, Inc.; Milliken Research Corporation; Noven Pharmaceuticals; Procter & Gamble Company; and Rhône-Poulenc, Inc.

Each professor is grateful to his mentor and his students for their contributions to his academic endeavors. The influence of my late professor, Jack H. Schulman, on everything that I have accomplished during the past 25 years cannot be overlooked. My involvement in the research on micelles, microemulsions, and monolayers, and many other areas of surface and colloid science, are directly linked to what I inherited as a student in Professor Schulman's laboratory.

I would like to thank all participants of the symposium, including plenary and invited speakers, as well as session chairs and cochairs for their invaluable contributions. I am also most grateful to my colleagues at the University of Florida and the members of the local arrangement committee for making all the detailed arrangements for this symposium. I am also grateful to our students, postdoctoral associates, research scholars, colleagues, and secretarial staff for enthusiastically completing numerous tasks related to this symposium.

I would like to conclude with a poem written by Nobel Laureate Indian poet Ravindranath Tagore (1861–1941).

> When the Sun went down in the evening,
> The Sun asked everyone on earth,
> "Who will take up my responsibility of providing light?"
> No one replied.
> However, a candle in a log cabin said,
> "My Lord, I can not light up the whole universe as you do,
> But I'll do my best to light up this small room."

This poem gives a profound message to everyone, including scientists. As researchers, all that we can do is to contribute to a small area of science or human knowledge. In our case, this is the very small area of surface and colloid science.

During the past quarter century, our group at the University of Florida has conducted extensive studies on micelles, microemulsions, and monolayers, and their technological applications. We hope that the scientific content of this volume is enthusiastically received by present and future researchers in the surface and colloid science community all over the world.

*Dinesh O. Shah*

# Contents

*In Memoriam*   iii

*Preface*   v

*Contributors*   xi

1. Micelles, Microemulsions, and Monolayers: Quarter Century Progress at the University of Florida  
   *Dinesh O. Shah*   1

### MICELLES

2. Quarter Century Progress and New Horizons in Micelles  
   *Fredric M. Menger*   53

3. Recent Advances in Aqueous Surfactant Phase Science: Coexistence Relationships of the "Sponge" Phase  
   *Robert G. Laughlin*   73

4. Surfactant Self-Assembly Structures at Interfaces, in Polymer Solutions, and in Bulk: Micellar Size and Connectivity  
   *Björn Lindman, Fredrik Tiberg, Lennart Piculell, Ulf Olsson, Paschalis Alexandridis, and Håkan Wennerström*   101

5. An Overview of Depletion and Surface-Induced Structural Forces in Thin Micellar Films  127
   *D. T. Wasan, A. D. Nikolov, and X. L. Chu*

6. Structure and Design of Abnormally Long Thread-Like Micelles and Their Relation to Vesicles and Liquid Crystals  145
   *C. Manohar*

## MICROEMULSIONS

7. Quarter Century Progress and New Horizons in Microemulsions  161
   *Krister Holmberg*

8. New Developments in Polymerization in Bicontinuous Microemulsions  193
   *Françoise Candau and Jean-Yves Anquetil*

9. Application of Microemulsions in Soil Remediation  215
   *K. Mönig, W. Clemens, F.-H. Haegel, and M. J. Schwuger*

10. The Importance of Surfactant Hydrophobe Structure in Microemulsion Formation  233
    *Robert S. Schechter and William H. Wade*

11. The Role of Surfactants in Enhanced Oil Recovery  249
    *Hisham A. Nasr-El-Din and Kevin C. Taylor*

12. Nanosized Particles: Self-Assemblies, Control of Size and Shape  289
    *M. P. Pileni, I. Lisiecki, L. Motte, C. Petit, J. Tanori, and N. Moumen*

13. Microemulsions as Tunable Media for Diverse Applications  305
    *Syed Qutubuddin*

14. Double Emulsions Stabilized by Macromolecular Surfactants  333
    *Nissim Garti and Abraham Aserin*

15. Interparticle Forces from SANS Measurements of Frozen Dispersions  363
    *David C. Steytler, Brian H. Robinson, Julian Eastoe, and Isabel MacDonald*

## MONOLAYERS

16. Phase Transitions in Lipid Monolayers at the Air–Water Interface  387
    *Harden M. McConnell*

17. Surfactant Monolayers in Relation to Foam Breaking by Particles  395
    *R. Aveyard, B. P. Binks, and P. D. I. Fletcher*

18. Dynamic Adsorption and Tension of Spread or Adsorbed Monolayers at the Air–Water Interface  417
    *Elias I. Franses, Chien-Hsiang Chang, Judy B. Chung, Karen Coltharp McGinnis, Sun Young Park, and Dong June Ahn*

19. What X-rays Tell Us About Langmuir Monolayers  437
    *Pulak Dutta*

20. Inorganic Extended Solid Langmuir–Blodgett Films  447
    *Daniel R. Talham, Houston Byrd, and Candace T. Seip*

21. Formation and Control of Unit Aggregates of Squaraines and Related Compounds in Langmuir–Blodgett Films  463
    *Huijuan Chen, Kangning Liang, Hussein Samha, Xuedong Song, David G. Whitten, Kock-Yee Law, and Thomas L. Penner*

22. Protein and Molecular Assembly Monolayer and Multilayer Film Studies with Scanning Probe Microscopy  481
    *J. A. DeRose and R. M. Leblanc*

23. Self-Assembled Amphiphiles on Surface: "Surface Rheology"  501
    *Yihan Liu and D. Fennell Evans*

24. Suprabiomolecular Architectures at Functionalized Surfaces  509
    *Wolfgang Knoll, Masahiko Hara, and Kaoru Tamada*

25. Langmuir–Blodgett Films of Condensation Polymers  519
    *Masa-aki Kakimoto and Toshio Imai*

26. Langmuir–Blodgett Film as Alignment Layers for Nematic Liquid Crystal Displays 543
A. Albarici, J. A. Mann, Jr., J. B. Lando, J. Chen, H. Vithana, D. Johnson, and Masa-aki Kakimoto

27. Dye-Sensitized Solar Cells Based on Redox Active Monolayers Adsorbed on Nanocrystalline Oxide Semiconductor Films 579
K. Kalyanasundaram and M. Grätzel

Index 605

# Contributors

**Dong June Ahn, Ph.D.** Professor, Department of Chemical Engineering, Korea University, Seoul, Korea

**A. Albarici, M.S.** Department of Macromolecular Science, Case Western Reserve University, Cleveland, Ohio

**Paschalis Alexandridis, Ph.D.**\* Assistant Professor, Department of Physical Chemistry 1, Center for Chemistry and Chemical Engineering, University of Lund, Lund, Sweden

**Jean-Yves Anquetil**† Charles Sadron Institute, Strasbourg, France

**Abraham Aserin, Ph.D.** Casali Institute of Applied Chemistry, The Hebrew University of Jerusalem, Jerusalem, Israel

**R. Aveyard, Ph.D.** Professor, School of Chemistry, University of Hull, Hull, England

**B. P. Binks, Ph.D.** Doctor, School of Chemistry, University of Hull, Hull, England

---

*Current affiliations:*
\*Assistant Professor, Department of Chemical Engineering, State University of New York–Buffalo, Buffalo, New York.
†Engineer, Clariant Chemie SA, Trosly Breuil, France.

**Houston Byrd, Ph.D.*** Department of Chemistry, University of Florida, Gainesville, Florida

**Françoise Candau, Ph.D.** Director of Research CNRS, Charles Sadron Institute, Strasbourg, France

**Chien-Hsiang Chang, Ph.D.**† Research Assistant, School of Chemical Engineering, Purdue University, West Lafayette, Indiana

**Huijuan Chen, Ph.D.**‡ University of Rochester, Rochester, New York

**J. Chen, Ph.D.** Department of Physics and Liquid Crystal Institute, Kent State University, Kent, Ohio

**X. L. Chu** Department of Chemical Engineering, Illinois Institute of Technology, Chicago, Illinois

**Judy B. Chung, Ph.D.**§ School of Chemical Engineering, Purdue University, West Lafayette, Indiana

**W. Clemens, Ph.D.**¶ Scientist, Research Center Jülich, Institute of Applied Physical Chemistry, Jülich, Germany

**J. A. DeRose, Ph.D.** Postdoctoral Fellow, Department of Chemistry, University of Miami, Coral Gables, Florida

**Pulak Dutta, Ph.D.** Professor, Department of Physics and Astronomy, Northwestern University, Evanston, Illinois

**Julian Eastoe, Ph.D., B.Sc.** School of Chemistry, University of Bristol, Bristol, England

**D. Fennell Evans** Department of Chemical Engineering and Materials Science, University of Minnesota, Minneapolis, Minnesota

---

*Current affiliations:*
*Assistant Professor, Department of Biology and Chemistry, University of Montevallo, Montevallo, Alabama.
†Associate Professor, Department of Chemical Engineering, National Cheng Kung University, Tainan, Taiwan.
‡Research Scientist, Research Laboratories, Eastman Kodak Company, Rochester, New York.
§Clorox Co., Pleasanton, California.
¶Professionelle Software GmbH, Bedburg, Germany.

# Contributors

**P. D. I. Fletcher, Ph.D.**   Professor, School of Chemistry, University of Hull, Hull, England

**Elias I. Franses, Ph.D.**   Professor, School of Chemical Engineering, Purdue University, West Lafayette, Indiana

**Nissim Garti, Ph.D.**   Professor, Department of Chemistry, and Head, School of Applied Science, Casali Institute of Applied Chemistry, The Hebrew University of Jerusalem, Jerusalem, Israel

**M. Grätzel, Ph.D.**   Professor, Institute of Physical Chemistry, Swiss Federal Institute of Technology, Lausanne, Switzerland

**F.-H. Haegel, Ph.D.**   Scientist, Research Center Jülich, Institute of Applied Physical Chemistry, Jülich, Germany

**Masahiko Hara, Dr.Eng.**   Deputy Head, Frontier Research Program, The Institute of Physical and Chemical Research (RIKEN), Wako, Saitama, Japan

**Krister Holmberg, Ph.D.**   Professor and Director, Institute for Surface Chemistry, Stockholm, Sweden

**Toshio Imai, Ph.D.**   Emeritus Professor, Department of Organic and Polymeric Materials, Tokyo Institute of Technology, Tokyo, Japan

**D. Johnson, Ph.D.**   Professor, Department of Physics and Liquid Crystal Institute, Kent State University, Kent, Ohio

**K. Kalyanasundaram, Ph.D.**   Institute of Physical Chemistry, Swiss Federal Institute of Technology, Lausanne, Switzerland

**Masa-aki Kakimoto, Ph.D.**   Professor, Department of Organic and Polymeric Materials, Tokyo Institute of Technology, Tokyo, Japan

**Wolfgang Knoll, Prof.Dr.**   The Institute of Physical and Chemical Research (RIKEN), Wako, Saitama, Japan, and Director, Department of Material Science, Max-Planck-Institute for Polymer Research, Mainz, Germany

**J. B. Lando, Ph.D.**   Professor, Department of Macromolecular Science, Case Western Reserve University, Cleveland, Ohio

**Robert G. Laughlin, Ph.D.** Research Fellow, Research and Development Department, Miami Valley Laboratories, The Procter & Gamble Company, Cincinnati, Ohio

**Kock-Yee Law, Ph.D.** Technical Manager, Wilson Center for Research and Technology, Xerox Corporation, Webster, New York

**R. M. Leblanc, Ph.D.** Professor and Chairman, Department of Chemistry, University of Miami, Coral Gables, Florida

**Kangning Liang, Ph.D.** Senior Research Associate, Department of Chemistry, University of Rochester, Rochester, New York

**Björn Lindman, Ph.D.** Professor, Department of Physical Chemistry 1, Center for Chemistry and Chemical Engineering, Lund University, Lund, Sweden

**I. Lisiecki, Ph.D.** Researcher, SRSI Laboratory, Pierre and Marie Curie University, Paris, and DRECAM-SCM, C.E.N. Saclay, Gif sur Yvette, France

**Yihan Liu** Department of Chemical Engineering and Materials Science, University of Minnesota, Minneapolis, Minnesota

**Isabel MacDonald, Ph.D.** Exxon Chemicals Ltd., Abingdon, England

**J. A. Mann, Jr., Ph.D.** Professor, Department of Chemical Engineering, Case Western Reserve University, Cleveland, Ohio

**C. Manohar, Ph.D.** Head, Interfacial Chemistry Section, Chemistry Division, Bhabha Atomic Research Center, Bombay, India

**Harden M. McConnell, Ph.D.** Professor, Department of Chemistry, Stanford University, Stanford, California

**Karen Coltharp McGinnis, M.S.*** Researcher, School of Chemical Engineering, Purdue University, West Lafayette, Indiana

**Fredric M. Menger, Ph.D.** Candler Professor, Department of Chemistry, Emory University, Atlanta, Georgia

---
*Current affiliation:*
*Process Engineer, Bayer Corporation, Addyston, Ohio.

**Contributors** xv

**K. Mönig, M.Sc., Ph.D.** Research Center Jülich, Institute of Applied Physical Chemistry, Jülich, Germany

**L. Motte, Ph.D.** Assistant Professor, SRSI Laboratory, Pierre and Marie Curie University, Paris, and DRECAM-SCM, C.E.N. Saclay, Gif sur Yvette, France

**N. Moumen, Ph.D.** Researcher, SRSI Laboratory, Pierre and Marie Curie University, Paris, France

**Hisham A. Nasr-El-Din, Ph.D.** Science Specialist, Laboratory Research and Development Center, Saudi Aramco, Dhahran, Saudi Arabia

**A. D. Nikolov, Ph.D.** Research Professor, Department of Chemical Engineering, Illinois Institute of Technology, Chicago, Illinois

**Ulf Olsson, Ph.D.** Associate Professor, Department of Physical Chemistry 1, Center for Chemistry and Chemical Engineering, Lund University, Lund, Sweden

**Sun Young Park, Ph.D.*** School of Chemical Engineering, Purdue University, West Lafayette, Indiana

**Thomas L. Penner, Ph.D.** Senior Research Associate, Research and Advanced Development Laboratories, Eastman Kodak Company, Rochester, New York

**C. Petit, Ph.D.** Researcher, SRSI Laboratory, Pierre and Marie Curie University, Paris, and DRECAM-SCM, C.E.N. Saclay, Gif sur Yvette, France

**Lennart Piculell, Ph.D.** Associate Professor, Department of Physical Chemistry 1, Center for Chemistry and Chemical Engineering, Lund University, Lund, Sweden

**M. P. Pileni** Professor, SRSI Laboratory, Pierre and Marie Curie University, Paris, and DRECAM-SCM, C.E.N. Saclay, Gif sur Yvette, France

---

*Current affiliation:*
*International Paper, Tuxedo, New York.

**Syed Qutubuddin, Ph.D.** Professor, Department of Chemical Engineering and Macromolecular Science, Case Western Reserve University, Cleveland, Ohio

**Brian H. Robinson, Ph.D.** School of Chemical Sciences, University of East Anglia, Norwich, England

**Hussein Samha, Ph.D.** Senior Research Associate, Department of Chemistry, University of Rochester, Rochester, New York

**Robert S. Schechter, Ph.D.** Professor Emeritis, Departments of Chemical Engineering and Petroleum and Geosystem Engineering, The University of Texas–Austin, Austin, Texas

**M. J. Schwuger** Professor, Research Center Jülich, Institute of Applied Physical Chemistry, Jülich, Germany

**Candace T. Seip, Ph.D.** Department of Chemistry, University of Florida, Gainesville, Florida

**Dinesh O. Shah, Ph.D.** Charles A. Stokes Professor of Chemical Engineering and Anesthesiology, and Director, Center for Surface Science and Engineering, University of Florida, Gainesville, Florida

**Xuedong Song, Ph.D.**[*] Research Associate, Department of Chemistry, University of Rochester, Rochester, New York

**David C. Steytler, Ph.D.** Doctor, School of Chemical Sciences, University of East Anglia, Norwich, England

**Daniel R. Talham, Ph.D.** Associate Professor, Department of Chemistry, University of Florida, Gainesville, Florida

**Kaoru Tamada, Ph.D.**[†] Senior Research Scientist, Frontier Research Program, The Institute of Physical and Chemical Research (RIKEN), Wako, Saitama, Japan

**J. Tanori, M.D.** SRSI Laboratory, Pierre and Marie Curie University, Paris, and DRECAM-SCM, C.E.N. Saclay, Gif sur Yvette, France

---

*Current affiliations:*
[*]Postdoctoral Fellow, National Laboratory, Los Alamos, New Mexico.
[†]Senior Research Scientist, Department of Molecular Engineering, National Institute of Materials and Chemical Research, Tsukuba, Ibaraki, Japan.

# Contributors

**Kevin C. Taylor, M.Sc.** Laboratory Scientist, Laboratory Research and Development Center, Saudi Aramco, Dhahran, Saudi Arabia

**Fredrik Tiberg, Ph.D.*** Assistant Professor, Department of Physical Chemistry 1, Center for Chemistry and Chemical Engineering, Lund University, Lund, Sweden

**H. Vithana, Ph.D.** Department of Physics and Liquid Crystal Institute, Kent State University, Kent, Ohio

**William H. Wade, Ph.D.** Professor, Department of Chemistry and Biochemistry, The University of Texas–Austin, Austin, Texas

**D. T. Wasan, B.S., Ph.D.** Motorola Professor, Department of Chemical Engineering, Illinois Institute of Technology, Chicago, Illinois

**Håkan Wennerström, Ph.D.** Professor, Department of Physical Chemistry 1, Center for Chemistry and Chemical Engineering, Lund University, Lund, Sweden

**David G. Whitten, Ph.D.**† Professor and Chairman, Department of Chemistry, University of Rochester, Rochester, New York

---

*Current affiliations:*
*Manager, Forest Products Section, Institute for Surface Chemistry, Stockholm, Sweden.
†Technical Staff Member, National Laboratory, Los Alamos, New Mexico.

# 1
# Micelles, Microemulsions, and Monolayers: Quarter Century Progress at the University of Florida

**Dinesh O. Shah**
*University of Florida, Gainesville, Florida*

## I. MICELLES

It is well recognized that a surfactant solution has three components: surfactant monomers in the aqueous solution, micellar aggregates, and monomers adsorbed as a film at the interface. The surfactant is in dynamic equilibrium among all these components. From various theoretical considerations, as well as experimental results, it can be said that micelles are dynamic structures whose stability is in the range of milliseconds to seconds. Thus, in an aqueous surfactant solution, micelles break and reform at a fairly rapid rate, in the range of milliseconds (1-3). Figure 1 shows the two characteristic relaxation times, $\tau_1$ and $\tau_2$, associated with micellar solutions. The shorter relaxation time, $\tau_1$, generally of the order of microseconds, relates to the exchange of surfactant monomers between the bulk solution and micelles, whereas the longer relaxation time, $\tau_2$, generally of the order of milliseconds to seconds, relates to the dissolution of a micelle after several molecular exchanges (4,5). It has been proposed that the lifetime of a micelle can be given by $n\tau_2$, where $n$ is the aggregation number of a micelle (6). Thus, relaxation time $\tau_2$ is proportional to the lifetime of the micelle. A large value of $\tau_2$ represents a high stability of the micellar structure.

Figure 2 shows the relaxation time $\tau_2$ of micelles of sodium dodecyl sulfate (SDS) as a function of SDS concentration (4,7,8). It is evident that the maximum relaxation time of micelles is observed at 200 mM SDS

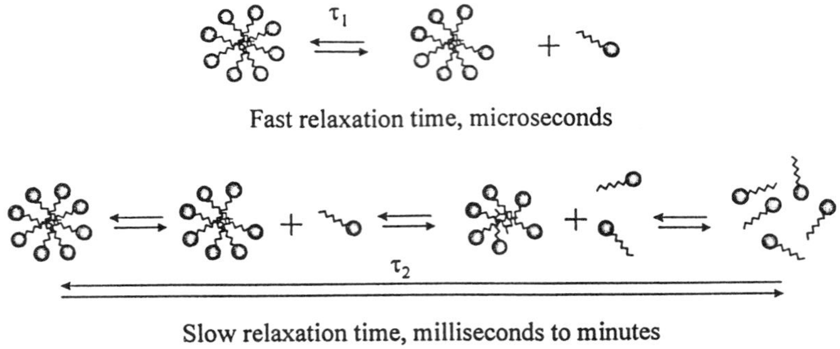

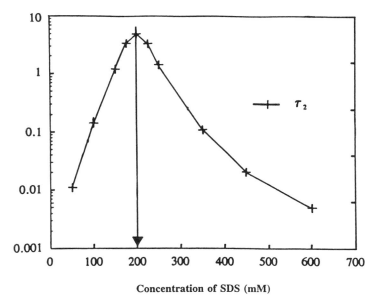

**Figure 1** Two relaxation times of micelles, $\tau_1$ and $\tau_2$, and related molecular processes.

**Figure 2** The relaxation time $\tau_2$ of SDS micelles as a function of SDS concentration.

concentration. This implies that SDS micelles are most stable at this concentration. For several years researchers at the Center for Surface Science and Engineering (CSSE) tried to correlate the relaxation time, $\tau_2$, with various equilibrium properties such as surface tension, surface viscosity, etc., but no correlation could be found. However, a strong correlation of $\tau_2$ with various dynamic processes such as foaming ability, wetting time of textile, bubble volume, emulsion droplet size, and solubilization of benzene in micellar solution was found (9).

Figure 3 schematically shows the effect of $\tau_2$ or micellar stability on the rate of adsorption of surfactant monomers at the air bubble surface. When a micellar solution has a large $\tau_2$ or high stability, it will provide less monomer to the newly created interface. Hence, the dynamic surface tension will be higher for micellar solutions having a large $\tau_2$. This will cause the volume of the bubble at the tip of the needle to be large. It is expected that a higher dynamic surface tension requires greater buoyancy force to break off the bubble from the tip of the needle. Thus, one expects a large bubble volume when the micelles are relatively stable, as represented by a large value of $\tau_2$ (10).

The foaming ability of a surfactant solution involves the sparging of air into a surfactant solution to create a large new interfacial area. The newly created surface consisting of foam lamellae has to be stabilized by surfactant molecules. When initial monomers present in a solution diffuse to the surface, micelles have to break down in order to provide the additional monomers. Thus, one expects that when micelles are relatively stable, less surfactant will diffuse to the interface and, hence, cause the film to

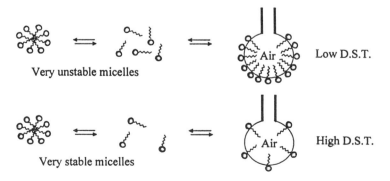

**Figure 3** The effect of $\tau_2$ or micellar stability on the flux of monomers and on the rate of adsorption of surfactant at the air bubble/water surface. D.S.T., dynamic surface tension.

break due to a lack of sufficient surfactant molecules at the interface (11). Figure 4 shows the effect of SDS concentration on $\tau_2$ and on foaming ability. It is evident that maximum $\tau_2$ corresponds to minimum foaming ability of SDS solution.

Figure 5 schematically illustrates the effect of $\tau_2$ on wetting time of fabric. For fabric wetting to be effective, the surfactant monomers have to adsorb on the hydrophobic sites inherent on the fabric surface and convert them to hydrophilic sites. Therefore, the wetting effectiveness depends upon the amount of surfactant monomers available for adsorption. A relatively stable micelle provides less surfactant monomers and, hence, causes poor wetting and a longer wetting time. For experimental evaluation of this phenomenon, 1 in.$^2$ pieces of various fabric were gently deposited on the surface of SDS solution. The amount of time before the fabric started sinking into the solution was measured. During this time, the solution penetrated into the fabric and replaced the air that had been trapped in the fabric (12). Figure 6 shows the effect of SDS concentration, and hence micellar relaxation time $\tau_2$, on the wetting time of cotton. It is evident that the maximum stability of micelles at 200 mM SDS concentration corresponds to the longest wetting time for cotton. Figure 7 shows the effect of

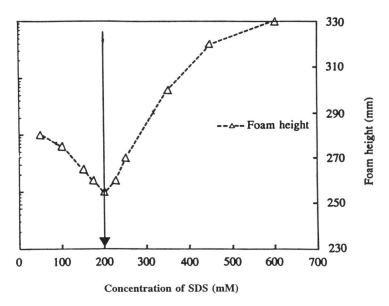

**Figure 4** The effect of $\tau_2$ on foamability of SDS solutions at various concentrations.

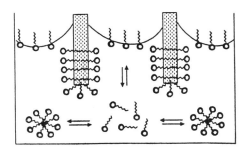

More stable micelles → Less monomer flux →
 Slower wetting process

**Figure 5** Schematic illustration of the effect of $\tau_2$ and hence micellar stability on wetting of fabric. Center for Surface Science and Engineering, University of Florida.

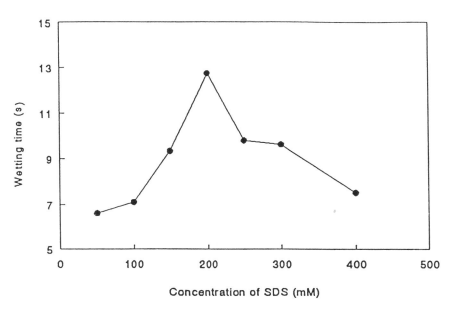

**Figure 6** The effect of SDS concentration on $\tau_2$ and on wetting time of cotton.

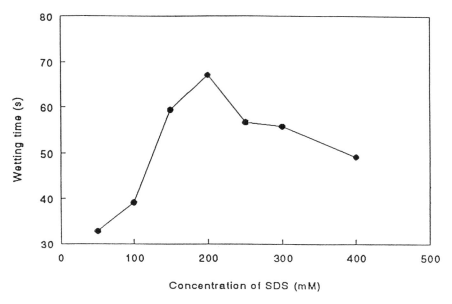

**Figure 7** The effect of SDS concentration on $\tau_2$ and on wetting time of rayon.

SDS concentration, and hence micellar relaxation time $\tau_2$, on the wetting time of rayon. Similar studies were carried out with several different fabrics (13). It is evident that, in each fabric, the maximum wetting time occurs at 200 mM SDS concentration, although the magnitude of the maximum wetting time varies.

Figure 8 schematically illustrates the effect of micellar stability on the flux of surfactant monomers from the aqueous phase to the oil–water interface during emulsification. The magnitude of this monomer flux determines the dynamic interfacial tension at the oil–water interface (14,15). In the emulsification process, the work done is dissipated in increasing the interfacial area of the oil–water interface. The work done, $W$, can be expressed by $W = \Upsilon \Delta A$, where $\Upsilon$ is the dynamic interfacial tension and $\Delta A$ is the change in the interfacial area upon emulsification (16). This implies that if the dynamic interfacial tension is high, the change in interfacial area will be low, causing the droplet size to be larger for a given amount of oil dispersed in water. Thus, a larger droplet size is expected at 200 mM SDS concentration, due to high micellar stability, in the hexadecane–water–SDS system. Figure 9 indicates the droplet size of various emulsions produced in the hexadecane–water–SDS system. It is evident that at 200 mM SDS concentration the average droplet size was a maximum.

# Quarter Century Progress at University of Florida 7

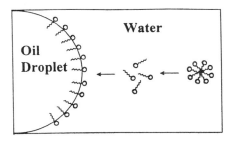

More stable micelles → Less monomer flux →
Higher interfacial tension → Larger droplet size

**Figure 8** Schematic illustration of the effect of micellar stability on the flux of surfactant monomers and on dynamic interfacial tension at the oil-water interface. Center for Surface Science and Engineering, University of Florida.

Figures 10 and 11 summarize the effects of SDS concentration on the phenomena already discussed, as well as on other related phenomena. Figure 10 shows typical phenomena in liquid-gas systems, while Figure 11 shows typical phenomena in liquid-liquid and solid-liquid systems. It is evident that each of these phenomena exhibit a maximum or minimum at

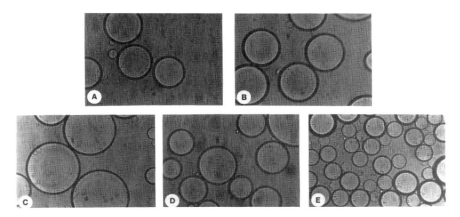

**Figure 9** The maximum size of droplets in hexadecane-water-SDS emulsions at (A) 50 mM, (B) 100 mM, (C) 200 mM, (D) 300 mM, and (E) 400 mM SDS concentrations.

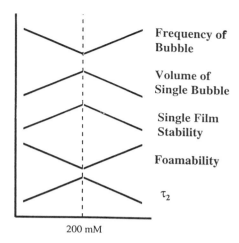

**Figure 10** Various liquid–gas system phenomena exhibiting minima or maxima at 200 mM SDS concentration.

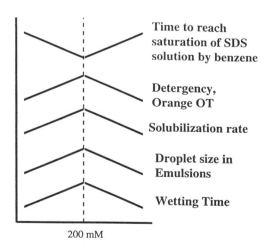

**Figure 11** Various liquid–liquid and solid–liquid system phenomena exhibiting minima or maxima at 200 mM SDS concentration.

## Quarter Century Progress at University of Florida

200 mM SDS concentration depending upon the molecular process involved. Thus, the "take-home message" emerging out of our extensive studies of the past decade is that the micellar stability can be the rate-controlling factor in the performance of various technological processes such as foaming, emulsification, wetting, bubbling, and solubilization (17).

Figure 12 shows the proposed explanation for the effect of SDS con-

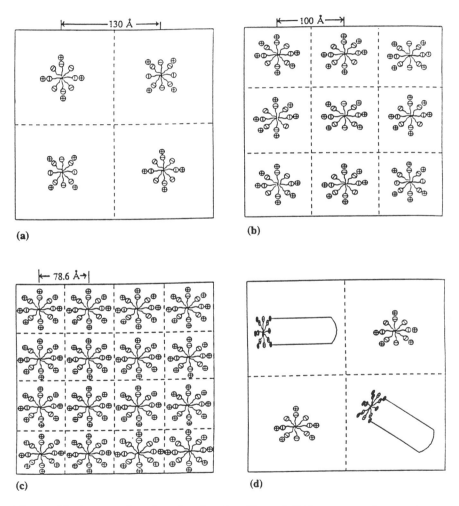

**Figure 12** Proposed explanation for the effect of SDS concentration on the intermicellar distance. (a) at 50 mM; (b) at 100 mM; (c) at 200 mM; (d) above 200 mM.

centration on the micellar stability. As the SDS concentration increases, the number of micelles increases, and thus, the intermicellar distance decreases. By knowing the aggregation number of the micelles, the number of micelles present in the solution can be calculated. The solution can then be divided into cubes such that each cube contains one micelle. From this, the distance between the center of the cubes can be taken as the intermicellar distance. Figure 12 shows the intermicellar distance at 50 mM, 100 mM, 200 mM, and 300 mM SDS concentrations (17-21). It is evident that the intermicellar distance is approximately 1 micellar diameter at 200 mM SDS concentration. A tremendous coulombic repulsion between the micelles is expected at such a short distance. One possible explanation for this phenomenon is that there is a rapid uptake of sodium ions as counterions on the micellar surface at this concentration, making the micelles more stable. Thus, the coulombic repulsion between micelles with the concomitant uptake of sodium ions allows the stabilization of micelles at this short intermicellar distance and, hence, maximum $\tau_2$ at 200 mM concentration. It should be mentioned that Per Ekwall proposed first, second, and third critical micelle concentrations (cmc) for sodium octanoate solutions (20). At the second cmc, he showed a sudden uptake or binding of sodium ions to the micellar surface and he proposed that, at the second cmc, there is a tight packing of surfactant molecules in the micelle. Thus, our 200 mM concentration could be equivalent to the second cmc as proposed by Ekwall. The phenomenon of a surfactant exhibiting a maximum $\tau_2$ appears to be a general phenomenon, and perhaps other anionic or cationic surfactants may form tightly packed micelles at their own characteristic concentrations. As with the first cmc, this critical concentration may also depend upon the physical and chemical conditions such as temperature, pressure, pH, salt concentration, and other parameters in addition to the molecular structure of the surfactant (22,23). Work is currently in progress at the CSSE on identifying a similar critical concentration for nonionic surfactants as well as mixed surfactant systems.

In addition to this work, it has been shown that, upon incorporation of short chain alcohols such as hexanol into the SDS micelles, the maximum $\tau_2$ occurs at a lower concentration of SDS (4,24,25). Thus, it appears that in a mixed surfactant system, one can produce the most stable micelle at a lower surfactant concentration upon incorporation of an appropriate cosurfactant (26). We also investigated the effect of long chain alcohols on the micellar stability, which showed similar results to short chain alcohols for all but dodecanol, which showed a significant increase in micellar stability over micelles containing only SDS due to the chain length compatibility effect (27). Figure 13 shows the effect of coulombic attraction between oppositely charged polar groups as well as the chain length compatibility effect on the $\tau_2$ of SDS plus alkyltrimethylammonium bromide solutions

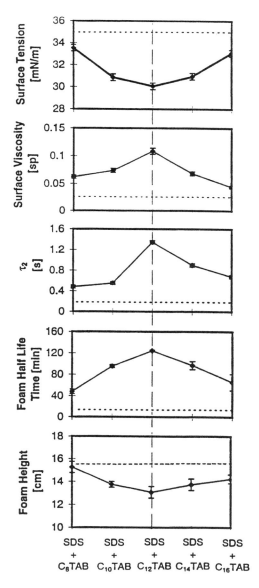

**Figure 13** Effect of coulombic attraction between polar groups of surfactants and chain length compatibility on $\tau_2$ or micellar stability and other interfacial properties of mixed surfactant solutions. The dotted line represents the specific property of 100 mM pure SDS solution.

(26). It shows that surface tension, surface viscosity, micellar stability, foaming ability, and foam stability are all influenced by the coulombic interaction as well as the chain length compatibility effect (26,28). It should be noted that the ratio of SDS to alkyltrimethylammonium bromide was 95 : 5 in this mixed surfactant system. However, even at this low molar ratio, the oppositely charged surfactant dramatically changed the molecular packing of the resulting micelles as well as the surfactant film adsorbed at the interface.

In view of the previously mentioned discussion, it is evident that micellar stability is of considerable importance to technological processes such as foaming, emulsification, wetting, solubilization, and detergency, since a finely tuned detergent formulation can significantly improve the cleaning efficiency as well as reduce the washing time in the laundry machine, resulting in significant energy savings at a national and global level. Micellar stability is thus a critical issue in any application in which the surfactants are present as micelles, and the subsequent monomer flux is utilized in the application.

## II. MICROEMULSIONS

As early as 1943, my teacher, Professor J. H. Schulman, published reports on transparent emulsions (29). From various experimental observations and intuitive reasoning he concluded that such transparent systems were microemulsions. Figure 14 illustrates the transparent nature of a microemulsion in comparison to a macroemulsion (30). He also proposed the concept of a transient negative interfacial tension to induce the spontaneous emulsification in such systems (31).

Considerable studies have been carried out on microemulsions during the past quarter century, during which time it has been well recognized that there are three types of microemulsions: lower-phase, middle-phase, and upper-phase microemulsions. The lower-phase microemulsion can remain in equilibrium with excess oil in the system, the upper-phase microemulsion can remain in equilibrium with excess water, whereas the middle-phase microemulsion can remain in equilibrium with both excess oil and water. As a result, the lower-phase microemulsion has been considered to be an oil-in-water microemulsion, and the upper-phase microemulsion has been considered to be a water-in-oil microemulsion, whereas the middle-phase microemulsion has been the subject of much research, and has been proposed to be composed of bicontinuous or phase-separated swollen micelles from the aqueous phase (32–43). Figure 15 shows the upper-, middle-, and

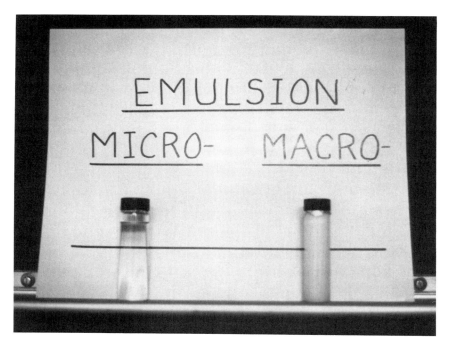

**Figure 14** Illustration of the transparent nature of a microemulsion, compared to a macroemulsion.

lower-phase microemulsions, as represented by the darker liquid in each tube (44–47).

The formation of lower-, middle-, and upper-phase microemulsions is related to the migration of surfactant from lower phase to middle phase to upper phase. Figure 16 illustrates that migration of the surfactant from the aqueous phase to the middle phase to the oil phase can be induced by changing a number of parameters, including adding salts (e.g., NaCl) to the system, decreasing the oil chain length, increasing the surfactant molecular weight, adding a cosurfactant, decreasing the temperature, and others (47,48). However, it has been reported that in oil–water–nonionic surfactant systems, the nonionic surfactant moves from lower phase to middle phase to upper phase as the temperature is increased (41,42), so each system must be carefully analyzed in order to determine the effects of certain parameters. Microemulsions exhibit ultralow interfacial tension with excess oil or water phases. Therefore, the middle-phase microemulsion is of spe-

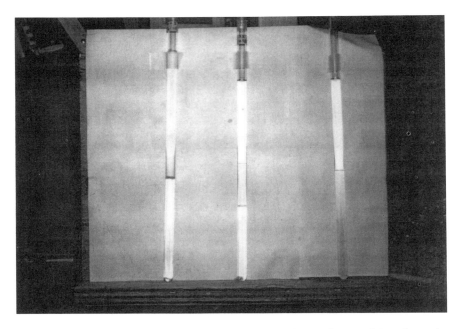

**Figure 15** Samples of lower-, middle-, and upper-phase microemulsions in equilibrium with excess oil, excess water and oil, or excess water.

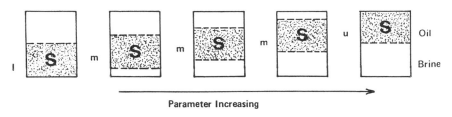

The transition l ⟶ m ⟶ u occurs by:

1. Increasing Salinity
2. Decreasing oil chain length
3. Increasing alcohol concentration ($C_4$, $C_5$, $C_6$)
4. Decreasing temperature
5. Increasing total surfactant concentration
6. Increasing brine/oil ratio
7. Increasing surfactant solution/oil ratio
8. **Increasing molecular weight of surfactant**

**Figure 16** The transition from lower- to middle- to upper-phase microemulsions can be brought about by the addition of salts or by varying other parameters.

# Quarter Century Progress at University of Florida

cial importance to the process of oil displacement from petroleum reservoirs.

## A. Microemulsions in Enhanced Oil Recovery

Figure 17 shows schematically a view of a petroleum reservoir, as well as the process of water or chemical flooding by an inverted five-spot pattern (32). Several thousand feet below the ground, oil is found in both tightly packed sand or sandstone in the presence of water as well as the natural gas. During the primary and secondary recovery processes (water injection method), about 35% of the available oil is recovered. Hence, about 65% of

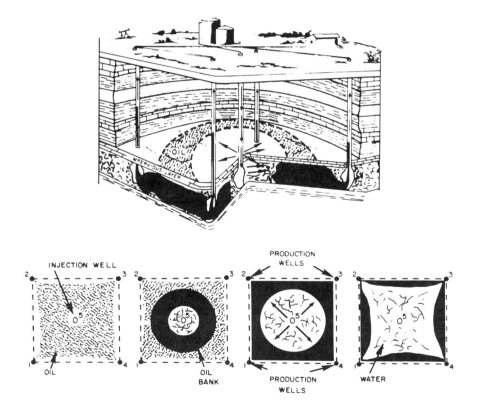

**Figure 17** Schematic view of a petroleum reservoir and the process of water or chemical flooding (five-spot pattern).

**Figure 18** Ultralow interfacial tension is required for the mobilization of oil ganglia in a porous medium.

original oil-in-place is left in the petroleum reservoir. This oil remains trapped because of the high interfacial tension (about 20–25 mN/m) between the crude oil and reservoir brine. It is known that if the interfacial tension can be reduced to around $10^{-3}$ mN/m, a substantial fraction of the residual oil in the porous media in which it is trapped can be mobilized. Figure 18 shows that ultralow interfacial tension is required for the mobilization of oil ganglia in porous media (49–51). Once mobilized by an ultralow interfacial tension, the oil ganglia must coalesce to form a continuous oil bank (Figure 19). The coalescence of oil droplets has been shown to be enhanced by a very low interfacial viscosity in the system (51). The incorporation of these two critical factors into a suitable surfactant system

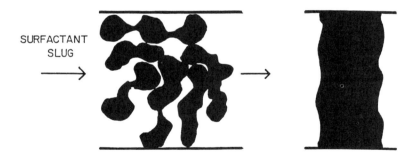

**Figure 19** Low interfacial viscosity is required for the rapid coalescence of oil ganglia and for the formation of an oilbank.

for oil recovery was crucial in developing the surfactant-polymer flooding process for enhanced oil recovery from petroleum reservoirs, which is shown schematically in Figure 20. Conceptually, a surfactant formulation is injected into the porous media in the petroleum reservoir so that, upon mixing with the reservoir brine and oil, the surfactant produces the middle-phase microemulsion in situ. This middle-phase microemulsion in equilibrium with excess oil and excess brine propagates through the petroleum reservoir. The design of the process is such that the oil bank maintains ultralow interfacial tension with reservoir brine until it arrives at the production wells.

One parameter that has been discovered to be crucially important in the successful implementation of the surfactant-polymer flooding process is the salinity of the aqueous phase. As discussed previously, the addition of salt to the microemulsion system induces the change from lower- to middle- to upper-phase microemulsion (Figure 21) (32). It was found that at a particular salt concentration, referred to as the optimal salinity, important processes become optimum for the oil recovery process. At the optimal salinity equal amounts of oil and brine are solubilized by the middle-phase microemulsion (52). As shown in Figures 22 and 23, the solubilized oil and brine are of equal volume, and the interfacial tension of the system is minimum at the optimal salinity (46,53).

In order to determine if the pressure drop across porous media, such as the sandpack or sandstone, was influenced by the salt concentration, an

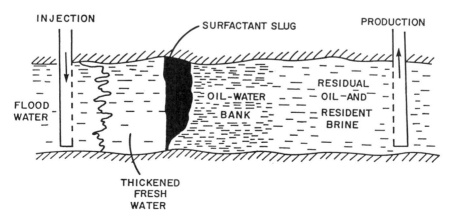

**Figure 20** Schematic representation of surfactant-polymer flooding process for enhanced oil recovery.

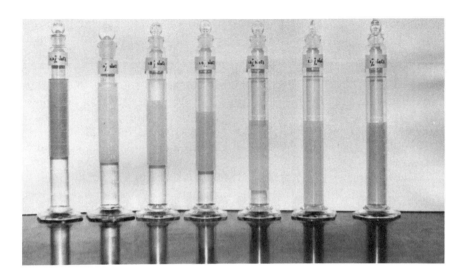

**Figure 21** The effect of salt (NaCl) concentration on transition from lower- to middle- to upper-phase microemulsions.

apparatus was developed to measure the pressure difference upon injection of a multiphase oil-brine-surfactant system into a porous medium, as shown schematically in Figure 24 (54). Figure 25 shows the pressure difference ($\Delta P$) when an oil-brine-surfactant-cosurfactant system was injected into a porous medium at various salt concentrations and flow rates. It is evident that the $\Delta P$ is minimum when the multiphase system is at the optimal salinity, where the middle-phase microemulsion exists in equilibrium with excess oil and brine phases (54). Thus, it appears that the major resistance for the flow of a multiphase system through porous media arises out of the high interfacial tension. Thus, when the interfacial tension is lowered to about $10^{-3}$ mN/m, the $\Delta P$ decreases strikingly. The high interfacial tension creates a substantial contribution to the $\Delta P$ for flow through the porous media.

Figure 26 summarizes various phenomena that are important in the surfactant-polymer flooding process as a function of salt concentration (32,53,55-57). It is evident that all of these parameters exhibit a maximum or a minimum at the optimal salinity. Thus, it appears that all of these processes are interrelated for the oil displacement in porous media by the surfactant-polymer flooding process. It also appears that the optimal salinity value is a critical parameter for oil recovery in this process.

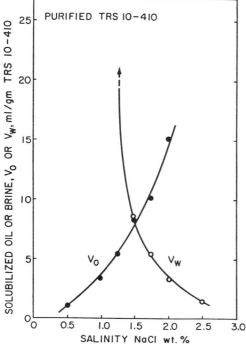

**Figure 22** Equal volumes of oil and water are solubilized in the middle-phase microemulsions at the optimal salinity.

## B. Formation of Nanoparticles Using Microemulsions

One very interesting use of microemulsions that has been investigated at the CSSE over the past decade is in the production of nanoparticles. Figure 27 schematically illustrates the formation of nanoparticles using water-in-oil microemulsions. For this process, two identical water-in-oil microemulsions are produced, with the only difference between the microemulsions being the nature of the aqueous phase, into which the two water-soluble reactants, A and B, are dissolved separately. Upon mixing the two almost identical microemulsions, the water droplets collide and coalesce, allowing the mixing of the reactants to produce the precipitate AB. Ultimately, these droplets will again disintegrate into two aqueous droplets, one containing

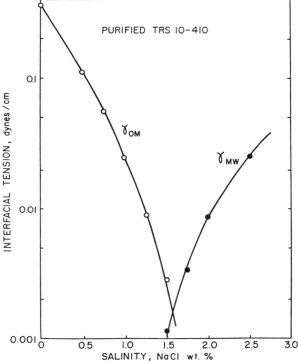

**Figure 23** The controlling interfacial tension is minimum at the optimal salinity.

the nanoparticle AB and the other containing just the aqueous phase (58–60). Thus, a precipitation reaction can be carried out in the aqueous cores of water-in-oil microemulsions, using the dispersed water droplets as nanoreactors. The size of the particles formed are mechanically limited by the size of the water droplets, which is determined by the structure and concentration of components. Since microemulsions are thermodynamically stable, the equilibrium size of droplets, which are fairly monodispersed, is determined by thermodynamic considerations. In this way, monodisperse particles in the 2–10 nm diameter range can be produced. Since homogeneity of particle size has inherently been a problem with other conventional methods of nanoparticle production, this method is an improvement over

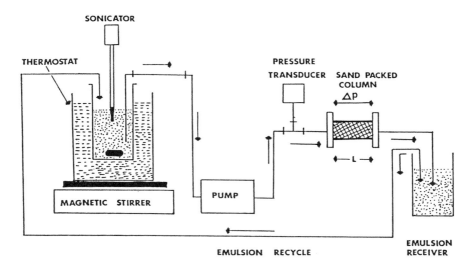

**Figure 24** Schematic representation of the injection of a multiphase system into a porous medium and the measurement of pressure difference, $\Delta P$, across it.

previous methods. Figure 28 is a transmission electron microscope (TEM) micrograph of silver chloride (AgCl) nanoparticles produced at the CSSE by the microemulsion method, illustrating the ability to produce ultrafine, monodisperse particles using this method (61–63).

Superconducting nanoparticles have also been produced at the CSSE using the microemulsion method. Table 1 shows the composition of two microemulsions used for synthesizing nanoparticles of YBCO superconductors (64–66). In this case, water-soluble salts of yttrium, barium, and copper were dissolved in the aqueous cores of one microemulsion and ammonium oxalate was dissolved in the aqueous cores of the other microemulsion. Upon mixing of the two microemulsions, precursor nanoparticles of oxalates of these cations were formed. The nanoparticles physically agglomerated due to interaction of the adsorbed surfactant upon centrifugation of the samples after the particles were precipitated. The particles were then washed with chloroform, methanol, or acetone to remove the surfactants and oil.

These nanopowders were then calcined at the appropriate temperature to convert the oxalate precursor into oxides of these materials. These oxides then were compressed into a pellet and sintered at 860°C for 24 h. The pellet was then cooled and the critical temperature was determined at which it exhibited zero electrical resistance. Table 2 summarizes various properties

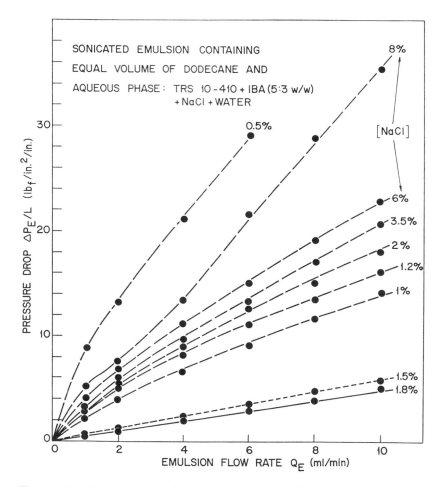

**Figure 25** The minimum $\Delta P$ across the porous medium is observed near the optimal salinity.

of superconducting samples prepared by microemulsions as compared to those prepared by the conventional aqueous coprecipitation method (64). It is evident that the critical temperature did not show any change for the two samples, but the fraction of the ideal Meissner shielding was strikingly different for the two samples prepared by different methods. It is the Meissner effect that is related to the levitational effect of the superconducting pellet on a magnetic field. Thus, it appears that the leakage of magnetic flux from the conventionally prepared sample was greater than that from

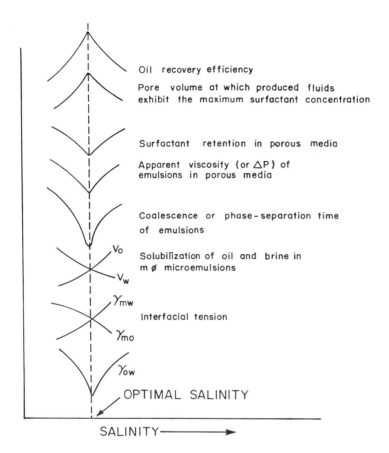

**Figure 26** Various phenomena occurring at the optimal salinity in the surfactant-polymer flooding process for enhanced oil recovery.

the sample produced by microemulsion. Figure 29 shows the scanning electron microscope (SEM) of the sintered pellets produced by the two different methods. It is evident that the pellets prepared from the nanoparticles produced by the microemulsion method showed 30–100 times larger grain size, less porosity, and higher density as compared to the samples prepared by conventional precipitation of aqueous solutions of these salts. A possible explanation for these effects is that nanoparticles, because of their extremely small size, can lose their identity very quickly and convert into atomic flux very rapidly, and this flux goes to the site of the growing grains and supports the grain growth. Therefore, samples prepared from

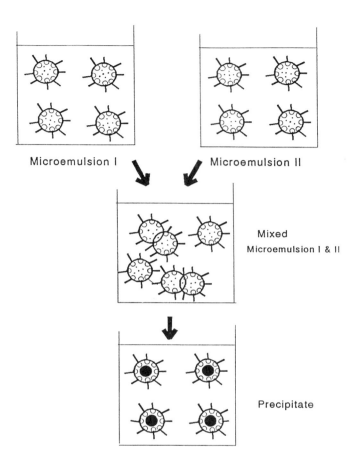

**Figure 27** Formation of nanoparticles using microemulsions (water-in-oil) as nanoreactors. The water droplets continually collide, coalesce, and break up upon mixing of two microemulsions containing reactants.

nanoparticles exhibit larger grain size and lower porosity (67). Thus, it appears that the nanoparticles or nanopowders may be useful to produce high-density ceramics.

Table 3 shows various nanoparticles synthesized by our research group over the past decade. Along with the superconducting materials, we have synthesized $\gamma$-$Fe_2O_3$ (68), $BaFe_{12}O_{19}$ (69-70), YBCO (64-66), BiSSCO (67), ZnO (71,72), $TiO_2$ (73), as well as polystyrene microlatexes (74), by using microemulsion processing. Since the pioneering work of Boutonnet et

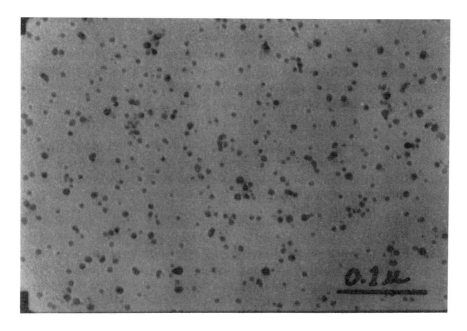

**Figure 28** TEM micrograph of silver chloride (AgCl) nanoparticles produced by reaction in microemulsions. Magnification = 200,000 X. Picture was taken 4 h after reaction. Composition: $(AgNO_3)$ = (NaCl) = 0.4 M in water, (AOT) = 0.14 M in $n$-dodecane, $N_{water}/N_{AOT}$ = 8.

**Table 1** Composition of Two Microemulsions for Synthesizing Nanoparticles of YBCO Superconductors

|  | Surfactant phase | Hydrocarbon phase | Aqueous phase |
| --- | --- | --- | --- |
| Microemulsion I | CTAB + 1-butanol | $n$-octane | (Y,Ba,Cu)nitrate solution, total metal conc. = 0.3 N |
| Microemulsion II | CTAB + 1-butanol | $n$-octane | Ammonium oxalate solution, 0.45 N |
| Weight fraction (for both I and II) | 29.25% | 59.42% | 11.33% |

**Table 2** Comparison of Physical Properties of YBCO Superconductor Prepared by Conventional Aqueous Phase Precipitation Versus Preparation by Microemulsion Reaction

| Physical property | Microemulsion reaction | Conventional reaction |
|---|---|---|
| ESD of (Y,Ba,Cu) oxalate precipitate | 47.4 nm | 380.6 nm |
| ESD of YBCO powder | 274.8 nm | 626.6 nm |
| Calcined at | 820°C, 2 h | 860°C, 6 h |
| Grain size of YBCO pellet (from SEM) | 15–50 $\mu$m | 0.5–2.0 $\mu$m |
| Sintered at | 925°C, 12 h | 925°C, 12 h |
| Density of sintered pellet (fraction of single crystal density) | 98 ($\pm$3)% | 90 ($\pm$2)% |
| Magnetic susceptibility of field-cooled sintered pellet (demagnetization corrected) | $-10.95 \times 10^{-3}$ (emu cm$^{-3}$) | $-3.06 \times 10^{-3}$ (emu cm$^{-3}$) |
| Magnetic susceptibility of zero-field-cooled sintered pellet (demagnetization corrected) | $-72.05 \times 10^{-3}$ (emu cm$^{-3}$) | $-11.43 \times 10^{-3}$ (emu cm$^{-3}$) |
| Fraction of ideal Meissner signal ($-1/4\pi$) | 90.5% | 14.4% |
| Superconducting $T_c$ | 93 K | 91 K |

al. on the formation of nanoparticles of heavy metals by the microemulsion method (75), we have added to the understanding of the reaction kinetics in microemulsions by investigating the interfacial rigidity of the microemulsion droplet. We have introduced the concept of chain length compatibility effect observed in the reaction kinetics of microemulsions, in which the interfacial rigidity is maximized by matching the chain length of the surfactant, oil, and cosurfactant alcohol, causing a minimum reaction rate (76). All of these contributions have led to a greater understanding of the method of nanoparticle production by microemulsions.

## III. MONOLAYERS

During the past quarter century, considerable studies have been carried out on the reactions in monomolecular films of surfactant, or monolayers. Figure 30 shows the surface pressure-area curves for dioleoyl, soybean, egg, and dipalmitoyl lecithins (77). For these four lecithins, the fatty acid composition was determined by gas chromatography. The dioleoyl lecithin has both chains unsaturated, soybean lecithin has polyunsaturated fatty

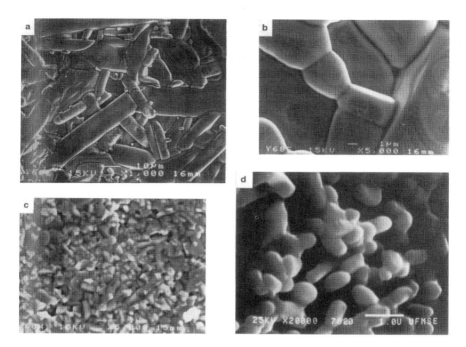

**Figure 29** SEM of superconducting pellets prepared from nanopowders (a, b) using microemulsions and conventionally prepared powders (c, d) (by precipitation of aqueous solutions).

acid chains, egg lecithin has 50% saturated and 50% unsaturated chains, and dipalmitoyl lecithin has both chains fully saturated. It is evident that, at any fixed surface pressure, the area per molecule is in the following order:

Dioleoyl lecithin > soybean lecithin > egg lecithin > dipalmitoyl lecithin

**Table 3** Nanoparticles Synthesized by Professor D. O. Shah's Group

- Silver halides — AgCl and AgBr
- Superconductors — YBCO (123) and BiSCCO (2223)
- Magnetic materials — $\gamma$-$Fe_2O_3$, $BaFe_{12}O_{19}$, and $CoFe_2O_4$
- Varistors — ZnO and ZnO + $Bi_2O_3$ with dopants
- Titanium dioxide ($TiO_2$)
- Polystyrene nanolatexes

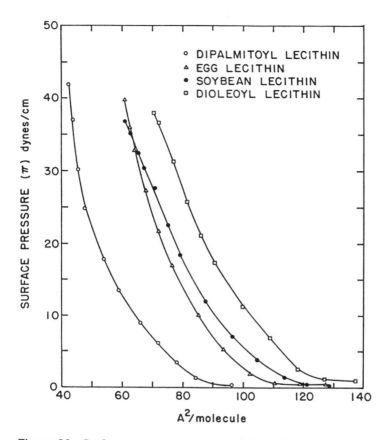

**Figure 30** Surface pressure-area curves of dioleoyl, soybean, egg, and dipalmitoyl lecithin.

It can be assumed that the area per molecule represents the area of a square at the interface. Thus the square root of the area per molecule gives the length of one side of the square, which represents the intermolecular distance. Figure 31 schematically illustrates the area per molecule and intermolecular distance in these four lecithins. The corresponding intermolecular distance was calculated to be 9.5 Å, 8.8 Å, 7.1 Å, and 6.5 Å at a surface pressure of 20 dyn/cm (78). Thus, one can conclude that a change in the saturation of the fatty acid chains produces subangstrom changes in the intermolecular distance in the monolayer.

In addition, it was desired to explore the effects that these small changes in intermolecular distance have on the enzymatic susceptibility of

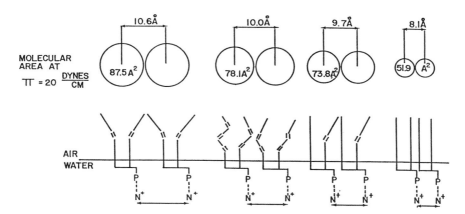

**Figure 31** Schematic representation of the area per molecule and intermolecular distance in dioleoyl, soybean, egg, and dipalmitoyl lecithin monolayers based on the data plotted in Figure 30.

these lecithins to hydrolytic enzymes such as phospholipase A (79-81), a potent hydrolytic enzyme found in cobra venom. Thus, microgram quantities of the enzyme phospholipase A were injected under this monolayer. Figure 32 illustrates the changes in surface pressure upon injection of the enzyme, and Figure 33 illustrates the changes in surface potential upon enzyme injection. The surface potential decreases immediately due to the hydrolysis of lecithin into lysolecithin and free fatty acids. Thus, by measuring the rate of change of surface potential, one can indirectly measure the rate of reaction in the monolayer. It is assumed that both of these quantities are proportional to each other. The kinetics of hydrolysis, as measured by a decrease in surface potential, was studied for each lecithin monolayer as a function of initial surface pressure and is shown in Figure 34 (77,82). It was found that the initial reaction rate increases as the surface pressure increases. Subsequently, as the surface pressure increases further, the reaction rate decreases until reaching a critical surface pressure where no reaction occurs. The critical surface pressure needed to block the hydrolysis of lecithin monolayer increased with the degree of unsaturation of fatty acid chains. Thus, it appears that, as the intermolecular distance increases due to the unsaturated fatty acid chains, a higher surface pressure is required to block the penetration of the active site of the enzyme into the monolayer to cause the hydrolysis. This also led to a suggestion that subangstrom changes in the intermolecular distance in the monolayer are significant for the enzymatic hydrolysis of the monolayers.

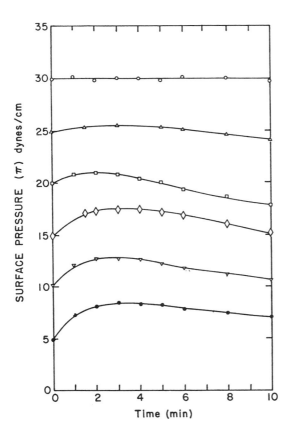

**Figure 32** Change in surface pressure upon injection of cobra venom phospholipase A under the lecithin monolayers at various initial surface pressures.

In addition to hydrolysis reactions, the enzymatic synthesis in monolayers was also studied. In this case, a stearic acid monolayer was formed on an aqueous solution containing glycerol. After compression to a desired surface pressure, a small amount of enzyme lipase was injected under the monolayer. The lipase facilitated the linkage of glycerol with fatty acid and produced monoglycerides, diglycerides, and triglycerides in the monolayer, as illustrated in Figure 35 (83). Since the amount of product that can be synthesized using a monolayer is in microgram quantities, this method is not attractive for large-scale enzymatic synthesis. Therefore, studies of enzymatic reactions in monolayers were extended to studies of enzymatic

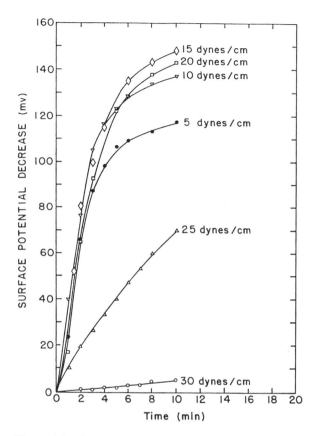

**Figure 33** Change in surface potential upon injection of cobra venom phospholipase A under the lecithin monolayers at various initial surface pressures.

reactions in foam. Foam provides a large interfacial area, and by continuous aeration, one can generate even larger interfacial area. A soap bubble is stabilized by a monolayer on both inside and outside surfaces of the bubble. The glycerol and enzyme can be put into the aqueous phase before producing the foam. Thus, it was shown that almost 88% of free stearic acid can be converted into di- and triglycerides in 2 h by reactions in foams (Figure 36). For surface active substrates (or reactants) and enzymes, the reactions in foams offer a very interesting possibility to produce large-scale synthesis of biochemicals.

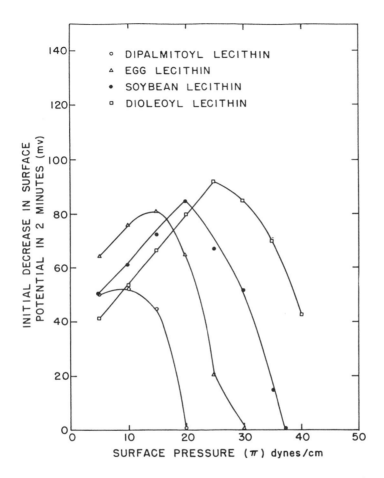

**Figure 34** The hydrolysis rate versus initial surface pressure of various lecithin monolayers.

### A. Phase Transition in Mixed Monolayers: Detection Using the Retardation of Evaporation of Water

Another interesting investigation at the CSSE was focused on the possible existence of phase transitions in mixed monolayers of surfactants. Figure 37 shows the area per molecule for cholesterol and $C_{20}$ alcohol in pure and mixed monolayers (84). The mixed monolayers exhibit the same area per molecule as predicted from the additivity rule. Thus, the conclusion can be

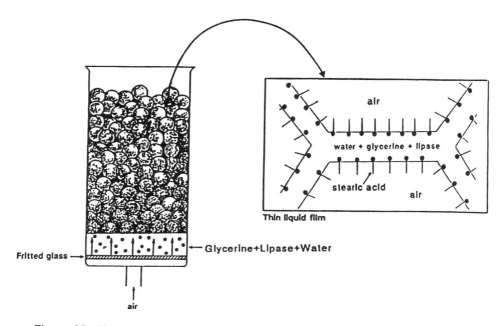

**Figure 35** Enzymatic synthesis of mono-, di-, and triglycerides in foam by lipase.

made that the molecules occupy the same area in the mixed monolayers as they occupy in their pure monolayers.

In contrast, Figure 38 shows the rate of evaporation from pure and mixed monolayers of cholesterol and $C_{20}$ alcohol (84). It is evident that the pure $C_{20}$ alcohol monolayer allows only one third of the water loss due to evaporation as compared to the pure cholesterol monolayer. This is presumably due to the fact that $C_{20}$ alcohol forms monolayers that are in the two-dimensional *solid* state. In contrast, the cholesterol monolayers are in the two-dimensional *liquid* state. However, when cholesterol is incorporated into a $C_{20}$ alcohol monolayer, the cholesterol mole fraction needs only to be about 20% to liquefy the solid monolayers of $C_{20}$ alcohol. The abrupt increase in evaporation rate of water at 20–25% cholesterol illustrates the two-dimensional phase transition in the mixed monolayers from a solid state to a liquid state. After the cholesterol fraction reaches about 25 mol %, the monolayer remains in the two-dimensional liquid state and, hence, there is no further change in the rate of evaporation of water. Thus, one can utilize the evaporation of water through film as a very sensitive probe for the molecular packing in monolayers. The existence of the solid state or liquid state for monolayers can be inferred from these experimental results.

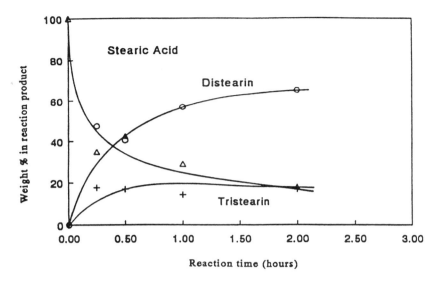

**Figure 36** The decrease in free fatty acids and synthesis of di- and triglycerides by lipase in foam.

Figure 39 shows a schematic explanation for the solid to liquid phase transition in the mixed monolayer of cholesterol and $C_{20}$ alcohol (84). The two components are of significantly different heights. Thus, when more cholesterol molecules are added to the $C_{20}$ alcohol monolayer, a chemical vacancy is produced above the cholesterol molecules that allows the rotation of the nearby $C_{20}$ alkyl chains in that space. The thermal motion of these alkyl chains causes the fluidization of the mixed monolayers.

It has been shown that mixed monolayers of oleic acid and cholesterol exhibit the minimum rate of evaporation at a 1 : 3 molar ratio of oleic acid to cholesterol. This is shown in mixed monolayers of oleic acid and cholesterol in Figure 40 as a function of surface pressure (84). In has further been shown that at a 1 : 3 molar ratio in mixed fatty acid and fatty alcohol monolayers, one observes the maximum foam stability, minimum rate of evaporation, and maximum surface viscosity in these systems (85).

## B. Chain Length Compatibility Effect in Foams and Microemulsions

Figure 41 shows the interrelationship among the molecular properties, surface properties, microscopic characteristics, and fluid displacement efficiency of steam or gas dry processes for enhanced oil recovery involving

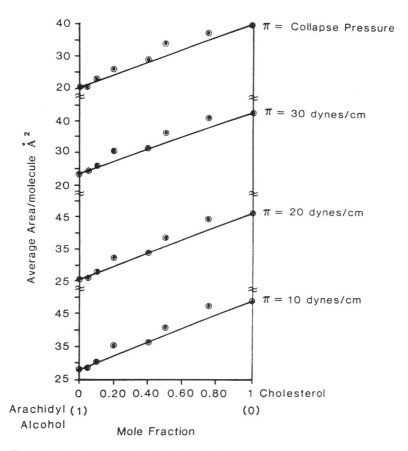

**Figure 37** Area per molecule for cholesterol and $C_{20}$ alcohol in pure and mixed monolayers.

surfactant systems (86–93). Figure 42 shows the effect of chain length compatibility on the bubble size in foams prepared from mixed surfactant systems (86,94). A valid question that can be asked is, Does the small bubble size observed in a test tube occur in porous media having a pore size of 5, 10, or 20 $\mu$m? Therefore, a micromodel was prepared by the etching process and pumped surfactant solution into the micromodel, followed by the subsequent pumping of air (93). By measuring reflected light from the surface of the micromodel, the formation and propagation of bubbles was studied in the micromodel containing a pore size of 20 $\mu$m. Figure 43 shows that even in microscopic spaces the chain length compatibility effect produces

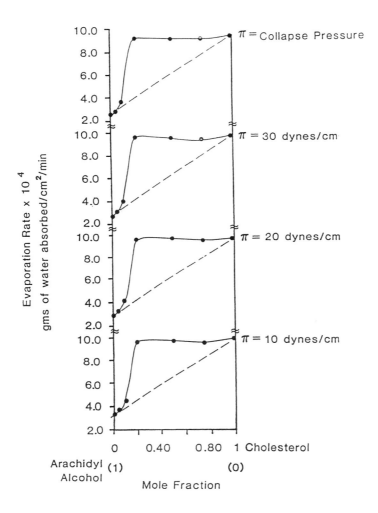

**Figure 38** Rate of evaporation from pure and mixed monolayers of cholesterol and $C_{20}$ alcohol.

the smallest bubble size in a pore (86,93). In summary, Figure 44 shows a correlation of molecular, surface, and microscopic properties with fluid displacement efficiency in porous media for various mixed surfactant systems (86). It is evident that when the chain lengths are equal, minima or maxima are observed with each of these properties.

Figure 45 schematically shows the proposed molecular mechanism for

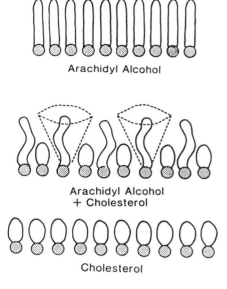

**Figure 39** Schematic explanation for the solid state → liquid state phase transition in mixed monolayers.

the chain length compatibility effect in mixed surfactant systems (86,94). The few atoms that stick out in the longer alkyl chains rotate freely and produce a thermal disturbance that propagates toward the polar headgroup in the surfactant, which causes a small expansion of the area per molecule, and hence increases the intermolecular distance (95).

It is not only the two component systems that were used in monolayers and foams but also the four component systems such as microemulsions that exhibit the chain length compatibility effect. In water-in-oil microemulsions, when the chain length of surfactant equals the combined chain length of alcohol and oil, the maximum solubilization of water is observed, as illustrated in Figure 46 (96). Of course, it should be emphasized that a system has to be appropriately designed to observe this chain length compatibility effect. Excessive amount of one component such as alcohol may eliminate such chain length compatibility effects. Nevertheless, the chain length compatibility relationship can be expressed as $L_S = L_O + L_A$ where $L$ is the molecular length and the subscripts S, O, and A refer to surfactant, oil, and alcohol, respectively. It was further confirmed that the interfacial film composed of surfactant, alcohol, and oil indeed has the highest rigidity when $L_S = L_O + L_A$, by studying the reaction rate in microemulsions using

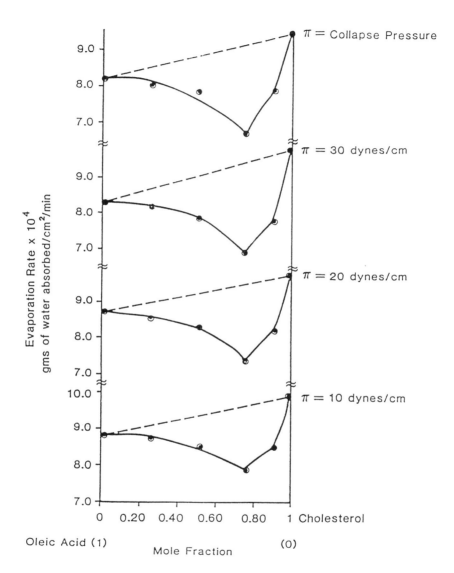

**Figure 40** The minimum rate of evaporation at a 1:3 molar ratio in mixed monolayers of oleic acid and cholesterol.

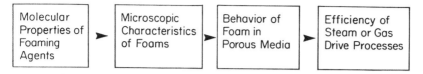

**Figure 41** The interrelationship among the molecular, surface, microscopic, and fluid displacement properties of steam or gas drive processes involving surfactant systems.

the stopped flow method (Figure 47) (76). We showed that the reaction rate was a minimum when octanol was the cosurfactant because the chain length of octanol plus octane is equal to that of cetyltrimethylammonium bromide surfactant.

## C. The Lipid Monolayer and Tear Film

Figure 48 shows various surface phenomena occurring in the tear film in the eye, including drainage, retardation of evaporation, wetting of the corneal surface, flow dynamics, and transfer of oxygen, salts, and drugs (97). A continuous aqueous film has been observed by the fluorescence of a water-soluble dye (sodium fluorescein) dissolved in the tear film. Dry spots are formed in the tear film between the blinks in patients suffering with dry eye

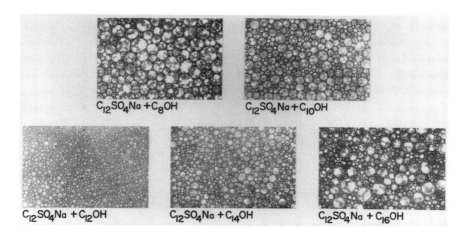

**Figure 42** The effect of chain length compatibility on the bubble size in foams prepared from mixed surfactant systems.

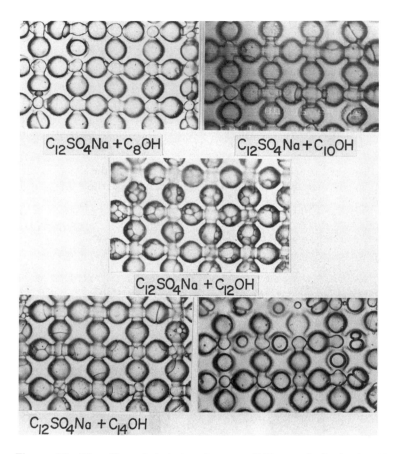

**Figure 43** The effect of chain length compatibility on the in situ bubble size upon injection of air into a micromodel filled with mixed surfactant solutions.

**Figure 44** A correlation of molecular, surface, and microscopic properties with fluid displacement efficiency of mixed surfactant systems.

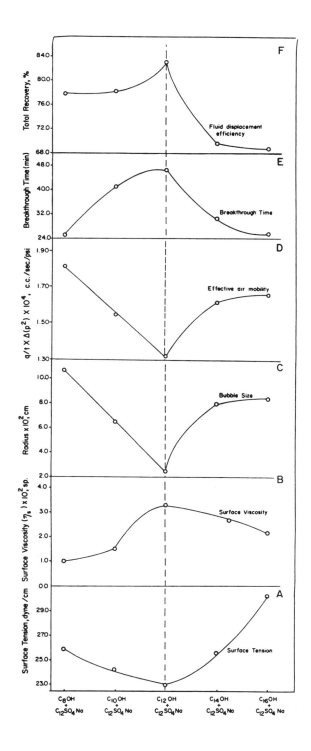

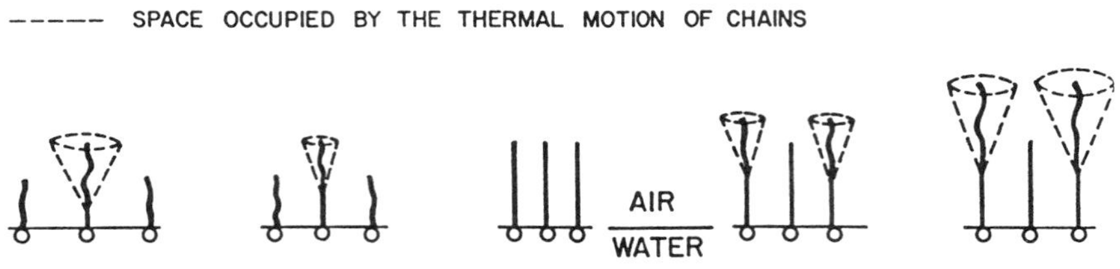

----- SPACE OCCUPIED BY THE THERMAL MOTION OF CHAINS

**Figure 45** A schematic presentation of the molecular mechanism of chain length compatibility effect in mixed surfactant systems.

syndrome. Figure 49 shows the instrumentation to measure the kinetics of thinning of the tear film between the blinks (98–102). For constant concentration of fluorescein in tear film, the fluorescent intensity is proportional to the thickness of the film. Thus, when the thickness of the tear film decreases between the blinks, the fluorescent intensity also decreases. Figure 50 shows the tracing of the fluorescent intensity between the blinks, indicating that the tear film indeed decreases in thickness between the blinks. However, if one adds a surface-active polymer solution in the eye (such as PVA, HPMC, etc.), the fluorescent intensity shows the opposite behavior, namely the thickening of the tear film between the blinks

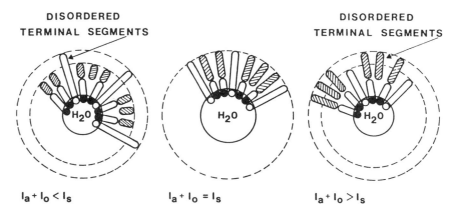

**Figure 46** The maximum solubilization of water-in-oil and the minimum reaction rate occurs in some microemulsions when the length of a surfactant molecule equals the sum of molecular length of oil and alcohol (i.e., $L_S = L_O + L_A$ where $L$ is the molecular length and subscripts S, O, and A refer to surfactant, oil, and alcohol, respectively).

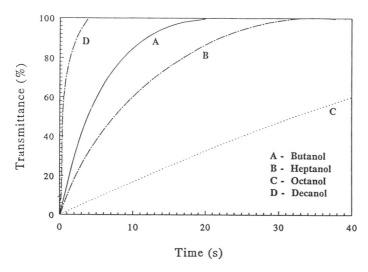

**Figure 47** Reaction kinetics in microemulsions, showing chain length compatibility for core C when $L_S = L_A + L_O$. The system contained $C_{16}$TAB and n-octane.

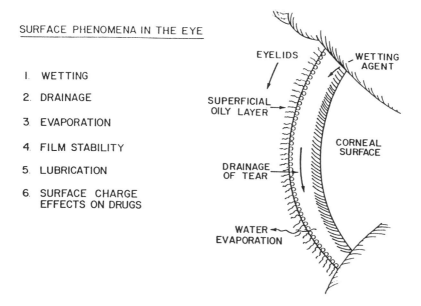

**Figure 48** Various surface phenomena occurring in the tear film on cornea.

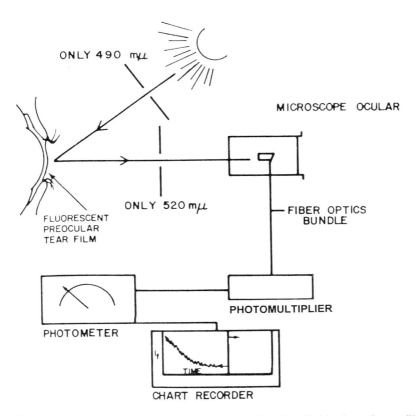

**Figure 49** Instrumentation to measure the kinetics of thinning of tear film between blinks.

(Figure 51) (98). Thus, the surface active film of polymer at the air-tear interface leads to thickening of the tear film.

Figure 52 shows a schematic explanation for the thickening of the tear film by spreading of a surface-active polymer monolayer at the air-tear interface by the Marangoni effect (98). As a result of the blinking process, the monolayer is compressed upon the eye. However, when one opens the eye, the compressed monolayer spreads upward and drags the boundary layer of water with it, and it is this process that thickens the boundary layer after installation of the surface active eye drop. Figure 53 shows a schematic explanation for the water layer dragged by a moving monolayer (98). If this mechanism is indeed operating in the eye, it should be able to be simulated on a glass surface. Figure 54 shows the experimental setup to determine the

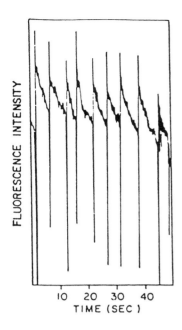

**Figure 50**  Thinning of tear film occurs between blinks.

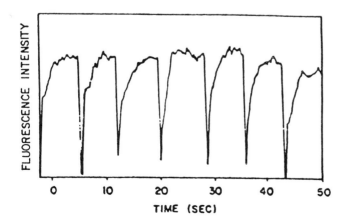

**Figure 51**  Thickening of tear film occurs between blinks when surface-active polymers, such as PVA, HPMC, etc., are added to the tear film.

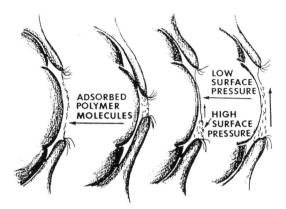

**Figure 52** A schematic explanation for the thickening of tear film by spreading of a surface-active polymer monolayer at the air-tear interface by Marangoni effect.

thickness of a water layer dragged by a moving monolayer on a vertical glass surface (98). As soon as the lower end of a wet glass slide is touched to the surface-active polymer solution, the film moves vertically and drags a substantial amount of water. The fluorescein present in the polymer solution then allows the measurement of the water layer dragged by the spreading film. From the calibration curve, the thickness of the boundary layer can be calculated. Table 4 shows the thickness of the water layer dragged by a moving monolayer of various polymer solutions or tear substitutes (98). It is also interesting that in 1938 my mentor, Professor J. H. Schulman, and Torsten Teorell showed that a moving monolayer of oleic acid can drag a 30-$\mu$m-thick water layer (103). Using a totally different technique involving fluorescence, we have shown that a moving monolayer of lipids or polymers can drag water layers that are 9 to 22 $\mu$m thick. Thus, both these techniques indicate that the hydrodynamic effect associated with a moving monolayer can drag a substantial amount of water. As long as enough surface-active molecules are present at the air-tear interface, the Marangoni effect will operate and will induce the upward movement of a thick layer of water.

Our research on the use of mixed polymer solutions, PVA and HEC, was granted a patent (104) from which subsequently a tear substitute was commercialized (Neo-tears). Thus, monolayers are fascinating systems with extreme simplicity and well-defined parameters. During the past 30 years of research, we have found the studies on monolayers to be rewarding in

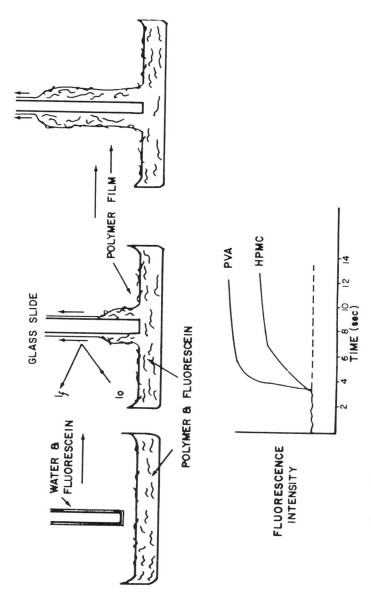

**Figure 53** A schematic explanation for the water layer dragged by a moving monolayer.

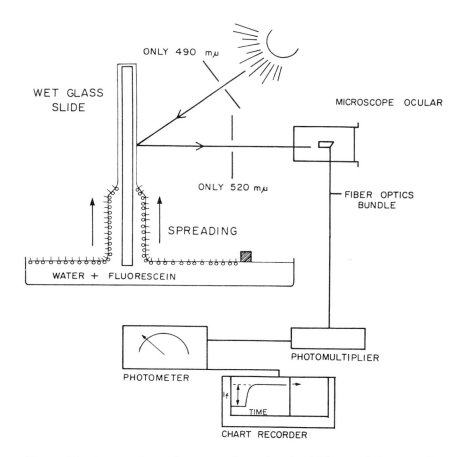

**Figure 54** An experimental setup to determine the thickness of the water layer dragged by a moving monolayer.

understanding the phenomena occurring at the gas–liquid, liquid–liquid, or solid–liquid interfaces in relation to foams, emulsions, lubrication, and wetting processes.

## ACKNOWLEDGMENTS

The author gratefully acknowledges the support provided by Mr. Byron Palla, whose enthusiastic and diligent assistance contributed significantly to the clarity and details of this chapter. Furthermore, the author wants to thank the Engineering Research Center for Particle Science and Technology at the University of Florida (ERC), the National Science Foundation (grant

**Table 4** The Thicknesses of Water Layer Dragged by a Moving Monolayer of Polymers or Various Tear Substitutes from the Dishes with 64 and 0.52 cm² Surface Area

| Polymer | Viscosity (centipoise) | Thickness ($\mu$m) of water layer dragged by polymers | |
|---|---|---|---|
| | | 64 cm² | 0.52 cm²[a] |
| Barnes-Hind wetting solution | –58 | 11 | 22 |
| Adapt | 70 | 16 | 17 |
| Presert | 18 | — | 16 |
| Lacril | 28 | 16 | 14 |
| Visculose | 130 | 30 | 11 |
| PVA | 120 | 19 | 18 |
| PVA | 20 | 19 | 12 |
| HPMC | 120 | 11 | 12 |
| HPMC | 20 | 9 | 9 |
| Monomolecular film of PVA | | 13 | |

[a]Surface area of trough.

#EEC-94-2989), and the ERC Industrial Partners for the financial support provided for these projects. Finally, the author acknowledges the support of the National Science Foundation (grant #CTS-9503607) for the symposium.

## REFERENCES

1. EAG Aniansson, SN Wall. J Phys Chem 78:1024, 1974.
2. EAG Aniansson, SN Wall. J Phys Chem 79:857, 1975.
3. EAG Aniansson, SN Wall, M Almgren, H Hoffmann, I Kielmann, W Ulbricht, R Zana, J Lang, C Tondre. J Phys Chem 80:905, 1976.
4. R Leung, DO Shah. J Colloid Interf Sci 113:484, 1986.
5. N Muller. In: Solution Chemistry of Surfactant. Vol. 1. Mittal, KL, ed. New York, Plenum Press, 1979.
6. WJ Gettings, JE Rassing, Wyn-Jones. In: Micellization, Solubilization and Microemulsions. Vol. 1. Mittal, KL, ed. New York, Plenum Press, 1977.
7. M Kahlweit. J Colloid Interf Sci 90:92, 1982.
8. T Inoue, Y Shibuya, R Shimozawa. J. Colloid Interf Sci 65:370, 1978.
9. SG Oh, DO Shah. J Dispersion Sci Tech 15:297, 1994.
10. SG Oh, SP Klein, DO Shah. AICHE J 38:149, 1992.

11. SG Oh, DO Shah. Langmuir 7:1316, 1991.
12. SG Oh, DO Shah. Langmuir 8:1232, 1992.
13. SG Oh. Ph.D. Dissertation, University of Florida, 1993.
14. VG Levich. Physiochemical Hydrodynamics. Englewood Cliffs, NJ, Prentice-Hall, 1962.
15. SG Oh, M Jobalia, DO Shah. J Colloid Inter Sci 155:511, 1993.
16. P Walstra. In: Encyclopedia of Emulsion Technology. Vol. 1. Becher, P, ed. New York, Dekker, 1983.
17. SG Oh, DO Shah. J Am Oil Chem Soc 70:673, 1993.
18. P Lianos, R Zana. J Colloid Interf Sci 84:100, 1981.
19. JB Hayter, J Penfold. J Chem Soc, Faraday Trans I 77:1851, 1981.
20. P Ekwall. In: Chemistry, Physics and Application of Surface Active Substance. Overbeek, JTG, ed. New York, Gordon and Breach Science Publisher, 1967.
21. F Reiss-Husson, V Luzzati. J Phys Chem 68:3504, 1964.
22. V Luzzati. In: Biological Membranes. Chapman, D, ed. New York, Academic Press, 1968, p 71.
23. L Mandell, K Fontell, P Ekwall. In: Ordered Fluids and Liquid Crystals. Advan Chem Ser 63:89, 1967.
24. SY Shiao, A Patist, ML Free, V Chhabra, PDT Huibers, AS Gregory, S Patel, DO Shah. Review Article, submitted for publication in Colloids and Surfaces.
25. T Inoue, Y Shibuya, R Shimozawa. J Colloid Inter Sci 65:370, 1978.
26. PDT Huibers, DO Shah. submitted for publication.
27. A Patist, V Chhabra, R Pagidipati, R Shah, DO Shah. Langmuir 13:432, 1997.
28. SY Shiao. Ph.D. Dissertation, University of Florida, 1976.
29. TP Hoar, JH Schulman. Nature (London) 152:102, 1943.
30. V Chhabra, ML Free, PK Kang, SE Truesdail, DO Shah. Proceedings of the 4th World Surfactants Congress, Barcelona, Spain, 1996, p 67.
31. JT Davies, EK Rideal. Interfacial Phenomena. New York, Academic Press, 1961, p 364.
32. DO Shah. In: European Symposium on Enhanced Oil Recovery. Bournemouth, England, Elsevier Sequoia S.A., Lausanne, Switzerland, 1981.
33. LE Scriven. Nature 263:123, 1976.
34. LE Scriven. In: Micellization, Solubilization and Microemulsions. Vol. II. Mittal, KL, ed. New York, Plenum Press, 1977, p 877.
35. ML Robbins. SPE 5839, SPE Improved Oil Recovery Symposium, Tulsa, OK, 1976.
36. RN Healy, RL Reed. Soc Pet Eng J 491, Oct, 1974.
37. RN Healy, RL Reed. Soc Pet Eng J 147, June, 1976.
38. CA Miller, R Hwan, WJ Benton, T Fort Jr. J Colloid Inter Sci 61:554, 1977.
39. R Hwan, CA Miller, T Fort Jr. J Colloid Inter Sci 68:221, 1979.
40. C Ramachandran, S Vijayan, DO Shah. J Phys Chem 84:1561, 1980.
41. K Shinoda. J Colloid Inter Sci 24:4, 1967.
42. K Shinoda, H Saito. J Colloid Inter Sci 26:70, 1968.

43. S Friberg, I Lapczynska, G Gilberg. J Colloid Inter Sci 56:19, 1976.
44. JC Noronha. Ph.D. Dissertation, University of Florida, 1980.
45. SI Chou. Ph.D. Dissertation, University of Florida, 1980.
46. WC Hsieh. Ph.D. Dissertation, University of Florida, 1977.
47. KS Chan. Ph.D. Dissertation, University of Florida, 1978.
48. WH Wade, E Vasquez, JL Salager, M El-Emory, C Koukounis, RS Schechter. In: Solution Chemistry of Surfactants. Vol. II. Mittal, KL, ed. New York, Plenum Press, 1979, p 801.
49. WR Foster. J Pet Tech 25:205, 1973.
50. VK Bansal, DO Shah. In: Micellization, Solubilization, and Microemulsions. Vol. I. Mittal, KL, ed. New York, Plenum Press, 1977.
51. CH Pasquarelli, DT Wasan. In: Surface Phenomena in Enhanced Oil Recovery. Shah, DO, ed. Proceedings of the Symposium on Enhanced Oil Recovery. New York, Plenum Press, 1979, p 237.
52. RL Reed, RN Healy. In: Improved Oil Recovery by Surfactant and Polymer Flooding. Shah, DO, Schechter, RS, eds. New York, Academic Press, 1977, p 383.
53. WC Hsieh, DO Shah. SPE 6594, International Symposium on Oilfield and Geothermal Chemistry, LaJolla, CA, 1977.
54. S Vijayan, C Ramachandran, H Doshi, DO Shah. In: Surface Phenomena in Enhanced Oil Recovery. Shah, DO, ed. Proceedings of the Symposium on Enhanced Oil Recovery. New York, Plenum Press, 1981, p 327.
55. MY Chiang. Ph.D. Dissertation, University of Florida, 1978.
56. SJ Satter. SPE 6843, 52$^{nd}$ Annual Fall Conference and Exhibition of SPE-AIME, Denver, CO, 1977.
57. MC Puerto, WW Gale. SPE 5814, SPE Improved Oil Recovery Symposium, Tulsa, OK, 1976.
58. PDI Fletcher, AM Howe, BH Robinson. J Chem Soc Faraday Trans. I 83:985, 1987.
59. HE Eicke, JCW Shepherd, A Steinemann. J Colloid Inter Sci 56:168, 1976.
60. Minero, E Pramauro, E Pelizzetti. Colloids Surfaces 35:237, 1989.
61. CH Chew, LM Gan, DO Shah. J Disp Sci Tech 11:593, 1990.
62. MJ Hou, DO Shah. In: Interfacial Phenomena in Biotechnology and Materials Processing. Moudgil, BM, Chander, S, eds. Amsterdam, Elsevier, 1988, p 443.
63. Dvolaitzky, R Ober, C Taupin, R Anthore, X Auvray, C Petipas, C Williams. J Disp Sci 4:29, 1983.
64. P Ayyub, AN Maitra, DO Shah. Physica C 168:571, 1990.
65. P Kumar, V Pillai, SR Bates, DO Shah. Mat Lett 16:68, 1993.
66. V Pillai, P Kumar, MJ Hou, P Ayyub, DO Shah. Adv Coll Inter Sci 55:241, 1995.
67. P Kumar, V Pillai, DO Shah. Appl Phys Lett 62:765, 1993.
68. V Chhabra, M Lal, AN Maitra, P Ayyub. Colloid Polymer Sci 273:939, 1995.
69. V Pillai, P Kumar, DO Shah. J Magn Mag Mater 116:L299, 1992.
70. V Pillai, P Kumar, MS Multani, DO Shah. Colloids Surfaces A Pysiochem Eng Aspects 80:69, 1993.

71. S Hingorani, V Pillai, P Kumar, MS Multani, DO Shah. Mat Res Bull 28: 1303, 1993.
72. M Singhal, V Chhabra, P Kang, DO Shah. Mat Res Bull 32:239, 1997.
73. V Chhabra, V Pillai, BK Mishra, A Morrone, DO Shah. Langmuir 11:3307, 1995.
74. A Jayakrishnan, DO Shah. J Polymer Sci Polymer Lett Ed 22:31, 1984.
75. M Boutonnet, J Kizling, P Stenius, G Maire. Colloids Surfaces 5:209, 1982.
76. V Pillai, DO Shah. Dynamic Properties of Interfaces and Association Structure. 1996, p 156.
77. DO Shah, JH Schulman. J Colloid Inter Sci 25:107, 1967.
78. DO Shah, JH Schulman. J Lipid Res 6:311, 1965.
79. K Saito, DJ Hanahan. Biochemistry 1:521, 1962.
80. J Murata, M Satake, T Suzuki. J Biochem (Tokyo) 53:431, 1963.
81. M Kates. In: Lipid Metabolism. Bloch, K, ed. New York, Wiley, 1960, p 185.
82. JH Moore, JH Williams. Biochim Biophys Acta 84:41, 1964.
83. CP Singh, DO Shah. Colloids Surfaces A Physico Eng Asp 77:219, 1993.
84. YK Rao, DO Shah. J Coll Inter Sci 137:25, 1990.
85. DO Shah. J Coll Inter Sci 37:744, 1971.
86. MK Sharma, DO Shah. SPE 10612, SPE Sixth International Symposium on Oilfield and Geothermal Chemistry, Dallas, TX, 1982.
87. AN Fried. R15866, USBM, 1961.
88. GG Bernard, LW Holm. Soc Pet Eng J 267, Sept., 1964.
89. JR Deming, M.S. Thesis, Pennsylvania State University, 1964.
90. LW Holm. Soc Pet Eng J 359, Dec, 1968.
91. MK Sharma, DO Shah, WE Brigham. Am Inst Chem Eng J 31:222, 1985.
92. MK Sharma, DO Shah, WE Brigham. I & EC Fundamentals 23:213, 1984.
93. MK Sharma, DO Shah, WE Brigham. Soc Pet Eng Reservoir Eng 253, 1986.
94. DO Shah, SY Shiao. Adv Chem Ser 144:153, 1975.
95. SY Shiao, A Patist, ML Free, V Chhabra, PDT Huibers, A Gregory, S Patel, DO Shah. Proceedings of Surfactants in Solution Conference, Jerusalem, Israel 1996. (In press)
96. VK Bansal, DO Shah, JP O'Connell. J Coll Inter Sci 75:462, 1980.
97. DO Shah. Chem Eng Educ J 14, Winter, 1977.
98. DA Benedetto, DO Shah, HE Kaufman. Invest Ophthalmol 14:887, 1975.
99. DM Maurice. Exp Eye Res 2:33, 1963.
100. SR Waltman, HE Kaufman. Invest Ophthalmol 9:247, 1970.
101. S Mishima, A Gasset, SD Klyce, et al. Invest Ophthalmol 5:264, 1966.
102. S Udenfriend. Fluorescence Assay in Biology and Medicine. New York, Academic Press, 1962.
103. JH Schulman, T Teorell. Trans Farad Soc 34:1337, 1938.
104. DO Shah, et al. U.S. Patent #4,131,651.

# 2
# Quarter Century Progress and New Horizons in Micelles

**Fredric M. Menger**
*Emory University, Atlanta, Georgia*

## I. INTRODUCTION

Micelles! Who could deny that these small, elusive, self-assembled systems are the flagship of colloid chemistry? No other member of the colloidal domain has received such attention by such an international group of scientists using such a diverse array of analytical tools. It is a source of pleasure that micelle research, pursued in seven continents, is never denied sunlight or summer. The resulting information has served well as the basis for understanding virtually every other member of the colloid family including vesicles, microemulsions, films, liquid crystals, foams, and polyelectrolytes.

The title assigned to this paper challenges me to describe a vast quantity of micelle research: past discoveries, present activity, future possibilities. This is, clearly, a delightful but impractical task. One feels like a person in a bountiful orchard who would like to sample a fruit from each tree. Fortunately, there is no cause to write a full-blown review; sampling only a few fruits will suffice. These few examples have been selected more or less arbitrarily. The only obvious bias lies in my citing a couple favorite micelle experiments from my own laboratory, a transgression for which I may be forgiven. By and large, however, I was guided by a desire to cover both diverse problems in micelle chemistry and a variety of methodologies that have been employed to solve them. Another person would doubtlessly have selected an entirely different set of publications. But no matter; a presentation such as this can employ many styles. Of course, in addition to reviewing the topic, I hope to inspire and provoke the reader (elements of good scientific writing, I am told), all in preparation for the exciting research papers that are to follow.

Before delving into chemical matters, it would be worthwhile to discuss briefly a bit of history. Colloid chemistry is now experiencing a remarkable revival after a sustained period of relative stagnation. How did it happen that a field so vital to almost every industrial enterprise in modern society (pharmaceuticals, paper, metals, petroleum, food, polymers, and paint, to name a few) could be shoved aside by the academic community as a second-class citizen? In my opinion, the culprit is a philosophy that has guided twentieth-century science, namely Western reductionism.

Reductionism teaches us that a complex system is best attached via its much smaller components. Information gathered on the components is then assembled to describe the system as a whole. Although this approach has had spectacular successes, the embattled colloid chemist is often unable to "divide and conquer." Experiments performed on monomeric surfactant molecules, for example, will never reveal the properties or structure of a micelle. The micelle itself must be investigated. Unfortunately, in modern times two major practitioners of reductionism, the quantum mechanicians and the molecular biologists, have substantially superseded the colloid chemist. Let us consider each in turn.

When physicists and physical chemists became enchanted with quantum mechanics, colloid chemistry was relegated to the position of stepchild. The excitement, the brains, the funding, the conferences, the hype focused on atomic particles rather than on colloidal particles. From the perspective of the colloid chemist, physicists and physical chemists had reduced their description of nature to a level too small to be useful. Information from atoms cannot be assembled to construct a colloidal entity. Electronic states of atoms cannot predict the solubility of glucose in water, let alone the properties of a polysaccharide gel. Reductionism had gone too far for many purposes, and yet it was receiving all the attention! This problem was brought home most poignantly by Nobel Prize winner Albert Szent-Györgyi in his *Personal Reminiscences*:

> In my hunt for the secret of life, I started research in histology. Unsatisfied by the information that cellular morphology could give me about life, I turned to physiology. Finding physiology too complex, I took up pharmacology. Still finding the situation too complicated, I turned to bacteriology. But bacteria were even too complex, so I descended to the molecular level, studying chemistry and physical chemistry. After twenty years' work, I was led to conclude that to understand life we have to descend to the electronic level, and to the world of wave me-

chanics. But electrons are just electrons, and have no life at all. Evidently on the way I lost life; it had run out between my fingers.[1]

Szent-Györgyi finally realized, in his increasingly reductionist career, that "the behavior of large and complex aggregates of elementary particles is not to be understood as a simple extrapolation of the properties of a few particles. Rather, at each level of complexity entirely new properties appear."[2]

Whereas quantum mechanicians attempt to describe chemistry in terms of electrons, molecular biologists attempt to explain life in terms of biomolecules (particularly DNA). Lest the reader doubt the extreme reductionist attitude prevalent in modern biology, let me quote from no less than Francis Crick of double-helix fame: "Your joys and sorrows, your memories and your ambitions, sense of personal identity and free will, are in fact no more than the behavior of a vast assembly of nerve cells and their associated molecules."[3] Or consider a quote from Carl Sagan in his well-known book, *Cosmos*:

> I am a collection of water, calcium and organic molecules called Carl Sagan. You are a collection of almost identical molecules with a different collective label. But is that all? Is there nothing in here but molecules? Some people find this idea . . . demeaning to human dignity. For myself, I find it elevating that our universe permits the evolution of molecular machines as intricate and subtle as we are.[4]

These quotes come from people who pay no attention to the colloidal nature of life. They do not seem to comprehend that life is composed of biomolecules that self-assemble according to the principles of colloid chemistry, many of which have not been unraveled. Even the total elucidation of the human genome will provide only the bricks of the building; the architectural plans will remain a mystery. Fortunately, my personal identity and ambitions will also remain a mystery, despite Francis Crick's clear implication that they are, at least potentially, within his grasp.

Ultimately, I am convinced, it is the colloid chemist who will have the upper hand in comprehending the products and processes of nature. This is because the cell cannot be described, Carl Sagan notwithstanding, as simply a bag of randomly mixed inorganic ions and organic compounds in water. Biomolecules as isolated entities have no more chance of revealing the essence of living systems than quarks have had in revealing the behavior of organic molecules. No, we must elevate ourselves to a higher level of complexity, the colloidal state. It is here where one can properly engage the mitochondrial membrane, the multienzyme complex, the ribosome, the microtubule, the protoplasmic gel. I cannot, of course, predict exactly when those who are well-versed in colloid chemistry will begin to dominate biology. All I can say is that the event is inevitable, so train your students well!

## II. THE EXPERIMENTS

In 1920 McBain and Salmon proposed the existence of micelles,[5] while in 1936 G. S. Hartley ventured to guess what the micelle might look like.[6] Hartley suggested that the micelle is spherical, with a radius approximately equal to the length of the surfactant's hydrocarbon chain. Although Hartley specifically expressed his belief that the surfactant molecules within a micelle are disordered, the scientific literature almost invariably portrays the "Hartley micelle" as having linear chains arranged radially, as in the spokes of a wheel (Figure 1). If there is any doubt as to the prevalence of this picture, the reader should consult four biochemistry texts published just this year.[7a-7d] Now, schematic diagrams can be useful as long as they are labeled as such and are thus not taken literally. The problem is that the Hartley representation has not been — and obviously is still not — labeled as schematic. Consequently, it has seeped into the scientific folklore as an accurate portrayal of reality.

The following anecdote illustrates how enamoured scientists can become with a model that has been relentlessly advertised with the aid of false but appealing pictures: At an international conference in 1977, I discussed the possibility that micelles are disorganized "brushheaps" of molecules with rough surfaces and an abundance of hydrocarbon-water contact. After the lecture, a person in the back of the room raised his hand to ask a

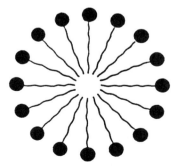

**Figure 1** The so-called "Hartley micelle" as portrayed in most modern biochemistry texts. Two decades of experiments utilizing a wide variety of methods have proven that micelles are far more disorganized than implied in the model.

question. He ceremoniously walked up the aisle (staring at me disdainfully the entire time), turned around, and told the audience "I don't believe any of this." He then walked back to his seat.

Nowadays, after a worldwide effort to define micellar structure using a host of experimental and theoretical methods,[8] informed chemists no longer regard the micelle as a radially symmetric oil droplet totally covered by a smooth ionic shell. One of our own contributions to the field focused on the question: Are surfactant chains in a micelle bent, or are they linear as implied in the Hartley model? To answer this question, we developed a new method for assessing chain conformations,[9] as described briefly in the next paragraph.

Our experiments were based upon the long-range nuclear magnetic resonance (NMR) coupling, $^3J$, between two $^{13}C$ atoms spaced four carbons apart ($*C_1$-$C_2$-$C_3$-$*C_4$). The coupling between $C_1$ and $C_4$ depends upon the dihedral angle about the $C_2$-$C_3$ bond in a typical Karplus-type function. Thus, if the chain at $C_2$-$C_3$ is linear (an "anti" conformation), then $^3J$ = 3.5–4.0 Hz. If the chain is bent (a "gauche" conformation), then $^3J$ = 1.5 Hz. By synthesizing surfactants bearing two $^{13}C$ atoms four carbons apart at known locations (a tedious chore!), we deduced the time-averaged conformations at various sites. This conceptually simple idea was applied to a variety of compounds, two of which are detailed below:

$$CH_3(CH_2)_6*CH_2(CH_2)_2*CH_2(CH_2)_3OSO_3^-$$
$$^3J = 2.6\,\text{Hz}$$
$$CH_3(CH_2)_6*CH_2(CH_2)_2*CH_2(CH_2)_2NH_3^+$$
$$^3J = 2.0\,\text{Hz}$$

The anionic and cationic surfactants are seen to have $^3J$ = 2.6 Hz and 2.0 Hz, respectively, when dissolved in $D_2O$ at concentrations manyfold higher than their critical micelle concentrations (cmc). In contrast, monomeric chains in chloroform have a $^3J$ = 3.8 Hz. Clearly, the majority of micellized surfactant chains are highly bent, definitive proof that the Hartley model is unacceptable.

Consider an aqueous micellar solution composed of cetyltrimethylammonium bromide plus added sodium chloride and butanol. In order to describe this system properly, one needs to know the concentrations of water, bromide, chloride, and butanol at the micelle surface (i.e., the Stern region). L. Romsted and colleagues (Rutgers)[10] have developed a particularly useful method for securing such hard-to-get information. The method utilizes a long-chain arenediazonium salt, $C_{16}H_{33}Ar-N^+_2\,BF^-_4$, bound to cetyltrimethylammonium bromide micelles. Arenediazonium salts are known to undergo slow C/N fission to lose nitrogen and create an extremely reactive aryl carbocation intermediate:

$$\text{Ar-}\overset{+}{\text{N}}_2 \rightarrow [\text{Ar}^+] + \text{N}_2$$

An aryl carbocation formed at a micelle surface reacts instantly with any nucleophile in its vicinity (i.e., water, halide, or butanol) to produce stable products that are quantitatively analyzed by HPLC:

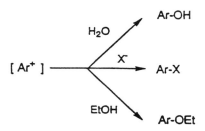

Assuming that inherent selectivities at the micelle interface are no different than they are for simple arenediazonium salts in pure water, it is possible to extract out the concentrations of all components in the Stern region. For example, 0.01 M cetyltrimethylammonium bromide (CTAB); (0.01 M HBr, 40°C) has the following interfacial concentrations: $[H_2O]_i = 45$ M and $[Br]_i = 1.5$ M. When 0.5 M butanol is added to the system, the concentrations are found to be $[H_2O]_i = 48$ M, $[Br]_i = 0.4$ M, and $[BuOH]_i = 8$ M. Thus, butanol binds efficiently to the micelle at the expense of bromide ion. The Romsted technique can be profitably applied to other colloidal systems, including vesicles and microemulsions.

Almost all aspects of micellar structure, including micellar fluidity, have engendered considerable debate. By the early 1980s, a variety of studies (including one of our own[11]) pointed to a "liquidlike" micelle interior. But it was a 1983 paper of P. Stilbs (Uppsala) and Lindman (Lund) that settled the matter beyond doubt.[12] The experiment is worth presenting in some detail, but before doing so it should be mentioned that the Stilbs et al. paper offers an unusually forthright exposé of the complications associated with the "probe method" as used in most fluidity studies. Among the problems, one finds (1) the probe molecules perturb the micellar structure; (2) the probe molecule may reside, to some extent, outside the micelle; (3) the probe molecule may assume many sites within a nonhomogeneous micelle; and (4) the motional behavior of a probe within a micelle may be difficult to translate into a macroscopic quantity such as "viscosity." The Stilbs approach eliminates, or at least minimizes, these difficulties.

Stilbs used deuterium NMR spin-relaxation methods to determine the correlation times for reorientation of *trans*-decalin-d[18] around its three principle rotational diffusion axes. *trans*-Decalin is small, rigid, and devoid of functionality that might otherwise "drag" it onto the micelle surface;

solubilization into the micelle interior seems assured. A good agreement was found between the correlation times for reorientation of *trans*-decalin in micelles and in hydrocarbon solvents with the same chain length. For example, the correlation times for *trans*-decalin in dodecane are 3.8, 10.3, and 3.3 ps; in dodecylbenzene, they are 6.7, 20.1, and 4.9 ps. *trans*-Decalin in dodecyltrimethylammonium bromide micelles have similar times of 6.3, 22.7, and 3.6 ps. Moreover, the reorientation rate for solubilized *trans*-decalin decreases with increasing surfactant chain length, a trend that is closely followed in alkane and alkylbenzene solvents. Increased viscosity with chain length reflects, no doubt, an enhanced probability of interchain entanglements. In summary, Stilbs and Lindman have proven, via model-independent measurements, that micelle interiors are rather normal, fluid-like environments as opposed to resembling "paraffin balls."

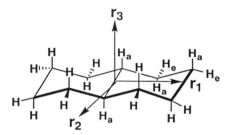

As just seen, NMR has played a key role in elucidating micelle structure in the past quarter century. Let me now cite another example where NMR provided valuable information, this time on the role of water. How much water surrounding a micelle, one may ask, differs significantly from bulk water? How does this "abnormal" water vary with the structure of the surfactant? At what rate does the bound water reorient? How long does such water remain in the vicinity of the micelle before diffusing back into the bulk? Questions such as this have been addressed by B. Halle and H. Wennerström (Lund) via the quadrupolar relaxation rates and splittings for water $^2$H and $^{17}$O nuclei.[13] Since the corresponding theory is complicated, I will merely summarize below their main results acquired from soap micelles.

It should be stated forthwith that the Halle–Wennerström approach assumes a "two-step" model of relaxation. In other words, only two components of water motion are considered: a fast anisotropic reorientation of bound water and a slower motion (as might occur during a micellar reorientation or a translational diffusion of water between two distinct environments). The NMR data verify the model and lead to the following conclusions: (1) bound water reorients quite rapidly (less than 10 times slower than in bulk water); (2) only one or two hydration layers are perturbed (in conflict with those who believe in long-range hydration structures[14]); (3) the lifetime of a hydration water molecule is on the order of 10 ns; (4) charged surface groups and their counterions impose larger structural perturbations on water than do the nonpolar groups; (5) less than two methylene groups in the alkyl chains of soap micelles are exposed to water.

With regard to the last point, I should note that soap micelles have structures that are radically different from the common surfactant micelles. For example, our double-$^{13}$C-labeling experiments have shown that the surfactant chains in soap micelles are linear (as in a bilayer), in sharp contrast to the highly bent configurations characteristic of surfactant micelles.[9] Thus, it is difficult to make sweeping generalizations about chain conformation, hydration, or any other aspect of micelle chemistry. Each system must be treated as unique and distinct.

We initially entered the micelle area because we naively believed that micelles were the simplest of the self-assembled systems. As seen above, micelles are in fact one of the most elusive of the colloidal particles because, being deformable, they readily respond to the conditions around them. This point was forcefully made by J. B. F. N. Engberts and colleagues (Groningen), and I quote from their work:

> Above a critical concentration, cooperative association sets in and the Gibbs energy of the system is minimized through a compromise of a variety of often opposing forces. These forces depend both on the molecular architecture of the surfactant and on the peculiar solvent properties of water. The urge for optimum aggregate stability is reflected in the rich variety of possible aggregate morphologies, each with its particular mode of alkyl chain packing and headgroup arrangement.[15]

Engberts used a set of nine related surfactants to show that minor structural variations lead to preferential formation of spherical micelles (SM), rodlike micelles (RM), or vesicles (V). Shown below are three of the surfactants (iodide salts) and their morphologies that illustrate the point:

**SM, RM**          **V**          **V**

Comparison of the first two compounds proves that the aggregate morphology is not determined solely by surfactant's total hydrophobic volume. Vesicle formation by the third compound is not completely understood. I will write no more about vesicles (lacking the wherewithal to properly cover even micelles) except to mention that the work of T. Kunitake (Fukuoka)[16] and J. Fendler (Syracuse)[17] was instrumental in ushering in the era of "organic" vesicular chemistry.

Engberts expressed the belief that systematic studies of chain-packing will become a major activity in surface chemistry,[15] and it is difficult not to concur with the sentiment. One hopes that ultimately the morphology of surfactant aggregates will become predictable with the aid of computer calculations.

1-Methyl-4-dodecylpyridinium iodide, the first of the preceding three surfactants, was also used to monitor the micropolarity at a micelle surface.[18] The method is based on the solvent-dependent charge-transfer band representing the transfer of an electron from iodide to the pyridinium ring. Monomeric compound has transition energies of 79.5, 99.6, and 117.1 kcal/mol in chloroform, ethanol, and water, respectively. Since the transition energy of the surfactant in water above its cmc equals 100.3 kcal/mol, it is clear that the effective polarity in the Stern region approximates that of ethanol. Adding large amounts of salt (e.g., 0.10 M NaI) lowers the cmc appreciably but does not much alter the surface micropolarity. One appealing feature of this experiment is its use of an intrinsic probe. Thus, the micropolarity is assessed via the surfactant headgroup itself rather than with an added reporter compound.

The solubilization capacity of micelles is a recurrent theme in the micelle literature. As I trek at an embarrassingly rapid rate through this literature, I will cite only one of the many important papers on solubilization, namely that of M. Almgren, F. Grieser, and J. K. Thomas (Notre Dame).[19] These authors studied the kinetics of micellar solubilization of phosphorescent aromatics (e.g., naphthalene and pyrene). In their method, the phosphorescent probes reside primarily within the micelles (i.e., $K_{assoc} = k_{enter}/k_{exit} \gg 1$). Moreover, a highly water-soluble quencher ($NO_2^-$) added to the medium was assumed to operate exclusively in the bulk water. Thus, as the quencher concentration is increased, the rate of phosphorescence loss ($k_{obsd}$) increases until the exit rate becomes rate-controlling. At that point, a further increase of quencher has no effect, and $k_{obsd} = k_{exit}$. The entry rate is obtained from $k_{enter} = K_{assoc} \times k_{exit}$.

It was found that $k_{exit} = 1.7 \times 10^3 \text{ s}^{-1}$ for pyrene in CTAB micelles, corresponding to a residence time $\tau$ of 588 $\mu$s. As might be expected, benzene has a much faster exit rate and, therefore, a much shorter residence time: $k_{exit} = 4.4 \times 10^6 \text{ s}^{-1}$ and $\tau = 1.3$ $\mu$s. Typical entry rates are 5–8 $\times$ $10^9 \text{ } M^{-1} \text{ s}^{-1}$ (within a factor of 2 from being diffusion-controlled). Exit rates are generally 2 to 4 times faster for sodium dodecyl sulfate (SDS) than for CTAB. There are strong indications from solubility data that pyrene is solubilized primarily at the surface of the CTAB micelles. Ion-dipole attraction[20] is a likely explanation for this preferred adsorption site.

A review of micelles' quarter century history, no matter how cursory, must not bypass the "double-relaxation" model initiated by N. Muller (Purdue),[21] experimentally tested by R. Zana and colleagues (CNRS, Strasbourg)[22] and theoretically developed by E. A. G. Aniansson and coworkers (Gothenberg).[23] The model invokes the presence of two distinct dynamic processes characterized by relaxation times $\tau_1$ and $\tau_2$. Accordingly, $\tau_1$ (representing rapid monomer exchange between the micelle and the bulk water) falls in the range of $10^{-6}$ to $10^{-3}$ s. The slower relaxation, $\tau_2$ (representing the total micellization–dissolution equilibrium) falls into the range of $10^{-3}$ to 1 s.

Monomeric exchange $(\tau_1)$: $A_n + A \Leftrightarrow A_{n+1}$

Micellization–dissolution $(\tau_2)$: $nA \Leftrightarrow A_n$

Consider a micellar system subjected, in a T-jump or p-jump experiment, to a sudden perturbation. Since the perturbation causes a small shift in the optimal size distribution, monomers will leave some micelles and enter others in order to restore an optimal size distribution. The relaxation time $\tau_1$, during which the total number of micelles remains unchanged, is given by the equation below (where $N$ is the most probable aggregation number, $\sigma$ is the standard deviation of the micelle size distribution, and $c$ is the surfactant concentration):

$$\tau_1 = \frac{k_{exit}}{\sigma^2} + k_{exit} \frac{(c - \text{cmc})}{(\text{cmc} \cdot N)}$$

By plotting the observed relaxation time $\tau_1$ vs. $(c - \text{cmc})$, one can extract both the monomer exit rate and J. For example, $k_{exit} = 1.0 \times 10^8$, $1 \times 10^7$, and $9.6 \times 10^5$ s$^{-1}$ for $NaC_8SO_4$, $NaC_{12}SO_4$, and $NaC_{14}SO_4$, respectively. The $\sigma$ for ionic micelles is about 10.

The micellization–dissolution process is somewhat more complicated. In order for micelle to disintegrate in toto, at some stage it must pass through a smaller and highly unfavorable aggregation number (a "bottleneck" that controls the efficiency of the process). Suffice it to state here without details, the $\tau_2$ has been expressed as a function of concentration and a "resistance" term reflecting the kinetically significant but least probable aggregation numbers. One-step dissolution of micelles is slow because micelles have a rather narrow size distribution, rendering the required, low-aggregation number "intermediates" unlikely. The slow relaxation time of micelles is an important process because, as shown by D. O. Shah (Gainesville),[24] it is a controlling factor in foaming, bubble size, wetting time, and solubilization efficiency.

Perhaps the time has come again to insert a bit of editorializing. As may already be apparent from this survey, there seems to be two groups of colloidal chemists. One group (in which I include myself) focuses on organic and biochemical aspects of the field. The other group is concerned more with physical and theoretical issues. Unfortunately, the camps do not communicate satisfactorily. They publish in different journals, go to different conferences, and they seldom cite each other. This is a pity, because each side has much to teach the other. The organic chemist has the advantage of being able to synthesize new compounds and, therefore, to carry out detailed studies of behavior as a function of structure. The more mathematically inclined have the advantage of an in-depth understanding at a fundamental level. It seems clear that in order for colloid chemistry to maximize its progress, collaborations must be established in which organic chemist, biochemist, physical chemist, physicist, and chemical engineer interact in a common effort to solve our extremely difficult problems.

Twenty-five years ago, people began grappling with micelles in the hope of catalyzing organic reactions with an efficiency and selectivity rivaling that of enzymes. Although this hope has not been realized, the hundreds of publications on the subject have produced a rich chemistry in which micelles are actually "doing something." Many of these publications analyzed their kinetic data using our "pseudophase" model where reactant A (with an intrinsic rate of $k_1$) binds to micelle M (within which it reacts with a rate $k_2$)[25]:

$$\begin{array}{ccc} \text{A} & + \text{ M} & \overset{K}{\Leftrightarrow} \text{AM} \\ \downarrow k_1 & & \downarrow k_2 \\ \text{P} & & \text{P} \end{array}$$

A plot of $1/(k_{obs} - k_1)$ vs. $1/[M]$ gives both the micellar rate constant and the reactant/micelle association constant:

$$\frac{1}{(k_{obs} - k_1)} = \frac{1}{(k_2 - k_1)} + \frac{1}{(k_2 - k_1)K[M]}$$

The preceding construct applies only to first-order and pseudo-first-order reactions. Second-order processes between neutral organic reactants are best handled according to the equation of Martinek and colleagues (Moscow).[26] The pseudophase model was further applied by Bunton and coworkers (Santa Barbara) to micellar ion–molecule reactions.[27] These latter treatments show (if I may be permitted another arid oversimplification) that micellar rate enhancements of biomolecular reactions, typically less than $10^2$-fold, are largely the result of a concentration effect. In other words, ionic and apolar reactants can assemble in the Stern region at concentrations higher than that in the bulk, thereby leading to an enhanced biomolecular reactivity. Catalysis by a micellar "medium effect" is, in most cases, small. When inert ions are added to the system, they can displace reactive ions bound to the micelle and thus inhibit the rate (much like competitive inhibitors in enzymology). To quote, finally, a review of the general subject, "Further progress depends not only upon the development of better kinetic models but also on increasing our understanding of colloidal structure and of ionic interactions with association colloids composed of ionic, nonionic, zwitterionic, and amphoteric surfactants and their mixtures, which are more typical of biological membranes and commercial applications."[28]

My brief foray into micelle-catalyzed reactions will include only two specific examples, both of which involve surfactants that have been functionalized to impart special reactivity. The first is that of R. Moss and colleagues (Rutgers), who synthesized a long-tailed iodosobenzoate[29]:

$$C_{16}H_{33}-\overset{CH_3}{\underset{CH_3}{\overset{+}{N}}}-CH_2CH_2O-\text{[iodosobenzoate ring]}$$

This compound, in the presence of cetyltrimethylammonium chloride (CTAC) micelles, hydrolyzes phosphotriesters with a remarkable $10^5$-fold rate acceleration. Our own favorite contributor to the field is shown below[30]:

$$C_{12}H_{25}-\overset{CH_3}{\underset{CH_3}{\overset{+}{N}}}-CH_2-\underset{O}{\overset{}{C}}-H \quad Br^-$$

Although not as fast as the Moss surfactant, it splits phosphotriesters (via a multistep mechanism that will not be detailed here) with an appealing enzyme-like turnover. Synthetic organic chemistry obviously allows access to a host of useful or interesting new colloidal materials.

Photochemists and photophysicists have been active contributors to the micelle discipline. When colloidal particles are present, an excited state (or electron-donor) can segregate itself from a quencher (or electron-acceptor) and thus delay its annihilation. Classical solution chemistry cannot hope to duplicate the ensuing benefits of this effect. As before, I will illustrate the use of micelles as "fluid nanoscopic photoreaction vessels" by means of only two representative examples.

## Progress and New Horizons in Micelles

M. Grätzel and coworkers (Lausanne) devised a lovely example of micelle-assisted photoinduced electron transfer.[31] The system contains micelles composed of dodecyl sulfate with $Cu^{2+}$ counterions. Adsorbed within the micelle is an electron-donor, N-methylphenothiazine, hereafter referred to as D. In the bulk water is placed $[Fe(CN)_6]^{3-}$, a species capable of oxidizing $Cu^+$ to $Cu^{2+}$. Now, consider the interesting series of events when the system is photolyzed: An electron is transferred, in less than a nanosecond, from D within the micelle to $Cu^{2+}$ counterion in the Stern region:

$$D + Cu^{2+} \rightarrow D^+ + Cu^+$$

Normally, such a transfer is impaired by the reverse process (i.e., return of the electron from $Cu^+$ back to $D^+$). In the present case, however, the $Cu^+$ escapes from the micelle into the bulk water where $[Fe(CN)_6]^{3-}$ lies in wait. Another redox reaction than takes place:

$$[Fe(CN)_6]^{3-} + Cu^+ \rightarrow [Fe(CN)_6]^{4-} + Cu^{2+}$$

The net effect of all this is an electron transfer from D to $[Fe(CN)_6]^{3-}$, with copper serving as the carrier. Since $[Fe(CN)_6]^{4-}$, being anionic, is repelled from the micelle surface, it is protected from a destructive encounter with micellized $D^+$. The result is a storage of light energy via redox-couple $D^+$ $[Fe(CN)_6]^{4-}$.

Turro and coworkers (Columbia) studied the emulsion polymerization of styrene as photo-initiated by dibenzyl ketone[32]:

It was found (and this is truly amazing) that the average molecular weight of the polystyrene varied with the magnitude of an applied magnetic field. For example, the molecular weight averaged $1 \times 10^5$ at zero external field; this increased to $5 \times 10^5$ at 1000 G. The effect is explained in the following manner:

Dibenzyl ketone, an oil-soluble photoinitiator, positions itself within the styrene-swollen micelles. Two things can happen upon photolysis and production of triplet radical-pairs (i.e., $Ph\dot{C}H_2 + CO + \dot{C}H_2Ph$). The radicals can either cage-combine or they can escape the micelle into the bulk water. The likelihood for escape is greater for the triplet state than for the singlet state, because the former has a longer lifetime. Now, a magnetic field is known to suppress triplet-to-singlet intersystem crossings. As the lifetime of the radicals thereby increases, their escape rate increases accordingly. Thus, the magnetic field creates micelles with fewer initiator radicals that, in the absence of partners with which to recombine, nonetheless promote multiple polymerization events. Since the probability of terminating reactions between propagating chains and initiator radicals is reduced, the molecular weight increases.

Photophysics has also been useful in characterizing micelles and micellar binding, as illustrated by experiments of Whitten and colleagues (Rochester) on 4-nitro-4'-methoxystilbene (NMS)[33]:

$$O_2N-\text{C}_6\text{H}_4-CH=CH-\text{C}_6\text{H}_4-O-CH_3$$

## NMS

The lifetime of the NMS triplet, produced by pulsed laser photolysis, depends upon the solvent polarity (e.g., $\tau = 0.077, 0.33$, and 4.9 $\mu$s in cyclohexane, methanol, and 50% MeOH/$H_2O$, respectively). Triplet decay in SDS, CTAB, and Brij-35 micelles is monoexponential with $\tau = 9.4, 1.5$, and 1.2 $\mu$s, respectively. Clearly, there is a single average binding site that is largely aqueous in nature, suggesting an interfacial adsorption. Binding in the Stern region occurs despite the fact that NMS is extremely water-insoluble ($<10^{-7}$ M in the absence of micelles). No doubt this behavior reflects the huge micellar surface area replete with hydrophobic domains that can latch onto apolar additives.

Colloid chemists interested in surfactant aggregation are now extending the subject from water to other polar solvents. Thus Evans and colleagues (Minnesota) have prepared micelles in hydrazine and ethylammonium nitrate,[34a,34b] while Rico and Lattes (Toulouse) have used formamide and $N$-methylsydnone as water substitutes.[35a,35b] Years ago, we ourselves once made micelles by adding a neutral compound, $C_{12}H_{25}-O-CH_3$, to concentrated $H_2SO_4$.[36] The strong acid protonates the ether and generates a surfactant in situ. Caution should be exercised in this area because, as shown by Das and coworkers (Lund), one cannot automatically assume that aggregates in water are even remotely similar to aggregates in nonaqueous solvents.[37] This makes nonaqueous solvents all the more interesting.

By now many readers will be asking, "What about reverse micelles?" Perhaps this is the point at which to quote Voltaire: "The secret of being tiresome is to tell everything." Actually, it is tempting to "tell everything" because, for one thing, I consider Aerosol OT (AOT) the single most remarkable compound in colloid chemistry (and one with which we are still tinkering). But space limitations force me to confine myself to "normal" micelles. Apologies are due to an important and active segment of micelle chemistry.

A good way to end this survey might be to describe new and unusual surfactants, seven of which are given in the scheme on page 68. The compounds in the scheme reaffirm my conviction that there are endless opportunities to vary the structure of amphiphiles and thereby create new and potentially useful colloidal systems.

> Compound #1 (S. H. Gellman, Wisconsin).[38] According to NMR, the onset of aggregation of #1 in water occurs in the range of 20–40 mM. At 60 mM, the aggregates begin to solubilize a water-insoluble dye (orange OT). If the ether side-chain is omitted from the molecule, it aggregates in the 5–10 mM range but loses the ability to solubilize dye even at 0.16 M. All evidence, including the absence of critical phenomena, points to small aggregates with unknown structure; whether or not one calls them "micelles" is a semantic issue.
> 
> Compound #2 (D. Kahne, Princeton).[39] This compound has been called a "facial amphiphile" to distinguish it from the linear head-tail amphilicity of conventional surfactants. An X-ray structure of #2 shows pairs of hydrogen-bonded glycosteroids stacked upon one another to create sugar-lined channels. Solution properties of these interesting amphiphilic compounds remain to be elucidated.
> 
> Compound #3 (J.-H. Fuhrhop, Berlin).[40] This water-soluble $\alpha,\omega$-bis(paraquat) amphiphile forms vesicles as the tetrachloride salt. The tetraperchlorate, however, organizes into monolayer sheets and into micelles of 100 Å diameter visible by rapid-freeze electron microscopy. It is proposed that the micelles have a cubic structure, shown in the structure on page 69. The experiments reveal a sensitive dependence of the aggregation number, aggregate curvature, and morphology upon the nature and stoichiometric ratio of counterions.

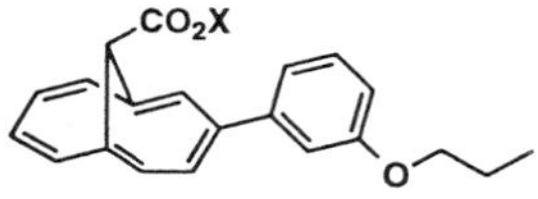

1

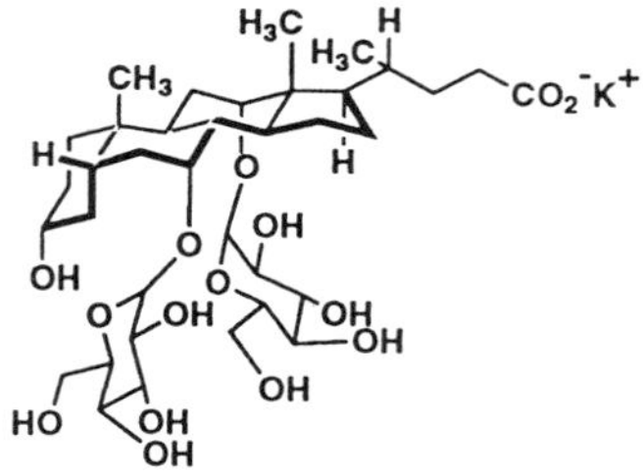

2

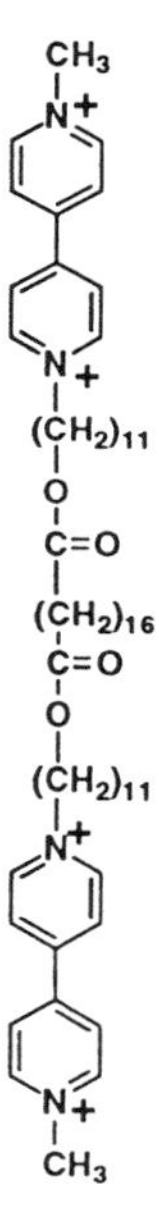

3

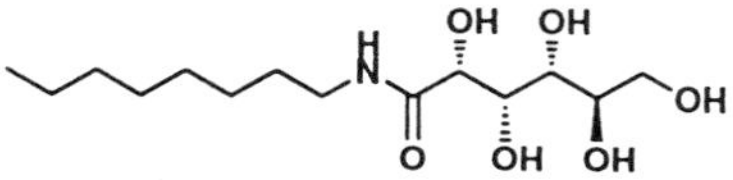

4

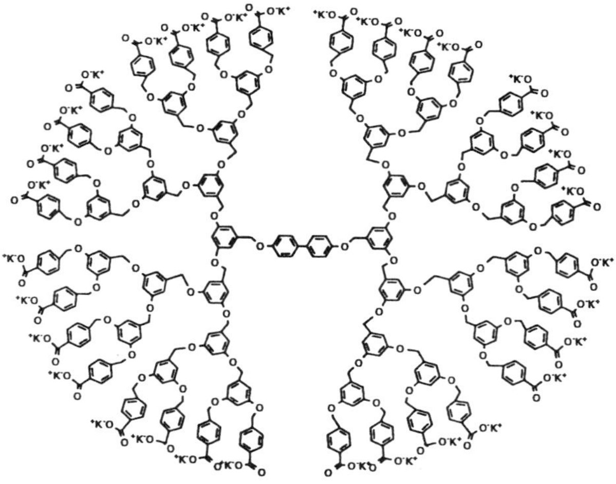

5

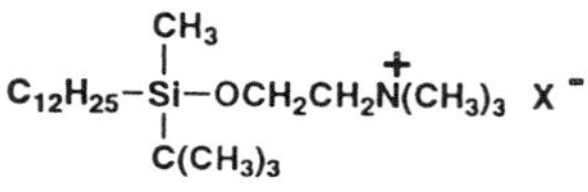

6

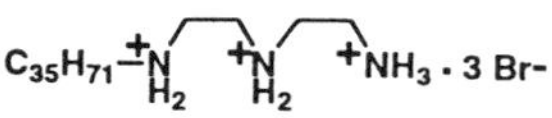

7

# Progress and New Horizons in Micelles

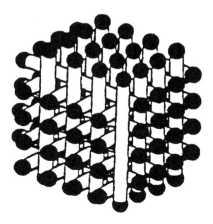

Compound #4 (J.-H. Fuhrhop, Berlin).[41] *N*-octyl-*D*-gluconamide forms micellar fiber bundles that have a quadrupole helix structure. Words cannot do these fibers justice; for full enjoyment, one must consult the original article and gaze upon its remarkable photographs.

Compound #5 (J. M. J. Fréchet, Cornell).[42] This water-soluble unimolecular dendritic micelle is my sole tribute to the growing area of polymer micelles. Macromolecules have an advantage over the surfactant micelle in that they possess no cmc, and therefore there are no concentration restrictions. Thus, compound #5 will solubilize pyrene even at concentrations as low as $5 \times 10^{-7}$ M.

Compound #6 (D. A. Jaeger, Wyoming).[43] This is a good example of a "destructible" surfactant. One merely has to treat the surfactant with aqueous acid or base in order to split off the $HOCH_2CH_2NMe_3^+$ unit:

$$C_{12}H_{25}\underset{\underset{C(CH_3)_3}{|}}{\overset{\overset{CH_3}{|}}{Si}}-OCH_2CH_2\overset{+}{N}(CH_3)_3\ X^- \xrightarrow[^-OH/H_2O]{H_3O^+/H_2O\ or} C_{12}H_{25}\underset{\underset{C(CH_3)_3}{|}}{\overset{\overset{CH_3}{|}}{Si}}-OH + HOCH_2CH_2\overset{+}{N}(CH_3)_3\ X^-$$

Since the resulting products are not amphiphilic, solutions can be readily made surfactant-free once the surfactant has performed a useful function.

Compound #7 (F. M. Menger, Emory).[44] We were curious to examine surfactants with "hyperextended tails" with as many as 35 carbons per tail. Imparting a reasonable solubility to such surfactants required headgroups with multiple charges. As it turned out, these compounds formed vesicles rather than micelles, despite their possessing only a single saturated tail. The current dogma that micelle vs. vesicle formation is determined by the headgroup vs. chain cross section (i.e., "cone" vs. "cylinder") seems to be a gross oversimplification. The interplay of specific attractive and repulsive interactions among the molecules is far more important than molecular shape *per se* in determining the morphology of the aggregates.

In conclusion, let me express the hope that I have not so much written a review as painted a picture. The picture is intended to portray the excitement and potential of colloid chemistry as exemplified by one small component of the field, micelles. Most colloid chemists are, at least in general terms, already familiar with the picture. So what is the purpose of my painting it? An answer might be that there is always the hope that a student might stumble across the work and say, "Hey, this is neat stuff." It was, in fact, students with similar temperament who have made possible our own colloid research, and I presume this is true for most everybody else in academia. Thus, this article has to be for the rest of the team.

## ACKNOWLEDGMENT

Research in colloid chemistry has been supported over the years by the National Institutes of Health, the National Science Foundation, and the Army Research Office.

## REFERENCES

1. A Szent-Györgyi. Personal Reminiscences.
2. WH Thorpe.
3. F Crick.
4. C Sagan. Cosmos. New York, Random House, 1982.
5. JW McBain, CS Salmon. J Am Chem Soc 43:426, 1920.
6. GS Hartley. Aqueous Solutions of Paraffin Chain Salts. Paris, Herman and Cie, 1936.
7a. RH Garrett, CM Grisham. Biochemistry. Fort Worth, TX, Saunders, 1995.
7b. D Voet, JG Voet. Biochemistry. New York, Wiley, 1995.
7c. GL Zubay, WW Parson, DE Vance. Principles of Biochemistry. Dubuque, IA, Wm. C. Brown, 1995.
7d. L Stryer. Biochemistry. New York, W. H. Freeman, 1995.
8. DF Evans, H Wennerström. The Colloidal Domain. New York, VCH, 1994.
9. FM Menger, MA Dulany, DW Carnahan, LH Lee. J. Am Chem Soc 109:6899, 1987.
10. A Chaudhuri, JA Loughlin, LS Romsted, JJ Yao. J Am Chem Soc 115:8351, 1993.

11. FM Menger, JM Jerkunica. J Am Chem Soc 100:688, 1978.
12. P Stilbs, H Walderhaug, B Lindman. J Phys Chem 87:4762, 1983.
13. B Halle, H. Wennerström. J Chem Phys 75:1928, 1981.
14. RP Rand, VA Parsegian. Biochim Biophys Acta 988:351, 1989.
15. J-JH Nusselder, JBFN Engberts. J Am Chem Soc 111:5000, 1989.
16. T Kunitake, Y Okahata. J Am Chem Soc 99:3860, 1977.
17. JH Fendler. Membrane Mimetic Chemistry. New York, Wiley-Interscience, 1982.
18. EJR Sudhölter, JBFN Engberts. J Phys Chem 83:1854, 1979.
19. M Almgren, F Grieser, JKJ Thomas. J Am Chem Soc 101:279, 1979.
20. DA Stauffer, DA Dougherty. Tetrahedron Lett 29:6039, 1988.
21. NJ Muller. J Phys Chem 76:3017, 1972.
22. J Lang, C Tondre, R Zana, R Bauer, H Hoffmann, W Ulbricht. J Phys Chem 79:273, 1975.
23. EAG Aniansson, SN Wall, M Almgren, H Hoffmann, I Kielmann, W Ulbricht, R Zana, J Lang, C Tondre. J Phys Chem 80:905, 1976.
24. SG Oh, DO Shah. J Dispersion Sci Technol 15:297, 1994.
25. FM Menger, CE Portnoy. J Am Chem Soc 89:4698, 1967.
26. K Martinek, AK Yatsimirski, AV Levashov, IV Berezin. In: Micellization, Solubilization, and Microemulsions. KL Mittal, ed. New York, Plenum, 1977, vol. 2, p. 489.
27. A Blaskó, CA Bunton, G Cerichelli, DC McKenzie. J Phys Chem 97:11324, 1993.
28. CA Bunton, F Nome, FH Quina, LS Romsted. Acc Chem Res 24:357, 1991.
29. RA Moss, KY Kim, SJ Swarup. J Am Chem Soc 108:788, 1986.
30. FM Menger, LG Whitesell. J Am Chem Soc 107:707, 1985.
31. Y Moroi, AM Braun, M Grätzel. J Am Chem Soc 101:567, 1979.
32. NJ Turro, M-F Chow, C-J Chung, C-H Tung. J Am Chem Soc 105:1572, 1983.
33. KS Schanze, DM Shin, DG Whitten. J Am Chem Soc 107:507, 1985.
34a. MS Ramadan, DF Evans, RJ Lumry. J Phys Chem 87:4583, 1983.
34b. DF Evans, A Yamaushi, R Roman, EZ Casassa. J Colloid Interface Sci 88:89, 1982.
35a. I Rico, A Lattes. J Phys Chem 90:5870, 1986.
35b. X Auvray, C Petipas, T Perche, R Anthore, MJ Marti, I Rico, A Lattes. J Phys Chem 94:8604, 1990.
36. FM Menger, JM Jerkunica. J Am Chem Soc 101:1896, 1979.
37. KP Das, A Ceglie, B Lindman. J Phys Chem 91:2938, 1987.
38. DG Barrett, SH Gellman. J Am Chem Soc 115:9343, 1993.
39. Y Cheng, DM Ho, CR Gottlieb, D Kahne. J Am Chem Soc 114:7319, 1992.
40. J-H Fuhrhop, D Fritsch, B Tesche, H Schmiady. J Am Chem Soc 106:1998, 1984.
41. J-H Fuhrhop, W Helfrich. Chem Rev 93:1565, 1993.
42. CJ Hawker, KL Wooley, JMJ Fréchet. J Chem Soc Perkin Trans-1 1287, 1993.
43. DA Jaeger, MD Ward, AK Dutta. J Org Chem 53:1577, 1988.
44. FM Menger, Y Yamasaki. J Am Chem Soc 115:3840, 1993.

# 3
# Recent Advances in Aqueous Surfactant Phase Science: Coexistence Relationships of the "Sponge" Phase

**Robert G. Laughlin**
*The Procter & Gamble Company, Cincinnati, Ohio*

## I. INTRODUCTION

Liquids are optically isotropic condensed phases which are fluids rheologically and show no physical evidence of sharply defined, long-range order. Defining what is a liquid and what is not has become progressively more difficult with the rapid development in recent times of the capacity to determine details of the microstructures of fluid phases, and to rationalize these structures on a theoretical basis. It is also difficult, on occasion, to experimentally distinguish between true one-phase equilibrium states and time-stable, colloidally structured biphasic mixtures (1).

It was recognized early in this century (2) that the water-rich liquid phases of surfactants are distinguished from those of nonsurfactants by the existence of micellar aggregates of substantial size, which first appear at the critical micelle concentration (cmc). In fact, micellar liquids are often regarded as two-phase mixtures (3a,3b), an approximation that can be very useful, while at the same time being incorrect in a strict thermodynamic sense. The liquid-phase chemistry of surfactant systems was further complicated in 1980, when Lang and Morgan published a preliminary phase diagram of the $C_{10}E_4$–$H_2O$ system (4). Yet another liquid phase which is neither a molecular solution nor a typical micellar solution is found in this system. This then-new liquid was termed the "anomalous" phase by these authors;

structurally related phases have since been termed "sponge" or "$L_3$" phases (5). The sponge phase has received considerable attention from both structural and theoretical chemists (6–16). It is claimed to exist not only in binary systems, but also in three- to five-component aqueous surfactant systems (8,13,17,18). The phase is presently regarded as having a structure that is qualitatively similar to a bicontinuous cubic liquid crystal phase, but more highly disordered. It has been suggested that still another structured liquid may exist between the $L_3$ and lamellar liquid crystal phases in a branched trisiloxane $E_5$ surfactant (19).

The $L_3$ phase has a viscosity much lower than that typically encountered among liquid crystal phases (especially among cubic phases). When first discovered by D. N. Rubingh (who laid the groundwork for the later studies of Lang and Morgan [4]), it was termed the "blue I" phase. It is optically isotropic, but may have a bluish appearance (viewed with backlighting) similar to that of any liquid near a critical point. The phase displays strong shear-induced birefringence (14), as do cubic phases.

Systems having the $L_3$ phase were probably encountered long before the $C_{10}E_4$ study was performed, but were not recognized as such at the time. For example, a number of semipolar phosphate and phosphonate ester surfactants investigated during the early 1960s were found to display a narrow liquid region which, on the basis of its shape and position within the diagram, would presently be regarded as being probably an $L_3$ phase (20,21). Considerable (unreported) work was later performed on diol polyether surfactants derived by reaction of small polyoxyethylene glycols with alkane 1,2-epoxides or glycidyl ethers (3-alkoxy-1,2-epoxypropanes), and some of these also display what is probably an $L_3$ phase (22).

From correlations between molecular structure and phase behavior, it is clear that the $L_3$ phase is comparatively rare in binary systems. These correlations are described more fully below, but it may be noted that the occurrence of the $L_3$ phase is strongly influenced by the relative position (within C-T space) of the liquid/liquid miscibility gap and the lamellar liquid crystal region (23). The existence of the miscibility gap and the positions of these regions, in turn, depend on the relative hydrophilicity and lipophilicity (HLB) of the surfactant (24). The $L_3$ phase has not been found so far in any surfactant having a strongly hydrophilic (zwitterionic or ionic) hydrophilic group, but it is often found among the less hydrophilic members of the single-bond and semipolar subclasses of the nonionic class. As noted above, it is also claimed to exist within a wide range of aqueous surfactant systems containing one or more additional components.

Our interest in this area was initially spurred by the fact that compounds that display the $L_3$ phase are also found in many commercially important surfactants, and by uncertainty as to the aqueous-phase behavior of these compounds. Surfactants that display this phase are typically low-melting, thermally stable, and easy to load and study within a diffusive interfacial transport (DIT) cell (25). Exploratory studies suggested that the published diagrams of some systems that display the sponge phase were in error. Thus, sponge-phase-forming surfactants constituted both worthwhile subjects of investigation in their own right, and also attractive compounds with which to refine the DIT-infrared (DIT-IR) phase studies method (26).

## II. CRITICAL ROLE OF METHODOLOGIES IN SURFACTANT PHASE SCIENCE

The uncertainties that presently exist regarding the phase behavior of $L_3$-forming surfactants are strongly correlated with limitations of the isoplethal and isothermal analytic phase study methods that have historically been used (27,28). Quantitative isothermal swelling methods (such as the diffusive interfacial transport or DIT experiment) were first reported in 1987 (25), and thus were not available at the time of the earlier studies. During a DIT experiment, an interface between the surfactant and water is created within a long, thin silica capillary having a rectangular cross section, and countercurrent swelling via diffusive transport produces all the phases that exist at the temperature of the experiment. Phase bands having one-dimensional concentration gradients, separated by clean, sharply defined interfaces, are typically formed. All the phases that exist are created in each study, and biphasic mixtures are not usually formed (29). Preliminary identification of phases may be inferred from optical textures in polarized light, and phase compositions are determined in situ. In the original DIT-NDX method, compositions were determined using refractive index data, but this approach has been found unreliable (1,30,31). The development of an alternative method of analysis (based upon near-FT-IR microspectroscopy) has been pursued for some years, and this DIT-IR method was utilized in the present work. The DIT-IR approach remains extremely promising.

Viewed from a rigorous perspective, it must be admitted that our knowledge of the phase behavior of systems that display the $L_3$ phase is primitive at this time. The flaws in the classical methods (27) and the scope and limitations of swelling methods have been addressed (28). A basic problem with traditional methods stems from the fact that composition is varied in a discontinuous (digital) fashion during their use, and narrow phase regions have been overlooked for this reason (32). Composition is varied in a continuous (analog) fashion during DIT experiments, on the other hand, so that the likelihood of missing phases is very small (33). The clean separation and analysis of coexisting phases in most aqueous surfactant systems are difficult to achieve (except when both phases are liquids of low viscosity), and errors have been made even by the best experimentalists for this reason. This was the case in the carefully studied $C_{10}E_4$-$H_2O$ system (4), and may also have been true in the $C_{12}E_5$-$H_2O$ system (6). Dispersions of one isotropic phase (liquid or cubic liquid crystal) in another of similar composition do not scatter light intensively (34), and phase boundaries can also easily be missed for this reason.

Much has been accomplished using these older methods—indeed the very foundations of this field were laid using these methods (35)—but it will be essential in the future to include isothermal swelling methods among the techniques utilized, both for efficiency and for reliability. It has been the experience of the author that errors are disclosed in published diagrams in a large fraction of the systems that have been reinvestigated using the DIT method. The problem of the literature diagram of $C_{12}E_3$ described in the studies reported below constitutes but one of several instances where this is true.

## III. COEXISTENCE RELATIONSHIPS OF THE $L_3$ PHASE

It is essential to determine both the range of compositions within which phases exist, and also their coexistence relationships with other phases, in order to fully know just the most elementary aspects of the thermodynamic state of a system. Both free energy and chemical potentials vary smoothly within a phase, while spatial discontinuities in free energy typically occur at the interfaces between phases (36). One would like to know whether or not such discontinuities exist in passing from one liquid phase to another. Experience suggests that liquid phases do not merge smoothly with other phases having a qualitatively different phase structure (crystal or liquid crystal), but two liquid phases can (and often do) merge with one another as temperature (or pressure) is changed. This happens at critical points in all systems having consolute boundaries, and is also true of the $L_1$ and $L_2$ phases in soluble surfactant systems at the cmc. A focus of the present work was to determine whether or not the $L_3$ phase in $C_{12}E_3$ and related systems merges smoothly with the other liquid phases that are present, or is separated from them by miscibility gaps.

In the $C_{10}E_4$ and $C_{12}E_5$ systems (Figure 1), it is evident by inspection that the $L_3$ phase is a discrete region that may coexist with the $L_1$ phase, the $L_2$ phase, or the D phase within appropriate ranges of composition and temperature. It is also evident that neither the $L_1$ nor the $L_2$ phase merges with the $L_3$ phase. However, the aqueous phase behavior of the closely

# Sponge Phase in Aqueous Surfactants

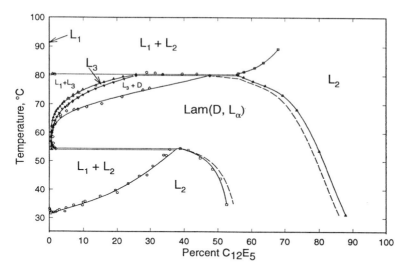

**Figure 1** The $C_{12}E_5$-water system showing the sponge phase and the sponge phase corridor, determined by R. Strey et al. (Ref. 6). The published diagram has been edited to suggest (with dashed lines) likely locations of the lamellar liquid crystal boundaries outside the corridor. The cusp in the $L_2$ region that lies between 0% and 50%, (where it touches the lower part of the corridor) may actually be an azeotropic point, but this has not been firmly established (private communication, R. Strey).

related surfactants $C_{10}E_3$ and $C_{12}E_3$ is much less clear (37). Serious violations of the Phase Rule exist in the published diagrams of these systems, so that the published data cannot be fully trusted.

Further, uncertainty has existed for some time with respect to the phase behavior of the low ethoxylogs of fatty alcohols. It has been firmly established that fatty alcohols do not exhibit surfactant-like phase behavior in aqueous systems (38), and that ethoxylogs containing three or more oxyethylene groups (such as $C_{10}E_3$ and $C_{12}E_3$) *do* exhibit surfactant phase behavior (37,39). The behavior of the respective $C_{10}E_1$, $C_{10}E_2$, and $C_{12}E_1$ ethoxylogs has not been reported, however, so it was uncertain as to whether surfactant phase behavior first appears within a $C_nE_x$ series at the $E_1$ or at the $E_2$ ethoxylog. In the $C_{12}E_2$-water system it has been reported that two cubic phases, the lamellar phase, the $L_1$, $L_2$, and $L_3$ liquid phases exist. Exploratory DIT studies suggest that the published diagram may not be correct in all details, but these data do indicate that surfactant behavior exists in this compound (40).

In the present work the $E_1$, $E_2$ and $E_3$ ethoxylates of the $C_{10}$ and $C_{12}$

alcohols were synthesized in a high state of purity, and their aqueous phase behavior was examined using the DIT-IR phase study method. Particular care was taken to address the influence of autoxidation on phase behavior, as it is known that this very real problem becomes progressively more serious as the level of impurities is decreased (41–43). Evidence that autoxidation does measurably influence the phase behavior of polyoxyethylene surfactants was again found during the present studies. Precautions to exclude oxygen were taken, but were not entirely successful.

## IV. MATERIALS AND METHODS

### A. Syntheses

Mono-, di-, and trioxyethylene glycol decyl and dodecyl ethers were synthesized from the redistilled pure glycols and alkyl bromides, using the modified Williamson synthesis of Gibson (in which sodium hydroxide replaces sodium as the reagent for forming the sodium alkoxide) (44). The melting points of the $C_{12}$ compounds fall between 10°C and 20°C, so it was possible to utilize recrystallization from pentane (with pressure filtration at $-10$°C to $-15$°C) as the final purification step for these compounds. Gas chromatography (GC) analysis of the trimethylsilyl derivative, using a Hewlett-Packard HP5880A Gas Chromatograph with autoinjector and DB-1 dimethylsilicone column (J&W Scientific Co.) gave the assays indicated in Table 1. After collecting and washing the precipitate, melting, and evaporating the solvent to constant weight, the samples were promptly subdivided and sealed in ampoules in vacuum. No apparent change in the GC assay occurred during the sealing process. The bulk of the sample was contained in a large ampoule, but a number of small ampoules containing 80–100 mg aliquots were also prepared. This procedure enables one to use a freshly opened ampoule in each experiment without excessive waste of sample. Since the DIT sample is contained within a silica capillary sealed by glass plates (cemented to the end of the cell using UV-cured resins), it too is protected from oxygen once loaded. Being saturated aliphatic ethers that

**Table 1**  Assays and Melting Points of $C_nE_x$ Surfactants

| Chain length | $E_1$ | $E_2$ | $E_3$ |
|---|---|---|---|
| $C_{10}$ | 99.5% | 99.6% | 99.5% |
| $C_{12}$ | 99.9%, 20.0°C | 99.9%, 18.3°C | 99.9%, 16.7°C |

are extremely stable thermally and transparent to light, these compounds are not photoreactive, and the only mode of decomposition that occurs within these vials is radiation-induced decomposition.

In spite of these precautions, variability in the physical data was observed that was traced to autoxidation. The temperatures of peritectic discontinuities could be reproduced within ±0.025°C during each determination, but the observed temperatures varied from one run to the next by almost 6°C (Table 2). This variation was particularly severe if the DIT cell had been exposed to high temperatures for long periods of time. GC analysis of a sample that had been stored at room temperature and then kept at high temperature indicated that measurable chemical degradation had occurred. The assay of the major component in this sample was originally 99.87%, decreasing to 99.37% after 26 days at room temperature in a septum-sealed gc vial (which does not rigorously exclude oxygen) and further to 98.67% after 3 days' additional storage at 58°C. GC analysis (Figure 2) indicated that numerous impurities ascribable to autoxidation had appeared as trace components. The positions of the lower ethoxylates that result are marked by dashed lines in Figure 2. The products of autoxidation of this class of compounds are well known, and their presence influences the physical behavior of the sample. The data suggest that either a rise or a fall in the transition temperatures may result from the presence of these impurities. The purity does not affect the sharpness of the transition, however.

## B. DIT-IR Studies

These were initiated in the manner described for DIT-NDX studies (25), after which the cell was placed in one slot of a two-slot thermostated oven designed for near-FT-IR transmission spectroscopic studies. During qualitative determination of the phase behavior, the cell was viewed using a Leitz

**Table 2** Influence of Autoxidation on $C_{12}E_3$ Peritectic Temperatures

| Days | $L_3$ Peritectic temperature | D Peritectic temperature |
|---|---|---|
| 0 | 57.2°C | — |
| 1 | 58.6°C | — |
| 5 | 59.3°C | 55.3°C |
| 22 | 58.5°C | — |
| 25 (3 days @ 58°C) | 54.3°C | 49.5°C |

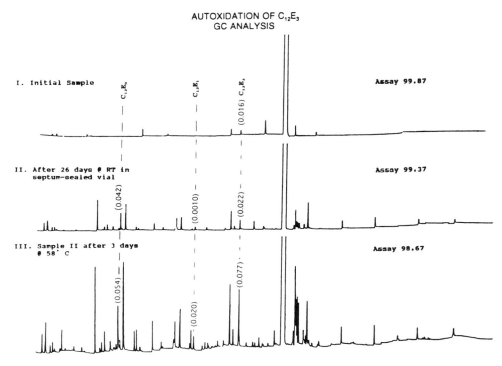

**Figure 2** Gas chromatographic traces of $C_{12}E_3$ as initially prepared (I); after 26 days at room temperature in a GC vial (crimp-sealed with a silicone septum coated on two sides with Teflon®, Supelco Cat. No. 2-7108) (II); and II after 3 days at 58°C (III). The injected sample volume and other GC parameters were identical in all three runs.

Orthoplan-Pol microscope fitted with Jamin–Lebedeff optics (25). After the qualitative behavior was established, quantitative analyses were performed using a Mattson Galaxy 4020 Spectrometer (modified for near-infrared studies) and SpectraTech IR-Plan Research Microscope with redundant adjustable apertures (26). The cell was positioned using a SpectraTech Motorized Micropositioning Stage (Version 0.95). A carbon-tetrachloride-filled reference cell was placed in the other slot, the oven was assembled, and the temperature was adjusted to the desired value. A few hours after the initially rapid swelling had subsided, a spectrum was taken to each side of each interface present using 50 × 250 μm collimated apertures oriented with the long dimension across the width of the cell chamber (450 m). The single-beam sample spectrum (4 cm$^{-1}$ resolution, 128 coadded scans) was ratioed to a carbon tetrachloride background spectrum (taken immediately before each sample spectrum) to obtain the absorbance spectrum. Water composition was estimated by integrating the area of the water combination band between 4600 and 5400 cm$^{-1}$. After normalizing the peak area data to 1 cm path length using cell thickness calibration data (25), the percentage of water was inferred using the CTA (composition–temperature–area) algorithm previously determined using octyldimethylphosphine oxide–water calibration data (45).

Data at more than one temperature can be obtained using the same cell—provided the temperature is programmed from one temperature to the next at a rate sufficiently slow for diffusive transport processes to keep up. If the programming rate is too fast, artifacts such as the emulsions that may be formed at peritectic discontinuities (described below) are introduced. For the $C_{12}E_3$ system, temperature programming rates of either 0.05 or 0.1 °C/min were used. The slower rate is preferred for quantitative studies, as the bands are longer, the gradients in composition are more shallow, and the validity of the data is greater.

## V. $C_{12}E_3$–$H_2O$ PHASE DIAGRAM

Exploratory studies of the $C_{12}E_3$–$H_2O$ system indicated that its phase diagram was qualitatively different from the published diagram, shown in Figure 3 (39). The phase diagram determined during the present work is shown in Figure 4, and data on the isothermal discontinuities that exist in this system are listed in Table 3. DIT-IR data were in good agreement with the published data at high temperatures, where a single interface corresponding to the $L_1/L_2$ miscibility gap (46) is found. Also, the composition of the $L_1$ phase was found to be zero, within experimental error, as expected. (The actual composition of this phase should closely resemble the cmc of $C_{12}E_3$, which is approximately 15–20 mg/l [47].)

As the temperature is programmed downward from above 60°C, the $L_3$ phase is inserted precisely at the $L_1/L_2$ interface. No such phenomenon is to be expected from the published diagram. Just before the $L_3$ phase appears, the $L_1/L_2$ interface is seen as a black line that is razor-sharp. A Becke line (a bright zone of constructive interference parallel to the interface) is also visible, its character and position depending on the focusing. The Becke line moves into the $L_2$ phase band (the phase of higher index) as the stage is dropped, as expected (48).

When the temperature is lowered from this point, an optical anomaly (a brightly colored zone about 10–15 $\mu$m wide) appears within the $L_2$ region exactly parallel to the interface. After 1–2 min a new interface develops,

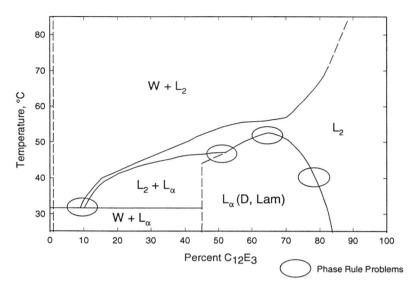

**Figure 3** The published phase diagram of the $C_{12}E_3$-water system, with Phase Rule problems circled. (From Ref. 39, used with permission.)

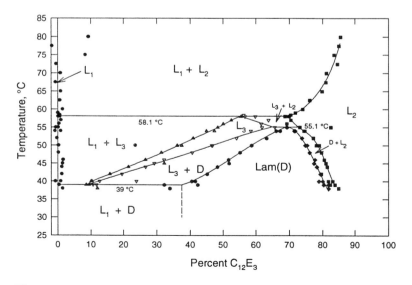

**Figure 4** The $C_{12}E_3$-water system, as determined using the DIT-IR method.

**Table 3** Isothermal Discontinuities in the $C_{12}E_3$-$H_2O$ System

| Name | Dilute phase | | Intermed. phase | | Conc. phase | | Temp. |
|---|---|---|---|---|---|---|---|
| $L_3$ Peritectic | $L_1$ | 0.002% | $L_3$ | 55.5% | $L_2$ | 69.9% | 58.1°C |
| D Peritectic | $L_3$ | 65.1% | DD | 69.3% | $L_2$ | 72.8% | 55.1°C |
| $L_3$ Eutectic | $L_1$ | 0.002% | $L_3$ | 8.6% | D | 37.4% | 39°C |

and the new $L_3$ phase band is formed. If the temperature is held at this value, the $L_3$ band will grow and display a substantial width.

If the temperature is raised to its original value before the new band length becomes too large, the new phase vanishes within seconds. If the temperature is again dropped, the insertion process is repeated, and if it is again raised the new phase is destroyed. Phase insertion and destruction can be repeated as many times as wished, always with the same result. A temperature change of 0.05°C is sufficient to cause phase insertion or phase destruction, so the temperature of the discontinuity can be established to within ±0.025°C during each determination.

It is established by such experiments that insertion of the $L_3$ phase is a reproducible, reversible process that occurs isothermally within the limitations of the temperature control, which is about ±0.03°C. The observation that the insertion temperature varies from one sample to another by a much larger amount (±3°C) is due to impurities introduced by autoxidation (see Section IV).

$L_3$ phase insertion occurs precisely at an interface (not within a phase band), which suggests that the composition of the new phase falls within the span of the $L_1/L_2$ miscibility gap. This premise was confirmed by DIT-IR data, and these two pieces of information establish the fact that this is a classical peritectic phase transition (49). In this particular case, the peritectic phase reaction involved is

$$0.21\, L_1 + 0.79\, L_2 \rightleftharpoons L_3 + q^{peri}$$

where $q^{peri}$ stands for the heat of the reaction. The reaction coefficients indicate the mass fractions of $L_1$ and $L_2$ phases required to form unit mass of the $L_3$ phase. They are dictated by the compositions of the three phases involved. When conducted reversibly, mass is conserved during this reaction as is required of all reversible processes in closed systems. Given the value of $q^{peri}$, a complete description of the isothermal calorimetric heat effects to be expected of the equilibrium process that occurs at the temperature of the peritectic, and within the composition span of the peritectic line, could be calculated (50).

If the $L_3$ phase band is allowed to grow too large before the temperature is raised, an increase in temperature causes an "emulsion" to form instead of regenerating the original $L_1/L_2$ interface. Discrete droplets (typically of the $L_1$ phase within the $L_2$) appear within the span of the cell chamber originally occupied by the $L_3$ band. This phenomenon simply reflects the situation that the span of distance covered by the $L_3$ phase has become too large for diffusion to reestablish the original interface, and nucleation occurs randomly within the $L_3$ band. When this happens, compositions no longer vary in a monotonic fashion along the length of the cell, and a new cell must be prepared if quantitative data are to be obtained.

It is of interest that the initial anomaly is visible only within the $L_2$ phase — never within the $L_1$. This observation suggests that the path by which the $L_3$ phase is formed is diffusion of water into the $L_2$ phase — not diffusion of surfactant into the dilute liquid phase. This result is not surprising, in hindsight, given the fact that the diffusion coefficient of water is substantially larger than is that of either surfactant monomers or micelles in aqueous mixtures, and the $L_1$ phase contains almost no surfactant (51).

If the temperature is lowered further after the $L_3$ band is formed, a similar process occurs again about 3°C below the $L_3$ peritectic temperature. This second discontinuity is also a peritectic phase transition, at which the phase reaction

$$0.45\,L_3 + 0.55\,L_2 \rightleftharpoons D + q^{peri}$$

occurs. While qualitatively similar to the $L_3$ transition (except for the phase structures involved), the details differ. The D phase appears more quickly than does the $L_3$ phase, and spans a wider range of x-coordinates within the chamber. Nevertheless, it may again be shown that a phase insertion process occurs within the span of the $L_3/L_2$ miscibility gap, that the insertion is essentially isothermal, and that it is readily reversible.

As the temperature is programmed downward below the D peritectic, the band sequence $L_1/L_3/D/L_2$ is preserved until about 44°C, at which point anomalies appear within the $L_3$ phase band and complexity occurs in the swelling processes. Below 39°C the $L_3$ phase no longer exists, and the band sequence is $L_1/D/L_2$. The data taken as a whole indicate that the $L_3$ phase vanishes at a eutectic discontinuity at about this temperature.

### A. Anomalous Swelling in $C_{12}E_3$

The swelling processes that occur above 44°C are straightforward and lead to the usual clean, sharp interfaces. The swelling of $C_{12}E_3$ at room temperatures is far more complex, and does not lead to the simple result found at higher temperatures. Instead, the phenomenon of "anomalous swelling" is observed (28). Anomalous swelling produces a striking complex, geometric, nonreproducible texture (Figure 5) (28). Widely differing surfactant molecules form similar textures, the characteristic short dimensions of which are similar to the chamber thickness. Anomalous swelling was originally discovered during studies of the polar lipids of myelin tissues, and the texture that results has been termed "myelinic texture." Myelinic texture has since been recognized to occur in many other systems, however. It has been suggested that myelinic texture results from the intrusion of bulk liquid water into a liquid crystal phase (52).

Anomalous swelling has been found to date only among surfactants that are poorly soluble in liquid water, and only at temperatures where the saturated liquid phase coexists with a lamellar liquid crystal. $C_{12}E_3$ below 39°C meets both of these criteria. In dilong-chain surfactants such as polar lipids or dialkyl quaternary ammonium salts, myelinic texture can be long-lived. In $C_{12}E_3$, however, this texture is not formed under certain conditions, and a simple $L_1/D$ interface results (Figure 6). The conditions which lead to this situation have not been fully defined. The myelinic textures of $C_{12}E_3$ thus appear to be kinetically unstable relative to those of dilong-chain

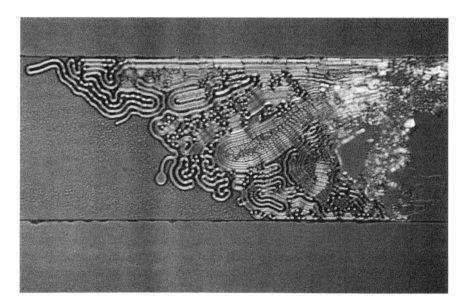

**Figure 5** The myelinic texture initially formed at the $L_1/D$ interface by the swelling of $C_{12}E_3$ at 28°C. Part of the textured D-phase band is seen to the right.

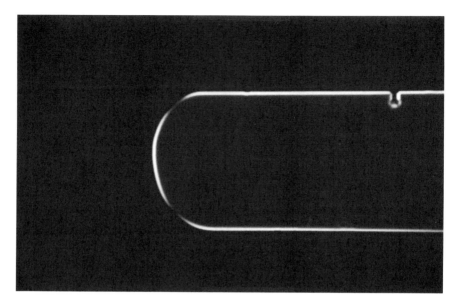

**Figure 6** The clean $L_1/D$ interface that results after programming slowly from higher temperatures down to 30°C, and then standing for 4 days. The $L_1$ phase is to the left. The existence of the pseudoisotropic (planar, homeotropic) featureless texture in the center of the band and brightness along the edge of the cell is characteristic of the D phase, but oily streak and mosaic textures are also typically seen elsewhere (see Figure 4). In this instance, brightness is also visible at the $L_1/D$ interface, but this is less common.

surfactants. The collapse of myelinic textures formed from single-chain surfactants which have a finite solubility by the process of ripening should be comparatively rapid (53a,53b). It is particularly noteworthy that the swelling of $C_{12}E_2$ (which does not have an $L_1/D$ miscibility gap) never produces myelinic textures.

Anomalous swelling aside, the phase sequence within the cell ($L_1/D/L_2$) and the equilibrium phase behavior at room temperature differ significantly from the phase behavior just below the D peritectic. The $L_3$ phase apparently disappears at a eutectic transition near 39°C, but the temperature of this transition cannot be ascertained with the same precision that is possible in the case of the two peritectics. A eutectic is the inverse of a peritectic in that a phase of intermediate composition is created by the isothermal *addition* of heat to a biphasic mixture (rather than by subtraction). The equation for this particular eutectic process is

$$0.77\, L_1 + 0.23\, D + q^{\,eu} \rightleftharpoons L_3$$

Straightforward eutectic insertion phenomena have been observed many times during other DIT studies, for example during the study of the *N*-dodecanoyl-*N*-methylglucamine–water system (30). In the case of $C_{12}E_3$ the observations are more complex and anomalous swelling confuses the issue, but the existence of this eutectic is nevertheless established. Anomalous swelling causes serious experimental difficulties in determining the compositions of the $L_1$ and D phases below the $L_3$ eutectic temperature, but is a colloidal phenomenon that is best treated separately from the equilibrium phase behavior (54).

## B. Liquid Crystal Boundaries

One of the curious facts of surfactant phase science is that the actual boundaries of liquid crystal regions are comparatively difficult to determine. A liquid boundary coexisting with a liquid crystal can easily be determined, but the liquid crystal itself cannot. This is mainly because the clean separation of liquid crystal phases during isothermal analytic studies is often incomplete. Separating coexisting liquid crystal phases is hopeless. These problems are irrelevant during swelling studies, however, as phases of high viscosity behave more simply than do phases of low viscosity (which are easily distorted by hydrodynamic forces). For this reason, more reliable information about the form of liquid crystal boundaries is obtained from DIT studies than from any other method, and it is of interest to consider this aspect of the $C_{12}E_3$ phase behavior. One important feature of their shape is the cusps that exist.

## C. Cusps in Phase Boundaries

Cusps in the boundaries of fluid phases are of special importance thermodynamically. Cusps reflect the crossing of lines described by different mathematical algorithms, and the two intersecting lines at a cusp cannot usually be defined using the same algorithm. Physically, *such cusps denote the existence of a discontinuity in the nature of the coexisting phase at the temperature of the cusp.* A familiar example is the sharp (about 90°) cusp in the boundary of the micellar liquid solution phase at the Krafft eutectic temperature of soluble surfactants (55). The slope of the Krafft boundary is slightly negative at this cusp, and the coexisting phase beyond the composition of this boundary is a crystal phase. The slope of the liquid crystal boundary above the Krafft eutectic is nearly vertical, and the composition of the coexisting liquid crystal beyond this boundary is similar to that of the liquid itself. It is the large thermodynamic difference between the crystal phase and the liquid crystal phase that is responsible for this cusp.

There are three cusps in the boundaries of the $L_3$ phase region. At the upper temperature limit is the cusp at the peritectic discontinuity, above which the phase vanishes as an equilibrium state. The data indicate that the slope of the dilute $L_3$ boundary below this peritectic is positive, while that of the concentrated $L_3$ boundary is negative. This shape was not evident from earlier data, but is more commonly encountered than if the two boundaries had both displayed positive (but different) slopes (as suggested in the tentative diagram of $C_{10}E_3$ [37]).

Below the peritectic another sharp cusp exists in the $L_3$ boundary at the D-phase peritectic temperature. Above this peritectic line the concentrated $L_3$ phase coexists with the $L_2$ phase, and the $L_3$ boundary has a negative slope. Below it the concentrated $L_3$ phase coexists with the D phase, and the $L_3$ boundary has a positive slope. The sharpness of this cusp signifies that a substantial difference exists in the thermodynamics of these two phases ($L_2$ and D), as well as in their composition and phase structure. Finally, the dilute $L_3$ phase vanishes at a eutectic and another sharp cusp exists here. Just as at the peritectic discontinuity, this cusp exists because the Phase Rule requires that the phase have a unique composition at these discontinuities.

There are two visible cusps in the liquid crystal phase boundary in this system. One occurs at the upper temperature limit of the phase, where (like the $L_3$ phase) the liquid crystal decomposes at a peritectic discontinuity. Typically the upper temperature limit of a liquid crystal phase is defined by an azeotropic point, and cusps do not exist at azeotropic points; they are the points at which two smooth boundaries are tangential (53a,53b). The difference in phase behavior at the upper temperature limit of the liquid crystal in the $C_{12}E_3$-water system (from that which is usually encountered) is presumably a direct consequence of the interference between the liquid crystal and liquid/liquid miscibility gap regions. This interference is also responsible for the existence of the $L_3$ phase. Had this not occurred, an azeotropic discontinuity (such as is found at the maximum temperatures of the liquid crystal regions in $C_{10}E_5$ or $C_{12}E_6$ [56]) would be expected.

Another cusp in the liquid crystal exists at the $L_3$ eutectic temperature. At this point the (dilute) coexisting phase above this temperature is the $L_3$ phase, while below this temperature the coexisting phase is the $L_1$ phase (nearly pure water). Other cusps likely exist at other points in the diagram where phase boundaries have not been determined.

## VI. OTHER $C_nE_x$ COMPOUNDS

Five other surfactants were studied to a degree sufficient to qualitatively define their phase equilibria, but quantitative phase diagrams have not been determined. Of particular interest is $C_{10}E_3$ (37). This surfactant is at high temperatures qualitatively identical in its phase behavior to $C_{12}E_3$ (the isothermal phase transition temperatures are lower), but at low temperatures these two surfactants differ. In this system the $L_3$ phase does not disappear at a perceptible lower eutectic temperature, but instead extends down to < 4°C (as shown in Figure 5.22 of Reference [23]). The temperature control limits of the present DIT system are only 25–80°C, so that well-controlled DIT studies down to 4°C are not possible. However, it was possible to program the temperature down to room temperature, transfer the cell to a 4°C room, and examine it using a microscope at this temperature. The same band sequence found at room temperature persists at 4°C. Had dramatic differences in phase behavior existed between 4°C and 25°C, this would not have been observed.

As the lipophilic chain length is decreased from 12 to 10, and as the number of oxyethylene groups is reduced from two to three in the $C_{10}$ and $C_{12}$ series, the upper temperature limits of the $L_3$ and D phases progressively decrease. Table 4 presents a summary of these data.

The phase behavior of $C_{12}E_2$ as revealed by exploratory DIT studies differs from that of $C_{12}E_3$ as determined in this work, and also from the published phase behavior (40). Starting from high temperatures, first an isotropic phase ($L_3$?) inserts at the $L_1/L_2$ interface, then the D phase inserts at the more concentrated of the two interfaces. Ultimately, as many as six different phase regions (five of them isotropic) are found between 25°C and 35°C; their structures have not all been assigned. No such complexity was observed in $C_{10}E_2$, but the phase behavior below room temperature cannot be examined by DIT and so is uncertain.

**Table 4** Isothermal Discontinuity Temperatures in $C_nE_x$ Surfactants

|  | $C_{10}E_2$ | $C_{10}E_3$ | $C_{12}E_2$ | $C_{12}E_3$ |
|---|---|---|---|---|
| $L_3$ Peritectic | 31.7°C | 52.4°C | 34.9°C | 58.1°C |
| D Peritectic | 27°C | 50.6°C | 33.6°C | 55.1°C |

## A. Nonsurfactant $C_nE_1$ Compounds

DIT studies of both $C_{10}E_1$ and $C_{12}E_1$ established the fact that surfactant phase behavior does not exist in aqueous mixtures of either of these compounds, using the previously defined criteria (57). No liquid crystal phases were observed — only the liquid/liquid miscibility gap. Penetration studies performed at 4°C also failed to disclose the existence of liquid crystal phases. These data demonstrate that within the $C_nE_x$ series ($n$ = 10 or 12), surfactant behavior first appears at $E_2$. Neither the fatty alcohols nor the $E_1$ ethoxylogs display surfactant phase behavior.

This result settles the issue of just how many ether and hydroxyl groups are required for surfactant phase behavior to exist among polyoxyethylene surfactants. One hydroxyl group alone (as in fatty alcohols) has been shown not to behave as a hydrophilic group, and now one hydroxyl groups plus one ether group has also been found not to be a hydrophilic group. The combination of two hydroxyl groups *does* constitute an operative hydrophilic group (monoglycerides and alkane-1,2-diols), however, as does the combination of one hydroxyl and two ether groups. These results are fully consistent with the considerable evidence that the intrinsic hydrophilicity of the ether group is somewhat less than is that of the hydroxyl group (58). Exactly how many ether groups in polyethers are required to create an operative hydrophilic group has still not been determined. Clearly one ether group is inoperative and five are operative (as in tetraoxyethylene glycol decyl methyl ether). The four ether groups in the modification of $C_{10}E_4$ in which a fluorine atom replaces the hydroxyl group are not operative (59), but just where surfactant behavior first appears within a simple aliphatic series of polyethers remains uncertain.

## VII. SURFACTANTS THAT DISPLAY THE $L_3$ PHASE OR THE PHASE CORRIDOR

The existence of the $L_3$ phase was first documented (in binary systems) in the $C_{10}E_4$-water and later in the $C_{12}E_5$-water systems. In these two systems interference between the liquid/liquid miscibility gap and the lamellar liquid crystal region is a prerequisite for its existence. Qualitatively similar phase behavior is also reported to exist in the $C_{12}E_4$-water system (39). In all these systems the $L_3$ phase vanishes on cooling, at a eutectic that lies above the critical point of the liquid/liquid miscibility gap, and the lower part of the miscibility gap (and also the critical point) is visible.

Viewing these diagrams from a broad perspective, a corridor is seen to exist which is tied to the lamellar liquid crystal region at its concentrated end and cuts through the liquid/liquid miscibility gap at its dilute end. Both the $L_3$ and lamellar phases are found within this corridor, as the swelling of the lamellar phase is highly exaggerated within the corridor (6). Also, a portion of the $L_2$ region is visible just below the corridor. These three compounds ($C_{10}E_4$, $C_{12}E_4$, $C_{12}E_5$) differ in this respect from $C_{10}E_3$, $C_{10}E_2$, $C_{12}E_3$, and $C_{12}E_2$ in that while the $L_3$ phase exists in all seven surfactants, the corridor is visible only in the first three. Thus, the existence of the $L_3$ phase and the existence of the corridor are not perfectly correlated. When the corridor exists the $L_3$ phase also appears to exist, but the $L_3$ phase is found in systems that do not display the corridor. Other data of uncertain quality exist which conflict with this generalization (22,39), and more work will be required to firmly establish the situation. It is possible that the corridor exists in all systems that form the $L_3$ phase, but its lower limit is obscured by interference from another boundary. This is not uncommon; for example, the freezing of ice or the separation of surfactant crystals is often superimposed upon another phase boundary, and obscures all or part of that boundary (60).

An attempt has been made to broadly survey the surfactant phase literature with respect to (1) the existence of an $L_3$ phase, and (2) the existence of the corridor. It is far less ambiguous to recognize the corridor than the $L_3$ phase, because identifying the $L_3$ phase with certainty requires incisive information as to phase structure. This survey included consideration of previously unreported data obtained by T. W. Gibson (21) using the stepwise dilution isoplethal method (34). These data were obtained on surfactants that are polyether diols in which a primary ($RCH_2OH$) hydroxyl group is present at the far end of a polyoxyethylene chain (as found in $C_nE_x$ compounds), and a secondary ($R_2CHOH$) hydroxyl group is positioned near the lipophilic chain. The two hydroxyl groups bracket all (or most of) the ether groups in these compounds. The molecular structures of the compounds studied are shown in Figure 7. Their important features with respect to hydrophilicity are the existence of two hydroxyl groups, plus a variable number of ether groups. The length of the lipophilic group was taken to be the longest continuous aliphatic chain attached either to an ether oxygen, or to the carbon bearing a secondary hydroxyl group. A useful acronym for these hydrophilic groups is $O_x(OH)_2$, in which $x$ designates the number of ether groups present. A parallel acronym for the hydrophilic groups in $C_nE_x$ compounds would be $O_x(OH)$. The results of this survey are tabulated in Tables 5 and 6.

It has been suggested that a "pivotal" structure exists in families of surfactants such as those listed in these tables (23). The pivotal surfactant along any row or column may be regarded as the most strongly hydrophilic (or least lipophilic) compound that displays a phase corridor. More strongly

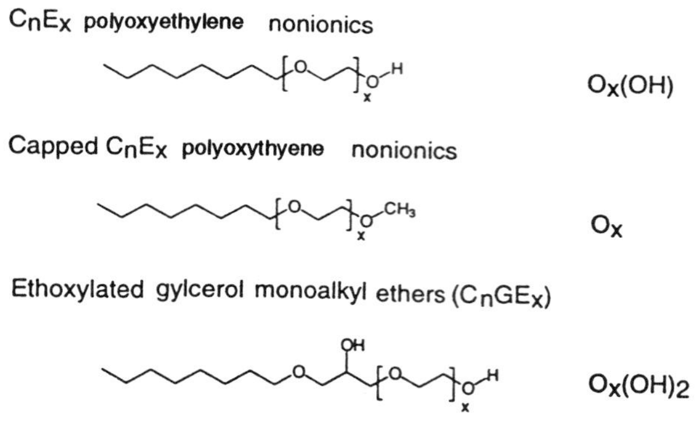

C$_n$E$_x$ polyoxyethylene nonionics

O$_x$(OH)

Capped C$_n$E$_x$ polyoxythyene nonionics

O$_x$

Ethoxylated gylcerol monoalkyl ethers (C$_n$GE$_x$)

O$_x$(OH)$_2$

**Figure 7** Molecular structures of single-bond nonionic surfactants that display the corridor and/or the L$_3$ phase.

**Table 5** Phase Behavior of C$_n$O$_x$(OH) Surfactants

| Chain length | O$_2$(OH) | O$_3$(OH) | O$_4$(OH) | O$_5$(OH) | O$_6$(OH) |
|---|---|---|---|---|---|
| C$_8$ | Corridor | — | Soluble | — | — |
| C$_{10}$ | Insoluble, L$_3$ | Insoluble, L$_3$ | Corridor | Soluble | Soluble |
| C$_{12}$ | Insoluble, L$_3$ | Insoluble, L$_3$ | Corridor, L$_3$ | Corridor, L$_3$ | Soluble |
| C$_{14}$ | — | — | — | Corridor, (C$_8$PhE$_5$) | Corridor |

**Table 6** Phase Behavior of C$_n$O$_x$(OH)$_2$ Surfactants

| Chain length | O$_2$(OH)$_2$ | O$_3$(OH)$_2$ | O$_4$(OH)$_2$ | O$_5$(OH)$_2$ | O$_6$(OH)$_2$ |
|---|---|---|---|---|---|
| C$_8$ | Corridor | Soluble | — | — | — |
| C$_{10}$ | — | — | — | — | — |
| C$_{12}$ | — | Corridor | Soluble | — | — |
| C$_{14}$ | — | Corridor | — | — | — |
| C$_{16}$ | — | — | — | Corridor | Soluble |
| C$_{18}$ | — | — | — | — | Corridor |

hydrophilic (or less strongly lipophilic) surfactants display "soluble surfactant" behavior, in which case the miscibility gap and the liquid crystal regions are separated by an $L_2$ region. Less strongly hydrophilic (or more strongly lipophilic) surfactants than the pivotal structure may also display the corridor, or may display only the $L_3$ phase.

In the $C_{10}$ row in the monohydroxypolyether family, for example, the $E_4$ ethoxylog is clearly the pivotal structure. In the $O_6(OH)$ column the $C_{14}$ homolog is evidently the pivotal structure, while in $O_6(OH)_2$ surfactants the $C_{18}$ homolog (four carbons longer) is the pivotal structure. Inspection of Tables 5 and 6 reveals that, in general, the pivotal lipophilic group chain length is longer for an $O_x(OH)_2$ functional group than it is for an $O_x(OH)$ functional group, for any value of $x$. Also, the pivotal hydrophilic group is more hydrophilic for longer lipophilic groups than it is for shorter lipophilic groups. It has been suggested that the qualitative concept of HLB (hydrophilic-lipophilic balance) is relevant to this aspect of surfactant phase behavior (24), and these data provide additional support for that premise. A perennial problem with HLB is the fact that hydrophilicity has yet to be determined on a solid thermodynamic basis (61); nevertheless, the concept is qualitatively valid.

It has been recognized that hydrophilicity can be varied in both an intensive and an extensive manner (62), and that a family of ethoxylogs constitutes variation in the extensive manner. Data also exist which suggest that the $L_3$ phase (and/or the phase corridor) exists among a variety of monofunctional semipolar compounds within which hydrophilicity is varied in an intensive manner (62). The most strongly hydrophilic of these are the phosphine oxides; the $L_3$ phase or corridor has never been seen (to date) in the more hydrophilic amine oxides, for example. The data which suggest these correlations have been published in various compilations (20,21), and those semipolar hydrophilic groups which behave in this manner are shown in Figure 8.

The qualitative picture with semipolar surfactants is remarkably similar to that found among single-bond surfactants. For each hydrophilic group a pivotal structure may be identified, and shortening the chain (increasing the HLB) results in "soluble surfactant behavior." An example of this may be seen in comparing $C_{12}PO$ (dodecyldimethylphosphine oxide, which displays soluble surfactant behavior) with $C_{14}PO$ (tetradecyldimethylphosphine oxide, which displays a corridor). In $C_{14}PO$ the Krafft boundary also cuts through the liquid/liquid miscibility gap just below the corridor, and complicates the phase behavior still further. The corridor is more prominent among phosphine oxides if the proximate substituents are lengthened to ethyl, as in $C_{12}(Et)_2PO$ and $C_{14}(Et)_2PO$.

Sulfoxides are somewhat less hydrophilic than are phosphine oxides,

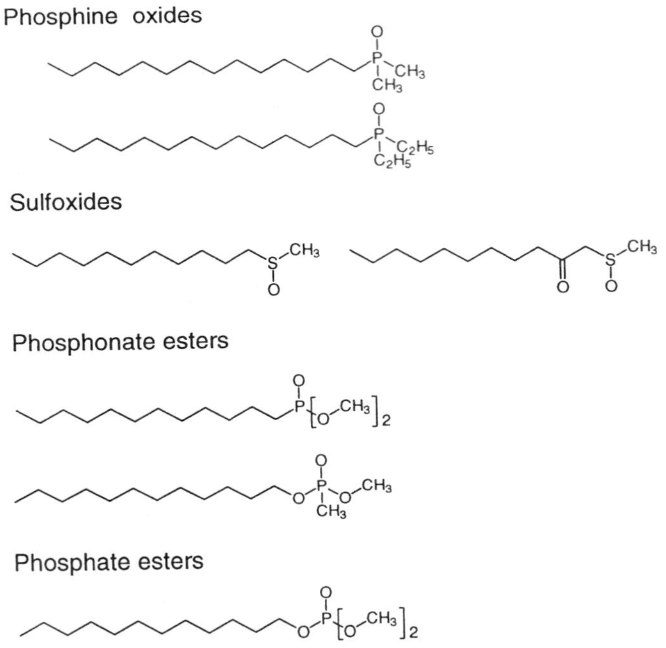

**Phosphine oxides**

**Sulfoxides**

**Phosphonate esters**

**Phosphate esters**

**Figure 8** Molecular structures of semipolar surfactants the display the corridor and/or the $L_3$ phase.

and it is interesting that the $L_3$ phase appears to exist in two sulfoxides (undecyl methyl sulfoxide, and 2-ketoundecyl methyl sulfoxide). Phosphonate esters are less hydrophilic than are sulfoxides, and in these compounds too the $L_3$ phase appears to exist. It is of interest that the two structurally isomeric phosphonate esters display virtually identical phase behavior (including the $L_3$ phase). As found elsewhere, the details of conformational structure and molecular shape are almost immaterial insofar as phase equilibria involving fluid states are concerned. Finally, the phosphate ester (which is the least hydrophilic semipolar group) appears to resemble the phosphonate esters, except that the thermal stability of the liquid crystal and the temperature range of the $L_3$ phase are lower. None of these compounds display the corridor.

## VIII. SUMMARY

It has become clear during the last 15 years that a relatively highly structured liquid phase (the $L_3$ or "sponge" or "anomalous" phase) exists in selected aqueous surfactant systems, in addition to the molecular and micellar solution phases. While considerable information exists regarding the structure and theoretical analysis of the $L_3$ phase, its exact position within the phase diagram and coexistence relationships with other phases are often much less clear. In this work it has been established that the $L_3$ phase in the $C_{12}E_3$-water system is a discrete phase, as had earlier been established in the $C_{10}E_4$ and $C_{12}E_5$ systems. The DIT-IR method has proven invaluable during these studies; its use has disclosed serious discrepancies between the actual and the published diagram of this and other surfactants.

Previously unreported information has been presented which extends and supports the premise, suggested earlier (4), that the existence of the $L_3$ phase is a consequence of interference between the liquid/liquid miscibility gap and the lamellar liquid crystal phase region.

The $C_{12}E_3$ DIT study revealed the existence of anomalous swelling (to form myelinic textures) at temperatures where the $L_1$ and D phase coexist, below the lower eutectic limit of the $L_3$ phase. While complicating the phase study, the existence of this phenomenon does not necessarily invalidate determination of the equilibrium phase behavior using this method. No anomalous swelling at all is observed in the $C_{12}E_2$-water system, which does not have an $L_1$/D miscibility gap.

## IX. A QUARTER CENTURY OF PROGRESS IN AQUEOUS SURFACTANT PHASE SCIENCE

A quarter century ago, aqueous surfactant phase science was a half century old. The first giant steps had been taken toward determining the complex phase behavior of these systems, not only in binary systems but also in ternary systems containing salts or oils. The intermediate phases that exist in aqueous soap systems had gone unrecognized during that entire time span, but their existence has since been documented.

A quarter century ago the number of surfactant structures whose aqueous phase behavior had been explored was limited. The number that have since been explored is vastly greater than was true then.

A quarter century ago surfactant chemists could not have stated exactly which polar functional groups were operative hydrophilic groups, and which were not. Little knowledge of the semipolar and zwitterionic subclasses existed. Now, all the polar functional groups that are operative as hydrophilic groups have been identified. It is even possible to scale a priori the relative intrinsic hydrophilicity of these groups.

A quarter century ago the fact that cloud point miscibility gaps were the lower part of a closed loop of coexistence was unknown, and the existence of the $L_3$ phase in weakly hydrophilic surfactants had not been recognized.

A quarter century ago reliance had to be placed entirely on isoplethal methods in executing binary phase studies—the same methods used by McBain during the early part of this century. Now, an isothermal phase studies method (the DIT method) exists which provides a level of information about surfactant phase behavior that was previously unattainable. It has yet to be widely used, however.

In spite of this considerable progress, great voids in our knowledge of this area exist (for example in the area of surfactant crystal chemistry). Misleading errors are common in published phase diagrams. Our knowledge of ternary phase behavior is truly primitive compared to the knowledge one would like to have; even the best studies are typically limited to one temperature. New methods applicable to ternary systems that are both more efficient and more incisive need to be developed, if this very important barrier to the practical utilization of surfactant phase science is to be overcome.

## ACKNOWLEDGMENTS

The author is greatly indebted to G. M. Bunke for syntheses of the samples used in this work, and to B. J. King for the careful execution of DIT-R studies. R. L. Munyon made substantial contributions of the development of this method during the early stages. R. E. Shumate provided access to the gas chromatograph and assistance during analyses of the samples. The various polyether diol surfactants were all synthesized by Dr. T. W. Gibson of the Procter & Gamble Miami Valley Laboratories (retired). Dr. Gibson also performed the preliminary screening of the phase behavior of these compounds.

## NOTES AND REFERENCES

1. RG Laughlin, RL Munyon, Y-C Fu, AJ Fehl. J Phys Chem 94:2546–2552, 1990.
2. GS Hartley. Aqueous Solutions of Paraffin Chain Salts. Paris, Hermann & Co, 1936.
3a. TL Hill. Thermodynamics of Small Systems. Vol. 1. New York, Benjamin, 1963.
3b. TL Hill. Thermodynamics of Small Systems. Vol. 2. New York, Benjamin, 1964.
4. JC Lang, RD Morgan. J Chem Phys 73:5849–5861, 1980.
5. The terminology of these liquid phases varies. In this paper $L_1$ will refer to a

molecular solution such as exists below the cmc, $L_2$ to a micellar liquid solution, and $L_3$ to the anomalous or sponge phase.

6. R Strey, R Schomaeker, D Roux, F Nallet, U Olsson. J Chem Soc Faraday Trans 86:2253–2261, 1990.
7. G Gompper, US Schwartz. Z Phys B: Condens Matter 97:233–238, 1995.
8. J Diacic, U Olsson, H Wennerstrom, G Jerke, P Schurtenberger. J Phys II 5:199–215, 1995.
9. J Appell, G. Porte, JF Berret, DC Roux. Prog Colloid Polym Sci 97:233–236, 1994.
10. W Helfrich. J Phys: Condens Matter 6:A79–A92, 1994.
11. G Gompper, M Schick. Phys Rev E 49:1478–1482, 1994.
12. RM Hill, M He, HT Davis, LE Scriven. Langmuir 10:1724–1734, 1994.
13. U Olsson, U Wuerz, R Strey. J Phys Chem 97:4535–4539, 1993.
14. EB Sirota, CR Safinya, RJ Plano, C Jeppesen, RF Bruinsma. Mater Res Soc Symp Proc 248:169–178, 1994.
15. ME Cates. Springer Proc Phys 66:275–280, 1992.
16. H Hoffmann, C Thunig, U Munkert, HW Meyer, W Richter. Langmuir 8:2629–2638, 1992.
17. SM Zourab, CA Miller. Collods Surf 95:173–183, 1995.
18. C Quilliet, M Kleman, M Bernillouche, F Kalb. C R Acad Sci Ser II: Mec Phys Chim Astron 319:1469–1474, 1994.
19. M He, RM Hill, Z Lin, LE Scriven, HT Davis. J Phys Chem 97:8820–8834, 1933.
20. RG Laughlin. In: Advances in Liquid Crystals. Brown, GH, ed. Vol. 3. 1978, pp 41–148.
21. J Sjoblom, P. Stenius, I Danielsson. In: Nonionic Surfactants, Physical Chemistry. Schick, MJ, ed. Vol. 23. 1987, pp 372–373.
22. TW Gibson, Miami Valley Laboratories, Procter & Gamble, unreported work.
23. RG Laughlin. The Aqueous Phase Behavior of Surfactants. London, Academic Press, 1994, pp 148–150.
24. RG Laughlin. The Aqueous Phase Behavior of Surfactants. London, Academic Press, 1994, pp 149–150, 354–355.
25. RG Laughlin, RL Munyon. J Phys Chem 91:3299–3305, 1987.
26. C Marcott, RG Laughlin, AJ Sommer, JE Katon. Fourier Transform Infrared Spectroscopy in Colloid and Interface Science, ACS Symposium Series No. 447, Scheuing, DR, ed. 1991, pp 71–86.
27. RG Laughlin. J Am Oil Chem Soc 67:705–710, 1990.
28. RG Laughlin. Adv Coll Interface Sci 41:57–79, 1992.
29. The phenomenon of anomalous swelling (28) does occur in some circumstances, and represents an exception to this generalization.
30. Y-C Fu, RG Laughlin, AS Glardon. Paper on the N-Dodecanoyl-N-methylglucamine-Water System, presented at the American Oil Chemists' Society Meeting, Anaheim, CA, April 25–29, 1992.
31. RG Laughlin, RL Munyon. Paper on the Dioctadecylammonium Cumenesulfonate-Water System, presented at the ICSCS Conference, Compiègne, France, July 7–13, 1991.

32. C Madelmont, R Perron. Colloid Polym Sci 254:6581–6595, 1976.
33. It is always possible to miss interfaces at isooptics, where the refractive indices of coexisting phases are the same. Sensitive optical methods must be utilized to detect any interface across which the index discontinuity is very small, as such interfaces may be nearly invisible using either bright field or polarized light illumination if both are isotropic. Jamin–Lebedeff interference optics have proven invaluable for the qualitative observation of interfaces. Isooptic phenomena have been encountered from time to time, e.g., during the study of the $N$-dodecanoyl-$N$-methylglucamine–water system (30). Fortunately, isooptic circumstances do not usually persist over a wide range of temperatures, and so become evident if temperature is varied during a study.
34. RG Laughlin. J Coll Interface Sci 55:239–241, 1976.
35. RG Laughlin. The Aqueous Phase Behavior of Surfactants. London, Academic Press, 1994, pp 441–459.
36. RG Laughlin. The Aqueous Phase Behavior of Surfactants. London, Academic Press, 1994, pp 54–56, 92–97.
37. RG Laughlin. The Aqueous Phase Behavior of Surfactants. London, Academic Press, 1994, p 148.
38. RG Laughlin., The Aqueous Phase Behavior of Surfactants. London, Academic Press, 1994, p 252.
39. DJ Mitchell, GJT Tiddy, L Waring, TA Bostock, MP McDonald. J Chem Soc Faraday Trans I 79:975–1000, 1983.
40. JP Conroy, C Hall, CA Leng, K Rendall, GJT Tiddy, J Walsh, G Lindblom. Prog Colloid Polym Sci 82:253–262, 1990.
41. JM Corkill, JF Goodman, RH Ottewill. Trans Faraday Soc 57:1627–1636, 1961.
42. M Donbrow, R Hamburger, E Azaz. J Pharm Pharmacol 27:160–166, 1975.
43. R Hamburger, E Azaz, M Donbrow. Pharm Acta Helv 50:10–17, 1975.
44. T Gibson. J Org Chem 45:1095–1098, 1980.
45. C Marcott, RL Munyon, RG Laughlin. Proceedings, 8th International Conference on Fourier Transform Spectroscopy, SPIE Vol. 1575, 1991, pp 290–291.
46. The nomenclature used to define miscibility gaps is as follows. A gap is defined by stating first the more dilute (water-rich) phase and then the more concentrated phase, separated by a slash, e.g., ($L_1/L_2$). The phase within this gap with which one is concerned is indicated before the parentheses. For instance, the dilute phase at the $L_1/L_2$ miscibility gap is $L_1(L_1/L_2)$, and the concentrated phase is $L_2(L_1/L_2)$. The phases that exist at three-phase discontinuities are similarly designated, with the more dilute phase being indicated first, e.g., ($L_1/L_3/L_2$) at that peritectic where the phase reaction $L_1 + L_2 - q \rightleftharpoons L_3$ occurs.
47. P Mukerjee, KJ Mysels. Critical Micelle Concentrations of Aqueous Surfactant Systems. National Standard Data Reference Series NBS-36. Washington, DC, US Government Printing Office, 1971, p 150.
48. NH Hartshorne, A Stuart. Crystals and the Polarizing Microscope. 2nd Ed. London, Edward Arnold, 1950, pp 224–229.
49. RG Laughlin. The Aqueous Phase Behavior of Surfactants. London, Academic Press, 1994, pp 61–62.

50. RG Laughlin, The Aqueous Phase Behavior of Surfactants. London, Academic Press, 1994, pp 80–82.
51. P Stilbs. Progress in NMR Spectroscopy 19:1–45, 1987.
52. RG Laughlin, RL Munyon, JL Burns, TW Coffindaffer, Y Talmon. J Phys Chem 96:374–383, 1992.
53a. AS Kabalnov, AV Pertsov, ED Shchukin. J Coll Inter Sci 118:590–597, 1987.
53b. AS Kabalnov, KN Makarov, AV Pertzov, ED Shchukin. J Coll Interface Sci 138:98–104, 1990.
54. RG Laughlin. The Aqueous Phase Behavior of Surfactants. London, Academic Press, 1994, pp 88–89.
55. RG Laughlin., The Aqueous Phase Behavior of Surfactants. London, Academic Press, 1994, pp 106–114.
56. RG Laughlin., The Aqueous Phase Behavior of Surfactants. London, Academic Press, 1994, pp 122, 124.
57. RG Laughlin., The Aqueous Phase Behavior of Surfactants. London, Academic Press, 1994, pp 248–252.
58. RG Laughlin., The Aqueous Phase Behavior of Surfactants. London, Academic Press, 1994, pp 300–308.
59. RG Laughlin., The Aqueous Phase Behavior of Surfactants. London, Academic Press, 1994, p 305.
60. RG Laughlin., The Aqueous Phase Behavior of Surfactants. London, Academic Press, 1994, pp 114–115.
61. RG Laughlin. J Soc Cosmet Chem 32:371–392, 1981.
62. RG Laughlin. The Aqueous Phase Behavior of Surfactants. London, Academic Press, 1994, pp 253–259.

# 4
## Surfactant Self-Assembly Structures at Interfaces, in Polymer Solutions, and in Bulk: Micellar Size and Connectivity

Björn Lindman, Fredrik Tiberg,* Lennart Piculell,
Ulf Olsson, Paschalis Alexandridis,† and Håkan Wennerström
*Lund University, Lund, Sweden*

## I. THE SPHERICAL MICELLE

It was G. S. Hartley who made the pioneering contributions to our understanding of the physical nature of globular micelles (1,2). He realized inter alia the liquid-like nature of the hydrocarbon interior of the micelle and the minimal contact between the solvent water and the surfactant hydrocarbon chains as the main driving force for the association. As illustrated in Figure 1, he pictured the interior of a micelle as essentially "a droplet of liquid paraffin" with the chains in "as near as can be done, a completely chaotic arrangement." He also argued strongly against some of his contemporaries drawing pictures with radially oriented straight alkyl chains (1): "The symmetrical 'asterisk' form so frequently drawn would imply a high degree of organisation and a density of great magnitude in the centre and of less magnitude near the circumference. It has no physical basis and is drawn of no other reason than that the human mind is an organising instrument and finds unorganised processes uncongenial."

The self-assembly of surfactant molecules into a sphere with minimal contact between the hydrocarbon chains and water would lead to a hydrocarbon core radius equal to the extended length of the surfactant alkyl chain; this is indeed observed experimentally (3,4). In fact, studies of the

---
*Current affiliations:*
*Institute for Surface Chemistry, Stockholm, Sweden.
†State University of New York–Buffalo, Buffalo, New York.

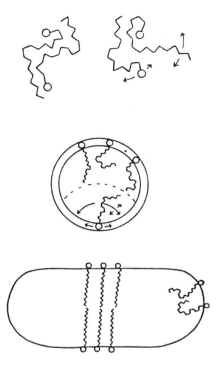

**Figure 1** Schematic of spherical and elongated micelles, and conformations of surfactant molecules therein, as proposed by Hartley. (From Ref. 2, used with permission; copyright Kolloid-Z, 1940.)

dimensions of aggregates in surfactant systems are generally quite informative regarding the mechanisms involved in self-assembly.

The Hartley picture has been repeatedly questioned, but there is now a consensus against any solid-like character of the surfactant alkyl chains and against proposals of an extensive water penetration far into the micelle core. Many have contributed to this picture, both experimentally and theoretically. Thirty years ago, P. Mukerjee provided an illuminating summary of our understanding, and has contributed with both thermodynamic and spectroscopic arguments (5–7). NMR relaxation experiments have provided quantitative information on motions and order of the alkyl chains and have indeed shown that their state is only very slightly different from that in a liquid hydrocarbon (except for the anisotrophy in motion dictated by the fixation of the surfactant headgroups at the interface) (8–10). Both neutron scattering (11,12) and NMR (13) techniques have demonstrated that only in the surface region of the micelle is there any significant hydrocarbon-chain water contact (14). (Previous confusion in this respect arose from misconceptions regarding the probe localization in techniques employing molecular probes, and incorrect interpretation of experimental data.) It has also been recently demonstrated by Monte Carlo simulations that the hydrocarbon–water boundary in a micellar system is extremely sharp (15,16). Even for a weakly self-assembling $C_7$ system, the water concentration changes from nearly 100% to almost 0% over a distance of 3 Å, comparable to the size of a water molecule. The accepted basic features of the spherical micelle have remained unchanged during quite some time, and the review by Gruen (17,18) as well as our own reviews (19,20) are still relevant.

## II. LARGER MICELLES

It was realized at an early stage that the nearly spherical micelle is not the only type of aggregate, but that at higher concentrations there may be a growth of the micelles (21–24). We now know, of course, that we can control micellar growth in many different ways, e.g., by electrolyte addition, addition of weakly surface-active solubilizates, temperature (19,20).

Hartley (1) correctly refuted a model by Lawrence for the formation of "secondary" micelles in terms of the association of "primary" (spherical) micelles by "cohesion of the polar end-groups which reside in the surface of the latter," and interpreted the observations in terms of a growth of the hydrocarbon interior (Figure 1). However, Hartley assumed that the surfactant hydrocarbon chains in these elongated micelles were in a very much more ordered straight state (except in the end-caps) (1,2). We now know that, on the contrary, there are only small differences in the states of the hydrocarbon chains for micelles of different shapes or sizes; we also know that this conclusion extends to all types of surfactant self-assemblies (above the Krafft point) (25).

Parallels to the Hartley versus Lawrence debate have, nevertheless, remained surprisingly vital, and several descriptions similar to that of Lawrence with notations such as "string-of-beads" (26) and "cross-linked micelles" (27) have appeared in literature. The arguments, however, could always be refuted (28) (a recent report invoking "cross-linked micelles" (27), for example, misinterprets the NMR relaxation data). This matter has a much broader significance than for simple micellar solutions and is, for example, very relevant also for microemulsions. There is nothing *a priori* that can lead us to exclude under all conditions the possibility of "micellar clustering" or "droplet clustering" as opposed to growth of the micelle or droplet core. If information on the shape of an aggregate can be obtained, however, the problem is often directly resolved. Observations of long, locally cylindrical micelles with a radius close to the surfactant molecule length imply that there is a growth of the core rather than an association of spherical micelles. We are not aware of any interaction or mechanism that can explain the aggregation into chain-like structures at equilibrium. Analogous arguments are applicable in the case where growth in two dimensions to disk-like structures is found with a thickness close to twice the length of the surfactant molecule or slightly less. Such structures cannot be understood in terms of clustering or association of spheres, where the result would be uniform growth in three dimensions.

## III. DISCRETE AND CONNECTED SURFACTANT SELF-ASSEMBLIES

The core growth picture is strongly supported by a broader view of self-assembled surfactant structures. Indeed, as we illustrate in Figure 2, there is a large number of possible structures, some of which are characterized by discrete "micellar" aggregates, while others contain nonpolar domains

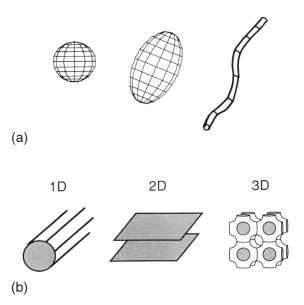

**Figure 2** Examples of discrete (a) and continuous (b) surfactant self-assembled structures; the latter can extend on 1 (cylinders), 2 (lamellae), or 3 (bicontinuous bilayer) dimensions.

# Surfactant Self-Assembly Structures

connected in one, two, or three dimensions (29). This applies both to systems that can be described in terms of monolayer structures and those that are of the bilayer type (Figure 3). In any surfactant system we have a segregation into water-rich and oil-rich domains as well as surfactant films. The surfactant films can be pairwise correlated into bilayer structures or be uncorrelated.

## IV. MICELLES, MICROEMULSIONS, AND SURFACTANT INTERFACIAL STRUCTURES

Micelles (19,20), microemulsions (30,31), monolayers and bilayers are central issues of the science of amphiphilic molecules (29). Perhaps we tend too often to consider them as separate matters, forgetting the strong connections and analogies there must be. For example, the very large number of self-assembled structures that have been identified in bulk surfactant–water systems, both isotropic solution and liquid crystalline phases, stand in strik-

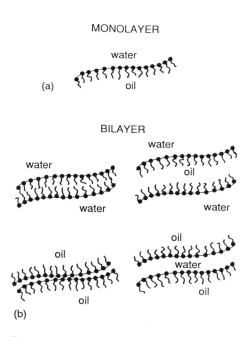

**Figure 3** Surfactant monolayers (a) and bilayers (b): in any surfactant system we have a segregation into water-rich and oil-rich domains, as well as surfactant films. The latter can be pairwise correlated (into bilayers) or uncorrelated.

ing contrast to the common monolayer or bilayer picture of surfactants as interfaces. Conversely, it took very long before connected-type structures penetrated into our thinking of micellar solutions. It was also long until it was generally accepted that the discrete spherical micellar-type structure is not the only one possible in microemulsions, and not even the typical one (32). The interrelations and analogies between the various modes of self-assembly would be fruitful for future developments of our understanding of surfactant systems. For example, while we will not expect identical surfactant structures at interfaces and in bulk, there is little reason to believe that aggregate polymorphism should be less rich for interfacial structures.

In this chapter we will by way of a few selected examples try to illustrate these problems. Our examples will be taken throughout from studies of nonionic surfactant systems, but there is no doubt that the general arguments apply equally well to other classes of surfactants.

## V. SELF-ASSEMBLY OF NONIONIC SURFACTANTS

Nonionic surfactants, also appropriately considered as short A-B diblock copolymers, have polar (usually polyethyleneoxide) and nonpolar parts of similar size. They exhibit the important feature that the spontaneous curvature ($H_0$) of the surfactant film is strongly temperature dependent (Figure 4) (33–36). At low temperature, when water is a good solvent for the poly(ethylene oxide) groups, the surfactant film is spontaneously curved toward the hydrophobic domains ($H_0 > 0$), while at high temperature, when the solvency conditions worsen, there is a spontaneous curvature toward water ($H_0 < 0$). The length of the polar headgroup affects the spontaneous curvature in a direction opposite to that of an increase in the temperature.

A consequence of this is that the self-assembled structures vary with temperature, as illustrated in Figure 5 (which shows only some of the identified structures). For the present discussion we should note the sequence spherical micelle → elongated micelle with circular cross section → vesicle → planar bilayers → sponge phase ($L_3$) as a function of increasing temperature or decreasing ethylene oxide (EO) chain length. The change in micelle size is illustrated in Figure 6, where the effective hydrodynamic radius, deduced from the micellar self-diffusion coefficient, is plotted against temperature (37–39). Note that at room temperature, the $C_{12}E_5$ micelles are very large and thread-like, while the $C_{12}E_8$ micelles are small and very nearly spherical.

The same types of considerations apply for ternary systems of surfactant, oil, and water. In a Shinoda section (40) of the phase diagram (at fixed surfactant concentration), we illustrate the fact that the spontaneous

# Surfactant Self-Assembly Structures

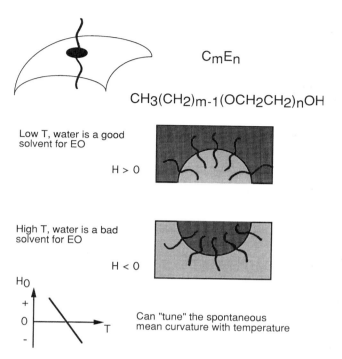

**Figure 4** The spontaneous mean curvature ($H_0$) of nonionic surfactant (alkyl-polyethyleneoxide, $C_mE_n$) films can be tuned with temperature. At low temperatures, when water is a good solvent for the poly(ethylene oxide) headgroup, the curvature is toward the hydrophobic domains ("oil-in-water"), while at high temperatures, when the solvency conditions worsen, the curvature is toward water ("water-in-oil").

curvature is primarily determined by temperature (Figure 7). At low or zero content of hydrocarbon (Figure 7a) the following sequence (as already illustrated in Figure 5) appears with increasing temperature: small spherical micelles — oil droplets that grow into elongated aggregates, lamellar liquid crystalline phase (and vesicles), and the bicontinuous sponge phase (based on a normal, oil-in-water bilayer). At the low-water-content side (Figure 7b) we have an analogous sequence of reversed structures moving toward lower temperatures: reversed spherical micelles — water droplets that grow into elongated aggregates, lamellar phase (and reversed vesicles), and the bicontinuous sponge phase (based on a reversed, water-in-oil bilayer).

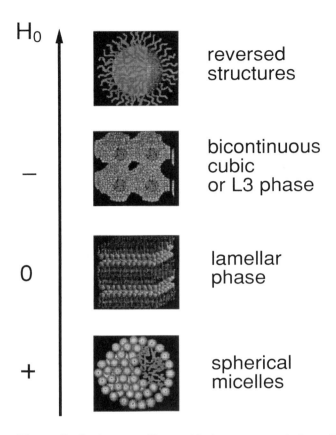

**Figure 5** Surfactant self-assembly in aqueous solutions. Schematic of different nonionic surfactant self-assemblies: the sequence of spherical micelles, planar bilayers, branched (bicontinuous) bilayers, and reverse ("water-in-oil") micelles is the result of increasing temperature or decreasing poly(ethylene oxide) (headgroup) chain length as reflected in the spontaneous curvature.

In the microemulsion channel, going from the lower left corner in the diagram to the upper right corner, there is also a change in connectivity from discrete oil droplets to discrete water droplets via the connected bicontinuous state (Figure 7). Finally, we note the channel stretching from the upper left corner of the phase diagram to the lower right corner where the structure is bicontinuous throughout.

These changes between connected and discrete structures in surfactant–oil–water microemulsions are, of course, not unique for nonionic sys-

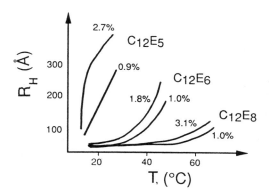

**Figure 6** Change in the micelle size effected by temperature for a number of nonionic surfactants (data extracted from References [37] and [38]). Nonionic micelles [$CH_3(CH_2)_{n-1}(OCH_2CH_2)_mOH$] grow with increasing temperature. Growth into elongated micelles is evident in the case of the $C_{12}E_5$ surfactant, while the $C_{12}E_8$ micelles remain spherical. Hydrodynamic radius = $kT/6\eta D_{mic}$.

tems, but can be found for all types of surfactants. However, different parameters tune the spontaneous curvature for different systems. While nonionics are tuned by temperature, ionics are tuned by electrolyte concentration (19,20,41).

Whether a phase is discrete or connected in the different components has consequences for several properties, but most notably transport coefficients. We illustrate this in Figure 8 by showing the relative oil and water self-diffusion coefficients ($D/D_0$) in a microemulsion system that undergoes a transition with temperature from oil-in-water to bicontinuous to water-in-oil (see Reference [35] for a full discussion).

## VI. CUBIC LIQUID CRYSTALS

Regarding lyotropic liquid crystalline phases, several of the most frequently occurring, in particular hexagonal and lamellar phases, were already at an early stage identified as being connected in one and two dimensions. Others, like the so-called nematic isotropic phases, were thought to be built up of discrete aggregates of different shapes, although the presence of continuous aggregates has been suggested recently also in this case (42). For other phases, in particular the cubic ones, the situation was confused for a long time and mistakes were made regarding the connectivity.

We now know that there are several types of cubic phases occurring in

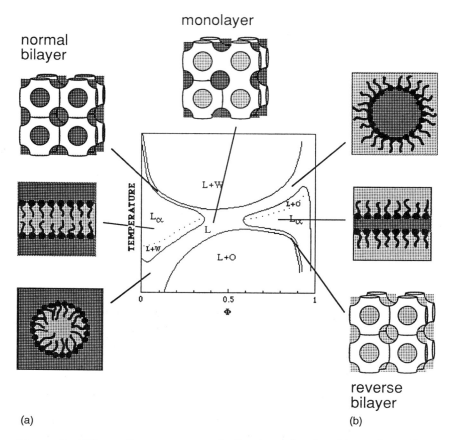

**Figure 7** Effects of temperature and oil volume fraction ($\phi$) on the surfactant monolayer spontaneous curvature and on the surfactant–oil–water phase behavior for a system containing a fixed surfactant concentration. At low or zero content of oil (a) at low water content (b). At similar water and oil contents, a bicontinuous structure with a surfactant monolayer separating the water and oil domains can be stable.

different regions of a phase diagram (43–45). Some cubic structures are built up of discrete (slightly anisometric) globular micelles of the normal or reversed type; there are also a number of structures consisting of infinitely connected surfactant aggregates. The micellar cubic phases may consist of either normal micelles ($I_1$ phase) or reverse micelles ($I_2$). The bicontinuous phases are often denoted $V_1$ when the curvature is toward oil and $V_2$ when the curvature is toward water. Consistent with a monotonic variation of the

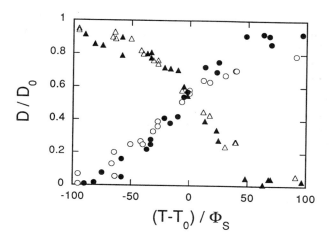

**Figure 8** The relative self-diffusion coefficients indicate the microemulsion structure change from oil-in-water to bicontinuous to water-in-oil with increasing temperature. Reduced self-diffusion coefficients, $D/D_0$, of water (triangles) and oil (circles) from two systems, $C_{12}E_5$-$D_2O$-tetradecane (filled symbols) and $C_{12}E_5$-$D_2O$-cyclohexane/hexadecane (unfilled symbols), are plotted as a function of the reduced temperature $(T - T_0)/\Phi_s$. $T_0$ is the temperature at which the spontaneous curvature is zero; by dividing the temperature scale with the surfactant concentration, $\Phi_s$, one takes into account the linear mean curvature dependence on $\Phi_s$. (Adapted from Ref. [35].)

interfacial mean curvature, the various cubic phases have different relative locations in the phase diagram according to the following generic sequence (44,45):

$$L_1-I_1-H_1-V_1-L_\alpha-V_2-H_2-I_2-L_2$$

which is symmetric around the lamellar phase ($L_\alpha$). The other phases are the normal ($H_1$) and reverse hexagonal ($H_2$) phases and the isotropic liquid solution phases with normal ($L_1$) and reverse ($L_2$) micelles, respectively. This phase sequence corresponds to a decreasing mean curvature ($H_0$) from the left ($H_0 > 0$) to right ($H_0 < 0$), where $H_0 = 0$ in the lamellar phase.

Progress in the field was assisted by the identification of more than one cubic phase in a given surfactant–water system. In this case, surfactant self-diffusion measurements allow one to distinguish between structures based on discrete units and those in which the surfactant films are connected over large distances. The self-diffusion in the latter is orders of magnitude more rapid than in the former (46).

Examples for nonionic surfactants with one cubic phase built up of discrete micelles and another of connected bilayers are shown in Figure 9, where a discrete micellar cubic region appears between the micellar solution and the hexagonal lyotropic liquid crystalline phase, and bicontinuous cubic region appears between the hexagonal and lamellar phases (47). Nonionic block copolymer self-assemblies can exhibit an even richer morphology (48–50). A recent example from a study of nonionic block copolymers in mixtures with oil and water is shown in Figure 10 (50), where two discrete and two bicontinuous cubic phases, one of each having a positive curvature and the other negative, have been identified in an isothermal system.

## VII. SURFACTANT SELF-ASSEMBLIES IN POLYMER SOLUTIONS

Examples of the role of aggregate size and type for surfactants mixed in polymer-containing systems are presented here; again we select the examples from nonionics. Phase diagrams for ternary mixtures of a nonionic surfactant, dextran and water (51) are shown in Figure 11. Micelles are characterized by large aggregation numbers and, therefore, exhibit typical polymer features, including "polymer incompatibility" in mixtures (51). Similar to the dextran–poly(ethylene oxide)–water system, there is a segregative phase separation in the nonionic-surfactant–dextran–water system. However, there are striking differences between different nonionic surfactants that illustrate the role of micellar size. For $C_{12}E_8$, which forms spherical (i.e., small) micelles, there is a rather weak segregation. Furthermore, the miscibility between nonionic micelles and polymer tends to increase with increasing temperature, as one would expect. For $C_{12}E_5$, on the other hand, the segregation is much more important and increases quite strongly with increasing temperature. Making an analogy with polymer solutions, one would assume that, in the latter case, there is a considerably higher degree of polymerization that also increases strongly with increasing temperature. This is exactly in line with our observations of large $C_{12}E_5$ micelles (Figure 6), which grow strongly as the temperature increases.

For a repulsive polymer–surfactant system, the diffusion of the surfactant micelles is mainly determined by simple excluded volume or obstruction effects. One of the most important features of micellar diffusion in a polymer network is the micellar size. As illustrated in Figure 12, there is a dramatic difference in the effect of a polysaccharide network on the diffusion of nonionic micelles of $C_{12}E_8$ and $C_{12}E_6$. The small spherical $C_{12}E_8$ micelles are obstructed to a much lower degree than the large elongated $C_{12}E_6$ micelles (53,54).

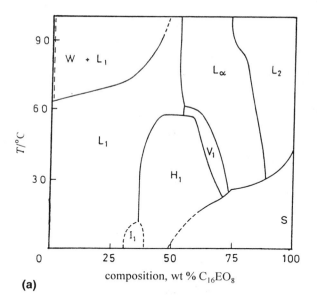

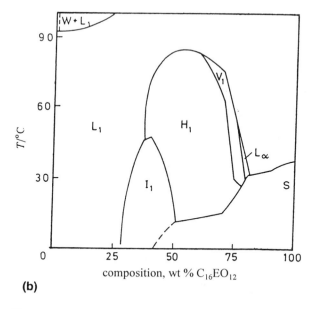

**Figure 9** Temperature-concentration phase diagrams of binary nonionic surfactant (alkyl-polyethyleneoxide)–water systems exhibiting both discrete ($I_1$) and bicontinuous ($V_1$) cubic regions; $L_1$ and $L_2$ denote isotropic solutions, while $H_1$ and $L_\alpha$ are hexagonal and lamellar lyotropic liquid crystalline regions, respectively. (a) Phase diagram of the $C_{16}EO_8$/water system over the temperature range 0–100°C. (b) Phase diagram of the $C_{16}EO_{12}$/water system over the temperature range 0–100°C. (From Ref. 47, used with permission; copyright Royal Chemical Society, 1983.)

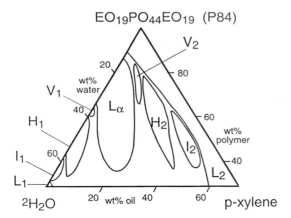

**Figure 10** Ternary phase diagram of an amphiphilic copolymer–water–oil system that exhibits two discrete ($I_i$) and two bicontinuous ($V_i$) cubic phases, one of which has a positive curvature (i = 1) and the other a negative (i = 2). (Adapted from Ref. [50].)

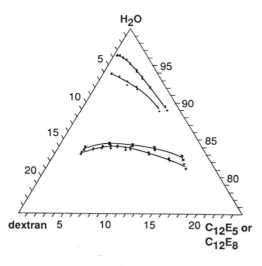

**Figure 11** Phase boundaries (compositions are expressed in wt%) obtained from cloud point measurements, for aqueous mixtures of dextran T70 (mol wt 70000) with (from top to bottom) $C_{12}E_5$ at 25°C, $C_{12}E_5$ at 10°C, $C_{12}E_8$ at 10°C, and $C_{12}E_8$ at 25°C. (Adapted from Ref. [51].)

# Surfactant Self-Assembly Structures

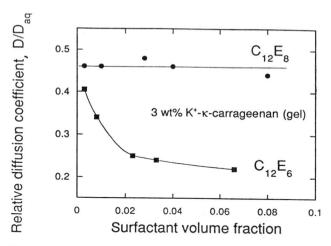

**Figure 12** The diffusion of surfactant micelles in a repulsive polymer–surfactant system is mainly determined by excluded volume or obstruction effects. In a polysaccharide (κ-carrageenan) network, the small spherical $C_{12}E_8$ micelles are obstructed to a much lower degree than the large elongated $C_{12}E_6$ micelles. (Adapted from Ref. [53].)

If the polymer contains hydrophobic groups, there will be a polymer–surfactant association rather than a segregation. Changing the temperature through the micelle–vesicle transition can lead to a thermal gelation (Figure 13) ascribed to a cross-linking of vesicles by the hydrophobically modified polymer (55).

For systems of infinite surfactant self-assemblies, such as bicontinuous microemulsions, any segregation or association effect will be particularly strong. While a regular nonionic cellulose derivative will not enter the middle-phase microemulsion of a three-phase surfactant–oil–water mixture, the corresponding polymer modified with a small number of hydrophobic grafts will associate with the surfactant films and hence dissolve in the microemulsion phase (56).

## VIII. STRUCTURE OF INTERFACIAL LAYERS

While a common description of surfactant adsorption on solid surfaces is given in terms of monolayers and bilayers (Figure 14) (29), there are very good reasons to assume that surfactant adsorption on hydrophilic surfaces without strong headgroup–surface interactions should more appropriately

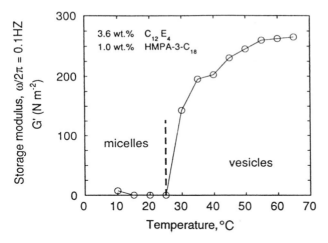

**Figure 13** An increase of the solution temperature through the surfactant micelle–vesicle transition leads to a thermal gelation (ascribed to a cross-linking of vesicles by the polymer) in a polymer–nonionic surfactant system where the polymer contains hydrophobic groups. (Adapted from Ref. [55].)

be considered in terms of surface-induced surfactant self-assembly. Depending on the surfactant, surface aggregates of different sizes are then implied (Figure 15). The first direct evidence for adsorbed layers built up of discrete micelles probably concerns nonionic surfactants adsorbed on silica surfaces studied by fluorescence quenching (57), but there is ample indirect evidence as well. We will here exemplify the surface-induced surfactant self-assembly by ellipsometric studies of the adsorption of nonionic surfactants on silica surfaces (58–60).

The equilibrium adsorption has the following characteristics (58):

1. There is little or insignificant adsorption of surfactant well below the bulk critical micelle concentration (cmc) value.
2. In a narrow concentration region leading to the assignment of a surface cmc, the adsorbed amount increases very strongly with increasing concentration. The degree of cooperativity increases with decreasing EO (headgroup) chain length of the surfactant. The surface cmc is lowered by a factor of 10 on increasing the number of carbon atoms in the surfactant alkyl chain by 2, but it is rather insensitive to changes in the number of EO segments in the headgroup. This behavior is the same as that for bulk micellization.
3. As the bulk cmc is attained, so is a plateau value of the adsorbed

# Surfactant Self-Assembly Structures

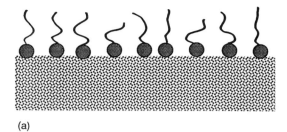

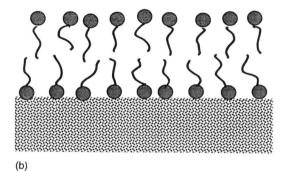

**Figure 14** Surfactant adsorption at (hydrophilic) solid surfaces in the form of monolayers (a) and bilayers (b).

   amount. The plateau adsorption varies strongly with the EO chain length, indicating that as the headgroup size increases, the surface is less and less covered by surfactant.
4. The thickness of the adsorbed layer jumps at the surface cmc from nearly zero to roughly 40–55 Å (depending on the surfactant type), corresponding to roughly twice the length of a surfactant molecule.

These observations demonstrate first that we never encounter monolayer adsorption and second that we do not see complete bilayers on adsorption. While the thickness of the adsorbed layer is consistent with both bilayers and discrete micelles of different sizes (since one of the dimensions of a micelle always corresponds to twice the surfactant molecule length), the different coverages and cooperativities for different nonionic surfactants can only be understood in terms of discrete micelles, which become smaller as the headgroup sizes increase; a transition from small surface

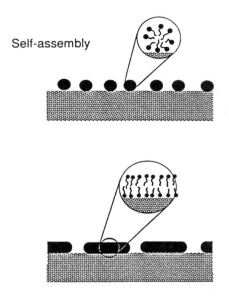

**Figure 15** The adsorption of surfactants on hydrophilic surfaces without strong headgroup-surface interactions can be viewed in terms of surface-induced surfactant self-assembly.

micelles to larger, disk-like micelles and, finally, bilayer-type aggregates is also possible. More in-depth analysis of the ellipsometric data provides quantitative estimates of the axial ratios of the surface micelles (assuming they are disk-like), which show exactly the same trends as for the bulk micelles, i.e., small micelles for $C_{12}E_8$, intermediate for $C_{12}E_6$, and large micelles for $C_{12}E_5$. We note that, although the weak attraction between the silica surface and the EO chains drives the adsorption, the adsorption decreases as the number of EO groups in a surfactant molecule increases. This demonstrates the role of surfactant self-assembly and that the number of EO groups in one micelle, rather than in one surfactant molecule, is the decisive factor.

The kinetic studies further illustrate the micellar picture of the adsorption (59). In Figure 16 we present a typical adsorption/desorption experiment. In the initial part (A1), the adsorption is determined by the diffusion of surfactant unimers and micelles (above the bulk cmc) through the stagnant layer adjacent to the surface. Transport limitations are imposed at high surface coverage, resulting in a slow-down of the adsorption rate (A2) and the attainment of the plateau value (P).

The initial desorption is also diffusion limited, with a slope which

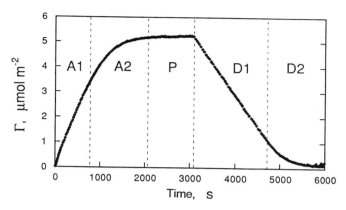

**Figure 16** Adsorption–desorption kinetics of $C_{16}E_6$ (C = 0.001 mmol/l) at hydrophilic silica surfaces. The surfactant is injected just prior to $t = 0$, while continuous rinsing with preheated water is initiated around $t = 3000$ s. A characteristic time-dependence of the adsorbed amount ($\Gamma$) is exhibited in each of the regimes, A1 (linear increase of $\Gamma$ with time), A2 (transition region), P (plateau region), D1 (linear decrease of $\Gamma$ with time), and D2 (exponential decay of $\Gamma$). (Adapted from Ref. [59].)

is proportional to the difference between the surface cmc and the bulk concentration, which is zero in these experiments. From the desorption rates, surface cmc values can be deduced that are in agreement with those obtained from the adsorption isotherms.

As the micelle concentration at the surface is reduced to a low value, the release of surfactant molecules from the micelles becomes the rate-controlling factor (rather than the diffusion) and there is a changeover from a linear to an exponential desorption. This provides further evidence for the micellar picture of the adsorbed layer. As can be seen in Figure 17, the rate in this exponential part is very strongly dependent on the surfactant alkyl chain length (but only moderately dependent on the headgroup size); the rate changes by a factor of 10 on changing the number of carbon atoms by 2, exactly as observed in kinetic studies of micelle disintegration in bulk solution.

We also note that the thickness of the adsorbed layer during adsorption and desorption experiments never takes on values consistent with a monolayer, not even when the adsorbed amount is very low.

The discrete, rather than connected, character of the adsorbed layer is also illustrated amply in adsorption/desorption experiments on mixed surfactant–hydrocarbon systems (60). The presence of a hydrocarbon leads

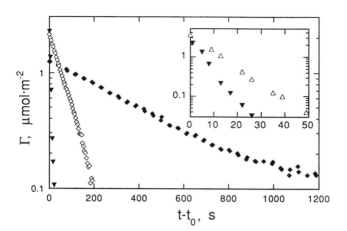

**Figure 17** Adsorbed amount of $C_{16}E_6$ (filled diamonds), $C_{14}E_6$ (open diamonds), and $C_{12}E_6$ (filled triangles) in regime D2 corrected for the small portion that remains adsorbed after long times, $\Gamma - \Gamma\infty$, versus the time from the commencement of the exponential regime, $t - t_{D2}$. The inset shows the corresponding curves obtained for $C_{12}E_5$ (open triangles) as well as that for $C_{12}6$ (filled triangles), reproduced on a shorter time-scale. (Adapted from Ref. [59].)

to slightly larger adsorbed amounts and large thicknesses due to the (well-known for bulk systems) swelling of the micelles. The most striking illustration is probably provided by the desorption experiments (Figure 18). During the initial part, the rate of desorption equals that observed in the absence of oil, i.e., desorption can be ascribed to surfactant desorption alone, with

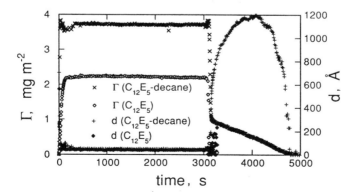

**Figure 18** Time evolution of the properties $\Gamma$ (adsorbed amount) and $d$ (adsorbed layer thickness) during the adsorption and desorption of $C_{12}E_5$-decane and neat $C_{12}E_5$ at silica. (Adapted from Ref. [60].)

# Surfactant Self-Assembly Structures

a rate controlled by the surface cmc value. During this desorption process there is a dramatic increase in the extent of the adsorbed layer from the surface. During desorption, the surface aggregates are depleted with respect to surfactant but not with respect to oil. This change in the composition leads to a changeover from surfactant-dominated aggregates (with a dimension limited by the surfactant molecule length in at least one dimension) to oil-dominated large droplet aggregates.

During the second, also diffusion-controlled, desorption step, oil is removed and the rate becomes controlled by the hydrocarbon solubility in water. Therefore, there are dramatic changes in the desorption rate as the number of carbon atoms in an alkane is varied (Figure 19).

## IX. FACTORS GOVERNING THE STABILITY OF SELF-ASSEMBLED STRUCTURES

Trying to understand the factors governing the stability of a certain aggregate, we can identify a number of relevant mechanisms.

1. The most important feature is the *spontaneous curvature* of the surfactant monolayer. This monolayer can reside at the border between aqueous and oil domains, but it can also rest on another oppositely oriented monolayer to form a bilayer.
2. The *amount of surfactant* determines the total monolayer or film area, while the *ratio between aqueous and apolar* volumes provides a constraint on the regions the film should enclose.
3. A third factor is *entropy,* which in particular at high dilutions favors the formation of finite aggregates relative to infinite ones. Entropic factors also give a preference to flexible disordered structures relative to ordered ones.
4. A fourth factor that becomes increasingly important at higher concentrations is the *interaggregate interaction* or the interaction with a surface.

Let us illustrate these different factors by considering the $C_{12}E_6$ micellar system at low surfactant concentrations. An optimally packed spherical micelle has a radius of ~3 nm. At low temperatures (20 °C), water is a good solvent for the EO headgroups, which then prefer a high fraction of gauche conformations leading to a repulsion between neighbouring EO chains in the palisade layer. The spontaneous curvature, $H_0$, is positive and high, of the order of 1/4 (nm$^{-1}$). Increasing the temperature leads to a gradual decrease in $H_0$. The surfactant film in a spherical micelle becomes slightly frustrated. There are a number of ways to release this frustration. The micelle can reduce its mean curvature by growing in size and becoming

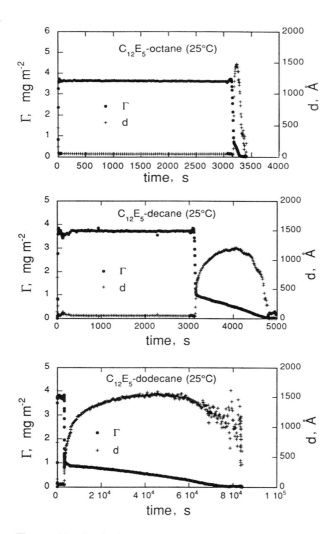

**Figure 19** Evolution of adsorption and desorption of $C_{12}E_5$–alcane systems with time. During the second desorption step, oil is removed with a rate controlled by the hydrocarbon solubility in water. These are dramatic changes in the desorption rate as the number of carbon atoms in an alkane is varied. (Adapted from Ref. [60].)

# Surfactant Self-Assembly Structures

somewhat elongated. It can interact with other micelles where the overlap of EO brushes leads to an increase in spontaneous curvature; the intrabrush interaction is partly replaced by an interbrush one. Yet, another alternative for the micelle to release the curvature frustration is to adsorb at an interface.

In all these cases there is an entropy penalty due to a decrease in the entropy of mixing. However, there is an interesting difference between the two bulk processes of micellar growth versus aggregation of intact micelles. In the latter case, aggregate size changes stepwise, while in the former the growth process is continuous, which reduces the entropy penalty for association since it allows for more possibilities.

There is, of course, also the possibility of a combination of aggregate growth and association. This is clearly an issue for micellar adsorption to a surface. The stronger the adsorption free energy, the more likely is it to find substantial differences between bulk and surface aggregate structure. For the weakly associated nonionic micelles on a silica surface the effect is small, while studies on hydrophobic surfaces like graphite show rather strong effects.

## X. CONCLUDING REMARKS

Our picture of surfactant self-assembly has in many respects remained rather unchanged during the last decades, but in some aspects our understanding has seen dramatic changes. This refers in particular to the problem of discreteness and connectivity.

Isotropic surfactant solutions were in the past generally considered in terms of smaller or larger discrete micelles. On the other hand, surfactant layers at interfaces were discussed in terms of homogeneous connected mono- or bilayers. For microemulsions, the discrete droplet picture was the only one considered until the 1970s, when it became questioned by Shinoda, Friberg, Scriven and others (61–63).

Now we generally accept that there are different types of bicontinuous surfactant solutions: they can be of the multiply connected (or branched) bilayer type, as found in the $L_3$ phase (64–66), or they can result from unlimited growth and branching of thread-like cylindrical micelles (67,68). Models of micellar growth into elongated aggregates due to chain-like association (string-of-beads) (26) can be rejected.

Our picture of microemulsion structure is summarized in Figure 7. We note that bicontinuous microemulsions are the rule for an efficient surfactant mixing of similar amounts of oil and water. There has been considerable confusion in this respect partly because of difficulties in distinguishing between changes in the size and shape of the droplets and droplet clustering.

The aggregate structure of surfactant layers at interfaces has been less discussed, mainly due to experimental difficulties (see, however, recent AFM studies [69,70]). However, as argued above, for the adsorption of water-soluble surfactants, i.e., surfactants forming micelles in bulk, on hydrophilic surfaces, there is generally an interfacial self-assembly of the surfactant into discrete aggregates. The surface micelles differ in size from those in bulk, but the size is influenced by the same factors as in bulk. Also regarding micelle kinetics, there are close analogies between interfaces and bulk.

The problem of finding a theoretical description of the balance between interactions and growth is a general one for many self-assembly processes, and it has only been solved in some particular cases.

## REFERENCES

1. GS Hartley. Aqueous Solutions of Paraffin-Chain Salts. A Study in Micelle Formation. Paris, Hermann & Cie, 1936.
2. GS Hartley. Kolloid-Z 22, 1940.
3. O Söderman, H Walderhaug, U Henriksson, P Stilbs. J Phys Chem 89:3693, 1985.
4. O Söderman, U Henriksson, U Olsson. J Phys Chem 91:116, 1987.
5. P Mukerjee. Adv Colloid Interface Sci 1:241, 1967.
6. P Mukerjee. J Phys Chem 76:565, 1972.
7. P Mukerjee. In: Micellization, Solubilization, and Microemulsions. Vol 1. Mittal, KL, ed. Plenum, 1977, p 171.
8. P Stilbs, H Walderhaug, B Lindman. J Phys Chem 87:4762, 1983.
9. B Lindman, O Söderman, H Wennerström. In: Surfactant Solutions. New Methods of Investigation. Zana, R, ed. Marcel Dekker, 1987, p. 295.
10. B Lindman, U Olsson, O Söderman. In: Dynamics of Solutions and Fluid Mixtures by NMR. Delpuech, J-J, ed. J Wiley, 1995, p 345.
11. B Cabane. J Physique 42:847, 1981.
12. B Cabane, R Duplessix, T Zemb. J Phys 46:2161, 1985.
13. B Halle, G Carlström. J Phys Chem 85:2142, 1981.
14. B Lindman, H Wennerström, H Gustavsson, N Kamenka, B Brun. Pure Appl Chem 52:1307, 1980.
15. J Shelley, K Watanabe, ML Klein. Int J Quantum Chem: Quantum Biol Symp 17:103, 1990.
16. JC Shelley, M Sprik, ML Klein. Langmuir 9:916, 1993.
17. DWR Gruen. Progr Colloid Polym Sci 70:6, 1985.
18. DWR Gruen. J Phys Chem 89:146, 1985.
19. H Wennerström, B Lindman. Phys Reports 52:1, 1979.
20. B Lindman, H Wennerström. Topics Current Chem 87:1, 1980.

21. F Reiss-Husson, V Luzzati. J Phys Chem 68:3504, 1964.
22. NA Mazer, GB Benedek, MC Carey. J Phys Chem 80:1075, 1976.
23. SJ Candau, E Hirsch, R Zana. J Physique 45:1263, 1984.
24. SJ Candau, E Hirsch, R Zana. J Colloid Interface Sci 105:521, 1985.
25. O Söderman, G Carlström, U Olsson, TC Wong. J Chem Soc, Faraday Trans 1 84;4475, 1988.
26. C Manohar, URK Rao, BS Valaulikar, RM Iyer. J Chem Soc, Chem Commun 379, 1986.
27. FM Menger, AV Eliseev. Langmuir 11:1855, 1995.
28. FAL Anet. J Am Chem Soc 108:7102, 1986.
29. DF Evans, H Wennerström. The Colloidal Domain: Where Physics, Chemistry, Biology, and Technology Meet. VCH, 1994.
30. DO Shah, ed. Macro- and Microemulsions. Washington, DC, American Chemical Society, 1981.
31. M Bourrel, RS Schechter. Microemulsions and Related Systems, Marcel Dekker, 1988.
32. B Lindman, K Shinoda, U Olsson, D Anderson, G Karlström, H Wennerström. Colloids Surf 38:205, 1989.
33. K Shinoda, H Arai. J Phys Chem 68:3485, 1964.
34. M Kahlweit, R Strey, P Firman. J Phys Chem 90:671, 1986.
35. U Olsson, H Wennerström. Adv Colloid Interface Sci 49:113, 1994.
36. R Strey. Colloid Polym Sci 272:1005, 1994.
37. PG Nilsson, H Wennerström, B Lindman. J Phys Chem 87:1377, 1983.
38. W Brown, R Johnson, P Stilbs, B Lindman. J Phys Chem 87:4548, 1983.
39. PG Nilsson, H Wennerström, B Lindman. Chem Scr 25:67, 1985.
40. K Shinoda, S Friberg. Emulsions and Solubilization. J Wiley, 1986.
41. M Kahlweit. J Phys Chem 99:1281, 1995.
42. MC Holmes, MS Leaver, AM Smith. Langmuir 11:356, 1995.
43. G Lindblom, L Rilfors. Biochim Biophys Acta 988:221, 1989.
44. K Fontell. Colloid Polym Sci 268:264, 1990.
45. RG Laughlin. The Aqueous Phase Behavior of Surfactants. Academic Press, 1994.
46. T Bull, B Lindman. Mol Cryst Liquid Cryst 28:155, 1975.
47. JD Mitchell, GJT Tiddy, L Waring, T Bostock, MP McDonald. J Chem Soc, Faraday Trans 1 79:975, 1983.
48. P Alexandridis, U Olsson, B Lindman. Macromolecules 28:7700, 1995.
49. P Alexandridis, U Olsson, B Lindman. Langmuir 13:X, 1997.
50. P Alexandridis, U Olsson, B Lindman. Langmuir, in press.
51. L Piculell, K Bergfeldt, S Gerdes. J Phys Chem 100:3675, 1996.
52. L Piculell, B Lindman. Adv Colloid Interface Sci 41:149, 1992.
53. L Johansson, P Hedberg, J-E Löfroth. J Phys Chem 97:747, 1993.
54. L Johansson, J-E Löfroth. J Chem Phys 98:7471, 1993.
55. K Loyen, I Iliopoulos, R Audebert, U Olsson. Langmuir 11:1053, 1995.
56. A Kabalnov, U Olsson, K Thuresson, H Wennerström. Langmuir 10:4509, 1994.
57. P Levitz, H Van Damme, D Keranis. J Phys Chem 88:2228, 1984.

58. F Tiberg, B Jönsson, J Tang, B Lindman. Langmuir 10:2294, 1994.
59. F Tiberg, B Jönsson, B Lindman. Langmuir 10:3714, 1994.
60. F Tiberg. J Chem Soc, Faraday Trans 92:531, 1996.
61. S Friberg, I Lapczynska, G Gillberg. J Colloid Interface Sci 56:19, 1976.
62. K Shinoda. Progr Colloid Polym Sci 68:1, 1983.
63. LE Scriven. Nature 263:123, 1976.
64. DM Anderson, H Wennerström, U Olsson. J Phys Chem 93:4243, 1989.
65. D Gazeau, AM Bellocq, D Roux, T Zemb. Europhys Lett 9:447, 1989.
66. J Daicic, U Olsson, H Wennerström, G Jerke, P Schurtenberger. J Phys II France 5:199, 1995.
67. M Monduzzi, U Olsson, O Söderman. Langmuir 9:2914, 1993.
68. I Harwigsson, O Söderman, O Regev. Langmuir 10:4731, 1994.
69. S Manne, JP Cleveland, HE Gaub, GD Stucky, PK Hansma. Langmuir 10:4409, 1994.
70. S Manne, HE Gaub. Science 270:1480, 1995.

# 5
# An Overview of Depletion and Surface-Induced Structural Forces in Thin Micellar Films

**D. T. Wasan, A. D. Nikolov, and X. L. Chu**
*Illinois Institute of Technology, Chicago, Illinois*

## I. MICELLAR STRATIFICATION PHENOMENON

One of the major areas of research in colloid science and interfacial engineering in the last quarter century has been the non-DLVO surface forces, such as depletion (i.e., volume exclusion effect) and structural forces, which affect structure formation and stability of colloidal dispersions such as foams and emulsions, as well as biological cells (1). The surfactant micelle-stabilized thin liquid films are the building blocks of concentrated colloidal dispersions. At high micellar concentrations, it has been observed that the thin liquid films become thinner in a stepwise fashion — that is to say, thinning foam and emulsion films formed from surfactant micellar solutions exhibit a number of thickness transitions before attaining an equilibrium film thickness (2-9). The process can be followed in Figure 1, which shows a photocurrent (film thickness) versus time interferogram of a horizontal flat film formed from the micellar solution of a nonionic surfactant (alcohol ethoxylate) (10).

We used the microinterferometric method to investigate thin micellar film behavior (6). As soon as the film is formed, it starts to thin. After it is thinner than 104 nm (the highest-order interferential maximum corresponding to the applied monochromatic 540 nm light reflected from the film), the film thickness is observed to change. The stratification process is temporally resolved, in that the film resides for a few seconds in each uniformly

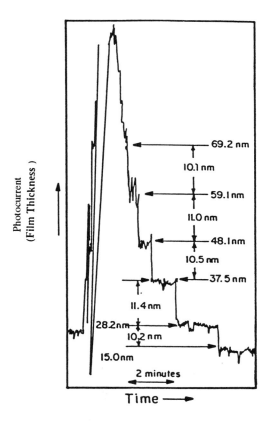

**Figure 1** Interferogram of a film formed from a solution of a nonionic detergent (Enordet AE1215, 5,2, $10^{-2}$ mol/l). As film thins, less light is reflected. Formation of metastable states of uniform thickness is revealed by "steps." Height of a step (black arrow) corresponds to micelle diameter, about 10 nm. Width of steps is proportional to lifetimes of respective metastable states.

thick state, prior to thinning to the next level. Dark spots, thinner than the rest of the film, appear and gradually grow in size, as shown in Figure 2. Eventually, the spots cover the entire film and the film "rests" for a time in a new state. Then, even darker spots appear, and after their expansion, a subsequent metastable state ensues. This process continues until the film finally reaches a stable state with no more stepwise changes. Each thickness state of the film appears in the interferogram (Figure 1) as an extended plateau which, after some time, drops step-like to the next plateau. The width of the plateau is indicative of the lifetime of the respective state. The

# Thin Micellar Films

**Figure 2** Stratification of film: 0.1 mol/l sodium dodecylsulfate surfactant solution.

height of the steps is nearly equal (about 10.6 nm), and it roughly corresponds to the bulk intermicellar distance.

The same phenomenon was observed for ionic surfactant solutions (11), such as $\alpha$-olefin sulfonates (12). For ionic surfactants the height of the steps was close to the effective micellar diameter, which includes the electric double layer around the micelles. Furthermore, foam films (i.e., thin symmetrical films between bubbles) formed from concentrated monodisperse suspensions of polystyrene latexes (13), silica particles (13), or sodium caseinate submicelles (14) stratify in a similar way, illustrating the general nature of the phenomenon.

Nikolov, Wasan, and Edwards (12) and Lobo and Wasan (15) observed the drainage and stability of pseudoemulsion films (i.e., asymmetrical aqueous micellar films between the oil and gas phases) at concentrations much above the cmc. Lobo and Wasan observed that, for a 4 wt % ethoxylated alcohol, the film thinned stepwise by stratification in a fashion similar to the symmetrical films formed from micellar solutions. Three thickness transitions were observed at 4 wt % concentration with $n$-octane as an oil phase, which was the same number of steps as observed by Nikolov et al. (10) in foam films at the same concentration. Similarly, Bergeron, Jimenez-Laguna, and Radke (16) observed stratified thinning in pseudoemulsion films. These observations on the micellar layering in the pseudoemulsion film confirms, again, the universality of the stratification phenomenon.

The phenomenon of stepwise thinning was observed not only in horizontal, but in large, vertical films, and foams and emulsions as well (17). We showed that vertical films formed from a latex suspension containing 150-nm particles in a vertical frame which contained stratification (18). A series of uniform stripes of different colors were observed at the upper part of the frame. The different colors are due to interference of the common (polychromatic) light reflected by the surface of the different stripes of uniform thickness. The boundaries between the stripes were very sharp, a consequence of the stepwise profile of the film in this region, and the liquid meniscus below the film appeared as a region with gradually changing colors. When observed in reflected light, the top strips had the following sharply distinguished colors: black, white, yellow, blue, red, and green-yellow. Following the stripes, a sequence of diffuse, alternate green and red stripes were observed that indicate the gradual change in film thickness where the order–disorder transition region is observed.

Similar, sharply defined stripes were found with vertical films from micellar solutions of nonionic surfactant (e.g., $C_{12-15}$ ethoxylated alcohol with 30 ethoxy groups) with a micellar diameter of about 10 nm (18). However, all stripes were very gray, though easily distinguished due to their different intensities, because the diameter of the micelles is small compared to the wavelength of the visible light.

Nikolov et al. (19) were the first to explain the stratification phenomenon as a layer-by-layer thinning of ordered structures of micelles or colloidal particles inside the film. According to the colloid crystal-like model first proposed by them, the different stripes in the stratifying (horizontal or vertical) films contain different numbers of micelle (or particle) layers. The micelles interact via screened electrostatic repulsion, forming an ordered structure because of the restricted volume of the film. The model permitted, for the first time, calculation of the structural contribution to the disjoining pressure of the film that arises from the presence of micellar structure within the films.

All the experimental and theoretical data for stratifying films show that stratification is a universal phenomenon and is due to the formation of a long-range colloid crystal-like structure within the foam film and a layer-by-layer thinning of such an ordered structure. This ordering occurs because highly charged Brownian particles (micelles) interact via repulsive forces and are forced into the restricted volume of the film. The classical DLVO theory of colloid stability (20,21), which explains order in colloidal systems as a balance of van der Waals attractive forces and electrostatic forces, cannot be used here because the intermicellar distances are too large for the van der Waals forces to be significant to balance the repulsive forces.

### A. Effects of Micellar Concentration and Size

The effect of important technological parameters on the film stratification phenomenon is discussed next. Figure 3 shows a plot of the effective volume fraction of the micelles as a function of the stepwise thickness transitions for anionic micellar solutions of sodium dodecyl sulfate (with a mean micellar diameter of 4.8 nm) and nonionic micellar solutions of $C_{12-15}$ ethoxylated alcohol with 30 ethoxyl groups (with a mean micellar diameter of 10 nm) (13). These curves show the effect of the effective concentration of micelle on film thickness transitions. The curves of the different systems can be

# Thin Micellar Films

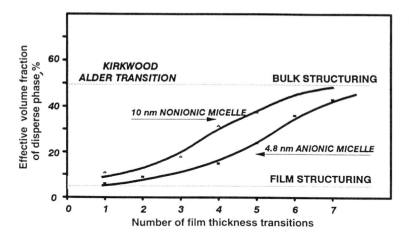

**Figure 3** Number of film thickness transition steps versus micellar concentration: dashed line, nonionic surfactant; solid line, anionic surfactant.

compared because the mechanism of film microstructuring is basically the same. The smaller-sized micelles induce more thickness transitions than the larger ones at the same effective volume fraction. With the smaller, anionic micelles, the film structuring process (i.e., order–disorder transitions) begins at a lower effective micellar volume percent (5 vol %) than with the larger, nonionic micelles (11 vol %). At those micellar concentrations, the intermicellar distance is about 2–3 times the micelle radius. This observation shows that an order–disorder transition can occur at a film thickness of several particle diameters and at a much lower concentration than 51 vol %, which was the concentration theoretically predicted by Kirkwood (22) and Alder and Wainwright (23) for structural transitions in bulk phase.

Kralchevsky et al. (24) described the film transition from one metastable state to the next one by a vacancy mechanism. The driving force behind the thinning process is the difference between the chemical potential of the micelles inside the film and inside the adjoining meniscus (Plateau border). Vacancies that initially appear at the film periphery move throughout the entire film area through micellar diffusion. For example, particles with a 19 nm diameter can pass through the film with a diameter of about $6 \times 10^{-2}$ cm in about $2 \times 10^{-2}$ s. The film-thickness transitions initiated by spot formation usually occur when there is a sufficiently large number of vacancies. As a result, a thickness transition occurs. The vacancies created at the film periphery diffuse and "condense" at the spot and gradually increase its area. Moreover, at high concentration of micelles, film structuring occurs at high film thickness and, depending on the film radius, the film can be several micelle layers thick.

Nikolov and colleagues (11,12) investigated the effect of surfactant chain length with alkyl sulfates and $\alpha$-olefin sulfonates. They found that the number of stepwise transitions increases with the chain length. For example, with $C_{16}$-$\alpha$-olefin sulfonate, two-step transitions were observed, but with the $C_{12}$-$\alpha$-olefin sulfonate only a one-step transition was observed (11). The effect can be explained by the lower cmc of the longer surfactant molecules. At the same concentration, more micelles are present in the solution of the longer chain surfactants, and the result is easier micellar ordering.

## B. Effect of Film Area

The film stability of horizontal flat film in the presence of 19 nm silica hydrosol was defined in detail regarding particle concentration and film diameter (13). From the practical point of view, the film stability can be improved by increasing the particle concentration (or concentration of micelles if they are the stabilizing colloid) and monodispersity in size or by decreasing the film diameter. With a $6 \times 10^{-2}$ cm film diameter, for example, a stable film with a thickness of 100 nm containing three particle layers inside could be formed. By increasing the film diameter to $10^{-1}$ cm, three more transitions were observed with no layers in the final film. In total, six film-thickness transitions were seen, which is the same number as that found in large macroscopic film. Foam films stabilized by micelles showed similar behavior (25). The effect of film area on film stability can also be explained by the vacancy mechanism (24). An obvious way of decreasing the film area in a foam system is to decrease the bubble size.

## C. Effect of Electrolytes

Adding electrolytes to the surfactant solution, the number of stepwise transitions decreases, and the thinning process becomes irregular in that some of the steps are fused together. This is especially noticeable with ionic surfactants in which the repulsive force between the micelles (or particles) is electrostatic (11). Moreover, above a threshold salt concentration, no stepwise transition occurs; that is, the electrolytes prevent ordering in the thinning film. In the $C_{16}$-$\alpha$-olefin sulfonate solution ($3.16 \times 10^{-2}$ mol/l), for example, two transitions were observed without added electrolyte (12). At one-to-one electrolyte concentrations higher than 1 wt %, the film thinned by a single-step transition. Addition of salt inhibits the ordering of ionic micelles, because the added ions compress the electric double layer around

# Thin Micellar Films

the micelles, resulting in a dramatic decrease in the effective volume of micelles. The threshold concentration of added NaCl for order–disorder increases with the micellar volume fraction (11) (see Figure 4).

## D. Effect of Temperature

In the application of foams, temperature is a very important parameter. The effect of temperature was studied with nonionic surfactants in which the intermicellar repulsion is the result of stearic forces (26). A decrease of temperature can prevent the occurrence of the last few stepwise transitions, and the stratification stops at a large film thickness. Thus, at lower temperatures, higher film stability results. In the Enordet AE1215-30 (ethoxylated alcohol), for example, two transition steps with a final thickness of 49 nm were observed at 26 °C. At 32 °C, six transitions and 14 nm final thickness was observed. At higher temperatures, the rate of stratification increased; that is, the drainage time decreased with the temperature. The surfactant with lower degree of ethoxylation is more sensitive to the temperature. Above 35 °C, the stepwise transitions become irregular, and at even higher temperatures, no transitions occur. Near the cloud point of the surfactant (but below it), the film ruptures without reaching a stable final thickness.

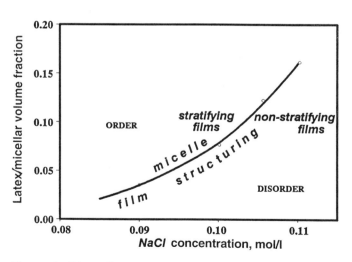

**Figure 4** Phase diagram of order/disorder transition. Volume fraction of micelles versus concentration of added NaCl. The curve represents the threshold concentration separating the regions with and without stratification in thinning foam films.

## E. Effect of Micellar Polydispersity

Comparing the interferograms of a foam film formed from the nonionic Enordet AE1215\30 (ethoxylate alcohol) surfactant solution and a similar solution in the presence of solubilized *n*-decane shows two significant changes occur. First, the number of stepwise transitions increases from five to six in the presence of the oil. Second, the film drainage time decreases from 6 to 4.3 min in the presence of solubilized oil. These phenomena can be explained by the effect of oil solubilization on the size and polydispersity of micelles and on the intermicellar interactions (27,28). Nilsson (29) and Lobo, Nikolov, and Wasan (28) found that the solubilization of decane leads to an increase in the micellar aggregation number and a reduction in the second virial coefficient (increased attraction between the micelles) of nonionic surfactants. Oil solubilization also increases the polydispersity of micelles (28). As already discussed, the number of film-thickness transitions is higher at higher micellar volume fractions. Therefore, oil solubilization is expected to result in an increased number of step transitions. At the same time, however, higher micellar attraction and polydispersity lead to the formation of a less packed in-layer structure in the thinning film that results in faster stepwise transitions, and lower foam stability in the presence of solubilized oil.

## II. CALCULATION OF EFFECTIVE PAIR INTERACTIONS BETWEEN MICELLES

From the classic DLVO theory, the interaction between two colloidal particles consists of two parts, the van der Waals interaction and the electrostatic interaction. In the frame of Debye-Huckle theory, the electrostatic force between two colloidal particles is purely repulsive. It was previously noticed by Asakura and Oosawa (30,31) that, in a system containing many particles, the particle volume exclusion effect may push two particles toward each other, the *depletion phenomenon*. Asakura and Oosawa (A-O) estimated the depletion force and potential for a dilute system. They concluded that the depletion force, which is purely attractive, could be significant enough to cause aggregation and flocculation of colloidal particles. Experimental evidence for the attractive depletion force as well as the phase separation has been observed in various systems.

It is known that in a many-body system, the pair potential of the mean force, or the effective pair interaction $u(r)$, is related to the packing structure of particles by

$$u(r) = -kT \ln g(r)$$

# Thin Micellar Films          135

where $g(r)$ is the radial distribution function. The A-O type depletion effect, which is attractive, is included in this expression.

Following the work by Henderson (30,32), the radial distribution function was calculated. The Ornstein-Zernike (O-Z) theory provides a theoretical approach to calculate the correlation function in a many-body system. It states simply that the total correlation function $h(r)$ between two particles, which is related to the radial distribution function $g(r)$ by $h(r) = g(r) - 1$, consists of two parts. The first part is the direct correlation between these two particles, denoted by $c(r)$, and the second part is contributed by all possible indirect correlations through the other particles in the system. The latter can be expressed exactly as the convolution of the direct correlation function and the total correlation function.

First, we calculated the effective pair interaction (i.e., the potential of the mean force) between hard spheres in a many particle system from the Ornstein-Zernike equation under the P-Y closure. A comparison of the effective potential calculated from the P-Y theory with that from the Asakura and Oosawa theory is shown in Figure 5 at low (8 vol %) and high concentrations (49 vol %). We notice that Asakura-Oosawa's result gives only the depletion effect, and is in reasonable agreement with P-Y theory in the low concentration limit. When the concentration is high, the two results are very different. The effective interaction potential becomes oscillatory, with both the period of oscillation and delay length equal to about particle size. The oscillatory effective interaction indicates the layering phenomenon of colloidal particles staying in between the two film surfaces at high concentration. The maxima at 1.5, 2.5, . . . indicate those distances are energetically less favored, while the minima at 1, 2, . . . suggest that the film surfaces are more likely separated by the distances in which 1, 2, . . . layers of particles can be packed.

## III. CALCULATION OF STRUCTURAL DISJOINING PRESSURE

Henderson suggested that one could calculate the structural forces and depletion from P-Y theory for a binary hard sphere system. We used the Derjaguin-Landau equation to convert the force to pressure (force/area) (30). The structural disjoining pressure is plotted in Figure 6 for different particle concentrations. A very sharp peak near the thickness of one particle diameter is apparent. The period of the oscillation is about the particle diameter—strong evidence that this oscillatory type pressure isotherm is caused by the formation of particle layers inside the film. The particle structures inside a thin film will be explored in detail by the Monte Carlo

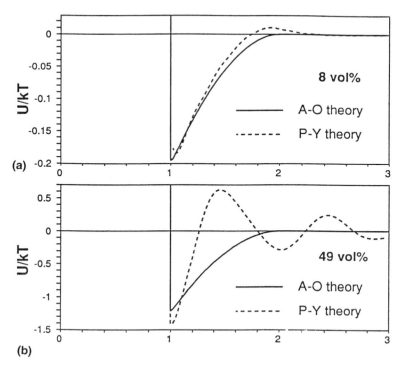

**Figure 5** Comparison of the results from PY theory and those from Asakura-Oosawa theory. (a) gives the comparison at low concentration (8 vol %) and (b) at high concentration (49 vol %).

simulations in the next section. From a thermodynamic point of view, the oscillatory disjoining pressure gives the necessary condition for the stepwise thickness transitions during the film thinning process.

In order to study the effect of the polydispersity on the structural disjoining pressure, a model film system was studied, in which the particles were polydisperse, and size distribution followed the gaussian distribution. Figure 7 shows the calculated structural disjoining pressure for three different cases. In all cases the average particle size was the same and the concentration in volume function was kept constant. It is clearly shown in the figure that the polydispersity has significant effects on the structural disjoining pressure. The higher the polydispersity, the weaker the oscillation of the disjoining pressure.

We compare our theoretical results with the experimentally measured force-distance curves by Parker et al. (30,33) for the colloidal system con-

# Thin Micellar Films

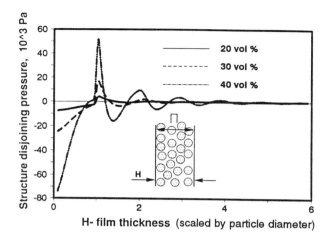

**Figure 6** Structural disjoining pressure exerted on a film. Particle concentration, 20, 30, and 40 vol %.

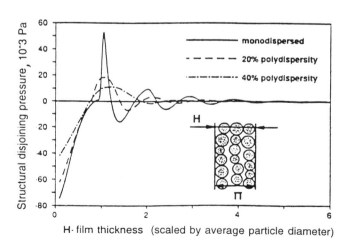

**Figure 7** The structural disjoining pressure isotherms versus distance for a model film system with polydispersed particles. (1) Particles are monodispersed, (2) particles are polydispersed (20% polydispersity), and (3) particles are polydispersed (40% polydispersity). The average particle size in all three cases are the same (10 nm) and the temperature is 298 K.

sisting of reversed micelle (water-in-oil) prepared from the AOT/water/ heptane microemulsion system. Due to the electroneutrality which holds for each individual micelle, the effect of electrostatic interaction can be neglected. Hence this system is very close to an ideal "hard sphere." Micelles are found to be spherical, with a polydispersity of about 25%. Parker et al. could measure the force–distance relation at all distances, including both the repulsive and the attractive parts. The result is found to be oscillatory; the period of oscillation is about the diameter of the micelles. In Figure 8, we compared the measured force with the calculation for a polydisperse micellar system, and found the theoretical prediction to be in good agreement with the experimental measurement.

## IV. MONTE CARLO SIMULATION OF MICELLAR STRUCTURING IN THE FILM

In order to reveal the in-layer structure formation inside a thin micellar film, we performed the grand canonical ensemble Monte Carlo (GCEMC) simulation in a model film, which consisted of two plane surfaces with

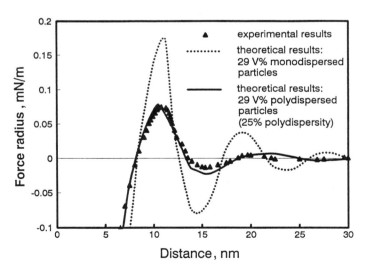

**Figure 8** Comparison of theory and the experiment. The theoretical results are calculated with the following parameters: dashed line, concentration 29 vol %, monodispersed, particle diameter 9 nm; solid line, concentration 29 vol %, polydispersity 25% (gaussian distributed) average particle diameter 9 nm.

# Thin Micellar Films

micelles/particles sandwiched between them. The details of our numerical results are given elsewhere (34).

In order to find the particle structure inside the layers parallel to the film surfaces, the radial distribution function (RDF) of the particles in a layer has been examined. The RDF gives the probability of finding particles around a reference particle in a shell of radius $r$. When the particle concentration is low, the particles inside a layer pack randomly to form a liquid-like structure. Only when the concentration is over a certain value do the particles in a layer start to form a more ordered structure. This value for the in-layer structure transition depends on the film thickness and the position of the layer inside the film. In Figure 9, we plot in-layer RDF of a surface layer versus the average particle concentration in the film and the in-layer distance. At the average effective particle concentration of 40 vol % (a solution of 0.052 mol/l of Enordet AE1215-30 corresponds to such a concentration) or lower, the in-layer RDF shows typical liquid-like 2-D structure without order, that is, damped peaks near the integers, 1, 2, 3, ... (in the unit of particle diameter). When the average effective concentration

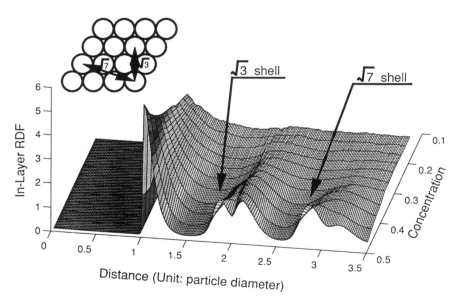

**Figure 9** Surface plot of inlayer RDF versus distance and particle concentration for a surface layer. New peaks at $\sqrt{3}$ and $\sqrt{7}$ can be seen at high concentration, including the formation of in-layer hexagonal particle structure. The film thickness is 15 particle diameters.

increases to somewhere between 40 and 45 vol %, new peaks begin to appear in the RDF, especially near $r = \sqrt{3}$ (the actual value is always somewhat larger than $\sqrt{3}$ due to nonvanishing spacing between particles) and $r = \sqrt{7}$, which indicates the formation of a 2-D hexagonal structure inside the layer. Notice that this peak grows gradually as the concentration increases, indicating a smooth structural transition from a liquid-like in-layer structure to the colloid crystal-like in-layer structure.

In Figure 10, we plotted the in-layer RDFs for different layers in a film of particle concentration of 46 vol %. One can see a peak near $\sqrt{3}$ for the surface layer, which indicates the formation of 2-D hexagonal packing. This peak becomes weaker in the next layer and disappears in the middle layer. The difference in the in-layer RDFs of different layers illustrates that there exists, in a single film, more ordered structure in the middle layer.

The in-layer packing structure depends not only on the position of the layer in the film but also on the film thickness. In Figure 10, we plotted the in-layer RDF of surface layers for different film thicknesses, $2\sigma$, $3\sigma$, $4\sigma$, and $5\sigma$, at the same concentration of 46 vol %. The degree of ordered 2-D hexagonal structure can be detected by the height of the peak near $\sqrt{3}$. It is

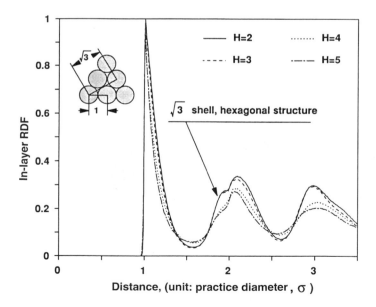

**Figure 10** In-layer radial distribution function for different layers in a film of the average particle concentration of 46 vol %, but different thicknesses. It shows that the thinner the film, the better formed is the ordered inlayer 2-D structure.

clearly seen that the in-layer particles in the thinner films of thicknesses $2\sigma$ and $3\sigma$ are better organized than those in the thicker films of thickness $4\sigma$ and $5\sigma$. The theoretical prediction of in-layer 2-D hexagonal particle structuring has been verified by our recent transmitted light diffraction experiments (35).

## V. CONCLUSIONS

The following conclusions can be drawn based on both the experimental observations and theoretical developments of micelle-layering phenomena in the confining geometry of the film with fluid surfaces such as those associated with a foam or emulsion system.

1. All of our experimental results with ionic and nonionic micelles, swollen micelles or microemulsions, globular proteins, and Brownian particles such as latex suspensions and silica hydrosols reveal that stepwise thinning due to micelle/particle organization into layers in a foam or emulsion type thin symmetric film or in a nonsymmetrical pseudoemulsion film is a universal phenomenon (36).
2. The organization of micelles into layers in the confined boundaries of the film induces structural forces stabilizing the film thickness at values corresponding to integral multiples of the average bulk intermicellar distance.
3. The distance between the micellar layers changes with an increase in the bulk micellar concentration.
4. The film thickness stability is affected by the effective micelle/particle concentration, added electrolyte, temperature, and film area.
5. Both the micelle/particle size and polydispersity in size affect the film stratification phenomenon.
6. The oscillations in structural disjoining pressure isotherms which have been measured experimentally and confirmed theoretically explain the micellar layering in thin films. The period of the oscillation and decay length is about the effective size of the micelles/particles. However, the micellar film thickness transition initiated by spot formation appears to be due to the "vacancies condensation" mechanism.
7. Monte Carlo simulations using the simple hard sphere model reveal that micelles/particles not only form layering structures normal to the film surface, but also self-organize in the layers

parallel to the film surfaces. At the micellar/particle concentration below the Kirkwood–Alder phase transition value (51 vol %) for the bulk phase, a 2-D hexagonal in-layer packing was found, which is supported by our recent transmitted light diffraction experiment (35).
8. The in-layer packing structure depends not only on the position of the layer inside the film but also on the film thickness.
9. The depletion and long-range structural forces, which are the non-DLVO forces, were calculated by using the "effective" pair interaction (i.e., the potential of the mean force) instead of the bare pair interaction. The numerical calculations reveal that the "voids" are formed due to the attractive depletion by polydispersed micelles.
10. The formation of in-layer ordered structures in thin micellar films offers a new mechanism for the stabilization of foams and emulsions. Stratification was photographed (37) in a real foam generated from 0.1 mol/l sodium dodecyl disulfate solution. The photograph (Figure 11) clearly captures the multibanded pattern

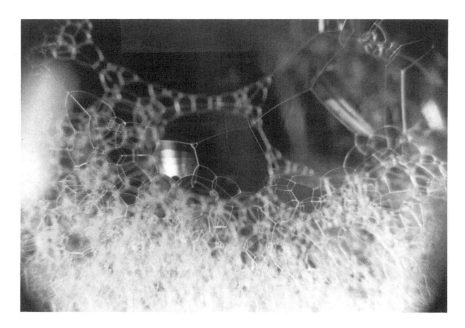

**Figure 11** Aqueous foam stabilized due to the stratification in the foam bubble lamellae (0.1 mol/l sodium dodecyl sulfate solution).

associated with stratification inside the curved lamella. The practical importance of the in-layer film microstructuring is that the lifetimes of foams and emulsions with stratifying films are much larger. The rheology of such dispersions will be quite different (38). The film, and thus foam or emulsion, stability can be improved by increasing the micellar concentration and decreasing polydispersity in micelle size, decreasing the individual film area (for example by decreasing the bubble or droplet size), decreasing the electrolyte concentration, or lowering the temperature.

## ACKNOWLEDGMENTS

This study has been supported by the National Science Foundation. One of the authors (Darsh T. Wasan) acknowledges Dr. Steve Christiano's assistance in the preparation of this manuscript.

## REFERENCES

1. PA Kralchevsky, ND Denkov. Chem Phys Lett 240:385, 1995.
2. ES Johonnott. Philos Mag 11:746, 1906.
3. RE Perrin. Ann Phys (Paris) 10:160, 1915.
4. HG Bruil, J Lyklema. Nature (London) 232:19, 1971.
5. JW Kenskemp, J Lyklema. Adsorption at Interfaces. Mittal, KL, ed. ACS Symposium Series 8. Washington, American Chemical Society, 1975, pp 191–198.
6. E Manev, SV Sazdanova, DT Wasan. Dispersion Sci Tech 3:435, 1982.
7. OM Krichevsky, J Stavov. Phys Rev Lett 74:2752, 1995.
8. D Langevin, AA Sonin. Adv Colloid Interface Sci 51:1, 1994.
9. V Bergeron, CJ Radke. Langmuir 8:3023, 1992.
10. AD Nikolov, DT Wasan, PA Kralchevsky, IB Ivanov. In: Ordering and Organization in Ionic Solutions. Ike, N., Sogami, I., eds. Singapore, World Scientific, 1988.
11. AD Nikolov, DT Wasan. J Colloid Interface Sci 133:1, 1989.
12. AD Nikolov, DT Wasan, DA Edwards. Presented at the 61st Annual Technical Conference and Exhibition of SPE, New Orleans, LA, October 5–8, 1986; SPE Preprint 15443.
13. AD Nikolov, DT Wasan. Langmuir 8:2985, 1992.
14. K Koczo, AD Nikolov, DT Wasan. J Colloid Interface Sci 178:694, 1996.
15. LA Lobo, DT Wasan. Langmuir 9:1668, 1993.
16. V Bergeron, AI Jimenez-Laguna, CJ Radke. Langmuir 8:3027, 1992.
17. ED Manev, SV Sazdanov, DT Wasan. J Dispersion Sci Tech 5:111, 1984.

18. DT Wasan. Chem Eng Educ 26:104–112, 1992.
19. AD Nikolov, PA Kralchevsky, IB Ivanov, DT Wasan. J Colloid Interface Sci 133:13–22, 1989.
20. B Derjaguin, L Landau. Acta Physicochim 14:633, 1941.
21. E Verwey, J Th G Overbeek. Theory of the Stability of Lyophobic Colloids. Amsterdam, The Netherlands, Elsevier, 1948.
22. JGJ Kirkwood. Chem Phys 7:919, 1939.
23. BJ Alder, TE Wainwright. Phys Rev 127:359, 1962.
24. PA Kralchevsky, AD Nikolov, DT Wasan, IB Ivanov. Langmuir 6:1180, 1990.
25. AD Nikolov, DT Wasan. Presentation at the American Institute of Chemical Engineers Meeting, Chicago, IL, November 1990.
26. AD Nikolov, DT Wasan, ND Denkov, PA Kralchevsky, IB Ivanov. Prog Colloid Polym Sci 82:87, 1990.
27. T Nakagawa, K Shinoda. In: Colloidal Surfactants. Shinoda, K, Nakagawa, T, Tamamushi, B, Isemura T, eds. Orlando, FL, Academic, 1963, p 139.
28. LA Lobo, AD Nikolov, DT Wasan. J Dispersion Sci Tech 10:143, 1989.
29. GJ Nilsson. Phys Chem 64:1135, 1957.
30. XL Chu, AD Nikolov, DT Wasan. J Chem Phys. 103:6653, 1995.
31. SA Sakura, FJ Dosawa. J. Chem Phys. 22:1255, 1954.
32. D. Henderson. J. Colloid Interface Sci 121:486, 1988.
33. JL Parker, P Richetti, P Keticheff, S Sarman. Phys Rev Lett 68:1955, 1992.
34. XL Chu, AD Nikolov, DT Wasan. Langmuir, 36:19, 1994.
35. DT Wasan, AD Nikolov. Particulate Two-Phase Flow, Roco, MC, ed. Boston, Butterworth-Heinemann, 1993, p 325.
36. DT Wasan, AD Nikolov, P Kralchevsky, IB Ivanov. Colloids Surfaces 67:139, 1992
37. DT Wasan, K Koczo, AD Nikolov. Foams: Fundamentals and Applications in the Petroleum Industry, Schram LL, ed. ACS Symposium, Washington, D.C. Ser. No. 242, 1994.
38. ES Basheva, AD Nikolov, PA Kralchevsky, IB Ivanov, DT Wasan. In: Surfactants in Solution. Vol. 11. Mittal, K, ed. New York, Plenum, 1990, p 467.

# 6
# Structure and Design of Abnormally Long Thread-Like Micelles and Their Relation to Vesicles and Liquid Crystals

**C. Manohar**
*Bhabha Atomic Research Center, Bombay, India*

## I. INTRODUCTION

It is fascinating to see the drastic changes in the flow characteristics when one adds a small amount of sodium salicylate (SS) to a dilute solution of cetyltrimethyl ammonium bromide (CTAB). The viscosity of CTAB solution, which is approximately that of water ~1 cP, increases drastically by orders of magnitude on addition of SS and the solution, more interestingly, shows elastic properties. An optimized demonstration has been prescribed by Hoffmann and Ebert (1).

### A. Hoffmann Prescription

Mix equal amounts of aqueous solutions of 50 mM CTAB and 30 mM SS. The resulting solution is highly viscous, and one can see elasticity by trapping some air bubbles in that solution, which recoils and oscillates if one suddenly spins the solution and stops. This solution also shows strong flow birefringence, as can be seen between two crossed polaroids.

Not only CTAB but most single-chain cationics show similar effects on mixing with SS. Figure 1 shows the zero shear viscosity measured by Rehage and Hoffmann on cetyl pyridinium chloride (CPC) and SS mixtures (2). One of the recent reviews gives a number of systems that have been

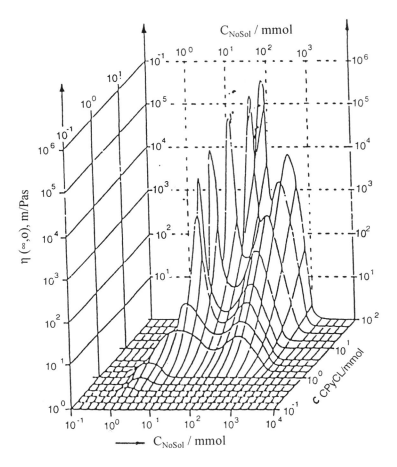

**Figure 1** Increase in viscosity of cetyl pyridinium chloride (CPC, abbreviated CPyCl in figure) on addition of SS. Note the double-peaked nature for all concentrations of CPC. Many cationic surfactants show this behavior on addition of SS and similar additives (see Ref. 6). (From Ref. 2, used with permission.)

investigated. Some of the systems that have been observed to show strongly viscoelastic properties are listed in Table 1.

Thus it is obvious that SS is not the only additive with cationics that produces strongly viscoelastic systems. It is convenient to classify the studies into two categories, namely rheological properties and structural investigations.

**Table 1** Surfactant Systems Exhibiting Viscoelastic Properties

| Surfactant | Counterion |
|---|---|
| Cetylpyridinium chloride | Na-salicylate |
| Cetyltrimethylammonium bromide | Na-salicylate |
| Cetyltrimethylammonium chloride | NaSCN |
| Cetyltrimethylammonium chloride | 2-Aminobenzenesulfonate |
| Cetyltrimethylammonium chloride | Perfluorobutyrate |
| Cetyltrimethylammonium chloride | 4-Methylsodiumbenzoate |
| $C_xF_{17}SO_3Na$ | $(C_2H_5)_4NOH$ |
| $C_9F_{19}CO_2Na$ | $(CH_3)_4NOH$ |
| | Surfactant |
| Tetradecyldimethylaminoxide | Sodium dodecylsulphate |
| Tetradecyldimethylaminoxide | $C_7F_{15}CO_2Na$ |
| $C_{14}H_{29}N^+(CH_3)_2\text{-}CO_2^-$ | Sodium dodecylsulphate |
| | Uncharged Compound |
| Cetyltrimethylammonium bromide | Chloroform |
| Cetyltrimethylammonium bromide | 1-Methylnaphthalene |
| Cetylpyridinium chloride | 4-Propylphenol |

## II. RHEOLOGICAL PROPERTIES

The interest in these types of systems has persevered because of the interesting rheological properties. It is clear now that additives like SS convert the spherical micelles into long worm-like micelles, whose contour length is larger than the persistence length $l$ (~ 150 Å), and they entangle to form a network producing high viscosities. Figure 2 shows a cryo-TEM, (transmission electron microscope) picture of such a system. Flow properties of these networks are extremely interesting, and the investigations have been pioneered by Candau et al. (3), Cates (4), and Hoffmann and colleagues (5). There are several excellent reviews on these aspects; therefore, we shall focus briefly on them (6,7).

The basic factors governing the rheological properties of a network are based on the concept of a tube (8). Each worm-like micelle, in a network, is conceived to be trapped inside a tube formed by the other micelles in the network. When a stress is applied, the micelle has to come out of the tube and adjust to the new conditions created by the stress. This relaxation of a worm-like micelles is decided by two times (4):

1. The lifetime of the worm-like micelle, namely, the time $\tau_b$ for which the micelle exists before a scission occurs somewhere along the length of the micelle. This is decided by the scission energy E,

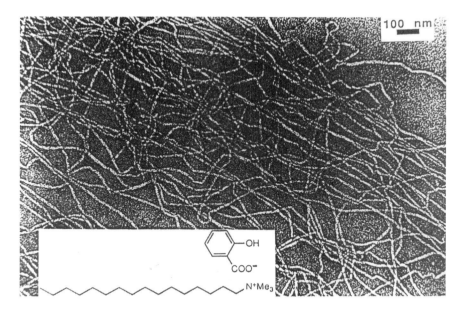

**Figure 2** Cryo-TEM pictures of network of worm-like micelles of CTAB and SS. Notice that the lengths of the worm-like micelles are in microns! (From Ref. 20, used with permission.)

namely the energy required to break the micelle into two pieces with end caps.

2. Characteristic time for reptation. This is the time required for a worm-like micelle to wiggle out of the tube formed by all the other worm-like micelles $\tau_r$.

Because of the existence of the breaking and remaking of the worm-like micelles, these have been termed as living polymers or equilibrium polymers and there is a length distribution whose number concentration $C(L)$ is given by

$$C(L) = C_0 \exp\left(-\frac{L}{\bar{L}}\right) \tag{1}$$

where $\bar{L}$ is the average length of the distribution and is given by

$$\bar{L} = \phi^\nu \exp\left(\frac{E}{kT}\right) \tag{2}$$

where $\phi$ is the volume fraction of the surfactant, $k$ is the Boltzmann constant, and $T$ is the temperature. $\nu$ is between 0.5–0.6 for dilute and semidilute regions.

If $\tau_b \gg \tau_r$ then the micelles live for a long time and the solution behaves like a conventional polymer solution and like a Maxwellian fluid. The Cole–Cole plot of the storage modulus $G'(\omega)$ and loss modulus $G''(\omega)$ would be a semicircle, as shown in Figure 3. Deviations from Maxwellian behavior occur when $\tau_b \ll \tau_r$ and the solution has one relaxation time,

$$\tau = (\tau_b \tau_r)^{1/2} \tag{3}$$

The typical dependence of $G'(\omega)$ and $G''(\omega)$ on $\omega$ is shown in Figure 4. At low frequencies the shear is applied slowly and the fluid has all the time to relax; the energy dissipation takes place and the solution does not behave like a elastic substance. However, when the frequency is increased, fewer modes have time to relax and the substance increasingly behaves like an

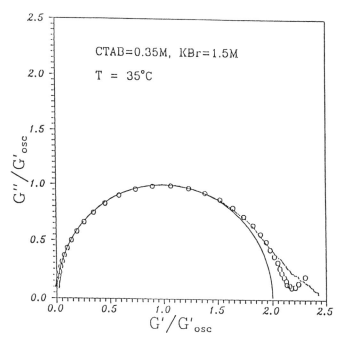

**Figure 3** Cole–Cole plot of loss and storage modulus showing a semicircle in conformity with Maxwell model. (From Ref. 9, used with permission.)

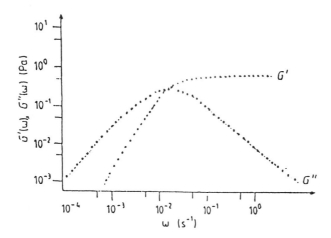

**Figure 4** A typical plot of the frequency dependence of loss and storage modulus. (From Ref. 7, used with permission.)

elastic solid. The plateau region of $G'(\omega)$ for high frequencies shows the rubbery region of the solution. The value $\omega = \omega_0$ at which the peak in the $G''(\omega)$ appears indicates roughly the characteristic relaxation time $\tau = \omega_0^{-1}$ of the systems. The dependence of these parameters on temperature, volume fraction $\pi$, salt content, etc., are complex questions and are being still explored, and this subject has been discussed in reviews (6,7,9,10).

## III. STRUCTURAL ASPECTS

The questions addressed here are

1. Why are the worm-like micelles formed when SS is added to CTAB?
2. Why there are two peaks in Figure 1?
3. What happens in between the two viscosity peaks?

The question of what shape a micelle takes has been discussed in the elegant works of Israelachivili et al. (11) and Tanford (12). To summarize their important conclusions, three parameters have to be defined. If a surfactant has $n$ carbon atoms in the chain and has a polar head area of $a$ at the water–micelle interface, and the length $l$ and the volume $v$ per hydrophobic chain are given by

$$l = 1.265\,n + 1.5\,\text{Å} \qquad (4)$$
$$v = 26.9\,n + 27.4\,\text{Å} \qquad (5)$$

then conditions for the formation of various shaped micelles can be summarized as (11)

$$\frac{v}{a\,l} < \frac{1}{3} \text{ spherical micelles}$$

$$\frac{1}{3} < \frac{v}{a\,l} < \frac{1}{2} \text{ cylindrical micelles}$$

$$\frac{v}{a\,l} > \frac{1}{2} \text{ vesicles or bilayers}$$

It is important to realize that these conditions are for a single individual surfactant molecule, and the assumptions that have gone into the derivation are primarily the hydrophobic effect, which attempts to avoid water contact and the packing considerations. Later, as we shall see, there are other possible features of surfactants, like the bending energy, which also seem to be relevant in deciding the supramolecular structure obtained.

It is most important to realize that molecules like SS are not conventional surfactants, but they do have surface activity (13a,13b). This is indicated by the orientation of SS on CTAB surface, as shown in Figure 5, which also shows that they take part in the packing. If there was no surface activity, then the SS molecule would have adsorbed with free portion of the benzene molecule sticking out of the micelle. In fact, shifting the OH group in SS to meta or para positions reduces the surface activity and thus the ability to induce the viscoelasticity (14). SS belongs to the class of hydrotropes (15) and has the ability to solubilize immiscible liquids. Once it is realized that SS acts like a mild anionic surfactant, it becomes beneficial to look at the more general class of mixtures of cationic–anionic surfactants. Beautiful experiments have been done by a group led by Kaler, and they have shown that spontaneous vesicles can be formed when one mixes suitably chosen mixtures on anionic and cationic surfactants (16). Figure 6 shows the phase diagram, and one can see that two vesicular phases are observed; in one phase the vesicles have a positive charge (cationic-rich side) and in the other they have a negative charge (anionic-rich side). AT higher concentrations there are two lamellar phases.

In view of the surface activity of SS it became tempting to check what would happen if one increases the hydrophobic part of the SS. The sodium hydroxynaphthalene carboxylate (SHNC) (the sodium salt of BON acid, a dye intermediate) was chosen for this purpose. Interestingly, on addition of SHNC to CTAB it was observed that there were two viscoelastic phases and

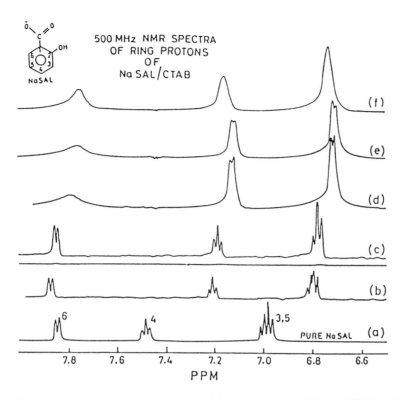

**Figure 5** NMR spectra of 8 mM SS (a) and solutions of 10 mM CTAB with varying amounts of SS (b, 1 mM; c, 2 mM; d, 4 mM; e, 6 mM; f, 8 mM). Notice from (a) and (b) that the positions 3, 4, and 5 have moved to a hydrophobic environment, showing that the SS molecule is oriented on micelle surface with carboxylic group sticking out in to water. This implies surface activity of SS. The difference in line widths of proton no. 6 and others shows that there is an asymmetric rotation of the molecule on micellar surface. (From Ref. 13, used with permission.)

two liquid crystalline phases (17). The viscoelastic phases were strongly flow birefringent. Figure 7 shows the polarized microscopic picture of a drop of SHNC solution in which a speck of CTAB powder has been placed in the middle, setting a concentration gradient from center to the edge. One can clearly see the two liquid crystalline phases and the dark band in the middle is nontransparent corresponding to the precipitation of the surfactant around 1 : 1 concentration of CTAB and SHNC. This suggested that as one goes away on both sides of the 1 : 1 composition in concentration, the supramolecular structures develop either positive charge (CTAB-rich side) or negative charge (SHNC-rich side), and effectively the two liquid crystal-

# Long Thread-Like Micelles 153

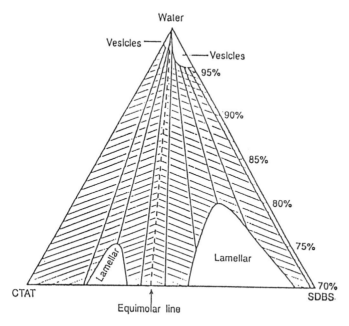

**Figure 6** Phase diagram of mixtures of CTA-tosylate and sodium dodecyl benzene sulfonate with water. Spontaneous two vesicular phases are formed in these anionic–cationic surfactant mixtures. Recently, hexagonal liquid crystals and worm-like phases also have been reported. (From Ref. 16, used with permission.)

line phases and the two viscoelastic phases were charge-induced effects. The liquid crystalline phases appeared to be nematic phases. The appearance of two viscoelastic phases in the neighborhood (on the higher charge side) of liquid crystalline phases suggested that the viscoelastic phases could be regarded as charge-induced isotropic phases, which normally have higher viscosity, of the liquid crystalline phase (18). Models of rods with charges and van der Waal attractions seemed to at least explain qualitatively the trends (19). More importantly, the appearance of double peaks in the viscosity of a large number of systems (eg., Figure 1) could be attributed to large micelles of positive and negative surface charges. Some of the electron microscopic pictures of these viscoelastic systems showed micron-long micelles.

Appearance of almost micron-long micelles, vesicles, and tubules at very low concentrations in a number of systems led to the suggestion that these systems might be made up of solid micelles (20). In these micelles the polar region of the supramolecular structures is supposed to be solid-like,

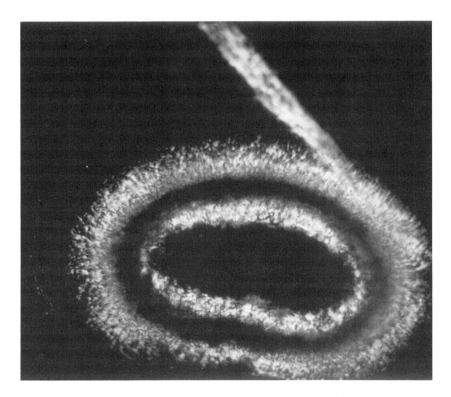

**Figure 7** Polarized microscopic picture of the CTAB–SHNC–water system. At center of the circular region is the speck of CTAB powder placed on a drop of SHNC solution. One can see two liquid crystalline phases separated by a band of dark, nontransparent precipitate of CTAHNC complex. Two viscous phases are also seen as dark bands.

and the structure formed is decided by the bend elastic constant rather than the packing considerations mentioned above. If the interface is solid-like, and is difficult to bend, therefore these form large structures with small curvatures. The comment by Fuhrhop and Helfrich (20) that in CTAB–SS systems the NMR line widths of SS were large and these could be on the verge of solidification led us to reexamine CTAB–SHNC systems. We expected that SHNC, being a larger molecule, had higher chances of freezing on CTAB surface. There was also other independent evidence that the SS molecule performed asymmetric rotations on the CTAB micellar surface, in contrast to aqueous solutions, leading to difference in line widths between various protons in different positions (21). This indicated that the packing of the molecules at the interface was tending to solid-type packing.

A preparation technique was developed (22) by which one could prepare anionic–cationic surfactant complexes without any smaller counter ions, and one such compound was cetyltrimethylammonium-3 Hydrooxy 2-Naphtalene carboxylate ($CTA^+HNC^-$). This sample was sparingly soluble in water and formed turbid solutions at room temperature which on heating became clear solutions, **but** the viscosity increased by an order of magnitude. This was puzzling, and a clue toward a possible solution came from the electron microscopic pictures at room temperature which showed vesicles of about 5 fm diameter. Figure 8 shows the scanning electron microscopic picture of the platinum-coated vesicles of CTAHNC. Figure 9 shows the light scattered at 90° and also the viscosity. One can see that as the turbidity decreases the viscosity increases, suggesting that one has a vesicle to worm-like micelle. An additional clue came from the proton NMR spectra of CTAHNC. Figure 10 shows the temperature dependence of NMR spectra of 0.6% solutions of CTAHNC. It can be seen that the spectra of $CTA^+$ and $HNC^-$ fall in two nonoverlapping regions, and one

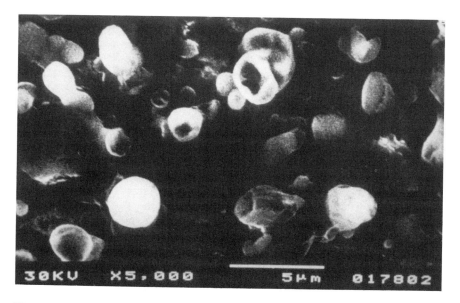

**Figure 8** SEM pictures of CTAHNC 0.6% solution vesicles are coated with platinum. The sizes are observed to be of 5 μm diameter and the vesicles are multilayered.

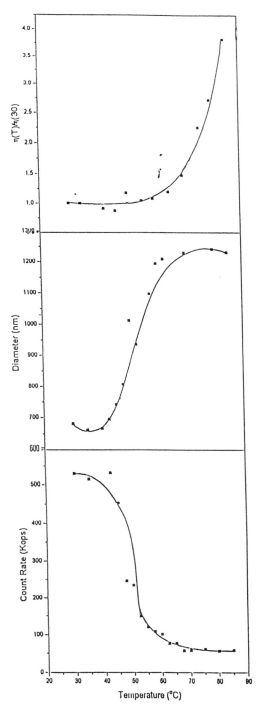

**Figure 9** Viscosity and effective diameter as measured by dynamic light scattering and intensity of light scattered at 90° as a function of temperature. The increase in viscosity with temperature indicates structural change to worm-like micelles on heating.

# Long Thread-Like Micelles 157

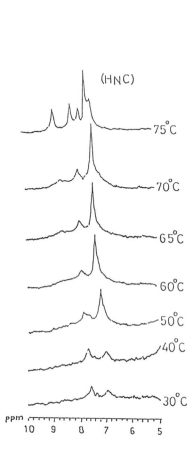
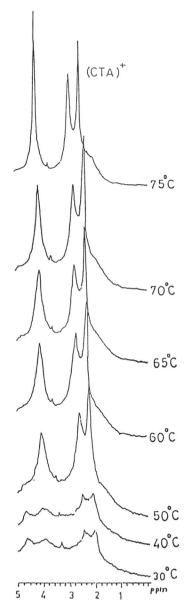

**Figure 10** Temperature dependence of the NMR spectra of HNC$^-$ and CTA$^+$ portions in a 0.6% solution. Note that the line widths are large at room temperature and decrease on heating in spite of the increase in viscosity of the solution. Room temperature spectra indicate the possibility of a solid type of ordering in the polar head region.

can follow them independently. The room temperature spectra are broad and are nonliquid-like. On heating, some portions of the HNC spectra become narrow indicating asymmetric rotations of the ion on the micellar surface. At 76°C the lines become sufficiently narrow that all the lines of HNC become noticeable, indicating comparatively free movement analogous to that in the solution. The room temperature widths are so large that the spectra could be regarded as solid-like. The spectra of the $CTA^+$ part, at higher temperatures, are very characteristic of the worm-like micelles and resemble that in CTAB-SS type system. This seems to suggest that we have a solid–fluid-type transition on the micellar surface that dictates what supramolecular structures should be formed.

There is another way to look at the cationic–anionic surfactant complexes that could be the beginning of an alternate, unified view toward the supramolecular structures. The cationic–anionic surfactant complexes, without any smaller ions, in aqueous solution create a two-dimensional (2-D) surface of polar heads (the micellar surface) that could be regarded as a coulomb gas in 2-D, which has been well studied (23) and known to exhibit the Bjerrum transition (24). According to this, at any temperature one should have a equilibrium

cation − anion ⇔ cation + anion

and since each ion has a hydrocarbon chain, the complex would behave like a double-chain lipid while each of the ions would behave as a normal detergent. Therefore, the system becomes comparable to a large class of lipid plus detergent systems that have been investigated in biochemical literature in connection with the membrane reconstitution studies (25,26). There it is observed that as one adds a detergent to vesicles formed by lipids there is a transition from vesicle to micelle, and these have been investigated by several techniques. In our cationic–anionic mixtures if one defines the effective radius of the ions is $R$ then the thermal energy required to break a ion pair would be

$$kT = \frac{e^2}{\epsilon 2 R}$$

where $e$ is the charge, $\epsilon$ is the dielectric constant of water, $T$ is the temperature, and $k$ is the Boltzmann constant. $T$ would be around 300 $K$ when ion $R$ is around 3.5 A. Therefore, it is expected that there is a considerable pairing in the polar head region which would be a strong function of temperature. In view of this, one should expect that the supramolecular structures formed are influenced by the organizations (phase transitions) occurring in the polar head region. Preliminary experiments and calculations performed seem to confirm these expectations of the strong analogies between lipid–detergent mixtures.

## IV. CONCLUSIONS

It appears that many of the dilute, strongly viscoelastic surfactant systems reported in the literature could be regarded as a mixtures of a cationic–anionic surfactants. The relative ratio controls the supramolecular structures formed, namely vesicles, liquid crystals, and worm-like micelles. It is likely that the phase transitions occurring in the polar head region appear to control the supramolecular structure formed, indicating that the curvature energy of the interface plays an important role in addition to the packing parameter ($v/al$). It is tempting to model the phase transitions in the polar head region of cationic–anionic surfactant systems by 2-D coulomb gas.

## ACKNOWLEDGMENTS

The work reported above was done in collaboration with S. D. Samant, B. K. Mishra, P. A. Hassan, Sushama Mishra, D. O. Shah, S. J. Candau, D. Langevin, W. Urbach, B. S. Valaulikar, R. A. Salkar, V. V. Kumar, and S. V. G. Menon under the Indo-French collaboration project no. 1007-1 sanctioned by the Indo-French Centre for the Promotion of Advanced Research.

## REFERENCES

1. H Hoffmann, G Ebert. Angewandte Chemie 27:902, 1988.
2. H Rehage, H Hoffmann. J Phys Chem 92:4712, 1988.
3. SJ Candau, E Hirsch, R Zana, M Delsanti. Langmuir 5:225, 1989.
4. ME Cates. Macromolecules 20:2289, 1987.
5. H Rehage, I Wunderlich, H Hoffmann. Prog Coll Polym Sci 72:1986, 1986.
6. H Rehage, H Hoffmann. Mol Phys 74:933, 1991.
7. M Cates, SJ Candau. J Phys Condens Matter 2:6869, 1990.
8. PG De Gennes. J Chem Phys 55:572, 1991.
9. F Lequeux, SJ Candau. In: ACS Symposium Series 578. Herb, CA, Prud'homme, RK, eds. Washington, DC, American Chemical Society, 1994, p51.
10. F Kern, R Zana, SJ Candau. Langmuir 7:1344, 1991.
11. JN Israelachivli, DJ Mitchell, BW Ninham. J Chem Soc Faraday II 72:1565, 1976.
12. C Tanford. The Hydrophobic Effect. 2nd Ed. New York, Wiley Interscience, 1980.

13a. C Manohar, URK Rao, BS Valaulikar, RM Iyer. J Chem Soc Chem Commun 379, 1986.
13b. URK Rao, C Manohar, BS Valaulikar, RM Iyer. J Phys Chem 91:3286, 1987.
14. S Gravsholt. J Colloid Int Sc 57:576, 1976.
15. D Balasubramanian, V Srinivas, VG Gaikar, MM Sharma. J Phys Chem 92:3865, 1989.
16. EW Kaler, A Kamalakaramurthy, BZ Rodriguez, JAN Zasadzinski. Science 145:1371, 1989.
17. BK Mishra, SD Samant, P Pradhan, SB Mishra, C Manohar. Langmuir 9:894, 1993.
18. M Doi, SF Edwards. J Chem Soc Faraday II 74:560, 1978.
19. J Narayanan, C Manohar. Phase Transitions 50:125, 1994.
20. JH Fuhrhop, W Helfrich. Chem Rev 93:125, 1994.
21. FAL Anet. J Am Chem Soc 108:7102, 1986.
22. RA Salkar, SD Samant. Private communication.
23. M Fisher. J Stat Phys 75:1, 1994.
24. N Bjerrum. Kgl Danske Videnskab Selskab 4:26, 1906.
25. A Helenius, K Simons. Biochim Biophys Acta 415:29, 1975.
26. D Andelman, MM Kozlov, W Helfrich. Europhys Lett 25:231, 1994.

# 7
## Quarter Century Progress and New Horizons in Microemulsions

**Krister Holmberg**
*Institute for Surface Chemistry, Stockholm, Sweden*

## I. THE TERM MICROEMULSION

Microemulsions are macroscopically homogeneous mixtures of oil, water, and surfactant, which on the microscopic level consist of individual domains of oil and water separated by a monolayer of amphiphile. They were scientifically described by Schulman in 1943 (1), but the concept had appeared in the patent literature before that (2,3). The traditional approach to preparing microemulsions was to titrate a milky emulsion with a medium chain alcohol such as pentanol or hexanol, which was later referred to as cosurfactant (4). In 1958 the term microemulsion was coined to describe the transparent, or slightly opaque, mixture that was formed after addition of a given amount of alcohol (5). Schulman's idea of the structure was that of an emulsion with very small droplets. The thermodynamic stability of microemulsions was from the beginning seen as the most distinguishing property, and it was postulated that the spontaneous formation was due to a transient zero, or negative, oil–water interfacial tension, $\gamma_{o/w}$. The cosurfactant was considered essential for obtaining zero interfacial tension; crowding of surfactant and cosurfactant molecules in the interfacial zone was seen as an important element in making the spreading pressure of the mixed film equal to or even exceeding $\gamma_{o/w}$.

Later research has modified Schulman's original views on microemulsions in some respects. It is now well established that cosurfactants are not needed to form microemulsions (although they are often used to simplify the formulation work, as will be discussed below). The concept of negative or zero surface free energy as a key element in microemulsion formation has been abandoned; it is true, however, that reduction of interfacial tension by several orders of magnitude is essential. However, the general description of the phenomenon, now half a century old, is still valid and, as Friberg (6) expressed it, "it is not only necessary but a pleasure to pay homage to Schulman's intuitive insight."

## II. PHASE BEHAVIOR OF OIL-WATER-SURFACTANT SYSTEMS

Pioneering work by Winsor, Shinoda, and others systematized the knowledge about phase behavior of oil-water-surfactant systems. The phase behavior of a three-component system can, at fixed temperature and pressure, best be represented by a phase diagram, as shown in Figure 1. At low surfactant concentration, there is a sequence of equilibria between phases, commonly referred to as Winsor phases (7). A microemulsion phase may be in equilibrium with excess oil (Winsor I, or lower phase microemulsion), with excess water (Winsor II, or upper phase microemulsion), or with both excess phases (Winsor III, or middle phase microemulsion). For nonionic surfactants the I → III → II transition (from left to right in Figure 1) may occur by raising temperature, while for ionic surfactant systems containing an electrolyte (i.e., a quaternary system), the transition may be induced by increasing salinity. System (a) in Figure 1, which represents a composition based on nonionic surfactant at low temperature, is indicative of the phase behavior of a very hydrophilic surfactant. Only trace quantities of oil can be solubilized into the oil-water (O/W) microemulsion which is in equilibrium with almost pure oil, as indicated by the tie lines. On raising temperature, the surfactant becomes less hydrophilic and more oil can be solubilized into the microemulsion, but the system is still of Winsor I type (b). Systems (c), (d), and (e) illustrate Winsor III systems with a three-phase triangle surrounded on two sides by two-phase regions. On increasing temperature, the microemulsion apex moves from left to right. When it is in a central position (d), i.e., at the point where the microemulsion contains equal amounts of oil and water, the system is referred to as balanced. The height of the microemulsion triangle at the point where the system is balanced can be seen as a measure of the surfactant efficiency. With a very efficient amphiphile, the microemulsion apex may appear at only a few percent surfactant; the importance of such systems is discussed below. Systems (f) and (g) may be seen as mirror images of (b) and (a).

The phase behavior depicted in the diagrams of Figure 1 can be visualized as test tube experiments, as shown in Figure 2.

# Progress and New Horizons in Microemulsions

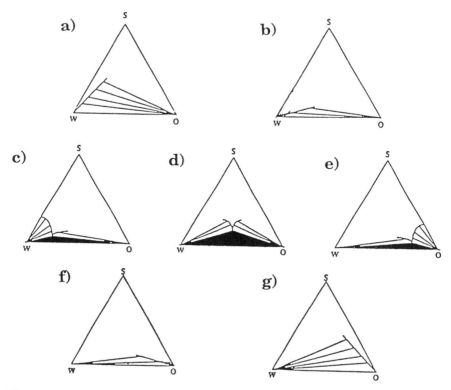

**Figure 1** A series of phase diagrams of a ternary system transformed from Winsor I [(a) and (b)] via Winsor III [(c), (d), and (e)] to Winsor II [(f) and (g)]. The dark triangles are three-phase regions. Tie lines indicate the compositions of the two-phase regions. S, W, and O stand for surfactant, water, and oil, respectively.

A phase prism, as shown in Figure 3, is a convenient way to illustrate phase behavior with temperature, but not pressure, as a variable. In order to simplify the work, the number of degrees of freedom is often reduced by one, either by keeping the oil to water ratio constant, usually at 1:1 (Figure 3a), or by using constant surfactant concentration (Figure 3b).

Figure 4 illustrates a section through the phase prism for a nonionic surfactant–oil–water system at a 1:1 oil to water ratio, equivalent to the plane cut out of Figure 3a. The three-phase region, W + L + O, exists between temperatures $T_1$ and $T_2$, and the width of the region is very dependent on surfactant concentration. At surfactant concentration $C^*$ and at the balanced temperature $T^*$, the three-phase region meets the one-phase microemulsion (the microemulsion apex of the three-phase region). At

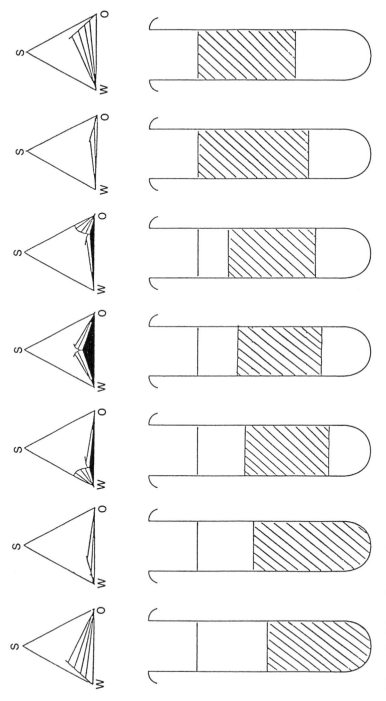

**Figure 2** Phase changes of a system containing equal amounts of oil and water and a given (low) amount of surfactant. For a nonionic surfactant system left to right transition may be induced by increasing temperature.

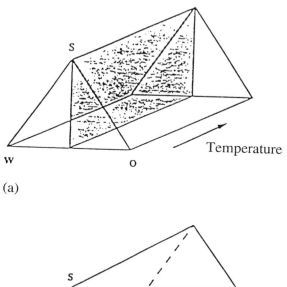

(a)

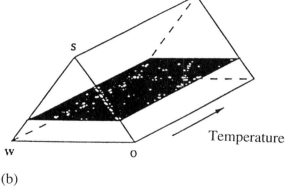

(b)

**Figure 3** Phase prisms illustrating cuts at constant oil to water ratio (a) and constant surfactant concentration (b).

higher surfactant concentration, the microemulsion is in equilibrium with a lamellar phase, $L_a$. Kahlweit and Strey have introduced this cut out of the phase prism to study phase behavior of systems based on nonionic surfactants (8).

The section through the phase prism representing constant surfactant concentration, equivalent to the plane cut out of Figure 3b, is also a useful tool for studying phase behavior of nonionic systems. This approach has been extensively used (9,10). A typical example is shown in Figure 5 in which a surfactant concentration slightly above $C^*$ has been used (11). The diagram illustrates the relationship between temperature and relative

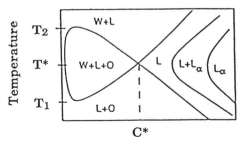

**Figure 4** Schematic phase diagram of a ternary system based on nonionic surfactant. L denotes microemulsion phase and $L_a$ lamellar liquid crystalline phase. (Redrawn from Ref. 9.)

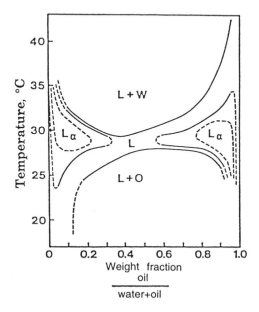

**Figure 5** Phase diagram of the system C12E5/$D_2$O/cyclohexane-hexadecane at a constant surfactant concentration of 7.0 wt %. (From Ref. 9, used with permission.)

amounts of oil and water. (Note that, by convention, weight fraction of oil rather than oil to water ratio is given on the x-axis.)

The phase diagram shows an isotropic solution phase forming a narrow channel which connects surfactant in water at lower temperature with surfactant in oil at higher temperature. The diagram illustrates the limited region of existence of the microemulsion phase typical of systems based on nonionic surfactants. The system is balanced at around 28°C. At higher temperatures, the surfactant is too oil soluble and an aqueous phase separates out. At lower temperatures the surfactant is too hydrophilic and oil separates out. The diagram also shows that lamellar liquid crystalline phases form at intermediate temperature both at high and low weight fractions of oil.

## III. MICROEMULSION STRUCTURE

Various techniques have been explored for studies of the internal structure (microstructure) of microemulsions. The pulsed-gradient spin-echo nuclear magnetic resonance (NMR) technique developed by Lindman and Stilbs (12) has proven particularly useful, since it gives direct information about the state of all three components—oil, water and surfactant—in the same experiment.

An illustrative example of the use of the diffusion NMR technique to study microemulsion internal structure is given in Figure 6. The system C12E5/water/hydrocarbon exhibits an isotropic channel extending from the lower left to the upper right corner of the temperature versus oil/(water + oil) diagram shown in Figure 5 above. The isotropic channel was studied with respect to the diffusion behavior of all components (11). Relative diffusion coefficients, $D/D°$, were recorded at several points along the channel. ($D$ is the observed diffusion coefficient and $D°$ is the diffusion coefficient of the neat liquid.)

Figure 6 shows the strong temperature and composition dependence of the diffusion coefficients of both water and oil. The full lines are drawn through approximate mean values of $D/D°$, with respect to temperature, at each composition point. The surfactant diffusion can also be monitored, and the values obtained can help in structure elucidations.

The very large variations in $D_{water}/D_{oil}$ on varying the weight fraction of oil is striking. Evidently, there is a smooth transition from O/W structure at low oil to water ratio to W/O at high ratio. At approximately equal amounts of oil and water, there is rapid diffusion ($D/D°$ of around 0.5) of both components, indicating a structural arrangement where neither component is confined into closed domains.

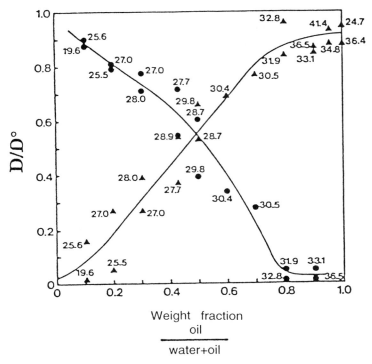

**Figure 6** Relative diffusion coefficients, $D/D°$, of water (●) and cyclohexane (▲) versus weight fraction of oil (11).

A similar type of smooth transition from O/W at low oil content to W/O at high oil content has been obtained by gradually changing the ratio of anionic to nonionic surfactant while keeping all other parameters constant (13). Also in this case both oil and water showed rapid diffusion in the region of approximately equal amounts of the components.

The isotropic region in which both oil and water show rapid diffusion is often referred to as bicontinuous (14). There has been considerable debate about the microstructure of bicontinuous microemulsions. Lamellar-like structures, interconnected rods, and structures resembling cubic liquid crystalline phases have been proposed. It now seems that the most general representation of a bicontinuous microemulsion is one of zero mean curvature structures with infinite curved channels of oil and water, as shown in Figure 7. The water on the hydrophilic side of the surfactant monolayer is continuously connected in three dimensions and the same is true of the oil on the hydrophobic side.

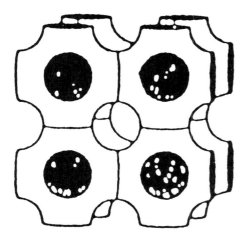

**Figure 7** Structure of a bicontinuous microemulsion composed of infinite and alternating oil and water channels.

The concept of curvature is a useful one in elucidating microemulsion microstructure. In principle, every point on a surface possesses two principal radii of curvature, $r_1$ and $r_2$. For a sphere, $r_1$ and $r_2$ are equal and positive. For a cylinder, $r_2$ is indefinite and for a plane, both $r_1$ and $r_2$ are indefinite. For the case of a monkey saddle, $r_1 = -r_2$, i.e., at every point the surface is both concave and convex with the same radii. Now, mean curvature and gaussian curvature are used to define the bending of surfaces. They are defined as follows:

Mean curvature: $H = 1/2(1/r_1 + 1/r_2)$
Gaussian curvature: $K = 1/r_1 \cdot 1/r_2$

It can easily be seen that for a plane $H = K = 0$ and for a monkey saddle surface $H = 0$ and $K < 0$. It seems that bicontinuous microemulsions can best be represented as isotropic systems with zero or close to zero mean curvature and with negative gaussian curvature of the dividing surface between hydrophilic and hydrophobic domains. As the system parameters are changed, such as temperature for microemulsions based on nonionic surfactants or electrolyte concentration for systems based on ionic surfactants, the spontaneous mean curvature will change and the system will gradually move away from bicontinuity. Figure 8 gives a schematic illustration of the gradual change in microstructure of a microemulsion based on nonionic surfactant as one moves along the isotropic channel in the temperature versus oil weight fraction diagram.

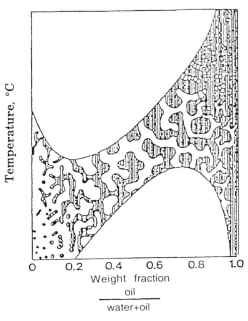

**Figure 8** Schematic illustration of microemulsion structure in a temperature versus oil weight fraction phase diagram. The microemulsion is based on nonionic surfactant. (From Ref. 15.)

The transition from Winsor I via Winsor III to Winsor II, which is very distinct on the macroscale (see Figure 2), seems not to be accompanied by a dramatic change in microstructure of the microemulsions. Typically, the lower phase microemulsion of a Winsor I system exhibits rapid water and slow oil diffusion. The $D_{water}/D_{oil}$ value decreases as the system approaches the three-phase region, but the transition into the Winsor III regime does not bring about a jump in the curve. When the system is balanced, i.e., when the middle phase contains equal amounts of oil and water, $D_{water}/D_{oil}$ is around unity. A further shift toward the Winsor II regime leads to a continuous decrease in water and increase in oil diffusion rates but, again, the transition from Winsor III to Winsor II, which is so distinct to the eye, is not reflected in a discontinuity in the $D_{water}/D_{oil}$ curve (16).

A problem in the interpretation of the diffusion data is to properly account for the reduction in water diffusion rate due to surfactant headgroup hydration and in oil diffusion rate due to oil penetration of the hydrocarbon region of the surfactant film. These effects tend to make it difficult to assign quantitative values of $D_{water}$ and $D_{oil}$ when the surfactant concentration is high. However, there should be no uncertainties about the magnitude of the values and the trends obtained (17).

Other techniques have also been used to study microstructure of isotropic systems. Of particular interest are recent results from freeze fracture electron microscopy (18). This technique must be used with caution since artifacts may mislead the interpretation of results. However, it now seems clear that very rapid freezing followed by fracture and replication of the fracture face gives reliable images of the microstructure. On studying a system based on nonionic surfactant, the same transition from predominantly water continuous via bicontinuous ($H = 0$) to predominantly oil continuous on temperature increase was seen, as had earlier been detected by self-diffusion NMR.

## IV. CHOICE OF SURFACTANT

As mentioned above, the curvature of the oil–water interface of a microemulsion may vary from highly curved toward oil to zero mean curvature to highly curved toward water. Unlike emulsions, the curvature of microemulsions is considerable at the scale of the surfactant. This implies that not only hydrophile–lipophile balance but also molecular geometry of the surfactant is an important factor in finding the optimum microemulsion surfactant.

A popular way of dealing with surfactant geometry is to use the packing parameter concept (19). The geometric or packing properties of surfactants depend on their optimal headgroup area, $a_o$, as well as on their hydrocarbon volume $V$ and critical chain length $l_c$ (Figure 9). The value of

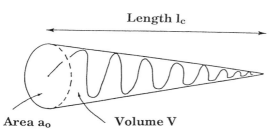

**Figure 9** A surfactant molecule with headgroup area $a_o$, hydrocarbon critical chain length $l_c$, and hydrocarbon volume $V$.

$a_o$ is governed by repulsive hydrophilic forces acting between the headgroups and attractive hydrophobic forces acting at the hydrocarbon-water interface. Steric chain-chain and oil penetration interactions determine $V$ and $l_c$.

Now, it has been shown that the value of the dimensionless packing parameter $V/(a_o \cdot l_c)$ will determine the type of aggregate that will spontaneously form in solution (Figure 10). Each of these structures corresponds to the minimum-sized aggregate in which the surfactants have a minimum free energy.

The packing parameter concept was originally developed to explain surfactant structures in binary surfactant-water systems. However, it can be extended to include oil, as is illustrated in Figure 11. The structures are similar to those in the surfactant-water system, but the existence of oil molecules adds new constraints on the surfactant molecules while lifting others. For example, highly stressed cylindrical micelles can turn into swollen micelles of spherical geometry by having oil fill up the interior regions. Oil penetration of hydrocarbon domains will often drive the system in the clockwise direction, i.e., from O/W to W/O, since the increase in $V$ leads to an increase of the value of the packing parameter $V/a_o \cdot l_c$ (20).

As illustrated in Figure 10, the packing parameter concept is intimately linked to the hydrophilic-lipophilic balance (HLB) concept. Increasing the length of the polyoxyethylene chain of nonionic surfactants increases the HLB number. It also increases $a_o$, thus decreasing $V/(a_o \cdot l_c)$. However, the packing parameter concept also takes into account bulkiness

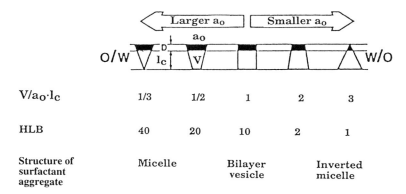

**Figure 10** The packing parameter $V/a_o \cdot l_c$ governs the shape of the surfactant aggregate formed in solution. The figure also shows correlation between packing parameter and HLB numbers. (Redrawn from Ref. 20.)

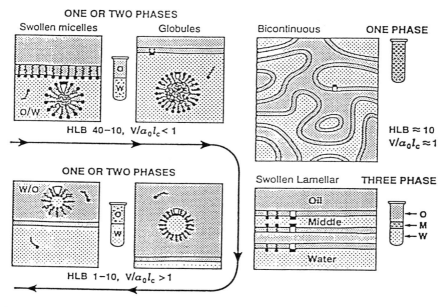

**Figure 11** Structure of surfactant aggregates in oil–water systems. (From Ref. 20.)

of the hydrophobic tail, and this is important in microemulsions and micellar systems that contain highly curved oil–water interfaces.

From geometrical considerations, it may be stated that surfactants with moderately long, straight-chain aliphatic hydrocarbon tails are best suited to prepare O/W microemulsions, surfactants with rather bulky hydrophobes are good for bicontinuous microemulsions, and surfactants with highly branched hydrophobic tails should be used for W/O microemulsions. Indeed, this has been found to be the case. As mentioned above, often a combination of surfactants are used in the formulation. The surfactant geometry will then be the mean geometry of the species involved. Consequently, a combination of a surfactant with a single straight-chain tail and one with two branched tails (for instance, see Figure 12c) may constitute an ideal mixture to formulate a bicontinuous microemulsion.

For some microemulsion applications, in particular in enhanced oil recovery (EOR—see below), there is a need for surfactants which by themselves, i.e., without cosurfactant or cosolvent, can solubilize large amounts of oil and water. By proper optimization of the surfactant geometry, including surfactant molecular weight, molecules with extreme solubilizing capacity have been obtained. An example is shown in Figure 13. The twin-

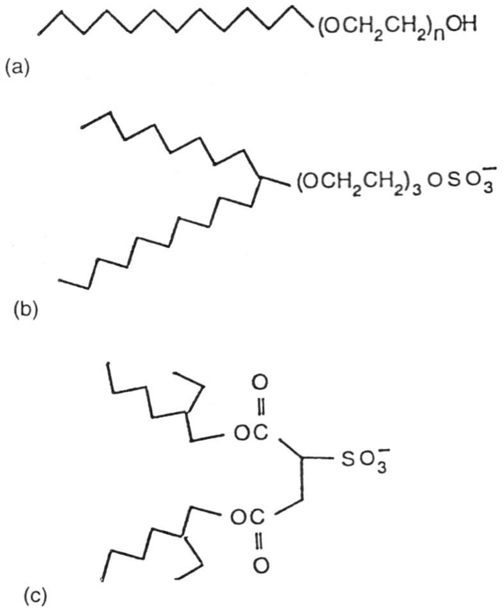

**Figure 12** Examples of surfactants for O/W microemulsions (a), bicontinuous microemulsions (b), and W/O microemulsions (c).

tailed sulfate surfactant used in 1.54 wt % can solubilize 49.2% aqueous NaCl solution and 49.2% hexane, representing 32 times as much water and hexane as surfactant (21).

The surfactant of Figure 13 has very low solubility in both oil and water. This is an important characteristic of a surfactant to be used with high efficiency in Winsor III systems. Very low saturation concentrations

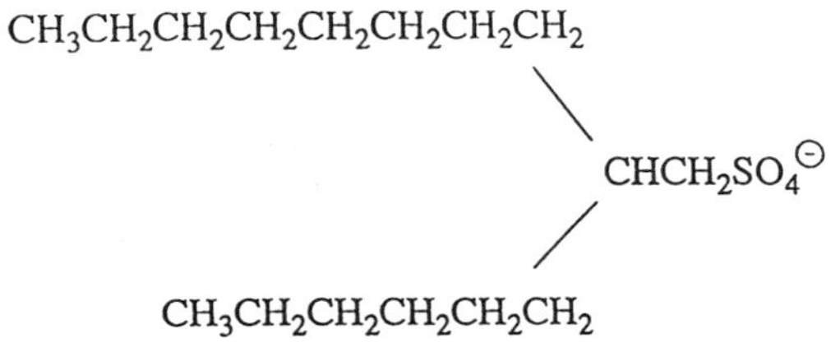

**Figure 13** Structure of a surfactant, sodium 2-hexyldecylsulfate, with very high solubilizing capacity for both oil and water.

in both excess phases are needed in order to have the surfactant confined to the location where it exerts its action, i.e., at the oil-water interfaces.

The interest in EOR for surfactants with high solubilizing capacity derives from the fact that there is a correlation between high solubilizing capacity and low oil-water interfacial tension. The surfactant of Figure 13 gives a $\gamma_{oil/water}$ of around 0.0005 mN/m, an extremely low value. Low oil-water interfacial tensions are believed to be crucial for oil mobilization in the reservoir (see below).

## V. APPLICATIONS OF MICROEMULSIONS

Microemulsions were in commercial use long before the term was coined. Early applications include floor polishes, cutting oils, and pesticide formulations. Today, microemulsions are used in a wide variety of products in both the industrial and the household sectors. In this section three very different applications will be discussed: enhanced oil recovery, cleaning, and reaction medium for organic reactions. The first one is of tremendous potential and has triggered a vast amount of research activity but remains to be commercialized; Winsor III systems are believed to be the key to success. The second application has an established position in the market and is usually of O/W type. The third application is an emerging one and is normally of W/O type.

### A. Microemulsions for Enhanced Oil Recovery (EOR)

Oil fields consist of porous rock, usually limestone or sandstone, in which the pores are filled with petroleum and brine. The porous rock formation is surrounded by impermeable rock. The permeability depends on pore size, which is typically 50–1000 nm. In a normal oil field 10–25% of the pore volume is occupied by brine, 55–80% by oil, and the rest is void volume. The pressure in the reservoir is normally high and the temperature typically is in the range 70–100°C.

When the first wells are dug, oil comes out under its own pressure. This spontaneous production is later supplemented by pumping action. Together, these two processes are referred to as the primary recovery. On average this stage leads to a recovery of 15–20% of the oil in place. In the next stage, the secondary recovery, water is used to sweep out remaining oil. Water is then pumped down the injection well and moves outward in a piston-like fashion, displacing the oil. The immobilized oil is recovered via production holes, as shown in Figure 14. The so called sweep efficiency is usually not very good, however, particularly when the oil is of higher vis-

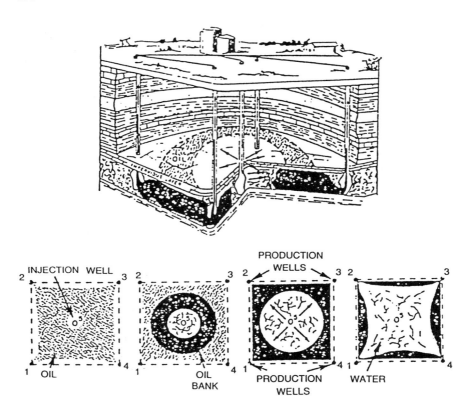

**Figure 14** Schematic picture of oil production by water flooding.

cosity than the displacing water. The secondary stage typically gives a recovery of 10% of the original oil in place. Hence, primary and secondary recovery together manage to recover less than a third of the total amount of petroleum in the reservoir.

Any oil recovery process following water flooding is referred to as enhanced oil recovery (EOR) or tertiary oil recovery. Surfactant flooding, sometimes called microemulsion flooding, is the technique of relevance to this paper. The interest in microemulsions for this purpose derives from their ability to reduce oil–water interfacial tension to ultralow values.

One of the main reasons for the inefficiency of water flooding through the reservoir rock is that oil is trapped by capillary action in the form of disconnected ganglia. In principle, increasing the water pressure is a way to mobilize such oil ganglia. However, in water flooding there is a limit to the pressure that can be used, because high pressures fracture the reservoir rock and large fractures cause considerable decrease in sweep efficiency. For an oil droplet trapped in a pore, as shown in Figure 15, it can be shown that

$$p_1 - p_2 = 2\gamma_{o/w}(1/r_1 - 1/r_2)$$

where $p_1$ and $p_2$ are the pressures on either side of the droplet with radii of curvature $r_1$ and $r_2$. As can be seen, the pressure drop $\Delta p = p_1 - p_2$ is proportional to the oil–water interfacial tension. It can be shown that, at least in water wet rocks, reduction of the interfacial tension to values on the order of $10^{-3}$ mN/m is needed in order to obtain substantial mobilization and recovery of oil.

As was first shown by Reed and Healy (22), oil–water interfacial tension has a deep minimum in the three-phase region, i.e., $\gamma_{o/w}$ decreases as a system goes from Winsor I to Winsor III, has a minimum in the middle of the Winsor III region, and increases as it moves on into the Winsor II regime (Figure 16). (The Winsor I > III > II transition can be achieved by raising temperature for a system based on nonionic surfactant and by increasing salinity for a system based on ionic surfactant.) The condition for which the hydrophilic and lipophilic properties of a surfactant are equally balanced is called the phase inversion temperature for nonionics, for which temperature is usually the most important variable. For ionic surfactants this state is often referred to as optimal conditions, e.g., optimal salinity. At this point there is equal volumetric solubilization of oil and

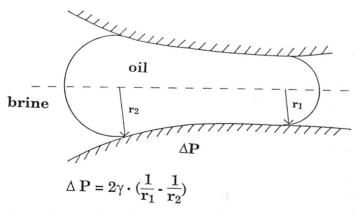

**Figure 15** An oil ganglion trapped in a pore.

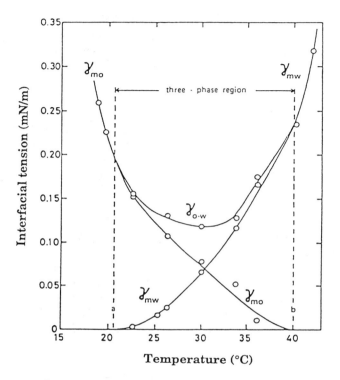

**Figure 16** Interfacial tensions as a function of temperature for the system C8E3–water–tetradecane. The symbols $\gamma_{o/w}$, $\gamma_{m/o}$, and $\gamma_{m/w}$ indicate interfacial tensions for oil–water, middle phase–oil, and middle phase–water, respectively. The system is balanced at 30°C (23).

water in the middle phase microemulsion and the interfacial tensions are at a minimum.

It could soon be demonstrated in laboratory flooding experiments that surfactant formulations capable of forming Winsor III microemulsions with the specific oil and brine present in the rock at the specific temperature of the reservoir could give a remarkable yield of oil recovered. Important contributions to the understanding of the mechanisms involved were made by the groups of Shah (24) and Schechter and Wade (25).

Since, particularly in off-shore reservoirs, the distances between injection and production holes are long, the formulations used should preferably contain as few surface-active components as possible. Mixtures of different surfactants or surfactant–cosurfactant combinations are probably less suitable, since they are likely to separate during their way through the porous rock (which can be expected to act like a long chromatography column). Therefore, surfactants have been sought that

1. Form Winsor III systems with the specific reservoir oil and brine at reservoir temperature
2. Are hydrolytically stable for an extended period of time under reservoir conditions
3. Do not precipitate in hard water
4. Do not adsorb extensively at the mineral surfaces of the reservoir.

In addition to the above requirements, "trivial" aspects such as cost, toxicity, and biodegradability need to be taken into account. Several laboratories have come up with branched ether sulfonates (or possibly sulfates) as a suitable choice of surfactant type (26–28). By tailor-making the hydrophobe branching, surfactants with extreme solubilizing capacity can be obtained (27). Representative compounds are schematically shown in Figure 17.

Figure 18 shows relative phase volumes as a function of surfactant concentration for a system containing a branched-tail ether sulfonate designed for specific reservoir conditions (29). Note that there is a Winsor I > Winsor III > Winsor II transition on dilution of the surfactant which must be taken into account in the optimization of the surfactant. The system shown in Figure 18, having a water to oil ratio of 4, goes from Winsor I to Winsor III at 1.0% surfactant and from Winsor III to Winsor

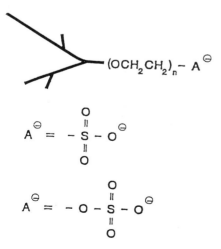

**Figure 17**  A branched-chain ether sulfonate surfactant suitable for EOR use.

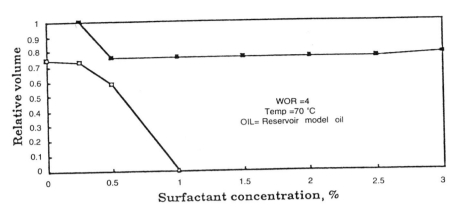

**Figure 18** Relative phase volumes versus surfactant concentration at a water to oil ratio (WOR) of 4.

II at around 0.25% surfactant. A very high displacement efficiency was obtained in core flooding experiments with reservoir oil as oil phase. A 0.5 pore volume slug containing 2 wt % surfactant in seawater produced 89% of the oil left in the core after water flooding (29).

### B. Microemulsions for Cleaning

Microemulsions, being microheterogeneous mixtures of oil, water, and surfactant, are excellent solvents for nonpolar organic compounds as well as for inorganic salts. The ability of microemulsions to solubilize a broad spectrum of substances in a one-phase formulation has been found useful for cleaning of hard surfaces—dirt is often a complex mixture of hydrophilic and hydrophobic components. Of particular interest from a practical point of view is the possibility of replacing formulations based on halogenated or aromatic hydrocarbons with microemulsions containing nontoxic aliphatic hydrocarbons. A typical example is given in Figure 19.

Microemulsions, mainly based on nonionic surfactants, have an established position in industrial cleaning of hard surfaces. They are usually sold as concentrated mixtures that are to be diluted with water before use. Hence, the isotropic domain should preferably extend to the water corner. A typical example of a suitable model system is that of C12E6, decane and water, shown in Figure 20. Nonionics are suitable surfactants in these formulations, since they can be formulated with ionic builders such as phosphate or citrate. Such systems suffer from the drawback of considerable temperature sensitivity, however, as is also illustrated in Figure 20.

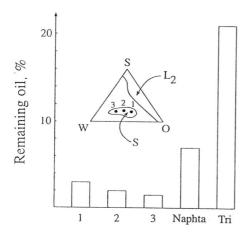

**Figure 19** Removal of lubricating oil by three microemulsions all lying within the surfactant phase (S phase) indicated in the phase diagram. The surfactant was a mixture of two ethoxylates with a mean HLB value of 10.7. The hydrocarbon was aliphatic hydrocarbon with bp 190–240°C. Oil removal by hydrocarbon only and by trichloroethane are shown as references. Remaining oil is measured as remaining fluorescence (30).

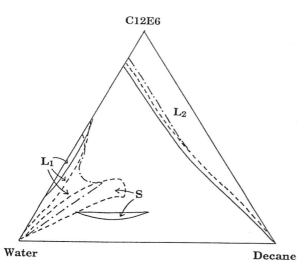

**Figure 20** Phase diagrams for the system C12E6/decane/water at 30, 40, and 50°C (31). (———) 50°C, (------) 40°C, (— · — ·) 30°C.

One way to increase the temperature interval is to use a mixture of nonionics, e.g., a blend of ethoxylates both below and above the degree of ethoxylation of the optimum compound. (Commercial ethoxylates have a very broad homologue distribution in themselves and give microemulsions with broader temperature regions than homologue pure model compounds. By mixing commercial nonionics, the temperature intervals can be further extended.)

The microemulsion concept can also be utilized in hydrocarbon-free cleaning formulations. It has been demonstrated that for nonionic surfactants, optimum detergency (in the case of oily soil) occurs around the phase inversion temperature of the system surfactant/hydrocarbon soil/water (32). The high level of soil removal at this temperature can be attributed to the ultralow interfacial tension achieved under these conditions and to the high rates of oily soil solubilization into middle-phase microemulsions.

The importance of a well-balanced system for soil removal is illustrated in Figure 21, which shows that for the two nonionic surfactants C12E4 and C12E5, optimum detergency occurs at the temperature at which the ternary system surfactant/hydrocarbon soil/water forms a three-phase system (33).

## C. Microemulsions as Reaction Medium for Organic Synthesis

In recent years microemulsions have been evaluated as a reaction medium for a variety of chemical reactions. In preparative organic chemistry, microemulsions may be used to overcome reactant solubility problems and may also increase the reaction rate, as will be discussed below. Microemulsions are also of interest as a medium for inorganic reactions. Microemulsions of W/O type have been found useful in preparing small particles of metal and inorganic salts.

Another use of microemulsions for particle synthesis is in the preparation of monodisperse latexes of very small droplet size. By polymerization in W/O microemulsions, ultrafine particles containing very high molecular weight polymers are obtained. In bioorganic synthesis, W/O microemulsions can be employed as "minireactors" both for condensation and for hydrolysis reactions. Most work has dealt with lipase catalyzed reactions using hydrophobic substrates. The enzyme is located in the water pools, and reaction is believed to take place at the oil–water interface.

The four types of reactions in microemulsions mentioned above, i.e., organic synthesis, formation of inorganic particles, polymerization, and bioorganic synthesis, can all be seen as emerging technologies of considerable current interest. In slightly different ways they all take advantage of

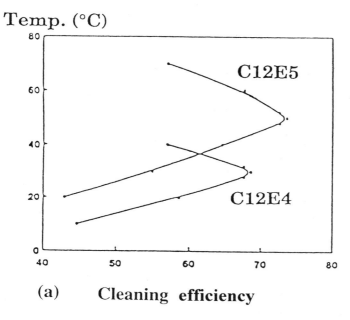

(a)  Cleaning efficiency

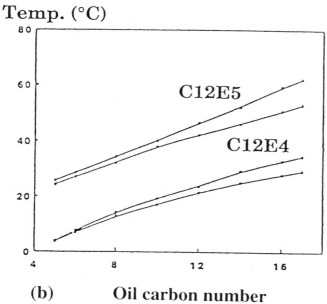

(b)  Oil carbon number

**Figure 21** Hexadecane removal (a) and extention of three-phase regions with different hydrocarbons (b) versus temperature for two nonionic surfactants (33).

the large interior interfacial area of microemulsions. Use of microemulsions in preparative organic chemistry will be discussed in some detail below.

## 1. Overcoming Reagent Incompatibility Problems

A common practical problem in synthetic organic chemistry is attainment of proper phase contact between nonpolar organic compounds and inorganic salts. There are many examples of important reactions where this is a potential problem: hydrolysis of esters with caustic, oxidative cleavage of olefins with permanganate-periodate, addition of hydrogen sulfite to aldehydes and to terminal olefins, preparation of alkyl sulfonates by treatment of alkyl chloride by sulfite or by addition of hydrogen sulfite to $\alpha$-olefin oxides. The list can be extended further. In all examples given there is a compatibility problem to be solved if the organic component is a large nonpolar molecule.

There are various ways to solve the problem of poor phase contact in organic synthesis. One way is to use a solvent or a solvent combination capable of dissolving both the organic compound and the inorganic salt. Polar, aprotic solvents are sometimes useful for this purpose, but many of these are unsuitable for large-scale work due to toxicity or difficulties in removing them by low vacuum evaporation.

Alternatively, reaction may be carried out in a mixture of two immiscible solvents. The contact area between the phases may be increased by agitation. Phase transfer reagents, in particular quaternary ammonium compounds, are useful aids in many two-phase reactions. Also, crown ethers are very effective in overcoming phase contact problems; however, their usefulness is limited by high price. (Open-chain polyoxyethylene compounds often give a "crown ether effect" and may constitute practically interesting alternative phase transfer reagents.)

Microemulsions are excellent solvents both for hydrophobic organic compounds and for inorganic salts. Being macroscopically homogeneous yet microscopically dispersed, they can be regarded as something between a solvent-based one-phase system and a true two-phase system. In this context microemulsions should be seen as an alternative to two-phase systems with phase transfer reagents. This is illustrated below by the use of microemulsions for detoxification of mustard, $ClCH_2CH_2SCH_2CH_2Cl$.

Mustard is a well-known chemical warfare agent. Although it is susceptible to rapid hydrolytic deactivation in laboratory experiments where rates are measured at low substrate concentrations, its deactivation in practice is not easy. Due to its extremely low solublity in water, it remains for months on a water surface. Addition of strong caustic does not increase the rate of reaction. Microemulsions have been explored as media for both hydrolysis and oxidation of "half-mustard," $CH_3CH_2SCH_2CH_2Cl$, a mustard model (Figure 22). Oxidation with hypochlorite turned out to be extremely rapid in both O/W and W/O microemulsions. In formulations based on either anionic, nonionic, or cationic surfactant, oxidation of the half-mustard sulfide to sulfoxide was complete in less than 15 s. The same reaction takes 20 min when a two-phase system, together with a phase transfer reagent, is employed (34).

## 2. Specific Rate Enhancements

By a proper choice of surfactants, rate enhancement analogous to micellar catalysis can be obtained. In microemulsion systems the effect may be referred to as microemulsion catalysis.

The importance of the choice of surfactant on reaction yield can be illustrated by the reaction between decyl bromide and sodium sulfite to give decyl sulfonate, a surface active agent.

$$C_{10}H_{21}Br + Na_2SO_3 \rightarrow C_{10}H_{21}SO_3Na + NaBr$$

Reactions were carried out in microemulsions based on decyl bromide dissolved in dodecane as oil component, aqueous $Na_2SO_3$ as water component, and either nonionic surfactant or nonionic surfactant plus a small amount of ionic surfactant as amphiphile. The phase diagram is shown in Figure

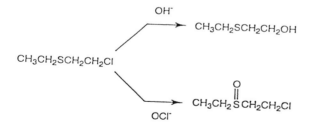

Reaction time for half-mustard destruction

| | |
|---|---|
| water + caustic | months |
| water + caustic + crown ether | 20 min |
| microemulsion | 15 sec |

**Figure 22** Transformation of 2-chloroethyl ethyl sulfide (half-mustard) into 2-hydroxyethyl ethyl sulfide (by alkali) or into 2-chloroethyl ethyl sulfoxide (by hypochlorite).

23. Reactions were run at three different compositions, one in the W/O microemulsion region (point A), one in the liquid crystalline phase, and one in the narrow isotropic channel which separates the liquid crystalline phase from the lower 2-phase region (point B). Point B represents a bicontinuous microemulsion (35,36).

The reaction was extremely slow in a surfactant-free mixture of the oil and water components, fairly sluggish in the liquid crystalline phase, and fast in the two microemulsion systems. Figure 24, which illustrates reaction at composition B of Figure 23, shows the influence of addition of a small amount of added ionic surfactant. Sodium dodecyl sulfate (SDS), which gives negative charge to the droplet surface, brings about a very low reaction rate. This is the expected result, since electrostatic double-layer forces will render approach of the sulfite ion into the interfacial region difficult.

It may seem surprising that the cationic surfactant $C_{14}TAB$ also de-

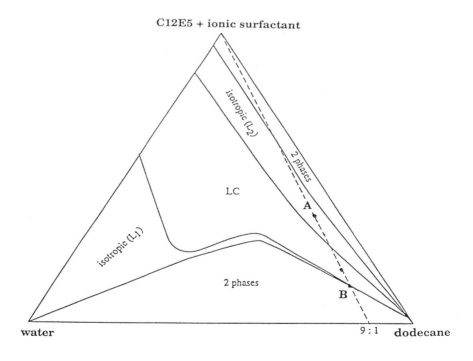

$$C_{10}H_{21}Br + Na_2SO_3 \longrightarrow C_{10}H_{21}SO_3Na + NaBr$$

**Figure 23** Pseudo-ternary phase diagram of the system described above. Reactions were run at compositions along the dotted line.

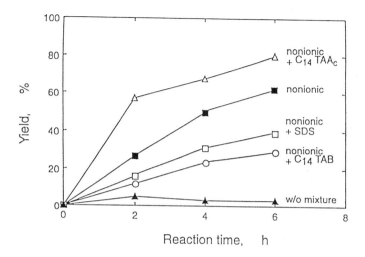

**Figure 24** Effect of ionic surfactants on rate of formation of decyl sulfonate from decyl bromide and sodium sulfite in microemulsion. $C_{14}TAA_c$ and $C_{14}TAB$ stand for cetyltrimethylammonium acetate and bromide, respectively.

creases reaction rate. The other cationic surfactant, $C_{14}TAA_c$, on the other hand, gives a considerable increase in reactivity. Evidently, the choice of counterion is decisive of the reaction rate. A large, polarizable counterion, such as the bromide ion, interacts so strongly with the surfactant palisade layer that approach of the anionic reactant, the sulfite ion, is prevented. Interaction with the acetate ion is much weaker, and sulfite ions are allowed to diffuse into the interfacial region were reaction occurs.

### 3. Effects on Regioselectivity

The presence of an oil-water interface may induce orientation of reactants in microemulsion systems, which in turn may affect the regioselectivity of organic reactions. This is nicely illustrated by nitration of phenol to nitrophenol (Figure 25).

Conventional nitration of phenol in aqueous media results in an approximate 2:1 ratio of para- to ortho-isomer. When the reaction is performed in a microemulsion based on sodium bis(2-ethylhexyl)sulfosuccinate (AOT), a selectivity for *o*-nitration in the order of 80% is obtained (37).

The preference for nitration in the ortho position when the reaction is performed in microemulsion is likely to be due to accumulation of phenol at the oil-water interface, with the phenolic hydroxyl group oriented into

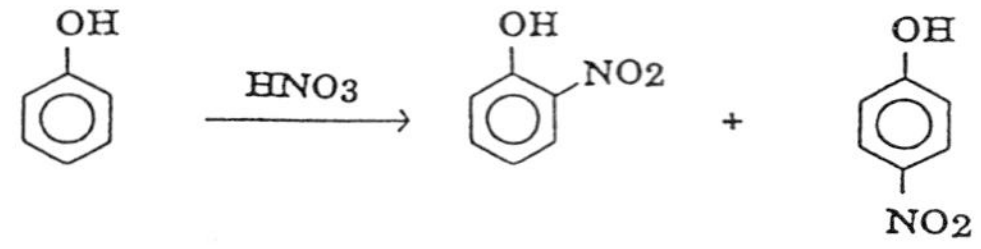

OH → HNO3 → OH-NO2 + OH (para-NO2)

|  | ortho | para |
|---|---|---|
| In water | 35 % | 65 % |
| In microemulsion | 80 % | 10 % |

**Figure 25** Nitration of phenol in water and in microemulsion. Increased regioselectivity due to interfacial ordering.

the water phase, as is illustrated in Figure 26. Since the reacting nitronium ion species resides entirely in the aqueous domain, an attack at the para position is less likely than at the ortho position. It is interesting that also a relatively non-surface-active molecule, such as phenol, exhibits such a marked orientational effect.

Use of microemulsions instead of aqueous solutions has been found to be a way to alter relative reactivities of different nucleophiles in substitution reactions relevant to protein immobilization (38). It was found that in a microemulsion proteins bind primarily via thiol groups to electrophilic groups at a surface. In water, coupling via amino groups dominate (Figure 27). The difference in relative reactivity of the functional groups is believed to be due to the difference in polarity between the two reaction media. Whereas polar, protic solvents, e.g., water, favor $S_N2$ reactions with uncharged nucleophiles (such as amino groups), solvents of low dielectric constant,

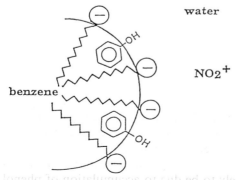

**Figure 26** Orientation of phenol at the oil-water interface.

# Progress and New Horizons in Microemulsions

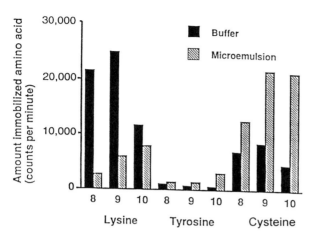

**Figure 27** Relative reactivities of lysine, tyrosine, and cysteine against epoxy groups in microemulsion and in aqueous buffer at pH 8, 9, and 10. The microemulsion consisted of 55% isooctane, 30% aqueous buffer, and 15% $C_{12}E_5$ (38).

e.g., microemulsions of limited water content, favor reactions with charged nucleophiles. (Thiol groups have $pK_a$ values at 9.0–9.5 and are thus partly deprotonized at typical immobilization conditions, i.e., pH 8–9.)

Many proteins, including antibodies and enzymes, have been immobilized to a solid support by the use of a microemulsion as reaction medium. For the majority of proteins the biological activity is not markedly affected by a few hours exposure to a microemulsion based on aliphatic hydrocarbon as oil component. The choice of surfactant seems not to be critical with regard to retention of biological function.

## 4. Immobilization of Proteins to Protein-Rejecting Surfaces

Hydrophobic surfaces which normally interact strongly with proteins can be made protein-rejecting by grafting with poly(ethylene glycol) (PEG) or other noncharged, hydrophilic polymers (39). These surfaces are of interest as such in biotechnical and medical areas where protein adsorption and cell adhesion need to be minimized. For several applications, e.g., solid phase immunoassay, therapeutic apheresis, and bioorganic synthesis using immobilized enzymes, protein immobilization to protein-repelling surfaces is also of interest. It has been shown that proteins attached to such inert surfaces retain their activity better than proteins immobilized to charged or hydrophobic surfaces. In addition, nonspecific adsorption of other biomolecules

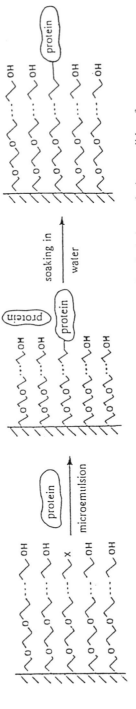

**Figure 28** Immobilization of protein to a reactive endgroup of poly(ethylene glycol) chains grafted to a solid surface.

will be minimized. However, covalent coupling of protein to such inert surfaces is not trivial.

Microemulsions of low water content are nonsolvents for hydrophilic polymers such as PEG. Thus, the pronounced protein-rejecting effect which is present in water and which is of entropic nature does not exist in the microemulsion. Coupling can be made to reactive groups on the hydrophilic polymer chains. When exposed to water, the chains will become hydrated and protein-repellent, as illustrated in Figure 28 (40).

## VI. CONCLUSIONS

Microemulsions have now reached the state where there is rather good understanding of the basic principles involved. There seems to be general agreement about the role of interfacial layer curvature and the importance of surfactant geometry. Several excellent experimental tools are available for studies of the microstructure of isotropic systems.

Many characteristic properties of microemulsions such as ease of formation, thermodynamic stability, high solubilizing capacity, and high penetrating ability into porous materials make these systems of interest for technical use. The three very different applications discussed above were chosen to illustrate the versatility of microemulsions for industrial use. Although much applied work has been done in this area, commercialization of microemulsions is still in its infancy. In particular, there is a large potential for these microheterogeneous systems in high-value processes where the small, well-defined, and easily adjustable dimensions of microemulsions can be taken advantage of. In the future, microemulsions are likely to play a role in the synthesis of nanosized materials of various kinds, as well as in organic and bioorganic processes where a high degree of selectivity is desired.

## REFERENCES

1. JP Hoar, JH Schulman. Nature 152:102, 1943.
2. VR Kokatnur. U.S. Patent 2,111,000 (1935).
3. D Bowden, J Holmstine. U.S. Patent 2,045,455 (1936).
4. LM Prince. Microemulsions: Theory and Practice. New York, Academic Press, 1977.
5. JH Schulman, W Stockenius, LM Prince. J Phys Chem 63:1677, 1959.
6. SE Friberg. J Disp Sci Technol 6:317, 1985.
7. PA Winsor. Solvent Properties of Amphiphilic Compounds. London, Butterworth, 1954.

8. M Kahlweit, R Strey. Angew Chem Intern Ed 24:654, 1985.
9. U Olsson, H Wennerström. Adv Colloid Interface Sci 49:113, 1994.
10. U Olsson, K Shinoda, B Lindman. J Phys Chem 90:4083, 1986.
11. U Olsson, K Nagai, H Wennerström. J Phys Chem 92:6675, 1988.
12. B Lindman, P Stilbs. In: Microemulsions. Friberg, S, Bothorel, P, eds. Boca Raton, FL, CRC Press, 1987, p 119.
13. B Lindman, K Shinoda, M Jonströmer, A Shinohara. J Phys Chem 92:4702, 1988.
14. LE Scriven. Nature 263:123, 1976.
15. M Kahlweit, R Strey, R Schomäcker. In: Reactions in Compartmentalized Liquids. Knoche, W, Schomäcker, R, eds. Heidelberg, Springer-Verlag, 1989, p 1.
16. P Guéring, B Lindman. Langmuir 1:464, 1985.
17. O. Söderman, P Stilbs. Prog NMR Spectroscopy 26:445, 1994.
18. R Strey. Colloid Polymer Sci 272:1005, 1994.
19. J N Israelachvili, DJ Mitchell, BW Ninham. J Chem Soc Faraday Trans 2 72:1525, 1976.
20. J Israelachvili. Colloids Surfaces A 91:1, 1994.
21. K Shinoda, Y Shibata. Colloids Surfaces 19:185, 1986.
22. RL Reed, RN Healy. In: Improved Oil Recovery by Surfactant and Polymer Flooding. Shah, DO, Schechter, RS, eds. New York, Academic Press, 1977, p 383.
23. H Kunieda, K Shinoda. Bull Chem Soc Jpn 55:1777, 1982.
24. DO Shah, RH Hamlin, Jr. Science 171:483, 1971.
25. M Bourrel, C Chambu, RS Schechter, WH Wade. Soc Pet Eng J 22:28, 1982.
26. DH Hoskin. U.S. Patent 4,446,079 (1984).
27. C Lalanne-Cassou, I Carmona, L Fortney, A Samil, RS Schechter, WH Wade, U Weerasooriya, V Weerasooriya, S Yiv. J Disp Sci Technol 8:137, 1987.
28. E Gilje, C Sonesson, P-E Hellberg, K Holmberg, S Svennberg. Norwegian Patents 170 411 and 170 972 (1992).
29. PE Hellberg, B Gustavsson, K Holmberg, L Karlsson, S Svennberg, E Gilje, T Maldal. In: Proceedings of the 3rd CESIO International Surfactants Congress, London, 1992.
30. P Herder, A Nyman, K Berglund. Unpublished work.
31. I Blute. Unpublished work.
32. CA Miller, KH Raney. Colloids Surfaces A 74:169, 1993.
33. K Stickdorn, MJ Schwuger, R Schomäcker, Tenside Surf Det 31:4, 1994.
34. FM Menger, AR Elrington. J Am Chem Soc 113:9621, 1991.
35. S-G Oh, J Kizling, K Holmberg. Colloids Surfaces A 97:169, 1995.
36. S-G Oh, J Kizling, K Holmberg. Colloids Surfaces A 104:217, 1995.
37. AS Chhatre, RA Joshi, BD Kulkarni. J Colloid Interface Sci 158:183, 1993.
38. K Holmberg, M-B Stark. Colloids Surfaces 47:211, 1990.
39. EW Merrill, EW Salzman. Am Soc Artif Intern Organs J 6:60, 1983.
40. K Bergström, K Holmberg. Colloids Surfaces 63:273, 1992.

# 8
# New Developments in Polymerization in Bicontinuous Microemulsions

**Françoise Candau and Jean-Yves Anquetil\***
*Charles Sadron Institute, Strasbourg, France*

## I. INTRODUCTION

Microemulsions are transparent or translucent systems of low viscosity containing large proportions of oil and water stabilized by amphiphilic compounds. In contrast with conventional emulsions, microemulsions form spontaneously, are thermodynamically stable, and exhibit a wide variety of structures: globular (oil-in-water, O/W or water-in-oil, W/O), bicontinuous, cubic, or even lamellar. They have been the subject of extensive research over the last two decades primarily because of their possible use in enhanced oil recovery but also for a large range of other applications, and several books have been issued (1–3). More recently, around 1980, the concept of polymerization in microemulsions appeared (4). Since then, the field has developed rapidly and the number of papers dealing with microemulsion polymerization has increased constantly. The very large interfacial area of microemulsions (up to 100 $m^2$/ml) compared to that of emulsions and the small size of the monomer-containing domains ($\sim 10^{-2}$ $\mu$m) give rise to large Brownian fluctuations, which lead to a polymerization mechanism very different from those observed in other heterophase polymerizations (5). One of the most striking features is a polymerization mechanism characterized by a continuous particle nucleation (6–8) in contrast with emulsion polymerization where the nucleation stage occurs at the early stages of the reaction. Such a mechanism results in the so-called uni-oligo molecular polymerization where only one to a few polymer chains are collapsed within a small-sized stable microlatex particle (5). Let us recall that

---
\**Current affiliation*: Clariant Chemie SA, Trosly Breuil, France.

in conventional emulsion polymers, several thousand polymer chains coexist within the same particle.

Polymerization reactions have been carried out in microemulsions of all types of structures and in particular in globular microemulsions (4, 9–30). The microlatexes thus formed have potential applications in microencapsulation and targeted drug delivery (24). The preparation of highly functional polymer particles of low size has recently been discussed by Antonietti, Basten, and Lohmann (31). These dispersions may be used as ligands, binding sites, and detoxination agents or in catalysis.

A major difference between emulsions and microemulsions comes from the amount of surfactant needed to stabilize the systems. This amount is much larger for microemulsions ($\sim$ 10% of the total mass). This is a drawback that can considerably restrict the potential uses of microemulsion polymerization, since high solid contents and low surfactant amounts are usually desirable for most applications. These requirements are far from being achieved at the present state of the art for polymerization in globular microemulsions, since in most cases the amount of monomer does not exceed a few percent and that of surfactants is around 15% (5). These values have to be compared to those used in conventional emulsion polymerization (polymer content $\cong$ 30–60%, surfactant concentration < 4%). On the other hand, several studies have shown that the formulation improves significantly when the globular monomer-containing microemulsion is replaced by a microemulsion with a bicontinuous character. The latter is an open-cell-like structure with randomly connected oil and water domains (5,32–34).

The aim of the present paper is to stress the relevance of a thorough and systematic research for an optimal formulation through two typical examples of polymerization in single-phase "bicontinuous" microemulsions which hopefully can serve as a guideline for "globular" microemulsions.

The first example deals with the synthesis of hydrophobic porous solid materials formed by polymerization of a bicontinuous microemulsion in which a cross-linking agent is incorporated (35–38). The second example, which is developed in more detail in the paper, deals with the synthesis of hydrophilic linear polymers of ultra-high molecular weights. The role of the various parameters involved in the formulation with special emphasis given to that of the monomer is thoroughly analyzed, allowing an optimization of the process.

## II. BICONTINUOUS MICROEMULSIONS BASED ON HYDROPHOBIC MONOMERS

Microemulsion polymerization yields in most case microlatexes, that is, polymer particles dispersed in a continuous medium (oil or water). The preparation of novel solid materials by this process is still at an exploratory stage. A few years ago, Qutubuddin and colleagues showed that polymerization of styrene in multiphasic microemulsions (Winsor I, II, III) led to porous solid materials with interesting morphology and thermal properties (39,40). More recently, an important contribution in this field was provided by Cheung and coworkers, who obtained porous polymeric structures by polymerization of monomers in single-phase microemulsions (35–38). The systems consisted of methyl methacrylate (MMA), acrylic acid (AA), a cross-linking agent, ethylene glycol dimethacrylate (EGDMA), water and sodium dodecyl sulfate (SDS) as the surfactant. Large amounts of monomers were used in the formulation (up to 70% in some cases). There was a close correlation between the microstructure of the synthesized material and the nature of the initial microemulsion. We briefly summarize the main results obtained by these authors.

Polymerization in microemulsions with a water/oil droplet structure yielded closed cell porous polymeric solids, having a morphology characterized by a disjointed cellular structure where the pores were distributed as discrete pockets throughout the solid.

Polymerization in microemulsions with a bicontinuous structure resulted in a polymer with an open-cell structure, i.e., an interconnected porous structure with water channels through the polymer.

The above morphologies were clearly evidenced by means of scanning electron microscopy. In addition, thermogravimetric analysis confirmed the results. The distinction between open-cell and closed-cell porous structures was made, based on the marked difference in the shape of the drying rate curve for the two structures. In closed-cell structures, the drying process is diffusion-limited, resulting in an exponential decrease in drying rate with decreasing moisture content. In open-cell structures, the drying process is dominated by transport of moisture from the interior of the solid to its surface (capillary forces), and a linear decrease in drying rate was observed up to a certain threshold of water content.

These results indicate that the morphology of the polymer keeps some memory of the initial structure, since this structure is retained to a certain extent. However, the length scale of the porous structure obtained (1–4 $\mu$m) is considerably larger than the length scale characteristic of microemulsions (<0.1 $\mu$m), due to phase separation effects during polymerization. The incorporation of the crosslinking agent EGDMA was found quite effective in reducing the occurence of phase separation. The porous polymers have a good mechanical stability, and their rigidity can be varied according to their composition and especially to the amount of cross-linking agent used. This process is a unique route for the formation of polymeric membranes without using an organic solvent, and there will be certainly interesting developments in this area in the near future.

## III. BICONTINUOUS MICROEMULSIONS BASED ON WATER-SOLUBLE MONOMERS

### A. Formulation

The example considered here concerns an original study of the microemulsion copolymerization of acrylamide (AM) and sodium-2-acrylamido-2-methylpropanesulfonate (NaAMPS) (41). In spite of their use in a large number of applications, these copolymers have been the object of very few fundamental studies. Beyond the fundamental aspect, the economical challenge is of prior importance, and this implies in a first step to find inexpensive formulation characterized by a high ratio of monomer-to-surfactant concentration, compatible with an economical process. The optimal formulation of the monomer-containing microemulsions was achieved by using a similar approach to the so-called cohesive energy ratio (CER) concept developed by Beerbower and Hill for the stability of classical emulsions (42). Its basic assumption is that the partial solubility parameters of oil ($\delta_0^2$) and emulsifier lipophilic tail ($\delta_L^2$) and of water and hydrophilic head are perfectly matched. When these conditions are met, one gets the following relationship:

$$HLB_0 = 20\delta_L^2/(K + \delta_L^2) \tag{1}$$

where $HLB_0$ is the optimum HLB (hydrophile-lipophile balance) of the surfactant when $\delta_L^2 = \delta_0^2$ and $K$ is a constant estimated at 230 for a W/O emulsion. It is thus possible to calculate the required HLB for a given oil.

These criteria led us to select the following experimental conditions for microemulsions based on NaAMPS/AM (41):

Oil : Isopar M (Exxon Corp.) ($\delta_0 = 7.79$ (cal/cm$^3$)$^{1/2}$
Aqueous phase/oil weight ratio = 1
Water/monomer ratio in the aqueous phase = 1
Surfactants: mixture of sorbitan sesquioleate (Arlacel 83, HLB = 3.7, $\delta_L = 7.87$ (cal/cm$^3$)$^{1/2}$) and of a polyoxyethylene sorbitol hexaoleate with 50 ethylene oxide residues (G 1096, HLB = 11.4, $\delta_L = 7.87$ cal/cm$^3$) (ICI).

The complex resulting from the association of the two nonionic surfactants at the W/O interface favors greater stability. In addition, blending surfactants allows the selection of an optimum HLB by varying their composition.

Figure 1 represents the percentage of surfactants required for the formation of a microemulsion as a function of the HLB number, for different compositions of the monomer feed. The curves delineate the transition between a turbid emulsion and an optically transparent microemulsion. The transition is sharp and can be easily detected by turbidimetry or visually. It can be seen that microemulsions are found in a HLB domain ranging between 8 and 11. This HLB domain is higher than that classically used in inverse emulsion polymerization (HLB ≅ 4–6) (43). The curves exhibit a minimum for an optimum HLB value, which increases upon increasing the content of the ionic monomer in the feed. Note also the low surfactant concentration needed for the formation of clear systems (5.5% < $S_{min}$ < 7.5%) in spite of the large proportions of monomers incorporated (~22%).

Values of HLB ~8–11 are indicative of systems located in a phase-inversion region (42). In the present case, this corresponds to microemulsions with a bicontinuous structure, in good agreement with previous studies performed on other monomer pairs (32–34,44,45).

The formation of microemulsions with a bicontinuous structure can be ascribed to the presence of monomer in large proportions in the system,

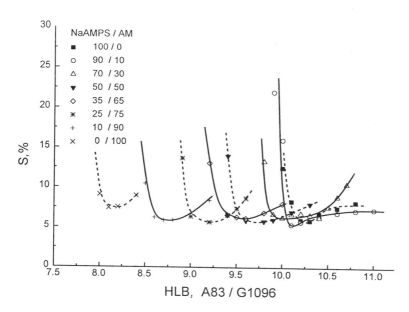

**Figure 1** Weight percentage of surfactants required for the transition emulsion → microemulsion versus the HLB value for various compositions of the monomer feed.

which affects its HLB and interfacial properties. The role of the monomers was thoroughly investigated by our group for other monomer pairs (5, 32–34,44–46). It was found that the water-soluble monomer usually acts as a cosurfactant, leading to a considerable extent of the microemulsion domain in the phase diagram. The cosurfactant role of the two monomers under investigation was confirmed by surface tension experiments (34,47). A direct consequence of this effect is an enhancement of the flexibility and fluidity of the interface, which favors the formation of a bicontinuous structure.

Confirmatory evidence of a bicontinuous structure was provided by an analysis of the phase equilibria occuring in the vicinity of the $HLB_{opt}$, for different monomer compositions. The experimental conditions were those defined above, but the surfactant concentration was fixed at 4 wt %, i.e., below the coexistence curves given in Figure 1. As the time required for the systems to reach equilibrium was long, an inhibitor of polymerization (hydroquinone) was added to prevent spontaneous polymerization. The corresponding phase equilibria obtained after 2 weeks are schematized in Figure 2. In all the cases, triphasic systems with a bicontinuous middle phase are obtained (Winsor III equilibria [48]). The amount of middle phase formed depends in a complex manner on the composition of the monomer feed. From these results, one can speculate that the monophasic systems formed in the same HLB range but at higher surfactant levels have retained their bicontinuous character (32). It is important to note that similar oil–water–surfactant systems do not lead to bicontinuous microemulsions in the absence of monomer.

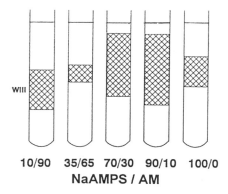

**Figure 2** Phase equilibria observed after 2 weeks as a function of monomer composition in the feed ($T = 20°C$) (surfactants: 4% wt/wt). ▨ bicontinuous middle phase; □ aqueous or organic phase in excess.

The cosurfactant effect of the monomer is not sufficient by itself to account for the high HLB values observed. These are also partly caused by a salting-out of the ethoxylated surfactants by NaAMPS, as previously shown by Corpart and Candau for polyampholytes prepared in microemulsions (34). It is well known that addition of electrolytes to aqueous solutions of ethoxylated surfactants modifies the cloud point $T_p$ (49). Some salts increase $T_p$ (salting-in), whereas some others depress it (salting-out). These effects are usually interpreted as a modification of the water structure around ions (50). Structure-maker ions such as $Na^+$ lower the hydration of the surfactant polar head (salting-out effect). Thus, at a given temperature, the HLB of the ethoxylated blend is made more lipophilic upon addition of a salting-out electrolyte in the system. Optimum microemulsification requires therefore a surfactant blend of higher HLB value. This salting-out effect accounts for the increase in the $HLB_{opt}$ observed upon adding increased amounts of NaAMPS in the feed (Figure 3).

On the other hand, the slight diminution of the surfactant content when the feed becomes richer in NaAMPS (Figure 4) can be explained as follows: the progressive addition of NaAMPS, while decreasing the surfactant solubility in water, tends to compress the hydrophilic heads in the aqueous phase. This favors the formation of a more compact film, which is a stabilizing factor (usually cosurfactants partly play this role), hence the

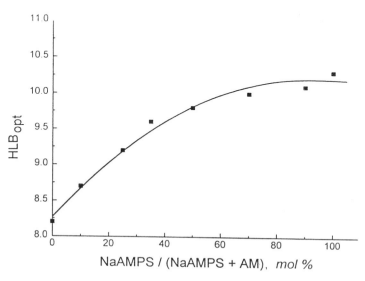

**Figure 3** Variation of the $HLB_{opt}$ with the composition of monomer mixture.

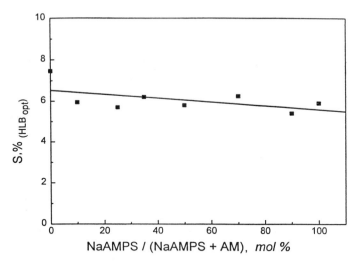

**Figure 4** Variation of the minimum amount of surfactants with the composition of monomer mixture.

lower amount of surfactants needed for the formation of the microemulsion. These results agree well with those previously obtained on the microemulsion copolymerization of sodium acrylate and acrylamide (32).

## B. Polymerization

The optimal conditions which were defined in the formulation of the polymerizable microemulsions are those needed to promote clear and stable microlatexes. The free-radical polymerization was initiated by UV irradiation (20°C) using AIBN as the oil-soluble initiator. As for the polymerization of other water-soluble monomers in microemulsions, the process is characterized by a very high rate of polymerization, since complete conversion is achieved in less than 30 min (5). As soon as the polymerization proceeds, the clear and fluid microemulsions become turbid and highly viscous. The end of the polymerization is characterized by a return to clarity and fluidity. These changes are indeed related to the progressive evolution of the bicontinuous structure toward a globular configuration, as shown from quasi-elastic light scattering experiments (see below). Although the mechanism at the origin of this structural change is not fully understood, one can propose the following qualitative explanations:

1. Microemulsions are characterized by flexible and fluctating interfaces that are continuously rearranging on the time-scale of microseconds due to thermal disruption.
2. The monomer acts here as a cosurfactant, thereby increasing the film flexibility (see above). The progressive monomer consumption from the interfacial layer induces a modification of its composition and of the film curvature energy.
3. The formation of very large macromolecules which are insoluble in the oily domains provokes the turbidity and viscosity increase observed during the reaction.

These factors taken together are likely responsible for the destabilization of the bicontinuous microemulsion, with a possible shift in the phase diagram. The conformation of random coils corresponding to the minimum free energy of polymeric chains would favor the formation of spherical particles.

## 1. Size of the Microlatex Particles

The determination of the size of the dispersed particles in the final systems was obtained by quasi-elastic light scattering (QELS) experiments in a way similar to that previously described for other monomer pairs (32,34). The latexes have been prepared from microemulsions containing 22% of monomers of variable composition and 12% surfactants (wt/wt). In all cases, the inverse microlatexes formed are perfectly transparent and stable and show no settling over months. The values of the hydrodynamic radii, $R_h$, as deduced from the diffusion coefficient values extrapolated to zero volume fraction (Stokes–Einstein equation) range roughly between 30 and 40 nm (Figure 5). The radius goes through a minimum for $\cong 70\%$ of NaAMPS in the monomer feed. Two opposite effects can explain this behavior.

1. The initial decrease of $R_h$ upon increasing NaAMPS content can be accounted for by a dehydration of the ethoxylated moities of the surfactant blend by NaAMPS monomer. This salting-out effect produces a decreased solubility of the surfactant in water and hence a diminution of the radius of curvature.
2. The upturn observed at very high NaAMPS content ($>90\%$) can be due to the enhancement of the osmotic pressure inside the particles resulting from the increasing number of counter-ions.

## 2. Copolymers Characterization

Most applications of water-soluble polymers, e.g., in paper manufacturing, sewage treatment, or enhanced oil recovery, require high molecular weights (51). The locally high monomer concentration in a microemulsion and the

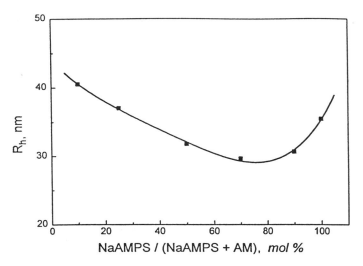

**Figure 5** Hydrodynamic radius of the latex particles versus the mole fraction of NaAMPS in the copolymer.

relative isolation of the growing radicals produce ultra-high molecular weights. This is illustrated in Table 1, where are listed the weight-average molecular weights, $M_w$, and the radii of gyration, $<R_G^2>^{1/2}$, determined in 0.1M NaCl aqueous solutions by static light scattering for a series of copolymers of variable composition. The experimental details concerning polymer characterization are given in Reference 52. It can be seen that inverse microemulsion polymerization leads to high molecular weights, ranging from $5 \times 10^6$ to $15 \times 10^6$ g · mol$^{-1}$, depending on the experimental conditions.

**Table 1** Characteristics of NaAMPS/AM Copolymers

| NaAMPS/AM | $10^{-6} * M_w$ (g · mol$^{-1}$) | $10^{-2} * <R_G^2>^{1/2}$ (nm) |
|---|---|---|
| 100/0 | 7.5 ± 1.5 | 2.3 ± 0.3 |
| 90/10 | 12 ± 1 | 2.9 ± 0.1 |
| 70/30 | 13 ± 3 | 2.8 ± 0.3 |
| 50/50 | 13 ± 3 | 3.5 ± 0.4 |
| 25/75 | 6 ± 1 | 2.3 ± 0.3 |
| 10/90 | 4.9 ± 0.5 | 1.6 ± 0.1 |

# Bicontinuous Microemulsion Polymerization

## 3. Titration of Residual Monomers

Most uses in paper, water treatment, or mining are based on the ability of polyacrylamide-based copolymers to flocculate solids in aqueous suspensions (51). For example, charged polyacrylamides are quite effective in clarifying potable water. However, residual monomer in the copolymer must be reduced to a very low concentration to meet the standard requirements.

Four syntheses of copolymers in microemulsion have been carried out at different monomer compositions (NaAMPS/AM: 90/10, 70/30, 50/50, 10/90). After polymer precipitation, the monomers were extracted with acetonitrile and titrated by liquid chromatography. Both monomers were titrated separately with different eluants. For the four samples investigated, the monomer content was in the detection limit of the apparatus, i.e., far below 20 ppm. This content is much lower than that usually found for direct or inverse polymer emulsions, proving the efficiency of the microemulsion polymerization technique.

## C. Optimization

### 1. Procedure

All the experiments described above were performed at 22% solid contents. We tested several methods in order to increase the solid contents in the final products.

*a. Increase of Monomer Concentration in the Aqueous Phase.* Experiments indicate that opaque microlatexes are obtained when the monomer content in the aqueous phase exceeds 55% wt/wt. The productivity gain remains limited, since this corresponds to a solid content of 24% of the total mass.

*b. Increase of the Aqueous-to-Oil Phase Ratio.* Experiments have been carried out in which the monomer aqueous phase to oil phase ponderal ratio was varied from 55 : 45 to 70 : 30, the other parameters being held constant (monomer to water ponderal ratio: 50 : 50 and surfactant level 12%). The corresponding volume fractions of the dispersed phase ranged from $\Phi = 50.3\%$ (55/45) to 64.6% (70/30). The final latexes show a slight opalescence but are very stable. The limiting factor here is the viscosity of the medium. For volume fractions of the dispersed phase higher than 55%, one observes a large divergence of the viscosity, in accordance with what is generally observed for conventional latexes (53). In practice, handling facilities limit the polymer content to 31%. This rather low content compared to that used for direct latexes (up to 60%) comes from the fact that the inverse polymer particles are also highly swollen by water (usually in a 1 : 1 ratio).

c. *Concentration of Microlatexes.* The polymer microemulsions may be concentrated by heating under vacuum to remove excess water and organic solvent by evaporation. This method did not allow us to increase the polymer content above 25% of the total mass. Above this limit, a strong decrease in the optical transmission, $T$, of the latex, is observed ($T = 40\%$ for a polymer content of 30%).

d. *Effect of HLB and Surfactant Content.* We have seen above that the HLB value is a critical parameter in the formulation of a monomer-containing microemulsion prior to polymerization. We now examine the effect of HLB and surfactant level on the latex stability. The surfactant content (G1096/Arlacel 83) was varied from 6 to 12 wt % and the HLB from 8.6 to 9.8. The results can be seen in Figure 6. The full line represents the minimum amount of surfactant required for the formation of a microemulsion as a function of the surfactant HLB. Four domains can be defined according to the HLB value:

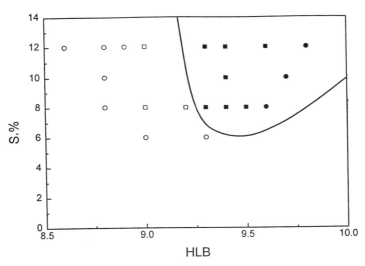

**Figure 6** Effect of the amount of surfactant and of the HLB value on the characteristics of the systems prior to and after polymerization. ○ opaque microemulsion, unstable latex; □ opaque microemulsion, stable latex; ■ transparent microemulsion, stable latex; ● transparent microemulsion, unstable latex.

8.5 < HLB < 8.9 (○). The initial system is opaque. The polymerization leads to a translucent latex with some settling after a few hours.

9 < HLB < 9.2 (□). The initial system is still opaque. During the polymerization, the medium clears up progressively to yield a stable and transparent microlatex. No significant change in the viscosity of the system is observed during the process.

9.3 < HLB < 9.6 (■). The initial system is a transparent bicontinuous microemulsion. During the process, the system becomes viscous and turbid (see above). The final product is optically clear and stable.

HLB > 9.6 (●). The initial system is still a clear bicontinuous microemulsion. During and after the reaction, the system becomes very viscous (formation of gels in some cases) and phase separates.

From Figure 6, one can deduce that stable microlatexes can be prepared from microemulsions containing surfactant contents as low as 8%. The domain of stability of microlatex is schematized in Figure 7 (dotted line). A comparison with the microemulsion domain (solid line) shows that the former domain is slightly shifted toward lower HLB values. It is likely that the best area for carrying out the polymerization reaction is located at the intercept of these two domains. In this area, the initial system is a

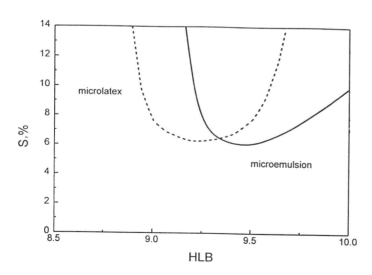

**Figure 7** Stability domains for the microlatex and the microemulsion.

transparent microemulsion, and the increase in viscosity during the reaction is negligible.

These few attempts show that it is not necessary to start with a transparent microemulsion to get a clear and stable latex. This result confirms previous data obtained on acrylamide–sodium acrylate copolymers (32). In this case, the turbidity observed for some systems was not due to the existence of a polyphasic domain in relation with insufficient amounts of surfactant but rather to the high electrolyte content (sodium acrylate), which lowered the cloud point of the surfactants (salting-out). In the course of polymerization, the monomers located partially at the interface diffuse into the particles and the salt saturated aqueous phase becomes progressively used up, accounting for the clarity of the systems at the end of the reaction.

The above studies show how subtle the interplay is between the composition of the initial microemulsion and the stability of the final microlatex, due to the great number of degrees of freedom involved in the formulation.

## 2. Effect of the Nature of Surfactants on the Particle Size

We have seen above that the best conditions of formation of monophasic microemulsions and subsequently clear microlatexes are obtained at the $HLB_{opt}$ value given by the minimum of the curves reported in Figure 1. Peculiar data obtained on latexes prepared at the optimal conditions ($HLB_{opt}$) but using different surfactant mixtures prompted us to analyze in more detail the effect of the influence of the surfactant on the final particle size.

In addition to Arlacel 83 (A83), and G1096, the following surfactants were studied:

1. Polyoxyethylene sorbitol monooleate with 40 ethylene oxide residues (G1086, HLB = 10.2)
2. Sorbitan monolaurate (Arlacel 20 (A20), HLB = 8.6)
3. Sorbitan trioleate (Arlacel 85 (A85), HLB = 1.8)
4. Sorbitan monooleate with 20 ethylene oxide residues (Tween 80 (TW80), HLB = 15).

The $HLB_{opt}$ of five different surfactant blends formed from those listed above were determined for microemulsions based on pure NaAMPS (Figure 8). The $HLB_{opt}$ is seen to vary from 8.2 to 10.9 according to the nature of the blend. With surfactant blends formed from G1096 and one of the Arlacel series, one can note an augmentation of the $HLB_{opt}$ with the hydrophobicity of the Arlacel series. The content of the ethoxylated surfactant (G1096) in the mixture remains roughly at the same level (between 77% and 86%; see Table 2).

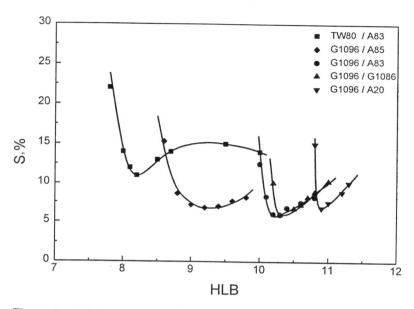

**Figure 8** Weight percentage of surfactants required for the formation of a microemulsion (NaAMPS/AM: 100/0). Effect of the nature of the surfactant pair.

For the A83/G1096 and A83/TW80 pairs, one observes a significant difference in both the composition of the ethoxylated surfactants (85.7% and 39.8%, respectively) and the values of the $HLB_{opt}$ (10.3 and 8.2, respectively) (Table 2). Again, the high $HLB_{opt}$ value observed with the predominantly ethoxylated A83/G1096 pair can be accounted for by the salting-out effect caused by the electrolytic monomer; the more pronounced the effect, the larger the number of ethylene oxide units. The surfactant blend becomes less hydrophilic and its $HLB_{opt}$ has to be shifted to higher values in order to counterbalance this solubility decrease.

For each surfactant blend, the polymerization reactions have been performed at the corresponding $HLB_{opt}$, the other experimental conditions being kept constant (22 wt % NaAMPS monomer, 12 wt % surfactant). All microlatexes are optically transparent and stable. Interestingly, the size of the latex particles exhibits a large variation, with an hydrodynamic radius ranging from 35 nm to 100 nm depending on the surfactant blend (Figure 9). Despite the fact that all experiments have been performed at the $HLB_{opt}$, there still exists a value of HLB = 10.3 (defined as $HLB^*_{opt}$) for which the droplet radius is minimum.

The large variation in $R_h$ can be understood by looking at the compo-

**Table 2** Hydrodynamic Radii Obtained for NaAMPS Microlatexes Prepared from Different Surfactant Blends

| Surfactant blend | HLB | Surfactant (wt%) | $HLB_{opt}$ | $R_h$ (nm) |
|---|---|---|---|---|
| A83[a] | 3.7 | 60.2 | 8.2 | 102 |
| TW80 | 15 | 39.8 | — | — |
| A83 | 3.7 | 60.2 | 8.2 | 80 |
| TW80 | 15 | 39.8 | — | — |
| A85 | 1.8 | 22.9 | 9.2 | 62.7 |
| G1096 | 11.4 | 77.1 | — | — |
| A20 | 8.6 | 17.9 | 10.9 | 64.5 |
| G1096 | 11.4 | 82.1 | — | — |
| A83 | 3.7 | 14.3 | 10.2 | 35.0 |
| G1096 | 11.4 | 85.7 | — | — |
| G1086 | 10.2 | 91.7 | 10.3 | 44.1 |
| G1096 | 11.4 | 8.3 | — | — |

[a]From Reference 34.

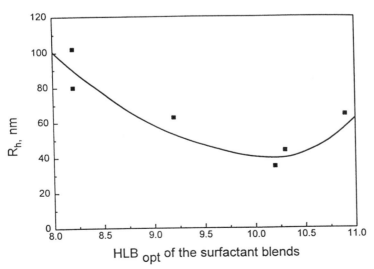

**Figure 9** Variation of the hydrodynamic radius of the latex particle versus the $HLB_{opt}$ of the surfactant blends.

sitions of the surfactant blends (Table 2). As a first approximation, the particle size is seen to decrease with the increasing ethoxylated surfactant content. For those surfactants, the salting-out effect tends to compress the ethoxylated units in the particles and lowers the water activity, which favors the bending of the interfacial film. This would account for the smaller particle sizes observed for ethoxylated-rich surfactant mixtures. Some other factors, although of less importance, also play a role. The molecular structures and geometrical packing of the surfactants as well as the complex they form at the interfacial film should also modify the radius of curvature of the particle. For example, the hydrodynamic radius varies from 35 nm to 65 nm for the pairs formed from G1096 and the Arlacel series, in spite of a fairly constant amount of ethoxylated surfactants.

## D. Comparison Between Copolymers Prepared by Emulsion and Microemulsion Polymerization

The efficiency of water-soluble polymers as flocculants should depend on their characteristics. The large differences observed between the kinetics and the mechanism of emulsion and microemulsion polymerizations are susceptible to modifying some polymer characteristics such as the molecular structure, the microstructure, or molecular weight. To our knowledge, the only comparative study available in the literature concerned the microstructure of various copolymers prepared by polymerization in microemulsion, emulsion, or solution (52,54). An interesting finding is that microemulsion polymerization seems to improve the structural homogeneity of the copolymers, with reactivity ratios tending toward unity. This result was accounted for by the marked differences in mechanism and microenvironment between the processes.

We have compared the molecular weights and intrinsic viscosities of NaAMPS/AM copolymers prepared in emulsion and microemulsion. The syntheses were carried out with raw and recrystallized monomers in order to check the possible effect of impurities on the molecular weight. Table 3 summarizes the experimental conditions used and the characteristics of the copolymers. Inspection of Table 3 leads to the following conclusions: (1) whatever the method of synthesis, higher intrinsic viscosities are obtained when purified monomers are used. (2) The Huggins coefficients, $K_H$, which characterize the solvent–polymer interactions, are significantly larger for emulsion polymers. Values higher than 0.5 are the signature of a poor solvent, the presence of aggregates, or some cross-linking of macromolecules, making them partly insoluble. This would explain the higher molecular weights obtained for emulsion polymers compared to those prepared in microemulsions and the apparent inconsistency between the values of $M_w$

**Table 3**  Compositions of the Initial Systems and Characteristics of the Copolymers

| Monomers polymer method | Raw microemulsion | Raw emulsion | Pure microemulsion | Pure emulsion |
|---|---|---|---|---|
| $HLB_{opt}$ | 9.6 | 6.5 | 9.6 | 6.5 |
| Monomers (wt %) | 22.01 | 24.45 | 21.97 | 23.95 |
| Surfactant (wt %) | 12.06 | 2.47 | 12.04 | 2.47 |
| $[\eta]$ ml · g$^{-1}$ | 2138 | 2154 | 2856 | 2728 |
| $K_H$ | 0.48 | 0.65 | 0.49 | 0.93 |
| $M_w \cdot 10^{-6}$ | 6.02 ± 0.6 | 11 ± 6 | 7.7 ± 0.7 | 12 ± 2 |

Note: (NaAMPS)/(AM) = 35/65; monomer/aqueous phase = 1 (wt/wt); surfactant blend: G1096/A83.

and $[\eta]$ measured by the two processes. This interesting result, which needs to be confirmed by further additional experiments, can be of great importance for applications and is certainly related to the difference in mechanism between the two processes.

In a microemulsion polymerization, the number of macromolecules per particle is extremely low, one in the limiting case, thus excluding the possibility of interchain cross-links. On the other hand, transfer reactions to polymer are likely to occur in an emulsion polymerization, since each final latex particle contains several hundred polymer chains. This property of microemulsion polymerization process was taken advantage of to produce Mannich polyacrylamides (i.e., substituted with tertiary aminoethyl groups) with dewatering properties superior to those obtained by inverse emulsion polymerization (55).

## IV. QUARTER CENTURY PROGRESS AND NEW HORIZONS

Twenty-five years ago, the process of polymerization in microemulsions was still unborn, since the first attempts in this area were reported in the early 1980s. At this time, the primary goal of the investigators was to produce thermodynamically stable latexes in the nanosize range, not attainable with the classical emulsion polymerization process. This was of particular relevance for inverse latexes, since inverse emulsion polymerization is known to yield unstable polymer particles with a broad size distribution (43). In addition, the optical transparency and thermodynamic stability of microemulsions were advantageous for photochemical or other reactions.

The process was successful in obtaining small-sized latex particles

# Bicontinuous Microemulsion Polymerization

($d < 30$ nm) from polymerization in globular microemulsions (9,14–21,25). However, some limitations rapidly appeared concerning the ratio of monomer over surfactant concentration, which was much lower than that used in emulsion polymerization. For this reason, only a few groups throughout the world pursued this type of research, which looked quite exotic to the rest of the community. In fact, it was not realized at that time that microemulsion polymerization was not a simple extension of emulsion polymerization and that the following questions needed to be answered:

1. Elucidation of the polymerization mechanism
2. Determination of appropriate formulation rules
3. Development of methods allowing production of products that can compete with other commercially available materials
4. Discovery of specific applications

Concerning the first three challenges, considerable progress has been achieved, whereas the search for new applications is still in infancy. However, the fast development of the last 3 years tells us that this will be one of the important targets in the near future.

Among the main achievements in the last decade, one can cite the following:

1. A polymerization mechanism has been observed very different from that observed in emulsion polymerization and which can be described by a *continuous* particle nucleation (5–8).
2. Each latex particle contains a very low number of polymeric chains, one in the limiting case (5–7,14,16,25,26), in contrast with emulsion polymerization where thousands of polymeric chains coexist within the same particle.
3. In the case of water-soluble polymers, it is more advantageous to start from bicontinuous microemulsions rather than globular systems (5,32–34).
4. The formulation has been optimized to such an extent (see present paper) that products prepared by microemulsion polymerization can be found on the market.

Considering the new outlook, it is obvious that there is a need for further fundamental mechanistic studies, but it is also clear that a more appealing aspect at the moment is to take advantage of the great variety of structures offered by microemulsions to produce novel compounds, as for example the porous solid materials described in this paper (35–38).

However, it must be kept in mind that any attempt to prepare a new material should be preceded by a careful determination of the phase diagram of the monomer-containing microemulsion.

## ACKNOWLEDGMENTS

One of us (J. Y. A.) acknowledges FONGECIF and Société Française Hoechst for the financial support provided during this work. We also thank G. Mattioda, A. Blanc, and P. Mallo (Société Française Hoechst, Stains) and J. P. Guette (CNAM) for initiating the project and continuous interaction.

## REFERENCES

1. DO Shah. Macro and Microemulsions: Theory and Applications. ACS Symposium Series 272, 1985.
2. SE Friberg, P Bothorel. Microemulsions, Structure and Dynamics. New York, CRC, 1986.
3. M Bourrel, RS Schechter. Microemulsions and Related Systems. New York, Marcel Dekker, 1986.
4. C Schauber. Thèse Docteur-Ingénieur, Université de Mulhouse, France, 1979.
5. F Candau. In: Polymerization in Organized Media. Paleos, CM, ed. Philadelphia, Gordon and Breach Science, 1992, pp 215–282.
6. F Candau, YS Leong, RM Fitch. J Polym Sci Polym Chem Ed 23:193, 1985.
7. MT Carver, E Hirsch, JC Wittmann, RM Fitch, F Candau. J Phys Chem 93:4867, 1989.
8. JS Guo, ED Sudol, JW Vanderhoff, MS El-Aasser. J Polym Sci Polym Chem Ed 30:691, 1992.
9. SS Atik, KJ Thomas. J Am Chem Soc 103:4279, 1981.
10. P Lianos. J Phys Chem 86:1935, 1982.
11. YS Leong, F Candau. J Phys Chim 86:2269, 1982.
12. PL Johnson, E Gulari. J Polym. Sci Polym Chem Ed 22:3967, 1984.
13. A Jayakrishnan, DO Shah. J Polym Sci Polym Lett Ed 22:31, 1984.
14. F Candau, YS Leong, G Pouyet, SJ Candau. J Colloid Interface Sci 101:167, 1984.
15. PL Kuo, NJ Turro, CM Tseng, MS El-Aasser, JM Vanderhoff. Macromolecules 20:1216, 1987.
16. JS Guo, MS El-Aasser, JM Vanderhoff. J Polym Sci Polym Chem 27:691, 1989.
17. VH Perez-Puna, JE Puig, VM Castano, BE Rodriguez, AK Murthy, EW Kaler. Langmuir 6:1040, 1990.
18. C Schauber, G Riess. Makromol Chem 190:725, 1989.
19. M Antonietti, W Bremser, D Müschenborn, C Rosenauer, B Schupp, M Schmidt. Macromolecules 27:3796, 1990.
20. JE Puig, S Corona-Galvan, A Maldonado, PC Schulz, BE Rodriguez, EW Kaler. J Colloid Interface Sci 137:308, 1990.
21. V Vaskova, V Juranicova, J Barton. Makromol Chem 191:717, 1990.
22. V Vaskova, V Juranicova, J Barton. Makromol Chem 192:989, 1991.

23. M Antonietti, W Bremser, D Muschenborn, C Rosenauer, B Schupp, M Schmidt. Macromolecules 24:6636, 1991.
24. C Larpent, ThF Tadros. Colloid Polym Sci 269(11):1171, 1991.
25. LM Gan, CH Chew, I Lye, T Imae. Polym Bull 25:193, 1991.
26. RA Mann. PhD dissertation, University of Sydney, Australia, 1991.
27. LM Gan, CH Chew, SC Ng, SE Loh. Langmuir 9:2799, 1993.
28. F Bleger, AK Murphy, F Pla, EW Kaler. Macromolecules 27:2559, 1994.
29. S Holdcroft, JE Guillet. J Polym Sci Polym Chem Ed 28:1823, 1990.
30. WM Brouwer. J Appl Polym Sci 38:1335, 1989.
31. M Antonietti, R Basten, S Lohmann. Macromol Chem Phys 196:466, 1995.
32. F Candau, Z Zekhnini, JP Durand. J Colloid Interface Sci 114:398, 1986.
33. C Holtzscherer, F Candau. Colloids Surf 29:411, 1988.
34. JM Corpart, F Candau. Colloid Polym Sci 1055:1067, 1993.
35. WRP Raj, M Sasthav, HM Cheung. Langmuir 7:2586, 1991.
36. M Sasthav, WRP Raj, HM Cheung. J Colloid Interface Sci 152:376, 1992.
37. WRP Raj, M Sasthav, HM Cheung. Langmuir 8:1931, 1992.
38. WRP Raj, M Sasthav, HM Cheung. J Appl Polym Sci 47:499, 1993.
39. S Qutubuddin, E Haque, WJ Benton, EJ Fendler. In: Polymer Association Structures: Microemulsions and Liquid Crystals. El-Nokaly, M, ed. ACS Symposium Series 384, 1989, pp 65–83.
40. E Haque, S Qutubuddin. J Polym Sci Polym Lett Ed 26:429, 1988.
41. JY Anquetil. Mémoire de Diplôme d'Ingénieur CNAM, Paris, 1994.
42. A Beerbower, MW Hill. In: McCutcheon's Detergents and Emulsifier Annual. Allured, Ridgewood, NJ, 1971, pp 223–235.
43. F Candau, RH Ottewill, eds. In: An Introduction to Polymer Colloids and Scientific Methods for the Study of Polymer Colloids and Their Applications. NATO ASI Series C 303, Kluwer, Dordrecht, 1990, pp 73–96.
44. C Holtzscherer, F Candau. J Colloid Interface Sci 125:97, 1988.
45. P Buchert, F Candau. J Colloid Interface Sci 136:527, 1990.
46. F Candau. In: Polymer Association Structures: Microemulsions and Liquid Crystals. El-Nokaly, M, ed. ACS Symposium Series 384, 1989, pp 48–61.
47. C Graillat, M Lepais-Masmejan, C Pichot. J Disp Sci Techn 11:455, 1990.
48. PA Winsor. Trans Far Soc 44:376, 1948.
49. M Schick. Colloid Sci 17:801, 1962.
50. JA Beunen, E Ruckenstein. Adv Colloid Interface Sci 16:201, 1982.
51. WM Thomas, DW Wang. In: Encyclopedia of Polymer Science and Engineering. Mark, H, Bikales, N, Overberger, CC, Menges, G, eds. New York, 1985, pp 169–235.
52. JM Corpart, J Selb, F Candau. Polymer 34:3873, 1993.
53. I Krieger, TS Dougherty. Trans Soc Rheol 3:137, 1959.
54. F Candau, Z Zekhnini, F Heatley. Macromolecules 19:1895, 1986.
55. JJ Kozakiewicz, SY Huang (American Cyanamic Co.). U.S. Patent 4,956,399, (1990).

# 9
# Application of Microemulsions in Soil Remediation

**K. Mönig, W. Clemens,\* F.-H. Haegel, and M. J. Schwuger**
*Institute of Applied Physical Chemistry, Jülich, Germany*

## I. INTRODUCTION

In current soil remediation processes, the contaminants are usually removed by abrasion from soil particles using mechanical energy and water as washing medium. In this manner, a fine-particle pollutant-enriched fraction is always formed during the washing process that cannot be remediated by present techniques. In particular, organic contaminants with very low water solubilities, like polycyclic aromatic hydrocarbons (PAH), are concentrated in this soil fraction.

Previous studies have shown that the presence of organic substances possessing sufficient amphipathic activity, e.g., humic acids and surfactants, enhances the solubility of PAH (1,2). For this reason, solutions of low surfactant content have been used in some remediation processes to improve the removal of organic matter (3). However, the main problem involved in this procedure is the complete loss of the surfactant during subsequent wastewater treatment. A new remediation concept applying microemulsions as the washing media has been developed (4) to take advantage of the enhanced solubilities of organic contaminants by using surfactants while simultaneously reducing their loss. In this case the surfactant will be recycled by phase separation of the microemulsion into a surfactant-rich aqueous phase and an oil phase after extraction. Suitable microemulsions have to be composed of biodegradable components in order to have

---

*\*Current affiliation:* Professionelle Software GmbH, Bedburg, Germany.

the chance of refilling the soil after decontamination and to avoid any risks due to toxicity.

The first part of this work is concerned with important properties of microemulsions regarding their use as extraction media in this new soil remediation concept. The second part contains some of the essential results on the phase behavior of ternary and pseudo-ternary mixtures of water, nonionic surfactants and rapeseed oil (RSO), rape oil methyl ester (RME), or castor oil (CO). The successful application of such microemulsion systems is demonstrated in the last section by presenting some extraction results.

## II. PROPERTIES OF MICROEMULSIONS

In mixtures of water, oil, and a nonionic surfactant, a region of composition and temperature might exist in which the system separates into three coexisting phases. The phase behavior of these ternary mixtures at constant pressure and temperature is described by a Gibbs triangle showing the one-, two-, and three-phase regions depending on composition. In order to consider the additional influence of temperature $T$ on the phase behavior at constant pressure, an upright phase prism with the Gibbs triangle as the base and $T$ as the applicate is used (Figure 1).

The three-dimensional representation can be favorably reduced to a two-dimensional pseudo-binary phase diagram. This can be obtained by a vertical section through the phase prism for a constant oil–water ratio (Figure 2). Intersecting the three-phase body, the phase boundaries form the shape of a fish, its body representing the three-phase region ($3\phi$) at the chosen oil–water ratio and the tail being the one-phase region ($1\phi$). If for a given mass fraction $\alpha$ of oil in the binary oil–water mixture the mass fraction $\gamma$ of the surfactant in the ternary mixture is raised, the volume of the middle phase of $3\phi$, to be termed microemulsion, grows at the expense of the remaining phases until all the water and oil is solubilized. The system directly undergoes the $3\phi$ to $1\phi$ transition only at one single temperature. At any other temperature between the lowest temperature $T_l$ and the uppermost temperature $T_u$ found for $3\phi$ at given $\alpha$, the system first turns from $3\phi$ into a two-phase mixture ($2\phi$) before it becomes $1\phi$. Once $1\phi$ is formed, it can be separated into two phases just by temperature variation.

The "fish" is surrounded by two different two-phase regions $\underline{2\phi}$ at lower and $\overline{2\phi}$ at higher temperatures. In $\underline{2\phi}$ the surfactant is much more soluble in the aqueous phase. As the temperature increases it becomes more oil-soluble. In $\overline{2\phi}$ most of the surfactant is dissolved in the oil phase. The

# Microemulsions in Soil Remediation

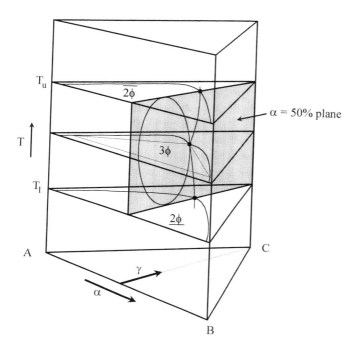

**Figure 1** Vertical section through the phase prism of a ternary system H$_2$O (A), oil (B), nonionic amphiphile (C) with temperature T as the applicate (schematic) at $\alpha = 50$ wt %.

extension of 3$\phi$ is described by an upper temperature $T_u$ and a lower temperature $T_l$. The mean temperature

$$\overline{T} = \frac{T_u + T_l}{2} \tag{1}$$

characterizes the position of 3$\phi$.

The position of the "fish" on the temperature scale and the extension of 3$\phi$ sensitively depend on the chemical nature of the components chosen (5). The more hydrophobic the oil for a given surfactant, the higher is the position of the fish on the temperature scale. The same dependence is valid for increasing hydrophilicity of the surfactant for a given oil. Growing hydrophobicity of the surfactants for a given oil shifts 3$\phi$ to lower temperatures. Often in such systems liquid crystalline phases (LC) appear in 1$\phi$ and reduce its extension or sometimes even prevent its existence. Such phases are preferentially formed in systems composed of long-chained oils (6).

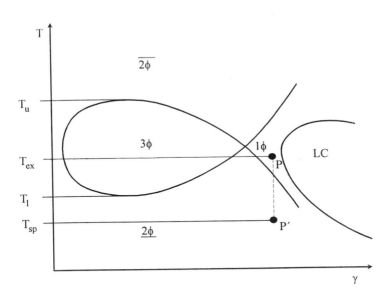

**Figure 2** The vertical section of Figure 1 yields a pseudo-binary phase diagram for measuring the position and extension of the three-phase body (schematic). $\alpha$ = const.

Besides, high surfactant contents are necessary in this case to solubilize those oils supporting again the formation of liquid crystals in $1\phi$.

Furthermore, the application of additives, e.g., alcohols, influences the hydrophobicity of the surfactant or the oil as a function of their chain lengths (7). In this manner they change the position of the "fish." Short-chained alcohols consisting of no more than three carbon atoms enhance the water solubility of the surfactant, shifting $3\phi$ to higher temperature values. Cosurfactants possessing a longer alkyl chain mainly dissolve in the oil phase, thus causing the opposite effect. The addition of electrolytes decreases the water solubility of the surfactant so that $3\phi$ is found at lower temperatures (8).

In order to find suitable surfactants for applying vegetable oil as the hydrophobic component, the influence of its chain length on phase behavior has to be investigated. Furthermore, the property of the surfactant has to be adjusted in such a way that the formation of liquid crystals is minimized.

A microemulsion used as extraction medium in soil remediation should be stable in a sufficient temperature and concentration range. A

# Microemulsions in Soil Remediation

mixture with the composition found at point $P$ in Figure 2 is appropriate. Such a microemulsion shows a submicroscopic bicontinuous structure with large internal interfaces combined with a very low interfacial tension between the water and oil domains (9). The nonrigid structure exhibits fast fluctuations of single component concentration and of domain size (about 5–500 nm). The good wetting properties of microemulsions enable their penetration even into micropores. Another important property for soil remediation achieved by microemulsion systems is their great solubilization capacity. But the most important reason for using a microemulsion as the extraction medium is its temperature-dependent phase behavior. As already shown in Figure 2, $1\phi$ can be left simply by temperature decrease from the extraction temperature $T_{ex}(P)$ to the splitting temperature $T_{sp}(P')$. In this way, phase separation can be achieved after the extraction process. By lowering the temperature, the one-phase system splits into a lower aqueous surfactant-rich phase and an upper oil-rich phase containing most of the contaminants extracted.

The process, schematically shown in Figure 3, uses the above-mentioned advantages of microemulsions for the remediation of contaminated soils, silt, and clay fractions (10,11). Within this concept, the contaminated soil is intensively mixed with a microemulsion of biodegradable components. The extraction process is followed by a solid–liquid separation yielding a decontaminated soil and the microemulsion containing the contaminants. After phase separation by lowering the temperature, the aque-

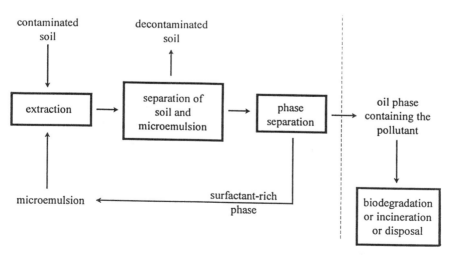

**Figure 3** Concept for soil remediation applying microemulsions.

ous surfactant-rich phase is recycled for further extraction. The contaminated oil phase has to be degraded, incinerated, or disposed of. In the case of PAH, degradation by microorganisms is possible (12). The residual components of the microemulsion on the soil can be removed by rinsing and biodegradation.

## III. EXPERIMENTAL

### A. Materials

Triply distilled water was used for these studies. The PAH investigated was pyrene (99% from Janssen Chimica, Brüggen-Bracht, Germany). Alkylpolyglucosides ($C_iG_j$; $i$ = chain length, $j$ = degree of glucosidation) and alkylpolyethoxylates ($C_iE_j$; $i$ = chain length, $j$ = degree of ethoxylation) were used as surfactants. $C_8E_5$, $C_{12}E_6$, and $C_{16}E_2$ were purchased from Fluka, Buchs, Switzerland with a purity >98%. $C_{10/12}G_{1.3}$ and $C_{12/14}G_{1.3}$, obtained from Hüls AG, Marl, Germany, and $C_{9/11}E_4$, supplied by ICI Surfactants, Essen, Germany, were of technical grade and used without further purification. Rapeseed oil of technical grade was obtained from GET, Aldenhoven, Germany, with mainly unsaturated $C_{18}$-fatty acids. Rape oil methyl ester was produced by base-catalysed transesterification of rapeseed oil with methanol and also was purchased from GET. Castor oil was from Heess, Stuttgart, Germany. The main component of this triglyceride is ricinoleic acid, which is a $C_{18}$-fatty acid with one double bond and a hydroxyl group at $C_{12}$.

### 1. Soil Samples

Extraction experiments were performed with two different soil materials.

1. Calcium bentonite, purchased from Süd-Chemie, Moosburg, Germany, was doped with 250 mg pyrene per kilogram of solid and was used as a model soil component. The material was air-dried and pulverized, yielding a particle size of <60 μm. The content of organic carbon was determined by IR-spectroscopy after combustion. It was 0.016 wt %.
2. The real contaminated sample was an air-dried fine silt fraction of the hydrocyclone outlet of a washing plant with a particle size <30 μm. It was contaminated with more than 3000 mg PAH per kilogram of soil (16 EPA PAH). The content of pyrene in two different charges was 430 mg/kg (A) and 400 mg/kg (B), respectively. The content of organic carbon was 13.5 wt %.

The particle size was determined by use of the Sedigraph 5100 lent by Micromeritics, Neuss, Germany. The soil materials were dispersed in 0.1 wt % sodium pyrophosphate solution in the autosampler by means of ultrasonication and stirring. Sedimentation in the sample cell is measured by X-ray detection.

## B. Methods

### 1. Preparation of the Artificially Contaminated Soil

Calcium bentonite (200 g) was suspended in about 2.8 l of a water–acetone mixture (1/1) by means of an Ultra-Turrax from Janke & Kunkel IKA Labortechnik, Staufen, Germany. A solution of 50 mg pyrene in 200 ml acetone was added stepwise. The whole suspension was stirred for 72 hours. Afterward, it was air-dried at room temperature for 1 week.

### 2. Determination of Phase Diagrams

Samples with different compositions of water, surfactant, and oil were weighed into test tubes with volume scale and thermostated in a water bath. After having been adapted to the bath temperature and shaken, the samples remained in the bath for 1 to 5 days to reach phase equilibrium. The number of coexisting phases could be detected visually. LC phases were identified by crossed polarizers. This procedure was performed at different temperatures, near the phase boundaries in steps of 2°C.

### 3. Distribution of Pyrene

For the determination of the distribution of pyrene between the polar and the nonpolar phase, the same procedures were chosen as described under "phase diagrams." The only differences were as follows:
1. The oil was artificially doped with a certain amount of pyrene.
2. Initially, an appropriate temperature was chosen to form a microemulsion, then the samples were cooled for phase separation.
3. In order to reach complete demulsification and equilibration, the samples were stored 1 day at constant temperature.

### 4. Extraction

The systems used for extraction are shown in Table 1. The surfactant mixture in systems 1 and 2 is to be understood as a pseudo-component.

   *a. Extraction with Systems 1 and 2 Performed at 50°C.* Soil (10 g) was stirred at 140 rpm with 60 ml of system 1 or 2 in a glass vessel positioned in a thermostated water bath. For kinetic studies, the extraction time

**Table 1** Composition of Microemulsions used as Extraction Media

| System | Surfactant (wt %) | | Oil (wt %) | Water (wt %) | α (wt %) |
|---|---|---|---|---|---|
| 1 | 56 $C_{16}E_2$ | 14 $C_{10/12}E_{1.3}$ | 1.5 RSO | 28.5 | 5 |
| 2 | 56 $C_{16}E_2$ | 14 $C_{10/12}G_{1.3}$ | 3 RSO | 27 | 10 |
| 3 | 22 $C_{9/11}E_4$ | | 39 RME | 39 | 50 |

was varied from 0.17 h (10 min) to 3 h. To achieve liquid–solid separation, filtration through a membrane filter with pores of 0.45 μm (Minisart SRP25; Sartorius, Göttingen, Germany) was performed at 50°C supported by manual pressure. The filtrate was analyzed for pyrene.

  *b. Extraction with System 3 Carried out at 43°C.* The contaminated soil (5 g and 10 g, respectively) was stirred with 30 g of the ternary mixture at 200 rpm in a glass vessel which was thermostated in a water bath. After 2 hours, stirring was stopped and the samples were left in the water bath overnight in order to achieve complete sedimentation of the soil. The pyrene concentration was determined in the supernatant.

5. Determination of Pyrene

The pyrene concentration was determined by means of UV derivative spectroscopy (Kontron Uvikon 860) at 333 nm. The samples were diluted by a factor of 100 with propane-2-ol and measured in 10 mm QS cuvettes from Hellma.

## IV. RESULTS

### A. Phase Diagrams

1. Determination of the Effective Chain Lengths of Native Oils

For characterizing the native oils, their effective chain lengths were determined by measuring the phase behavior of the ternary systems consisting of native oil, water, and pure surfactant at $\alpha = 50$ wt %. Figure 4 shows the pseudo-binary phase diagrams of water/RME/$C_8E_5$, water/CO/$C_8E_5$, and water/RSO/$C_{12}E_6$. The position of the three-phase bodies on the tempera-

# Microemulsions in Soil Remediation

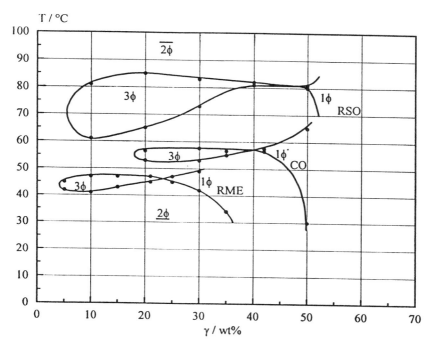

**Figure 4** Three-phase behavior as a function of the surfactant content $\gamma$ at $\alpha = 50$ wt % for the following systems: rape seed oil (RSO)/$C_{12}E_6$/water, castor oil (CO)/$C_8E_5$/water, rape oil methyl ester (RME)/$C_8E_5$/water.

ture scale is characterized by the mean temperature (see definition in the introduction). Comparing these mean temperatures with the corresponding ones of the three-phase bodies of systems containing alkanes (5), a certain carbon number of alkane can be assigned to the native oil. Thus, RSO corresponds approximately to the alkane with 18 carbon atoms, for RME an effective chain length of about 4 was found, and for CO it was about 6–7. Using this principle, appropriate surfactants can be chosen to obtain phase diagrams in the desired temperature range. However, at a certain temperature, depending on the system, LC phases appear and cover parts of the phase diagrams. For that reason the extraction temperature cannot be decreased at will.

The three-phase body of the RSO system does not show the symmetrical form of its methyl ester and alkanes. At low surfactant concentrations, the body shows a slight deformation to lower temperatures due to interfacially active components, e.g., di- and monoglycerides in the oil. Using triglycerides like RSO and CO, a higher surfactant content is necessary to form microemulsions in comparison to corresponding systems with alkanes. In contrast to this, much less surfactant is necessary for complete solubilization of the methyl ester of RSO.

For technical applications, commercially available low-cost surfactants have to be used. These are complex mixtures of components with different chain lengths and degrees of ethoxylation or glucosidation. As a consequence, technical surfactants do not show the ideal phase behavior exhibited by the pure component systems.

## 2. The $C_{16}E_2$ + $C_{10/12}G_{1.3}$/RSO/Water System

To solubilize RSO, a mixture of two nonionic surfactants of technical grade was used leading to a quaternary system. For representing the phase behavior of such systems in a phase prism at constant pressure, the number of independent variables has to be reduced from four to three. This can be done by keeping either the temperature or one of the composition variables fixed. In this case, a fixed mixing ratio of surfactants was used as a pseudo-component leading to a pseudo-ternary system.

Figure 5 shows the boundaries of 3ϕ at different temperatures of the pseudo-ternary system $C_{16}E_2$ + $C_{10/12}G_{1.3}$ at a ratio of 8/2, RSO and water.

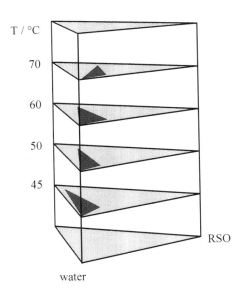

**Figure 5** Phase prism of the pseudo-ternary system $C_{16}E_2$ + $C_{10/12}G_{1.3}$ at a ratio of 8/2, RSO and water. The scaling of the applicate (T-axis) is not equidistant.

# Microemulsions in Soil Remediation

The position and extension of the 3ϕ body deviate strongly from those of real ternary mixtures. This deviation is the consequence of the reduced number of variables of state. For this reason only a narrow three-phase region shifted to the edge of water has been detected. The surfactant-rich corner of 3ϕ shows a distinct dependence on temperature in respect to its position, indicating the existence of a microemulsion. Nearly the whole three-phase area is surrounded by liquid crystals and two-phase regions. Neither an upper temperature $T_u$ nor a lower temperature $T_l$ was found between room temperature and 80°C. At lower temperatures, liquid crystalline and solid phases were formed instead of 2ϕ.

Liquid crystals also limit the extensions of 1ϕ in the mixture of $C_{16}E_2$ + $C_{10/12}G_{1.3}$ (7/3), RSO, and water at $\alpha = 10$ wt %. The pseudo-binary phase diagram of this system is presented in Figure 6. The dimensions and the deformations of 3ϕ are remarkable, but they are the consequence of using technical products for the oil as well as for the surfactant. The mixture becomes 1ϕ at a surfactant content of about 45 wt %. At this surfactant concentration, liquid crystals are formed at temperatures lower than about 50°C. In order to suppress the formation of LC phases and decrease the temperature range available for extraction, other oils and surfactants have to be used.

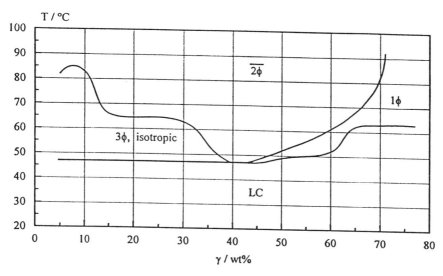

**Figure 6** Pseudo-binary phase diagram of the system $C_{16}E_2$ + $C_{10/12}G_{1.3}$ (7/3), RSO and water at $\alpha = 10$ wt %.

## 3. The $C_{9/11}E_4$/RME/Water System

The phase behavior of the ternary system $C_{9/11}E_4$/RME/water is presented in Figure 7. The oil content $\alpha$ was kept at 50 wt %. The "fish" body is strongly skewed to higher temperatures with decreasing $\gamma$ in contrast to systems with pure surfactants, as shown in Figure 2. This can be explained by the different solubilities of more or less hydrophilic components of the surfactant in oil and water (13).

The microemulsion region is reduced by the appearance of liquid crystals at surfactant concentrations larger than 26 wt %. For that reason the extraction and subsequent phase separation must be performed between 20 and 25 wt %, as shown in Figure 7 for 22 wt %.

## B. Extractions

### 1. Extraction Kinetics of Systems 1 and 2 at 50°C

Extraction with both systems led to fast removal of the contaminants. After about 30 min the extraction was finished and about 100% of the

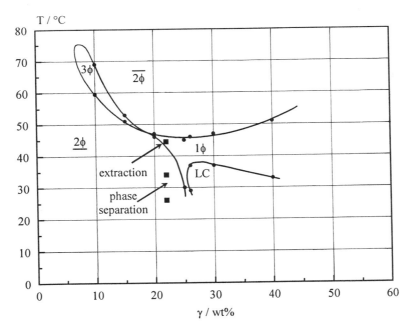

**Figure 7** Phase behavior of the ternary system $C_{9/11}E_4$/rape oil methyl ester/water at $\alpha$ = 50 wt %. LC represents the region of liquid crystalline phases.

contamination, determined by hot extraction with toluene, was removed. Figure 8 presents the extraction kinetics of the fine fraction (A) with systems 1 and 2 (Table 1). $\Delta Q_{Py}$ denotes the amount of extracted pyrene relative to the mass of soil. It is proportional to the concentration $c$ of pyrene in solution. The concentrations at different times were fitted to the function

$$c(t) = c_\infty (1 - e^{-kt}) \tag{2}$$

representing the solution of the linear adsorption kinetics model with the rate constant $k$ and the plateau value $c_\infty$.

Comparing the kinetics determined for different $\alpha$ on the fine fraction, only slight differences were observed. The necessary time to reach equilibrium for an extractant/soil ratio (e/s) of 6 is about half an hour in both cases.

Another important factor influencing the extraction process is the content of organic components, e.g., humic acids, in the soil. Organic contaminants strongly interact with the humic fraction (14). Despite the great difference in organic content, the artificially contaminated calcium bentonite and the real contaminated fine fraction show no remarkable difference in extraction rates (Figure 9).

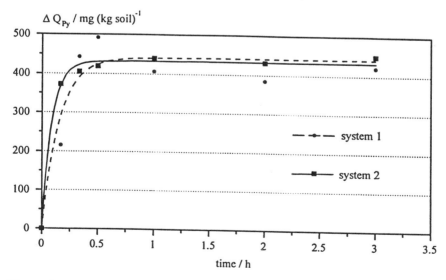

**Figure 8** Extraction kinetics of the real contaminated fine fraction (A) with systems 1 and 2 (Table 1).

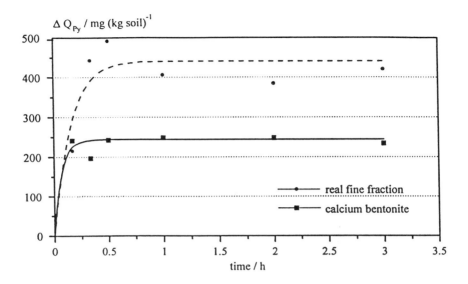

**Figure 9** Extraction kinetics of the doped calcium bentonite and the real fine fraction (A) with system 1 (Table 1).

## 2. Extraction with System 3 at 43°C and Subsequent Phase Separation of the Microemulsion

The $C_{9/11}E_4$/RME/water system forms bicontinuous microemulsions between 38 and 45°C at an oil content $\alpha$ of 50 wt % and a surfactant concentration $\gamma$ of at least 20 wt % (see Figure 7). In this region extractions were performed at $\gamma = 22$ wt % and at e/s = 6. The concentration of pyrene as a representative PAH was determined in the microemulsion. Figure 10 shows the results of the multistep extraction of the real contaminated soil. The total amount of pyrene $\Sigma\Delta Q_{Py}$ removed from the soil after each extraction step is presented. In the first extraction step about 10% more pyrene was extracted than with a Soxhlet extraction using toluene as the solvent. For the second extraction step, half the extraction liquid was replaced by fresh microemulsion. After this extraction step, within the error limits, no significant additional amount of pyrene was removed from the soil. In the further steps, 83% of the liquid was exchanged. No further dissolution of pyrene was observed.

The splitting of the microemulsion was investigated at 26 and 34°C for the surfactant concentration $\gamma$ of 22 wt %. About 30% of pyrene for the higher temperature and 50% for the lower one was separated with the

# Microemulsions in Soil Remediation

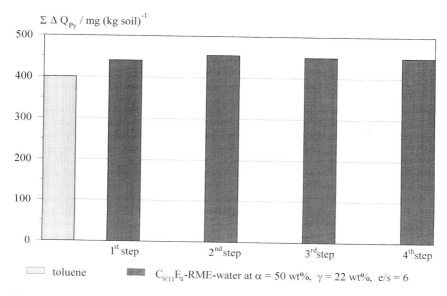

**Figure 10** Total amount of pyrene extracted from the real contaminated soil (B) in a multistep process for an extractant/soil ratio of 6. After the first step, half the solution was removed and replaced by fresh microemulsion. After the second and third steps, 83% of the liquid was exchanged.

oil phase. Leaving the samples in the thermostat, nearly complete separation was reached after about 1 h at 34°C and about 12 h at 26°C. Sophisticated processing and the use of technical equipment supporting coalescence and creaming will certainly decrease the time necessary for demulsification. The pyrene concentrations in the polar and unpolar phases were nearly the same for both temperatures. The separated amount of oil, however, increased with decreasing temperature, thus being the decisive factor for the separated amount of contaminants.

## V. DISCUSSION

The investigations show that the formation of bicontinuous microemulsions with native oils like RSO is possible at slightly elevated temperatures. In order to shift the microemulsion areas to lower temperatures, very hydrophobic surfactants are necessary. The major part of these surfactants is dissolved in the oil compartments so that the oil becomes more hydrophilic and microemulsions can be formed at lower temperatures.

For CO a much smaller effective chain length was found compared to that of RSO. This can be attributed to the higher polarity of CO because of its hydroxyl group. The lowest effective chain length was found for RME. Because of the polarity of the trigylcerides and the ester, their effective chain lengths are significantly reduced in comparison with their real number of carbon atoms.

Furthermore, these characterization experiments have shown that more surfactant is necessary for solubilizing triglycerides like RSO and CO compared with the methyl ester. The application of high surfactant concentrations, however, is unfavorable, since surfactant loss is inevitable during extraction owing to enhanced adsorption or induced formation of LC phases on the solid surface.

The RME system is most appropriate for application in soil remediation. Microemulsions are formed in a temperature range of 38 and 45°C at moderate surfactant contents $\gamma$ between 20 and 25 wt %. The area of liquid crystals is much smaller than for the presented RSO system. This is due to the smaller chain lengths of $C_{9/11}E_4$ and of RME in comparison with $C_{16}E_2$ and RSO.

The extraction experiments showed that the contaminants were removed with the RSO microemulsion to the same extent as was achieved with organic solvents. With the RME system even 10% more pyrene was extracted than with Soxhlet extraction by toluene. This result was already achieved in the first extraction step. Additional treatment with partially fresh microemulsion did not show any further dissolution of pyrene. It can thus be concluded that nearly all contaminants were removed in the first step. But since the separation of solution and solid is never complete and some liquid remains in the filter cake, at least a second extraction step must follow.

Since the subsequent phase separation only yields a partial segregation of the contaminants, a multistep process is necessary anyway. With the results described in this paper, two extraction steps will lead to sufficient remediation (15).

## VI. CONCLUSIONS

Appropriate microemulsions consisting of biodegradable components were found for application in soil remediation. These microemulsions are excellent extraction media for removing pyrene as a representative contaminant from the soil. The extraction rates are high and the contaminants can be dissolved in the microemulsion in one extraction step. The separation of pyrene with the oil phase is satisfactory and sufficient for application in a technical multistep process.

## REFERENCES

1. WD Clemens, F-H Haegel, MJ Schwuger. Langmuir 10:1366, 1994.
2. I Kögel-Knabner, P Knabner, H Deschauer. In: Advances in Soil Organic Matter Research. The Impact on Agriculture and the Environment. Wilson, WS, ed. Cambridge, The Royal Society of Chemistry, 1991, pp 121-128.
3. I Gotlieb, JW Bozzelli, E Gotlieb. Sep Sci Techn 28:793, 1993.
4. WD Clemens, F-H Haegel, MJ Schwuger, K Stickdorn, G Subklew, L Webb. In: Contaminated Soil '93. Vol. 2. Arendt, F, Annokkée, GJ, Bosman, R, van den Brink, WJ, eds. Dordrecht, Kluwer Academic Publishers, 1993, pp 1315-1323.
5. M Kahlweit, R Strey, P Firman. J Phys Chem 90:671, 1986.
6. SE Friberg, L Gan-Zuo. J Soc Cosmet Chem 34:73, 1983.
7. M Kahlweit, R Strey, G Busse. J Phys Chem 95:5344, 1991.
8. M Kahlweit, R Strey, P Firman, D Haase, J Jen, R Schomäcker. Langmuir 4:499, 1988.
9. MJ Schwuger, K Stickdorn, R Schomäcker. Chem Rev 95:849, 1995.
10. W Clemens, F-H Haegel, MJ Schwuger, C Soeder, K Stickdorn, L Webb. Process and Plant for Decontaminating Solid Materials Contaminated with Organic Pollutants. WO 94/04289 3.3.1994.
11. K Bonkhoff, W Clemens, F-H Haegel, G Subklew. In: Cost Efficient Acquisition and Utilization of Data in the Management of Hazardous Waste Sites. Lewis, RA, ed. Waste Policy Institute, Air & Waste Management Association, Pittsburgh, PA, 1994, pp 198-207.
12. M Ortmann, C Eschner, J Gerheim, T Muckenheim, H Kneifel, L Webb, K Bonkhoff, F-H Haegel, G Subklew, HP Rohns. In: In-situ-Sanierung von Böden. Kreysa, G, Wiesner, J, eds. DECHEMA, Frankfurt am Main, Germany, 1996, pp 217-221.
13. H Kunieda, M Yamagata. Langmuir 9:3345, 1993.
14. TD Gauthier, RW Seitz, CL Grant. Environ Sci Technol 21:243, 1987.
15. K Bonkhoff, F-H Haegel, K Mönig, MJ Schwuger, G Subklew. In: Contaminated Soil '95. Vol II. van den Brink, WJ, Bosman, R, Arendt, F, eds. Dordrecht, Kluwer Academic Publishers, 1995, pp 1157-1158.

# 10
# The Importance of Surfactant Hydrophobe Structure in Microemulsion Formation

**Robert S. Schechter and William H. Wade**
*The University of Texas–Austin, Austin, Texas*

## I. INTRODUCTION

Some years ago a considerable joint research effort that involved a number of industrial laboratories as well as university groups was mounted to better understand the surfactant parameters that contribute to the interfacial tension at an oil–water interface. Motivating this unusual surge of research was the perceived need to improve the efficiency of water-flood processes routinely used to displace oil from subterranean host rocks. Conventional wisdom held that by adding surfactants that reduced the interfacial tension to ultralow values (LIFT), the effectiveness of water-flooding would be greatly enhanced. At the time almost nothing was known about how such low interfacial tensions are developed. Our group joined the research effort with one of the tasks being the determination of the role of the surfactant structure. Our findings with respect to the structure of the surfactant's hydrophobic moiety were particularly interesting, and, we believe, indispensable to a practitioner faced with the selection of a surfactant for any one of the myriad of applications that require them. In this brief paper, we will summarize our findings. Even in retrospect, the results seem fascinating. For example, branching the hydrophobes yields a more lipophilic product, one less prone to form liquid crystals, one that produces less stable emulsions, one with a greater tolerance for electrolytes, but one that is less capable of reducing the interfacial tension.

Although the main thrust of the research was to understand ultralow interfacial tensions, it was decided to carry out a concurrent investigation of microemulsion phase behavior. This approach was adopted because a seminal paper by Reed and Healy (1) showed the existence of a correlation between phase behavior and interfacial tension. Indeed, the phase behavior of mixtures of oil, water and surfactants is the most efficient, reproducible, incisive method for characterizing these complex systems. At the most fundamental level, it is important to identify the type of microemulsion that tends to form under various conditions of temperature, added electrolyte, or cosolvent. A widely used classification is as follows: oil-in-water microemulsions are called either Winsor I or Type I systems, water-in-oil microemulsions are Winsor II or Type II systems, and middle-phase systems are Type III. The beauty of these classifications is that considerable information may be gleaned by simply preparing a series of volumetric pipettes containing equal volumes of oil and water together with all of the added components. One of the components (as, for example, the electrolyte) is systematically varied from tube to tube and the phase behavior observed. The varied additive is sometimes called the scanning variable. While temperature may also be a scanning variable, the electrolyte concentration is used primarily here.

## II. RESULTS AND DISCUSSION

Figure 1 presents a typical set of experimental results. At low salinity, a Type I system is obtained. At a critical salinity, $S_1$, there is a transition to a Type III system followed at $S_2$ by a transition to a Type II system. One notes the trend in interfacial tensions, $\gamma$, and that low interfacial tension is associated with Type III systems. At $S^*$, often termed the optimum salinity, the two interfacial tensions are equal and frequently both low. Also, one notes that the solubilization parameter, $\sigma$, expressed as cubic centimeters of water or oil solubilized per cubic centimeter of neat surfactant, are, at least qualitatively, mirror images of the tension curves. This reciprocal behavior was first quantified by Chun Huh (2). Using thermodynamic arguments, he predicted that at $S^*$,

$$\gamma\sigma^2 = \text{constant} \tag{1}$$

With $\gamma$ in dynes per centimeter and $\sigma$ in cubic centimeters per cubic centimeter, the constant is approximately 0.35 for all of the system variables we have explored (3-7). One also notes in Figure 1 that the $\sigma$ curves cross at $S^*$. This is an universal observation.

Other than salinity, the system variables studied were temperature ($T$), alcohol cosolvent molecular weight and concentration ($A$), and al-

## Surfactant Hydrophobe Structure

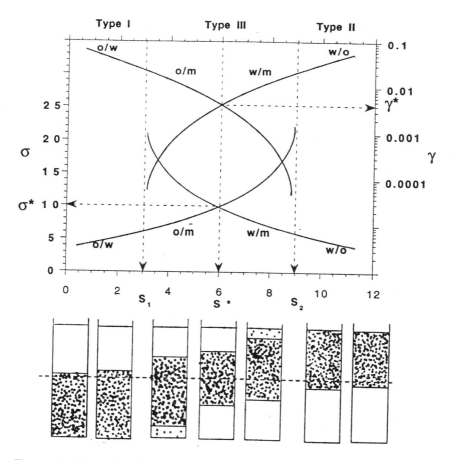

**Figure 1** Winsor I ⇔ II ⇔ III phase behavior for a properly designed oil-water-surfactant system with typical values for solubilization parameter, $\sigma$ (cc/cc), and interfacial tension, $\gamma$ (dyn/cm).

kane—or equivalent thereof—carbon number (ACN or EACN). These variables were found to be interrelated by the following relationship for optimum formulations (8):

$$\ln(S^*/S_o^*) = k\{EACN - N_{MIN}(MIX)\} + bA + c(T - 25) \qquad (2)$$

$S_o^*$ is the standard state $S^*$, always chosen to be 1 wt %. The value of $k$ is $0.16 \pm 0.01$ and is invariant, at least for sulfonate or sulfate surfactants. The constant $b$ depends on alcohol molecular weight and structure. The value of $c$ shows some variation with surfactant, but for anionics without added ethylene oxide, the $c$ values are so low that the temperature term is rather unimportant. Examining Equation 2, one sees that $N_{MIN}(MIX)$ is the EACN required for an optimum system at 25°C, 1.00 wt % salinity, and with no alcohol in the system. We assign $N_{MIN}$ as a characteristic property of a surfactant as an alternative to hydrophilic–lipophilic balance (HLB), whose group additivity calculated values (9) for anionics are of questionable accuracy.

Just as the EACN of a multicomponent oil phase is related to the EACNs of the individual components (10) by

$$EACN_{MIX} = \Sigma_i X_i EACN_i \tag{3}$$

so is the $N_{MIN}$ of a mixed anionic surfactant system (11) by

$$N_{MIN}(MIX) = \Sigma_i X_i N_{MIN\,i} \tag{4}$$

where $X_i$ is the mole fraction of each oil or surfactant. This behavior implies both oil phases and surfactant surface phases behave ideally, and no exceptions have been found to the above two equations. Further, from Equation 2, one notes that to remain at optimum with constant salinity, temperature, and alcohol composition, an increase in EACN of the oil phase requires an equivalent increase in $N_{MIN}(MIX)$. For a given headgroup, this should simply require an increase in hydrophobe molecular weight. One would expect $N_{MIN}(MIX)$ to increase linearly with average molecular weight of a mixture of surfactants. Figure 2 (12) displays such behavior for mixtures of four surfactants: two alkyl ortho-xylene sulfonates, and TRS 10-80 and Martinez 470, two petroleum sulfonates that are complex mixtures in themselves. For binary mixtures $N_{MIN}(MIX)$ varies linearly with molecular weight. Three systems show the expected positive slope; however, one notes that despite having nearly equal molecular weights, mixtures of TRS 10-80 and $C_{15}$ o-xylene $SO_3Na$ vary greatly in $N_{MIN}(MIX)$. Even more disturbing is the observance of *negative* slopes for two systems. Observations such as this lead inescapably to the conclusion that $N_{MIN}$ for a given surfactant is not a simple function of the molecular weight of the hydrophobe, but must be related in some complex fashion to its precise molecular arrangement.

Once this was realized, we embarked on a surfactant synthesis program which continues to the present time. Figure 3 (13) is a graphic demonstration of the importance of hydrophobe branching. A series of six monoisomeric alkane sulfonates were synthesized where a $NaSO_3^-$ was added to the second, fourth, fifth, sixth, seventh, and ninth carbon in a linear octadecane chain. Measurements were made at 40°C with decane for the oil phase. The salinities shown by each data point are values at optimum and

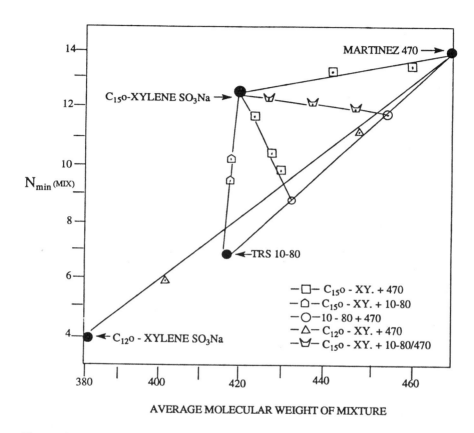

**Figure 2** Variation of surfactant $N_{MIN}$ with average molecular weight of four binary and one ternary systems. (From Ref. 11, used with permission.)

one notes a *twenty-fold* shift from the #2 isomer to the #9. Obviously the single-tailed (#2) species is quite hydrophobic, whereas the twin-tailed (#9) species is relatively lipophilic. This unexpected result remains the general observation for all surfactants subsequently studied. The ordinate of Figure 3 shows the minimum alcohol concentration required to obtain optimum Type III microemulsion systems rather than more complicated liquid crystal systems. The alcohol requirements vary greatly with isomer number, reaching 0 for the near midchain #6 and #7 species. Once again in all subsequent studies, it was found that straight-tailed species tend to form liquid crystal systems and thus require large cosolvent concentrations for breaking these relatively rigid structures. For practical applications, it is important to rec-

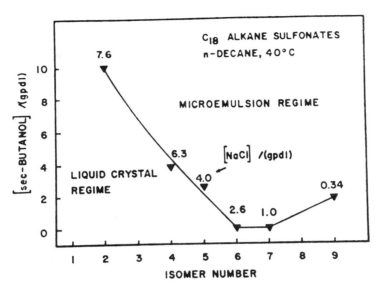

**Figure 3** Alcohol concentration in grams per deciliter and optimum salinity as related to hydrophobe structure. (From Ref. 12, used with permission).

ognize that twin-tailed species require smaller quantities of a cosolvent. Commercial samples of α-olefin sulfonates behave like the #2 isomer species (14a,14b) and vinylidine sulfonates (6a,6b), with their near twin-tailed structures, mimic the #6 and #7 isomer species, thereby emphasizing the practical importance of these observations.

The sodium salts of a number of monoisomeric alkyl benzene sulfonates were also synthesized (3). Optimum middle-phase systems were prepared by varying the ACN of the oil phase, employing Equation 3 to interpolate intermediate values of EACN when necessary. Salinity, alcohol composition, and surfactant concentration were kept constant. The nomenclature $x\phi C_y$ means that the benzene ring is attached to the $x$th carbon of a chain $y$ carbons long. All of these alkyl benzenes were sulfonated in the para position and converted to the sodium salt. In Figure 4, one notes the uniform spacing between the lines for the $x\phi C_{12}$, $x\phi C_{14}$, $x\phi C_{16}$, and $x\phi C_{18}$ species. Increased branching leads to higher optimum ACN, as can be inferred from the alkane sulfonate isomers in concert with Equation 2. One also notes that increased branching leads to significantly reduced optimum solubilization parameters. This is also a quite general observation. Considering Equation 1, it is furthermore a general observation that increased branching results in higher optimum interfacial tensions. There is also the

# Surfactant Hydrophobe Structure

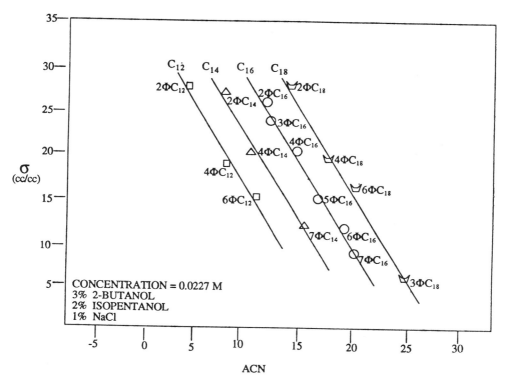

**Figure 4** Solubilization parameter, $\sigma$, and optimum ACN for various monoisomeric alkyl benzene sulfonates for system conditions as noted.

general observation that increased branching increases the range of the scanned variable over which Type III systems persist (3). A further observation on tail branching is that Winsor (15) and Evans (16) have shown for alkane sulfates, and our group (3) for alkyl benzene sulfonates that increased branching leads to a logarithmic increase in critical micelle concentration (cmc).

These various results are summarized in Table 1, along with nonequilibrium observations that both emulsion and foam stability are reduced by increased branching. The message of these observations is that tradeoffs are inevitable when selecting a surfactant to achieve a desired property.

With this information in hand, the objective was set to develop a suite of model surfactants that would yield Winsor Type III systems at relatively low temperatures (40°C) with no alcohol and for a variety of salinities (17). The results of these efforts are displayed in Figure 5. The nomenclature

**Table 1**  Difference of Properties for Single and Twin-Tailed Surfactants

| HYDROPHOBE STRUCTURE \\ PROPERTY | (STRAIGHT TAILED) ⊖ Na+ | (LIMITING STRUCTURES) ←→ | (MIDCHAIN BRANCHED) ⊖ Na+ |
|---|---|---|---|
| 1) PARTITIONING CHARACTERISTICS | PREFERS TO MICELLIZE IN WATER | | PREFERS TO MICELLIZE IN OIL |
| 2) INTERFACIAL TENSIONS | LOW | $\sigma^2 \gamma$ = CONST | HIGH |
| 3) SOLUBILIZATION PARAMETERS | HIGH | | LOW |
| 4) SENSITIVITY TO SYSTEM COMPOSITION | HIGH | | LOW |
| 5) COSOLVENT REQUIREMENTS | HIGH | | LOW |
| 6) MACROEMULSION STABILITY | HIGH | | LOW |
| 7) FOAM STABILITY | HIGH | | LOW |
| 8) CMC | | ------ INCREASES ------→ | |

$xC_y(EO)_n$ means that an ethylene oxide (EO) chain $n$ units long was added to a carbon chain $y$ carbons long at the $x$th carbon. The terminus of the EO chain has a $NaSO_3^-$ added in all cases. The number by each data point is the ACN of the oil phase, and the salinities are for optimum systems. The model surfactant requiring the smallest salinity, as previously discussed, is the near midchain alkane sulfonate, $7C_{18}$. At first glance one might be tempted to simply add more EO groups to obtain higher $S^*$'s, but this increased structural length was found to lead to a requirement for alcohol or higher temperatures to avoid the formation of liquid crystals. Rather, one must, while maintaining the near midchain branching, reduce hydrophobe molecular weight. The most hydrophilic surfactant synthesized was $6C_{12}(EO)_3$ having $S^*$'s varying from 16 to 21% depending on the oil phase alkane. One also notes that in going from $7C_{14}(EO)_2$ to $6C_{14}(EO)_2$ to $5C_{14}(EO)_2$, $S^*$ increases quite significantly, as was earlier found for the sulfonates.

Obviously the model compounds used in the studies described above are not commercially viable, but a number of species have been identified

# Surfactant Hydrophobe Structure

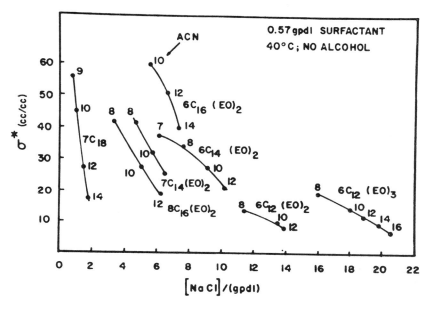

**Figure 5** Solubilization parameters and optimum salinity for a variety of branched-tail $(EO)_n SO_3 Na$ for system conditions as shown. (From Ref. 16, used with permission.)

that will perform at low temperature with little or no alcohol that either are, or could be, commercially available. They are shown in Figure 6. The multiple methyl groups in the hydroformulated tridecyl alcohol (TDA) reduces tail branching density, and hence, liquid crystal formation. The addition of propylene oxide (PO) or EO to the TDA and end capping with sulfate or sulfonate produces surfactants with a wide variety of $S^*$'s (18). The dialkyl sulfosuccinates have the usual properties of twin-tailed surfactants and are the only anionic species identified that will give Winsor phase behavior with chlorinated solvents for the oil phase (19a–19c). The species $C_{16}EXSO_4Na$ uses an Exxon internally methyl-branched Guerbet alcohol for the hydrophobe and could have EO added prior to sulfation (20). The species designated $C_8(PO)_xSO_4Na$ uses 2-ethyl hexanol coupled to PO for the hydrophobe (21). The species designated $C_{12}GA(PO)_xSO_4Na$ utilizes the more common $C_{12}$ Guerbet alcohol coupled to PO. The rationale for tail-branching promoting microemulsion formation is that it reduces attractive van der Waals intrahydrophobe forces to the point that the system is above the liquid crystal melting point.

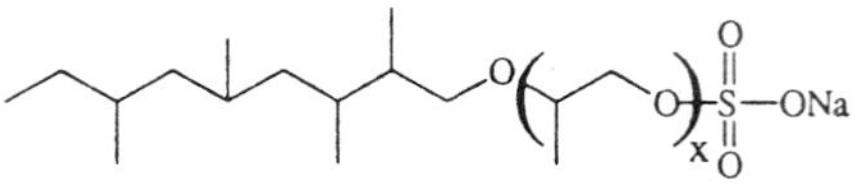

TDA(PO)$_x$SO$_4$Na

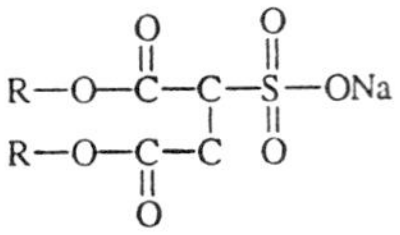

R = C$_6$H$_{13}$, C$_7$H$_{15}$, C$_8$H$_{17}$

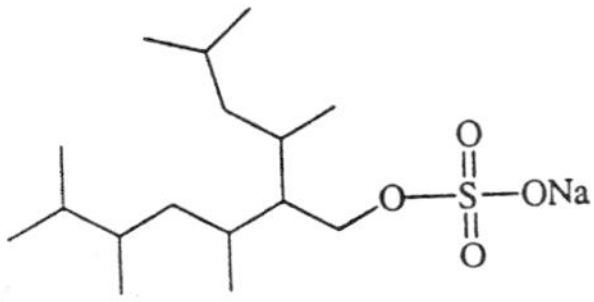

C$_{16}$EXSO$_4$Na

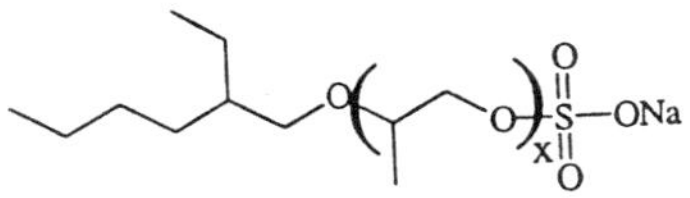

C$_8$(PO)$_x$SO$_4$Na

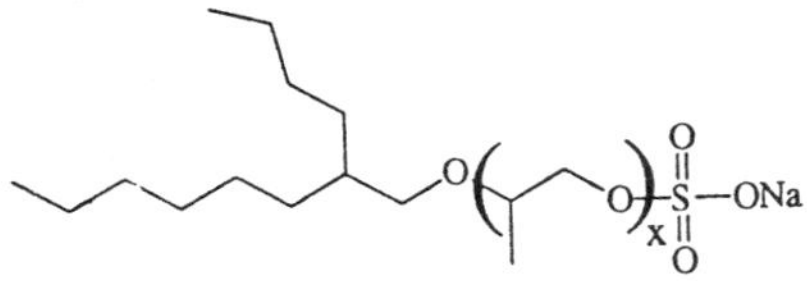

C$_{12}$GA(PO)$_x$SO$_4$Na

**Figure 6** Optimized surfactant hydrophobe structures for microemulsion formation.

The C$_{14}$ and C$_{16}$ Guerbets coupled to EO have also been studied (20). Results obtained for C$_{16}$EX(EO)$_n$SO$_4$ with seven different values of $n$ are shown in Figure 7. *sec*-Butanol was added to allow optimum systems at unusually low temperatures. The numbers beside each data point are optimum solubilization parameters. For surfactants with $n$ = 0–3, $S^*$ increases with temperature — the characteristic behavior of anionics. With $n$ = 6–10,

# Surfactant Hydrophobe Structure

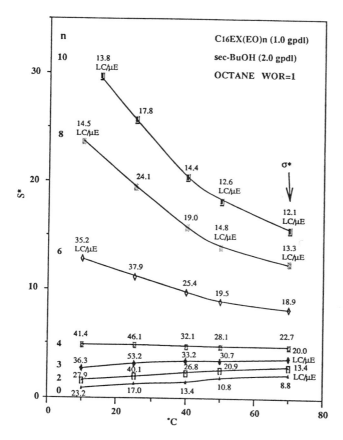

**Figure 7** Optimum salinity versus temperature for different levels of ethoxylation for $C_{16}EXSO_4Na$. (From Ref. 1, used with permission.)

$S^*$ decreases with increased temperature—the characteristic of a nonionic surfactant. With $n = 4$, $S^*$ is independent of temperature, thus the nonionic and anionic temperature characteristics are balanced.

As noted earlier, there is evidence that anionic surfactant mixtures behave ideally with regard to $N_{MIN}(MIX)$ and hence with respect to the formation of micelles. A systematic study was performed for a number of surfactant mixtures where different mole ratios of straight and twin-tailed species were varied to identify systems that would function alcohol-free (22). One such system is shown in Figure 8. The straight-tailed surfactant oleyl alcohol with one EO sulfated, $O(EO)_1S$, and the twin-tailed species

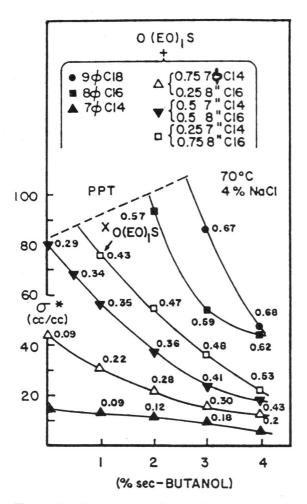

**Figure 8** Mixtures of straight and branched-tail species with system conditions as noted which show systematic variation in solubilization parameter and alcohol requirements. (From Ref. 21, used with permission.)

are $7\phi C_{14}$, $8\phi C_{16}$, and $9\phi C_{18}$. The number beside each data point is the mole fraction of $O(EO)_1S$. Temperature is constant at 70°C and salinity is constant at 4 wt %. With $7\phi C_{14}$ as the twin-tailed species, very little $O(EO)_1S$ is required for an optimum formulation. This system requires no alcohol and has relatively small values of $\sigma^*$. When $8\phi C_{16}$ is used as the twin-tailed species, greater values of $\sigma^*$ are obtained, because the mole ratio of $O(EO)_1S$ is much higher, and using $9\phi C_{18}$, the effect is further enhanced. Unfortunately, one finds that these two systems will not function without alcohol. However, if one makes mixtures of $7\phi C_{14}$ and $8\phi C_{16}$ for the twin-tailed species, then one generates intermediate systems. For instance, a 50/50 mole ratio will, with a 0.29 mole fraction of $O(EO)_1S$, produce an alcohol-free optimum system with a large $\sigma^*$. A sufficiently large number of such single/twin mixture studies have been done so that one can safely consider this to be a general way of tailoring surfactant formulation for microemulsions.

## III. SUMMARY

In summary, our studies of the last quarter century have led us to an understanding of the very significant role of hydrophobe structure in surfactant performance for microemulsion formation. We have synthesized monoisomeric model compounds as guides for more commercially viable species, we have identified a number of species that are, or potentially are, commercially viable mimics of the model species, and we have demonstrated the systematic behavior of mixtures of the two limiting structures.

## ACKNOWLEDGMENTS

The authors wish to recognize the following entities without whose financial assistance this research effort would not have been possible:

| | |
|---|---|
| Amoco Production | ARCO |
| British Petroleum | Celanese |
| Chevron Oil Field Research | Conoco |
| Dowell Division of Dow Chemical U.S.A. | Elf-Aquitaine |
| Energy Research and Development Administration | El Paso Products |
| Exxon Production Research | GAF |
| Getty Oil | Gulf Research & Development |
| Imperial Chemical Industries | Marathon Oil |
| Mobil Research & Development | National Science Foundation/RANN |
| Norsk Hydro Research Centre | Rhône-Poulenc Petrole Services |

Shell Development
Stephan Chemical

Suntech
Texaco
Union Oil
Robert A. Welch Foundation

SOHIO
Sun Exploration & Production
Tenneco Oil
TOTAL
U.S. Department of Energy
Witco Chemical

## REFERENCES

1. RL Reed, RN Healy. In: Improved Oil Recovery by Surfactant and Polymer Flooding. Shah, DO, Schechter, RS, eds. San Diego, CA, Academic Press, 1977.
2. C Huh. J Colloid Interface Sci 71:408, 1979.
3. Y Barakat, LN Fortney, RS Schechter, WH Wade, SH Yiv. J Colloid Interface Sci 92:561, 1983.
4. Y Barakat, LN Fortney, C Lalanne-Cassou, RS Schechter, WH Wade, U Weerasooriya, V Weerasooriya, SH Yiv. SPE preprint 10679 (1982) and Soc Petr Eng J 913, Dec 1983.
5. A Graciaa, L Fortney, RS Schechter, WH Wade, SH Yiv. Soc Petr Eng J 743, Oct 1982.
6a. C Lalanne-Cassou, RS Schechter, WH Wade. Am Chem Soc Div Petrol Chem 29:1187, 1984.
6b. C Lalanne-Cassou, RS Schechter, WH Wade. J Dispersion Sci Tech 7:479, 1986.
7. RS Schechter, WH Wade, U Weerasooriya, V Weerasooriya, S Yiv. J Dispersion Sci Tech 6:223, 1985.
8. JL Salager, E Vasquez, JC Morgan, RS Schechter, WH Wade. Soc Petr Eng J 19:107, 1979.
9. JT Davies. Proc 2nd Int Congr Surf Act, Schulman, JH, ed. London Butterworth 1:426, 1957.
10. JL Cayias, RS Schechter, WH Wade. Soc Petr Eng J 16:351, 1976.
11. JL Salager, M Bourrel, RS Schechter, WH Wade. Soc Petr Eng J 19:271, 1979.
12. JK Jacobson, JC Morgan, RS Schechter, WH Wade. Soc Petr Eng J 17:122, 1977.
13. M Abe, D Schechter, RS Schechter, WH Wade, U Weerasooriya, S Yiv. J Colloid Interface Sci 114:343, 1986.
14a. I Carmona, RS Schechter, WH Wade, U Weerasooryia. Soc Petr Eng J 25:351, 1985.
14b. I Carmona, RS Schechter, WH Wade, U Weerasooryia, V Weerasooryia. J Dispersion Sci Tech 4:361, 1983.
15. PA Winsor. Trans Faraday Soc 44:376, 1948.
16. HC Evans. J Chem Soc 579, 1956.

## Surfactant Hydrophobe Structure

17. M Abe, D Schechter, RS Schechter, WH Wade, U Weerasooriya, S Yiv. J Colloid Interface Sci 114:343, 1986.
18. M Aoudia, WH Wade, V Weerasooriya. J Dispersion Sci Tech 16:115, 1995.
19a. J Baran, GA Pope, WH Wade, V Weerasooriya. Langmuir 10:1146, 1994.
19b. J Baran, GA Pope, WH Wade, V Weerasooriya, A Yapa. Environ Sci Tech 8:1361, 1994.
19c. J Baran, GA Pope, WH Wade, V Weerasooriya, A Yapa. J Colloid Interface Sci 168:67, 1994.
20. C Sunwoo, WH Wade. J Dispersion Sci Tech 13:491, 1992.
21. WH Wade, unpublished data.
22. C Lalanne-Cassou, I Carmona, L Fortney, A Samii, RS Schechter, WH Wade, U Weerasooriya, V Weerasooriya, S Yiv. J Dispersion Sci Tech 8:137, 1987.

# 11
# The Role of Surfactants in Enhanced Oil Recovery

**Hisham A. Nasr-El-Din and Kevin C. Taylor**
*Saudi Aramco, Dhahran, Saudi Arabia*

## I. INTRODUCTION

Processes that rely on the injection of surfactants or surfactant-forming materials into a reservoir form a key part of chemical flooding. Micellar-polymer, MP, and alkali-surfactant-polymer, ASP, flooding are two examples in which surfactants serve a specific purpose. In these processes, it is necessary to understand the behavior of a surfactant as it is injected into a reservoir, as it travels through that reservoir over a period of weeks or months, and as it flows out of the reservoir through a producing well. This chapter discusses the basics required for an appreciation of these processes.

The progression from MP to ASP flooding has special significance in this chapter. Micellar-polymer flooding is technically well developed, relatively well understood, and has undergone many technically successful field trials (1). However, this process is inherently expensive because of the large surfactant concentrations that must be injected into the reservoir. Alkali-surfactant-polymer flooding is a much newer technology, is more complex, and is not technically well developed. Many lessons learned from micellar-polymer flooding can be applied to the ASP process. Alkali-surfactant-polymer flooding is inherently much less expensive than the micellar-polymer process, primarily because the surfactant concentration is significantly lower. Field trials are in progress, although many details remain confidential. This technology is at the stage that the micellar-polymer process was in during the early 1970s. As more is learned, this process may come into much more widespread use.

This chapter reviews the role of surfactants in micellar-polymer and alkali-surfactant-polymer processes. The forces that trap oil in porous media and the viscous forces that tend to mobilize this trapped oil are introduced. This is followed by a discussion of the traditional micellar-polymer process. Problems associated with field application of classical alkaline flooding are examined in a separate section. Finally, application of the alkali-surfactant-polymer process in the lab and the field is examined in detail.

## A. Capillary Forces in Oil Mobilization

The elimination or reduction of capillary forces is critical in enhanced oil recovery (EOR) processes using surfactants. Capillary forces are responsible for oil entrapment. They are influenced by pore geometry, interfacial tension, and rock wettability. Viscous forces, which act to displace oil, are functions of the velocity and viscosity of the displacing fluid.

When an aqueous phase flows through a porous medium containing oil, some oil will be readily displaced, but capillary forces will act to trap oil in some of the pore spaces. No matter how much aqueous phase flows through the material, a certain amount of oil, known as the residual oil, will remain trapped. The residual oil saturation is generally expressed as a percentage of the original oil in place (%OOIP), and can be upward of 40%. It is this oil that is the target of many enhanced oil recovery techniques.

The capillary number, $N_c$, is a dimensionless ratio of viscous to capillary forces, which provides a measure of how strongly trapped residual oil is within a given porous medium (2). Various definitions have been used for capillary number, but the following equation is common for water-wet porous media:

$$N_c = \frac{V\mu}{\sigma} \qquad (1)$$

where $V$ is Darcy velocity, $\mu$ is viscosity of the displacing phase, and $\sigma$ is the interfacial tension between displaced and displacing phases.

Figure 1 shows a typical capillary number curve for water-wet Berea sandstone. The shape of the curve is affected by wettability and pore size distribution of the porous medium. For the oil phase, mobilization of residual oil usually begins at a capillary number of about $10^{-5}$ (the critical capillary number), while complete oil recovery occurs at $N_c$ of about $10^{-2}$. For a variety of water-wet sandstones, Chatzis and Morrow (3) found a critical capillary number of $10^{-5}$ and complete recovery of oil at $N_c$ of

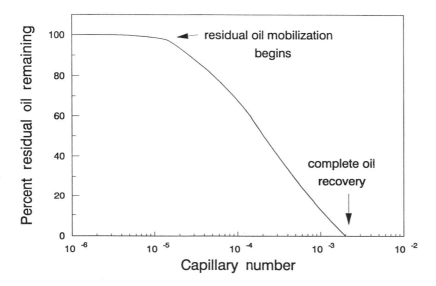

**Figure 1** Capillary number correlation.

$10^{-3}$. To greatly increase capillary number, it is most practical to decrease interfacial tension by the addition of surfactants.

It is important to realize that if interfacial tension between two phases becomes zero, then the two phases become miscible. This is the ultimate aim of many types of EOR: to make oil-water interfacial tension equal to 0, so that a displacing fluid can miscibly displace oil trapped in the porous medium. In practice, it is difficult to make interfacial tension equal to 0 for liquids of such different characteristics as oil and water.

## B. Micellar-Polymer Flooding

A surfactant is used to lower interfacial tension to values as low as $10^{-3}$ mN/m in MP flooding. Capillary number is increased, allowing the recovery of residual oil from porous media. The term micellar is used because injected surfactant solutions are always above their critical micelle concentration. Petroleum sulfonates are the most commonly used surfactants (4,5). Other surfactants such as ethoxylated alcohol sulfates (6) and nonionic surfactants mixed with petroleum sulfonates have also been used (7).

In micellar-polymer flooding, a microemulsion is either injected into the formation, or the surfactant solution is designed to create a microemulsion once it contacts oil in the reservoir. The structure of microemulsion systems has been reviewed (8). Bicontinuous and droplet-type structures are two types that can occur in microemulsions. The droplet-type structure is conceptually more simple, and is an extension of the emulsion structure that occurs at high values of interfacial tension. In this case, very small thermodynamically stable droplets occur, typically smaller than 10 nm (9). Each droplet is separated from the continuous phase by a monolayer of surfactant. In bicontinuous microemulsions, oil and water layers in the microemulsion may be only a few molecules thick, separated by a monolayer of surfactant. Each layer may extend over a macroscopic distance, with many layers making up the microemulsion.

Compositions of injected micellar fluids can vary greatly. They include aqueous solutions of surfactant as well as complex mixtures containing components such as cosurfactants, cosolvents, or stabilizers in addition to surfactant, oil, and brine. Whatever the composition of the injected fluid, once in the reservoir the fluid system consists primarily of oil, water, and surfactant.

Figure 2 displays the general sequence of a micellar-polymer flood. An initial fresh water preflush is sometimes used to lower salinity and divalent ion concentration in the reservoir. This is followed by the surfactant slug, containing a surfactant–polymer mixture designed to produce a microemulsion with the crude oil mixture. A polymer drive follows, which prevents "fingering" of the brine into the surfactant–polymer slug. The

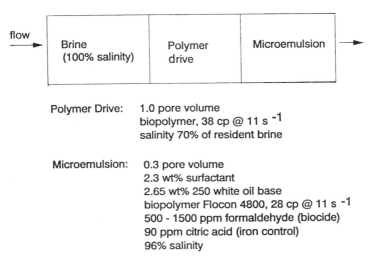

Polymer Drive: 1.0 pore volume
biopolymer, 38 cp @ 11 s $^{-1}$
salinity 70% of resident brine

Microemulsion: 0.3 pore volume
2.3 wt% surfactant
2.65 wt% 250 white oil base
biopolymer Flocon 4800, 28 cp @ 11 s $^{-1}$
500 - 1500 ppm formaldehyde (biocide)
90 ppm citric acid (iron control)
96% salinity

**Figure 2** Typical microemulsion injection scheme.

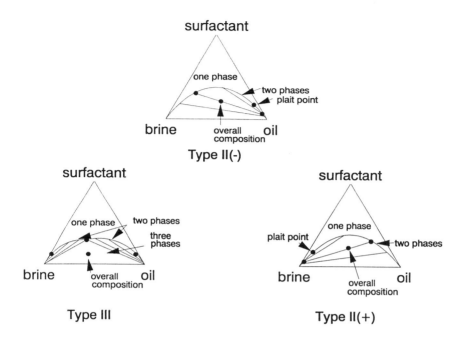

**Figure 3** Pseudo-ternary diagram of oil–water–surfactant system with three compositions of interest.

polymer drive is often injected with a concentration gradient, with polymer concentration decreasing as the injection progresses (graded viscosity scheme).

The phase behavior of the fluid system can be quite complex, but may be approximately described by means of pseudo-ternary diagrams in which the pseudo-components are surfactant, brine, and oil (Figure 3). Depending on the system being studied, the pseudo-components can range from pure substances to complex mixtures. For example, the oil may be a pure hydrocarbon or a crude oil. The surfactant can include cosurfactants and cosolvents, and the brine may contain a variety of ionic constituents. The pseudo-ternary diagram is separated by a multi-phase boundary into a single phase region above and a multi-phase region below the phase boundary.

Nelson and coworkers (10–12) and Healy, Reed, and Stenmark (13) have extensively studied phase behavior in micellar flooding. In Nelson's methodology, there are three different phase behavior environments (Figure 4). These are the type II(−), the type II(+), and the type III phase environments.

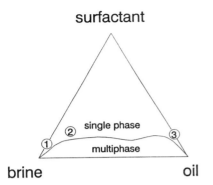

**Figure 4**  Phase behavior environments.

Type II(−) and II(+) phase environments have a maximum of two phases. In both cases, one phase only may be present at high surfactant concentrations. Type II(−) and II(+) phase environments are distinguished when the phases are plotted on a pseudo-ternary diagram. The tielines in the two-phase region give either a negative slope for type II(−) behavior, or a positive slope for type II(+) behavior. Winsor (14) assigns microemulsions occurring in the two-phase region of a type II(−) diagram as type I, and defines them as a microemulsion in equilibrium with excess oil. The microemulsion contains mostly brine and surfactant, while any oil present is solubilized in micelles. As such it is a water-external emulsion. The plait point of the pseudo-ternary diagram tends to be close to the oil apex. Microemulsions in the two-phase region of a type II(+) phase environment correspond to a Winsor Type II emulsion. The microemulsion is in equilibrium with excess brine, and contains mostly oil and surfactant. The brine is solubilized in micelles, making this an oil external emulsion. In this case, the plait point of the pseudo-ternary diagram tends to be close to the brine apex.

The type III phase environment may contain a maximum of three phases. When this is the case, the emulsion present corresponds to Winsor type III, in which a microemulsion is in equilibrium with pure oil and pure brine phases. However, type II(−) behavior and type II(+) behavior may also be observed under certain conditions. In practice this occurs when all of the brine or oil can be incorporated into the microemulsion, or when insufficient surfactant is present to produce a measurable microemulsion.

There are several factors that can affect the phase type that is observed. These factors generally act by changing the partitioning of the surfactant between the brine and oil phases. In general, any change in the

# Surfactants in Enhanced Oil Recovery

surfactant–oil–brine system that increases the solubility of surfactant in oil compared with brine will cause the phase environment type to shift from II($-$) to III to II($+$).

Increasing salinity generally decreases the solubility of the surfactant in the brine phase. This shifts the phase behavior from II($-$) to III to II($+$). Changing the oil type or the structure of the surfactant so that the surfactant is more soluble in the oil will also shift the phase behavior in a similar manner. The effect of surfactant structure on phase behavior has been discussed in detail (15).

The following changes will also shift phase behavior from type II($-$) to III to II($+$):

1. Decrease in temperature for anionic surfactants
2. Increase in temperature for nonionic surfactants
3. Increase in alcohol concentration (alcohols of less than four carbons)
4. Decrease in alcohol concentration (alcohols of more than four carbons)
5. Increase in the divalent ion concentration of the brine.

## C. Solubilization Parameter

Solubilization parameter is defined as

$$\frac{V_o}{V_s} \quad \text{or} \quad \frac{V_w}{V_s} \tag{2}$$

where $V_o$ is the oil volume, $V_s$ is the volume of surfactant, and $V_w$ is the volume of water, all three measured in the microemulsion phase. Interfacial tension, $\eta$, can be measured between the microemulsion and oil phases ($\eta_{mo}$) or between the microemulsion and water phases ($\eta_{mw}$). As either measure of interfacial tension decreases, the solubilization parameter increases. Healy and Reed (16) first showed that low interfacial tension correlates with high solubilization parameter, while Huh (17) showed it to be theoretically valid. Graciaa et al. (15) and Glinsmann (18) have validated the concept experimentally. This correlation is very useful, because it enables the results of phase behavior experiments to partially replace the experimentally more difficult measurement of interfacial tension.

## D. Field Application of the Micellar-Polymer Process

The Loudon field in Illinois, operated by Exxon, is an interesting example of micellar-polymer flooding design (19,20). The reservoir is a moderate permeability sandstone with excellent properties for micellar-polymer flooding, except one: the salinity of the formation brine is very high, approximately 10.5 wt % total dissolved solids and 4000 ppm divalent ions. Exxon has been studying the micellar-polymer process in this field for over 10 years, and to date has completed two pilot projects with two others in progress. The sequence of the injected microemulsion and polymer drive is outlined in Figure 5. A microemulsion of 0.3 pore volume was used, containing a relatively low surfactant concentration, 2.3 wt %. The formula of the surfactants used was a sulfate of a propoxylated ethoxylated tridecyl alcohol, of the following structure: $i\text{-}C_{13}H_{27}O(C_3H_6O)_m(C_2H_4O)_nSO_3Na$. The values of $m$ and $n$ for the two surfactants used were 4,2 and 3,4, respectively. In practice, a mixture of these two surfactants was used, so that the slug composition could easily be varied in the field to correct for any changes in injection brine composition. This type of surfactant is particularly suited to high salinities. If the number of ethoxyl groups is increased, optimal salinity increases, while if the number of propoxyl groups increases, the optimal salinity decreases. The mixture used had optimal salinity at the high salinity and hardness levels associated with this reservoir. The synthesis of this surfactant and its use in micellar-polymer flooding has been patented by Exxon (21).

A particularly interesting part of the pilot involved the treating of produced emulsions. Over the life of the pilot, 93% of the injected surfactant was produced, which led to serious emulsion problems. Heating the

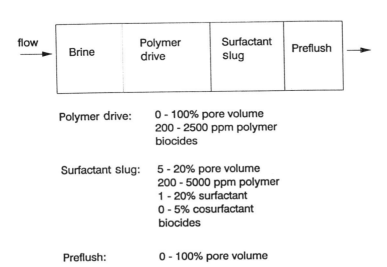

**Figure 5** Loudon micellar-polymer flood.

emulsion to a specific, but unreported, temperature caused the surfactant to partition completely into the aqueous phase, leaving the crude oil with very low levels of surfactant and brine. The resulting oil was suitable for pipeline transportation. The critical separation temperature had to be controlled to within 1°C. At higher temperatures, surfactant partitioned into the oil, while at lower temperatures significant quantities of oil remained solubilized in the brine. Recovered surfactant was equivalent to the injected surfactant in terms of phase behavior, and had the potential for reuse.

The pilot area used for this test was small, 0.71 acre. However, the test was a technical success, recovering 68% of the waterflood residual oil. The pilot began in 1982 and ended in November 1983. Since then, Exxon has initiated two other micellar-polymer floods in the Loudon field, one a 40-acre pilot and the other an 80-acre pilot.

## E. Classical Alkaline Flooding

Classical alkaline flooding uses alkaline solutions only to recover waterflood residual oil. Carboxylate soaps are formed when a crude oil with acidic components reacts with hydroxide ion in an alkaline solution. These petroleum soaps are capable of adsorbing at the oil-water interface and lowering interfacial tension. Crude oils suitable for alkaline flooding generally have a total acid number (TAN) of 0.1 to 2 mg potassium hydroxide per gram of oil. The injection of alkaline solutions into a reservoir can improve oil recovery by several mechanisms. Emulsification and entrainment, emulsification and entrapment, decreasing interfacial tension, and wettability reversal have been proposed (22,23).

Despite the success of alkaline flooding in laboratory studies, most field applications were not as successful as anticipated. Mayer et al. (24) showed that only 2 of 12 projects had significant incremental oil recovery: North Ward Estes and Whittier with 6-8% and 5-7% pore volume, respectively. Estimated recovery from the Wilmington field was 14% with a classical alkaline flooding method (25). However, postproject evaluation of that field suggested no improvement over conventional waterflooding (26).

Some of the important reasons for the poor field results of classical alkaline flooding are that petroleum soaps are sensitive to ionic strength, there is low acid content in the oil, there is alkali consumption due to rock/fluid interactions, and there is lack of mobility control. To overcome these problems, several injection schemes have been suggested.

The petroleum soaps produced from the reaction of the alkali with the organic acids are very sensitive to increases in the ionic strength of the aqueous phase. In 1975, Burdyn and colleagues began to address this problem with the addition of synthetic surfactants and polymers to the alkaline flooding process (27). The surfactant provides low interfacial tension (IFT) behavior over a broader range of salinity than can be obtained with alkali alone.

The petroleum soaps produced in alkaline flooding have an extremely low optimal salinity. For instance, most acidic crude oils will have optimal phase behavior at a sodium hydroxide concentration of approximately 0.05 wt % in distilled water. At that concentration (about pH 12) essentially all of the acidic components in the oil have reacted, and type III phase behavior occurs. An increase in sodium hydroxide concentration will not produce more petroleum soaps and, as a result, the ionic strength increases. As salinity increases, the petroleum soaps become much less soluble in the aqueous phase than in the oil phase, and a shift to over optimum or type II(+) behavior occurs.

The addition of a synthetic surfactant can modify the properties of the petroleum soaps that are produced from the reaction of the organic acids with hydroxide ion (27,28). This process has been termed "surfactant-enhanced alkaline flooding." The added synthetic surfactant is chosen to have a very high optimal salinity, and the resulting petroleum soaps/synthetic surfactant mixture produces optimal phase behavior at intermediate salinities.

The mixing of a synthetic surfactant and petroleum soaps can be explained in terms of the surfactant mixing rules. Taylor (29) found that at a fixed salinity and alkali concentration, a specific mole fraction of synthetic surfactant to petroleum soaps was required to produce optimal phase behavior as the water-to-oil ratio was varied. This relationship is useful in understanding the effect of the variables of water-to-oil ratio and synthetic surfactant concentration in alkali, surfactant, and polymer systems.

Synthetic surfactants are added to alkali/oil systems to either lower their IFT or to raise the salinity at which optimal phase behavior occurs (28). Many investigators have reported very low IFT values when a synthetic surfactant was added to alkali/oil systems (30–35). Nelson et al. (28) examined the effect of adding a synthetic surfactant on the salinity range at which optimum phase behavior exists. They obtained a significant increase in the salinity range for optimum phase behavior by adding a small amount of a surfactant having a higher salinity tolerance than the petroleum soaps. Their experiments showed a significant improvement in oil recovery using such formulations. Although Nelson et al. used anionic surfactants in their work, Saleem and coworkers (36,37) have obtained significant improvement in oil recovery by using a nonionic surfactant (Triton X-100).

Recently, Al-Ghamdi and Nasr-El-Din (38) examined the effects of oilfield chemicals on the cloud point of nonionic surfactants. They found that alkalis and water-soluble polymers depress the cloud point of these surfactants, thereby limiting the temperature range where these surfactants can be used.

Polymers are usually used in alkaline flooding to reduce the mobility of the aqueous phase, especially with viscous oils (39). Three different schemes have been proposed to use polymers in alkaline flooding, including injection of a polymer slug after the alkali slug, A/P; co-injection of polymer with the alkali slug followed by a chase brine, AP; and co-injection of polymer with the alkali slug followed by another polymer slug having a lower polymer concentration, AP/P (graded viscosity scheme). Table 1 lists some of the proposed schemes to mobilize residual oil using alkaline flooding together with the total acid number of the oil, the ratio of the slug to oil viscosity, $\mu_s/\mu_o$, and the geometry of the core examined.

A significant increase in oil recovery was observed when a polymer was co-injected with alkali, A (40–46) or alkali/surfactant slugs, AS (30,31, 47–49).

Alkali consumption in sandstone reservoirs occurs due to silica dissolution, clay transformation, precipitation in hard brines, reaction with organic acids present in the crude oil, and ion-exchange reactions with clay minerals (50–52). As a result of the alkali loss, the concentration of the

**Table 1** Various Proposed Schemes for Alkaline Flooding

| Reference | Oil TAN (mg KOH/g oil) | $\mu_s/\mu_o$ | Flood scheme | Core geometry |
|---|---|---|---|---|
| Sloat and Zlomke (40) | ≤0.05 | ≈1 | (AP)/P | Radial |
| Ball and Pitts (41,42) | 2.2 | ≈1 | (AP)/P | Radial |
| Nelson et al. (28) | 1.0 | ≈1 | (AS)/P | Linear |
| Krumrine et al. (32) | 2.8 | 0.44 | (AP) | Linear |
| Martin et al. (34) | 0.05 | ≈1 | (AS)/P | Linear |
| Mihcakan and van Kirk (43) | 1.75 | ≈1 | (AP)/P | Linear and radial |
| Burk (44) | 2.8 | 0.75 | (AP) | Linear |
| Potts and Kuehne (45) | 1.5 | 1.25 | (AP) | Linear |
| Shuler et al. (30) | 0.5 & 1.4 | ≈1 | MP/(AP) | Linear |
| Clark et al. (49) | NR | NR | (ASP)/P | Radial |
| Lin et al. (31) | NR | ≈1 | (ASP)/P | Radial |
| Manji and Stasiuk (48) | 0.45 | ≈0.34 | (AP) | Radial |

*Note:* NR = not reported; TAN = total acid number; A = alkali; AP = alkali + polymer; AS = alkali + surfactant; ASP = alkali + surfactant + polymer; P = polymer; S = surfactant; M = micellar.

hydroxide ion significantly decreases and the efficiency of alkaline flooding diminishes (53).

One way to compensate for alkali loss is to increase its concentration in the slug (43,45). However, using a strong alkali at high concentrations may have an adverse effect on the interfacial tension between the oil and the chemical slug. Also, it enhances rock dissolution (54) and may cause scale formation in producing wells (55). This problem can be resolved by using buffered alkalis, e.g., sodium carbonate or a mixture of sodium bicarbonate/sodium carbonate.

## F. Phase Behavior

The use of phase behavior diagrams in surfactant-enhanced alkaline flooding is more complex than in micellar-polymer flooding, for several reasons. One is that phase behavior is very sensitive to the oil to water ratio employed. From surfactant mixing rules (29), varying the amount of oil will affect the amount of petroleum soaps present, and the nature of the mixed surfactant will change. Another complication is that stable emulsions with high values of interfacial tension are much more likely to occur with the heavier oils used in the process, which can lead to improper evaluation of phase behavior. The third problem is that total surfactant concentration is much lower than is seen in micellar-polymer flooding. Sometimes, interfacial tension may be very low, and phase behavior can be in the type III environment, but a middle phase may not be readily apparent because of the low surfactant concentration.

Figures 6 and 7 depict phase behavior diagrams for David Lloydminster crude oil (Alberta, Canada) and the surfactant Neodol 25-3S, an ethoxylated alcohol sulfate, in the presence of 1 wt % sodium carbonate. The properties of this oil are listed in Table 2. Phase behavior measurements were carried out according to the method of Nelson et al. (28). The region of optimal phase behavior is shown at a surfactant concentration of 0.1 wt % in Figure 6. The region of optimal phase behavior is shaded. Above this region, type II(+) behavior occurs, while type II(−) behavior occurs below. Volume percent oil refers to the amount of oil present in the phase behavior tube used. For a given oil to water ratio, a transition from type II(−) to type III to type II(+) occurs as salinity is increased. As the amount of oil increases relative to the amount of aqueous phase, the same trend in phase behavior is seen.

Figure 7 displays the same system, but with a lower synthetic surfactant concentration. The type III phase behavior region is shifted to lower salinities and lower oil to water ratios. This is a direct result of changes in the petroleum soaps/synthetic surfactant ratio as oil to water ratio varies.

# Surfactants in Enhanced Oil Recovery

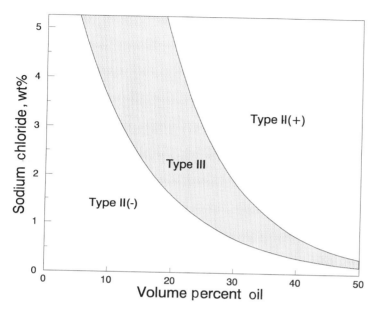

**Figure 6** Activity map with David Lloydminster crude oil, 0.1 wt % Neodol 25-3S, and 1 wt % sodium carbonate.

Nelson et al. (28) indicated that these phase behavior diagrams can be used in an equivalent fashion to the salinity requirement diagrams used for micellar-polymer flooding. Nelson (56) claims that a surfactant-enhanced alkaline flood should be designed so that the flood begins at the optimum-overoptimum phase boundary. The residual oil saturation is used to determine the oil to water ratio in the diagram. This assumes that equilibration is rapid, and does not address the possibility of petroleum soaps being extracted and concentrated in the flood front. A salinity gradient is applied when the alkaline agent is removed from the drive fluid. Nelson's results have been very promising in published laboratory core flood experiments.

## G. Dynamic Interfacial Tension

Crude oil–alkali systems are unusual in that they exhibit dynamic interfacial tension. A solution of 0.05 wt % sodium hydroxide in contact with David Lloydminster crude oil initially produces ultralow values of IFT (Figure 8). A minimum value is reached, after which IFT increases with time by nearly three orders of magnitude, when measured in the spinning drop tensiome-

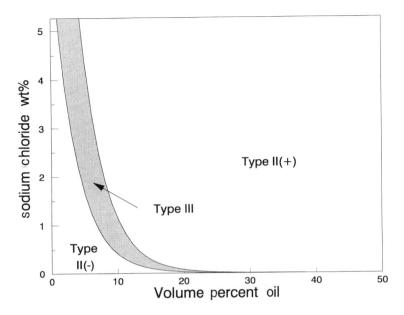

**Figure 7** Activity map with David Lloydminster crude oil, 0.02 wt % Neodol 25-3S, and 1 wt % sodium carbonate.

ter. This minimum is significantly less than that observed between David oil and David softened brine (28.5 mN/m at 32°C). The composition and properties of this brine are listed in Table 3. Taylor and colleagues (57) showed that dynamic interfacial tension can also occur in crude-oil/alkali/surfactant systems. Figure 8 displays interfacial tension versus time for a

**Table 2** Properties of Various Samples of David Crude Oil, Lloydminster "A" Pool (dead oil)

| Property | Value |
| --- | --- |
| Viscosity | 144 mPa·s at 23°C |
|  | 75.9–89.8 mPa·s at 32°C[a] |
| Density | 922 kg/m$^3$ at 23°C |
|  | 915.5–919.5 kg/m$^3$ at 32°C[a] |
| Total acid number | 0.35–0.45 mg KOH/g oil |
| Water content | 0.3 wt % |

[a]Reservoir temperature.

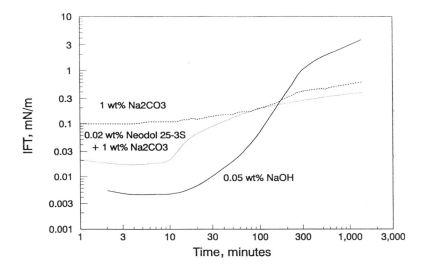

**Figure 8** Interfacial tension versus time, David Lloydminster crude oil.

**Table 3** David Synthetic Softened Brine Properties and Composition

| Property | Value |
|---|---|
| Viscosity | 0.78 mPa·s at 32°C |
| Density | 1000 kg/m$^3$ at 32°C |
| pH | 8.4 |
| Conductivity | 11.3 mS/cm at 23°C |
| Composition | |
|   Ion concentration (mg/l) | |
|     $Cl^-$ | 3464 |
|     $Na^+$ | 2447 |
|     $HCO_3^-$ | 368 |
|     $SO_4^{2-}$ | 56 |
|     $K^+$ | 4 |
|   Total dissolved solids | 6339 |

solution containing 1 wt % sodium carbonate, and 0.02 wt % of Neodol 25-3S. The addition of the synthetic surfactant greatly reduces the minimum IFT value obtained.

Dynamic IFT arises from the reaction of acidic components in the crude oil with sodium hydroxide in the aqueous phase to form petroleum soaps. This reaction is assumed to occur rapidly at the interface, but desorption of these active species from the oil–brine interface is taken to be slower. This leads to a maximum in the concentration of surface active species at the interface at some point in time where a minimum interfacial tension occurs. Subsequently, IFT increases as equilibrium is approached (58).

Nasr-El-Din and Taylor (59,60) and Taylor and Nasr-El-Din (61) have extensively studied various factors affecting dynamic interfacial tension in systems containing crude oil, sodium carbonate, high salinity tolerant surfactants, and water-soluble polymers. They showed that IFT of these systems depends on the concentration of alkali, synthetic surfactant, and polymer.

## H. Apparent Viscosity

It is well known that the apparent viscosity of dilute aqueous solutions of partially hydrolyzed polyacrylamide is constant at low shear rates, and at a critical shear rate, it decreases as the shear rate is increased (62,63). At a given shear rate, the apparent viscosity is also a function of polymer concentration, degree of hydrolysis, ionic strength, pH, and temperature (64). Figure 9 depicts the effect of sodium carbonate concentration on the apparent viscosity–shear rate relationship of various ASP slugs containing 0.1 wt % surfactant and 0.1 wt % polymer (Alcoflood 1175L). At a given shear rate, the apparent viscosity diminished as the alkali concentration was increased, especially at low shear rates. Also, the critical shear rate increased with sodium carbonate concentration. The critical shear rate is the shear rate at which there is a transition from Newtonian to shear-thinning behavior. Over the shear rate range tested, the apparent viscosity, $\mu$, can be predicted using the Carreau model (65) given as:

$$\mu = \frac{\mu_n}{[1 + (\tau_r \gamma)^2]^m} \tag{3}$$

where $\mu_n$ is the low-shear Newtonian viscosity, $\tau_r$ is a rotational relaxation time (inverse of the critical shear rate) and $\gamma$ is the shear rate. The exponent $m$ is related to the power-law exponent $n$ ($m = (1 - n)/2$).

The effect of sodium carbonate on the flow curves of partially hydrolyzed polyacrylamide shown in Figure 9 is mainly due to the variation of

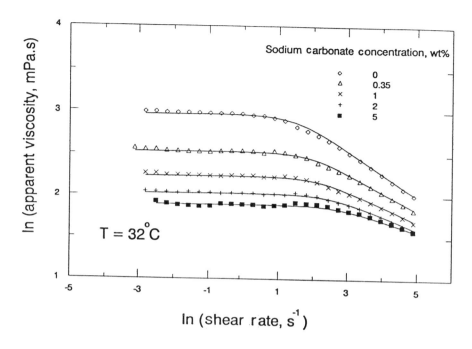

**Figure 9** Flow curves of alkali–surfactant–polymer solutions.

sodium ion concentration. At low sodium ion concentrations, a polymer chain is stretched because of the repulsive forces between the negative charges of the carboxylate groups on the chain (66). This means that the hydrodynamic radius of the polymer chain is large and, consequently, polymer solution viscosity is high. As the concentration of sodium ion is increased, the repulsive forces within the polymer chain decrease due to the charge screening effect of the counterions, and the chain coils up. This change in the polymer chain conformation reduces its hydrodynamic radius, which in turn causes the polymer solution viscosity to diminish.

It is important to note that partially hydrolyzed polyacrylamide may undergo further hydrolysis at high pH values, especially at elevated temperatures (67). As a result of polymer base hydrolysis, the viscosity of the polymer solutions will increase at low ionic strength only (64). It is worth noting that xanthan gum also undergoes base hydrolysis at high pH values. Nasr-El-Din and Noy (68) examined the effect of such hydrolysis on the apparent viscosity of two different types of xanthan gum. Base hydrolysis can cause significant drop in the apparent viscosity of xanthan solutions.

## II. COREFLOOD EXPERIMENTS

Previous corefloods using alkali-surfactant-polymer flooding were conducted in linear (69-71) and radial cores (72). Linear core floods tend to give more optimistic oil recoveries due to higher sweep efficiency (33,48), while radial core floods, due to their flow geometry, give more realistic oil recoveries, especially around injection wells. For these reasons, results obtained using both core geometries are discussed in the present review.

### A. Coreflood Assembly

The experimental setup used for the linear corefloods is shown in Figure 10. It consists of a core holder, a syringe pump, oil, brine, and chemical reservoirs. To avoid chemical and mineral changes in the Berea sandstone that result from firing (33,73,74) unfired cores were employed. Cylindrical core plugs 2.6 cm in diameter, 9 cm (short core) and 40.6 cm (long core) in length were cut from 90 cm lengths of cylindrical Berea sandstone (500 mD nominal permeability). All linear core floods were conducted at the reservoir temperature (32°C) and a confining pressure of 6900 kPa (69,70). A

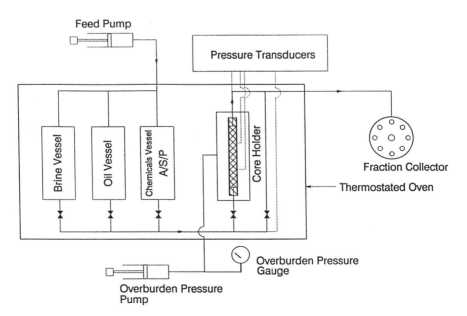

**Figure 10** Experimental setup—linear core assembly.

complete description of the experimental work conducted with the radial core geometry is given by Nasr-El-Din and Hawkins (72).

## B. Coreflood Procedure

The procedure for the radial and linear coreflood experiments included the following basic steps (69-72): Brine saturation and determination of brine permeability; oil flood and determination of the initial oil saturation, $S_{oi}$, and the effective permeability to oil at the irreducible water saturation, $k_o$ ($S_{wi}$); brine flood and measuring residual oil saturation, $S_{or}$, and the effective permeability to brine at the residual oil saturation, $k_w (1 - S_{or})$; chemical flood and determining the residual oil saturation after the chemical flood, $S_{orc}$, was determined. In some experiments with the long core setup, small ASP slugs (less than 0.5 pore volume — PV) were chased with a polymer slug, which was followed by a chase brine flood.

During each coreflood run, the pressure drop across the core was monitored. The core effluent was collected in approximately 0.2 PV increments and oil cut was determined. The aqueous phase in the effluent was analyzed for the presence of the tracer, alkali, synthetic surfactant, and polymer. Tritium (tracer) concentration was measured by liquid scintillation counting using a Beckman LS-100C system. Alkali concentration was determined using acid/base titration. Surfactant concentration in the aqueous phase was determined by the two-phase titration method (75) using Hyamine 1622 as titrant and dimidium bromide/disulphine blue indicator. Polymer concentration was measured by the starch triiodide method using flow injection analysis (76).

## III. RESULTS

## A. Effect of Alkali Concentration

Nasr-El-Din and coworkers (69) examined the effect of alkali concentration on recovery of residual oil using short linear core floods. The polymer used in these corefloods was Alcoflood 1175L. The cores had an average porosity of 0.218 (Table 4), which is in good agreement with previously published values for unfired Berea sandstone (77).

Figure 11 displays the oil cut, cumulative oil recovery, and the pressure drop across the core as a function of the cumulative core effluent for a 2 PV surfactant/polymer, SP, slug having 0.1 wt % surfactant and 0.1 wt % polymer. The oil cut was very low ($\leq 4$ vol %) during the injection of the SP slug and the chase brine. Cumulative oil recovery increased from 40%

**Table 4** Effect of Alkali Concentration—Summary of Short Linear Corefloods

| | | Sodium carbonate concentration (wt %) | | | | |
|---|---|---|---|---|---|---|
| Parameter | Units | 0.0 | 0.35 | 1.0 | 2.0 | 5.0 |
| Pore volume | cm$^3$ | 9.81 | 9.67 | 9.41 | 9.66 | 9.56 |
| Porosity | — | 0.221 | 0.218 | 0.215 | 0.220 | 0.218 |
| $S_{wi}$ | — | 0.185 | 0.199 | 0.184 | 0.196 | 0.195 |
| $S_{or}$ | — | 0.488 | 0.453 | 0.480 | 0.473 | 0.461 |
| $S_{orc}$ | — | 0.447 | 0.392 | 0.271 | 0.217 | 0.158 |
| $S_{orf}$ | — | 0.433 | 0.374 | 0.254 | 0.182 | 0.136 |
| $K_o(S_{wi})$ | mD | 589 | 564 | 546 | 593 | 508 |
| $k_w(1 - S_{or})$ | mD | 31.7 | 30.9 | 32.1 | 24.7 | 21.1 |
| $k_w(1 - S_{orf})$ | mD | 5.6 | 4.9 | 13.1 | 20.7 | 23.2 |

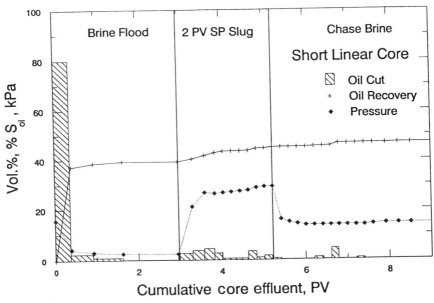

**Figure 11** Flood history for a slug having 0.1 wt % polymer, 0.1 wt % surfactant, and 0 wt % sodium carbonate—short linear core.

## Surfactants in Enhanced Oil Recovery

$S_{oi}$ at the end of the brine flood to 47% $S_{oi}$ at the end of the chase brine flood.

A sharp increase in the pressure drop across the core was observed upon the injection of the SP slug. This increase was due to the higher viscosity of the slug compared with that of the brine. The pressure drop across the core remained high throughout the SP flood, and sharply dropped once the chase brine was injected. The pressure drop across the core at the end of the chase brine flood was higher than that at the end of the brine flood. This high pressure drop was due to polymer entrapment and the unswept polymer left in the core as a result of poor mobility control during the chase brine.

Figure 12 depicts the flood history for a 2 PV ASP slug having 0.1 wt % surfactant, 0.1 wt % polymer, and 5 wt % sodium carbonate. Oil cuts up to 37 vol % were obtained during the chemical flood. Cumulative oil recovery significantly increased during the chemical flood and reached 83.1% $S_{oi}$ at the end of the chase brine flood. The pressure drop increased to a maximum of 17 kPa upon the injection of the ASP slug. Then, it gradually decreased during injection of the rest of the ASP slug and the chase brine.

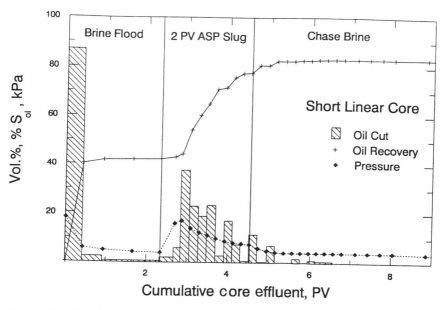

**Figure 12** Flood history for a slug having 0.1 wt % polymer, 0.1 wt % surfactant, and 5 wt % sodium carbonate – short linear core.

**Table 5** Properties of ASP Slugs as a Function of Sodium Carbonate Concentration—Radial Cores

| Parameter | Sodium carbonate concentration[a] (wt %) | | | | | | |
|---|---|---|---|---|---|---|---|
| | 0.0 | 0.17 | 0.37 | 1.0 | 2.0 | 3.9 | 5.2 |
| $\rho_s$, kg/m$^3$ | 1003.4 | 1004.7 | 1007.1 | 1013.8 | 1023.7 | 1044.1 | 1053.8 |
| $\mu_n$, mPa·s | 46.8 | 35.4 | 28.6 | 19.6 | 16.0 | 13.2 | 13.4 |
| $n$[b] | 0.68 | 0.69 | 0.71 | 0.73 | 0.77 | 0.79 | 0.79 |
| $K$[b], mPa·s$^n$ | 47.8 | 39.5 | 33.9 | 27.2 | 21.2 | 18.2 | 17.9 |
| pH | 8.1 | 10.2 | 10.4 | 10.8 | 11.0 | 11.1 | 11.1 |

[a]All solutions contain 0.1 wt % surfactant and 0.1 wt % Dow Pusher 1000E.
[b]$n$ and $K$ are the power-law parameters.

## B. Effect of Core Geometry

Nasr-El-Din and Hawkins (72) examined the effect of alkali concentration on oil recovery using radial Berea sandstone cores. The ASP slugs contained 0.1 wt % polymer, 0.1 wt % surfactant, and from 0 to 5.2 wt % sodium carbonate (Table 5). It is worth noting that this is the range of alkali concentrations that is used in most laboratory and field tests dealing with alkaline flooding (34,44). Table 6 summarizes core properties and oil saturations at the end of various floods for each experiment.

To examine the effect of alkali concentration further, the incremental oil recovery is plotted in Figure 13 as a function of alkali concentration. A steep increase in incremental oil recovery occurred by increasing sodium

**Table 6** Effect of Alkali Concentration—Summary of Radial Corefloods

| Parameter | Sodium carbonate concentration[a] (wt %) | | | | | | |
|---|---|---|---|---|---|---|---|
| | 0.0 | 0.17 | 0.37 | 1.0 | 2.0 | 3.9 | 5.2 |
| Pore volume (cm$^3$) | 11.0 | 10.9 | 10.8 | 11.0 | 11.1 | 11.3 | 11.2 |
| Porosity | 0.256 | 0.253 | 0.250 | 0.254 | 0.257 | 0.261 | 0.261 |
| $S_{wi}$ | 0.414 | 0.407 | 0.441 | 0.365 | 0.398 | 0.408 | 0.364 |
| $S_{or}$ | 0.343 | 0.34 | 0.284 | 0.367 | 0.342 | 0.334 | 0.365 |
| $S_{orc}$ | 0.3 | 0.267 | 0.178 | 0.192 | 0.145 | 0.144 | 0.152 |
| $S_{orf}$ | 0.284 | 0.261 | 0.152 | 0.163 | 0.118 | 0.095 | 0.096 |

[a]All solutions contain 0.1 wt % surfactant and 0.1 wt % Dow Pusher 1000E.

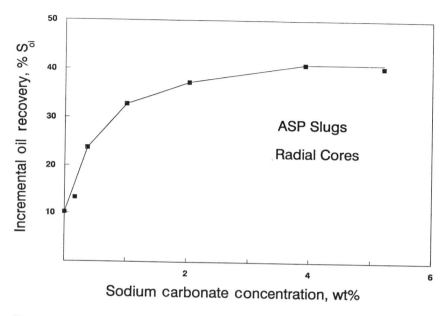

**Figure 13** Effect of sodium carbonate concentration on tertiary oil recovery using radial cores. All solutions contained 0.1 wt % polymer and 0.1 wt % surfactant.

carbonate concentration up to 1 wt %. However, incremental oil recovery was almost constant at sodium carbonate concentrations ≥2 wt %. The influence of alkali concentration on oil recovery was also examined in linear cores (69). The trend obtained was similar to that shown in Figure 13.

From oil recovery data shown in Figure 13 and the properties of the slugs given in Table 5, it can be concluded that ASP slugs at a pH of 10–11 can be effectively used to recover waterflood residual oil. This result confirms those obtained by Burk (44) and Lorenz and Peru (55) that buffered alkalis can be used to recover waterflood residual oil. Also, a significant amount of oil was recovered at $\mu_s/\mu_o$ of 0.1–0.2. This means that formulating the viscosity of an alkali slug such that its viscosity matches the oil viscosity (Table 1) overestimates the amount of polymer needed to recover residual oil and therefore is not economical.

## C. Effects of ASP Slug Size and Core Length on Oil Recovery

The results discussed so far were obtained using ASP slugs of large size (=2 PV) and employing either radial or short linear cores. In this section, the effects of ASP slug size and core length on oil recovery are examined. The ASP slugs employed in these experiments contained 0.1 wt % polymer, 0.1 wt % surfactant and 2 wt % alkali. This slug composition was chosen because it gave very low IFT with David oil, had no injectivity problems and, most importantly, effectively recovered waterflood residual oil (69). Table 7 summarizes the coreflood tests conducted to examine the effect of ASP slug size.

Figure 14 displays the effects of ASP slug size and core length on tertiary oil recovery. The oil recovery increased significantly with the slug size up to 0.5 PV. Increasing the slug size further resulted in a smaller increase in the oil recovery. These trends are similar to those theoretically predicted by Islam and Chakma (78). It is worth noting that increasing the core length up to 40.6 cm had no significant effect on the tertiary oil recovery for the system examined by Nasr-El-Din and colleagues (70). Figure 15 shows the flood history for a 0.2 PV ASP slug chased by 0.3 PV polymer slug, long core.

## D. Effect of a Chase Polymer

The results discussed so far indicate that small ASP slugs were not successful in recovering waterflood residual oil. One way to improve the efficiency of a small ASP slug is to use a chase polymer slug. The chase polymer will provide a better mobility control at the end of the ASP slug, which will maintain its integrity.

**Table 7** Effect of ASP Slug Size—Summary of Long Linear Corefloods

| Parameter | Units | Slug size (PV) | | |
|---|---|---|---|---|
| | | 0.2 | 0.5 | 2.0 |
| Pore volume | cm$^3$ | 45.43 | 47.38 | 45.31 |
| Porosity | — | 0.210 | 0.225 | 0.218 |
| $S_{wi}$ | — | 0.162 | 0.192 | 0.235 |
| $S_{or}$ | — | 0.461 | 0.423 | 0.426 |
| $S_{orc}$ | — | 0.461 | 0.381 | 0.185 |
| $S_{orf}$ | — | 0.350 | 0.258 | 0.178 |
| $K_o(S_{wi})$ | mD | 703 | 767 | 518 |
| $k_w(1 - S_{or})$ | mD | 24.3 | 32.0 | 16.6 |
| $k_w(1 - S_{orf})$ | mD | 8.0 | 18.6 | 16.3 |

*Note:* ASP slugs contained 2 wt % alkali, 0.1 wt % surfactant, and 0.1 wt % polymer.

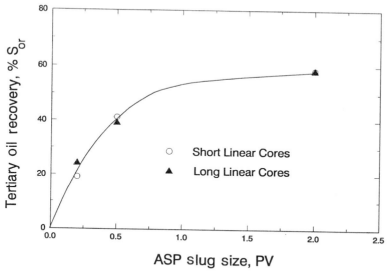

**Figure 14** Effect of slug size and core length on tertiary oil recovery.

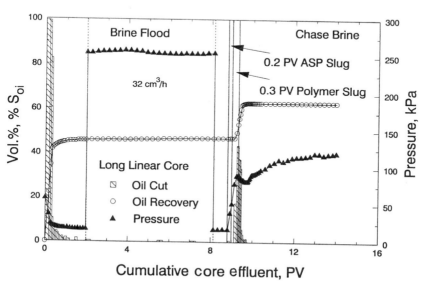

**Figure 15** Flood history for a 0.2 PV ASP slug chased by 0.3 PV polymer slug — long core.

Table 8 lists coreflood data and tertiary oil recovery obtained with small ASP slugs followed by a chase polymer. The 0.5 PV ASP slug followed by a chase brine gave a higher oil recovery than those obtained using small ASP slugs chased with a polymer slug. The concentration of the synthetic surfactant in the aqueous phase using small ASP slugs was nearly 0. As a result, the IFT of these slugs was not low enough to mobilize residual oil. This result emphasizes the importance of having low IFT during the chemical flood. It also demonstrates the importance of maintaining a high concentration of the synthetic surfactant in the aqueous phase.

### E. Effect of Alkali Concentration on the Propagation of Synthetic Surfactant

The effect of sodium carbonate concentration on the propagation of the synthetic surfactant is of interest because adsorption depends on salinity and pH (64,79). To examine the effect of alkali concentration on the propagation of the synthetic surfactant (Neodol 25-3S) in short linear cores, the normalized concentration of the synthetic surfactant, $C/C_o$, is plotted in Figure 16 as a function of the normalized cumulative core effluent. In this figure, $C$ is the concentration of the synthetic surfactant in the aqueous core effluent and $C_o$ is its concentration in the injected solution. At a sodium carbonate concentration of 0 wt %, $C/C_o$ reached 0.8 at a normalized cumulative core effluent of nearly 0.4. It then increased at a much

**Table 8** Effect of a Chase Polymer – Summary of Long Linear Corefloods

| Parameter | Units | Slug size and sequence | | | |
|---|---|---|---|---|---|
| | | A/S/P P | 0.5/0 | 0.2/0.3 | 0.1/0.4 |
| Pore volume | cm$^3$ | 47.38 | | 45.30 | 47.78 |
| Porosity | — | 0.225 | | 0.218 | 0.221 |
| $S_{wi}$ | — | 0.192 | | 0.247 | 0.231 |
| $S_{or}$ | — | 0.423 | | 0.401 | 0.398 |
| $S_{orc}$ | — | 0.381 | | 0.323 | 0.337 |
| $S_{orf}$ | — | 0.258 | | 0.281 | 0.321 |
| $K_o(S_{wi})$ | mD | 767 | | 585 | 458 |
| $k_w(1 - S_{or})$ | mD | 32.0 | | 21.4 | 7.1 |
| $k_w(1 - S_{orf})$ | mD | 18.6 | | 2.92 | 1.34 |
| Tertiary oil recovery | % $S_{or}$ | 39.0 | | 29.9 | 19.4 |

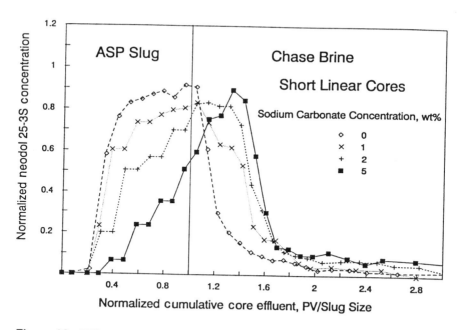

**Figure 16** Effect of sodium carbonate concentration on surfactant propagation—short linear core.

slower rate and reached its maximum value of 0.9 at the end of the ASP slug. $C/C_o$ significantly decreased upon injection of the chase brine. As the alkali concentration in the ASP slug was increased, a delay in the surfactant breakthrough and a slower rate of surfactant propagation during the chemical flood were observed. The influence of alkali concentration on the propagation of the synthetic surfactant in short linear cores was similar to that obtained using radial cores (72).

## F. Effect of Slug Size on Chemical Propagation

Figure 17 shows the propagation of the polymer, synthetic surfactant, alkali, and tracer of a 0.2 PV ASP slug using the short linear core. The breakthrough of the chemical species occurred during the chase brine flood. The normalized tracer concentration reached a maximum of only 0.3. The sodium carbonate profile was lower than that of the tracer. The loss of sodium carbonate was due to the reaction with the rock matrix, the organic acids present in the oil, and the polymer (partially hydrolyzed polyacrylamide). The normalized polymer concentration was also lower than that of

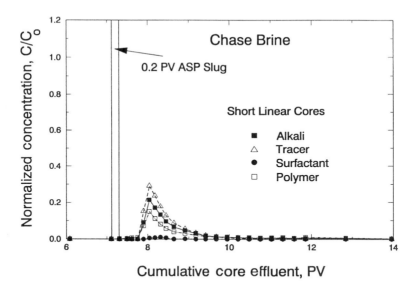

**Figure 17** Chemical propagation for a 0.2 PV ASP slug—short linear core.

the tracer. The loss of the polymer was due to adsorption and mechanical trapping. The normalized concentration of the synthetic surfactant was significantly lower than the tracer profile. This difference was due to surfactant adsorption and partitioning into the oil phase.

Unlike the results obtained with the 0.2 PV ASP slug, the maximum normalized concentration of the polymer was higher than the tracer for the 2 PV ASP slug in a short linear core, as shown in Figure 18. This result was due to the inaccessible pore volume, which reduced the aqueous pore volume available for the polymer. The maximum normalized concentration of all chemical species increased for the 2 PV ASP slug, especially the normalized surfactant concentration, which reached its maximum value during the chase brine injection.

## G. Effect of Core Length on Chemical Propagation

Figure 19 depicts chemical propagation in the long core for a 0.2 PV ASP slug. The features of chemical propagation in the long cores were parallel to those observed in the short cores, including the following: the breakthrough of the chemicals occurred during the chase brine flood, the normalized polymer concentration was lower than the tracer, and the normalized

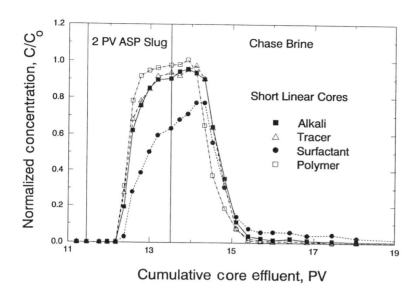

**Figure 18** Chemical propagation for 2 PV ASP slug — short linear core.

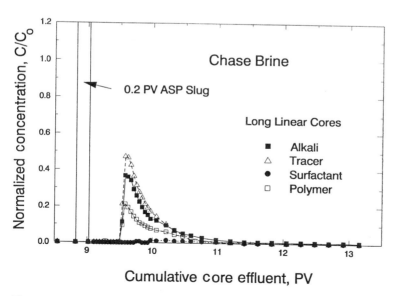

**Figure 19** Chemical propagation for a 0.2 PV ASP slug — long linear core.

surfactant concentration was significantly less than the tracer. A close examination of the results shown in Figures 17 and 19 indicate that the normalized concentrations of all chemical species are less dispersed for the long core floods. These results can be explained in terms of the core Péclet number, Pe, defined as:

$$\text{Pe} = \frac{VL}{D} \tag{4}$$

where $V$ is the velocity, $L$ is the core length, and $D$ is the dispersion coefficient. The velocity and the dispersion coefficient were the same for the short and long cores and, as a result, the core Péclet number was higher by a factor of 4.5 for the long core. According to Bidner and Vampa (80), increasing the core Péclet number will minimize the dispersion of various chemical species in porous media. Accordingly, the response of these species in the core effluent will be sharper for the case of the long core.

## H. Mechanisms of Oil Recovery

For ASP slugs, oil recovery was found to be a function of alkali concentration and slug size. For slugs having low sodium carbonate concentrations ($\leq 0.5$ wt %), oil recovery was relatively low. The IFT of these slugs was not low enough to mobilize the trapped oil. Although apparent viscosity of these slugs was high, oil recovery was low. This result indicated that slug viscosity was not a determining factor in oil recovery if the IFT was high. Phase behavior studies (69) showed that at low sodium carbonate concentrations ($\leq 0.35$ wt %) and an oil volume fraction of 0.5, a lower-phase emulsion is formed. Examining salinity changes during ASP floods of low alkali concentrations, it can be seen that a lower-phase emulsion is the only possible type of emulsion that can be formed. Previous studies have shown that the conditions that correspond to this type of emulsion do not recover significant amounts of oil (56,81,82).

For slugs having sodium carbonate concentrations $\geq 1$ wt %, oil recovery was high, which can be explained by one or more of the following mechanisms: low IFT, salinity changes, and wettability reversal. At sodium carbonate $\geq 1$ wt %, the IFT was two orders of magnitude lower than that observed at 0 wt % sodium carbonate concentration (Tables 9 and 10). This means that the capillary forces that trap the oil diminish and consequently, more oil can be recovered at high sodium carbonate concentrations (1 to 5.2 wt %).

Salinity changes could also play an important role in determining tertiary oil recovery obtained at high sodium carbonate concentrations.

# Surfactants in Enhanced Oil Recovery

**Table 9** Variation of Oil Recovery with the Capillary Number — Radial Cores

| Parameter | Sodium carbonate concentration (wt %) | | | | | | |
|---|---|---|---|---|---|---|---|
| | 0.0 | 0.17 | 0.37 | 1.0 | 2.0 | 3.9 | 5.2 |
| $\mu^a$ (mPa·s) | 28.4 | 24.1 | 21.4 | 17.5 | 14.5 | 12.8 | 12.8 |
| IFT$^b$ (mN/m) | 2.50 | 0.649 | 0.153 | 0.038 | 0.031 | 0.013 | 0.021 |
| $N_c \times 10^5$ | 2.89 | 9.47 | 35.47 | 117.6 | 119.4 | 250.5 | 154.9 |
| Tertiary recovery$^c$ (%) | 17.2 | 23.3 | 46.5 | 55.6 | 65.5 | 71.6 | 72.7 |

$^a$Measured at a shear rate of 5 s$^{-1}$ and 22°C.
$^b$Minimum IFT.
$^c$Tertiary oil recovery = $(1 - S_{orf}/S_{or}) \cdot 100$.

Examining the flood history of the slug having 5 wt % sodium carbonate, the salinity gradients at the front and the rear of the slug are very high. At both ends, significant dilution of the slug occurs and, as a result, the salinity of the slug approaches the optimum salinity values during most of the flood (69). It is well known that oil recovery under these conditions is very high (81).

**Table 10** Effect of Sodium Carbonate Concentration on Oil Recovery — Short Linear Cores

| Parameter | Sodium carbonate concentration (wt %) | | | | |
|---|---|---|---|---|---|
| | 0.0 | 0.35 | 1.0 | 2.0 | 5.0 |
| $\mu^a$ (mPa·s) | 18.5 | 12.4 | 9.2 | 7.5 | 6.6 |
| IFT$^b$ (mN/m) | 3.36 | 0.2 | 0.026 | 0.023 | 0.026 |
| $N_c \times 10^5$ | 0.624 | 7.0 | 39.7 | 36.6 | 29.1 |
| Tertiary recovery$^c$ (%) | 11.3 | 17.6 | 47.1 | 61.5 | 70.6 |

$^a$Measured at a shear rate of 5 s$^{-1}$ and 22°C.
$^b$Minimum IFT.
$^c$Tertiary oil recovery = $(1 - S_{orf}/S_{or}) \cdot 100$.

Wettability reversal is another mechanism to recover residual oil at high sodium carbonate concentrations (i.e., high ionic strength). High salinity (high ionic strength) drives the anionic surfactant (Neodol 25-3S) out of the aqueous phase (83). Therefore, a significant amount of the synthetic surfactant is adsorbed on the rock surface during the ASP flood. Hence, the wettability of the core is changed from water-wet to oil-wet. The trapped oil becomes mobile and more oil is recovered (84).

The tertiary oil recovery significantly increased with the ASP slug size up to 0.5 PV (Figure 18). The low oil recovery obtained with the small ASP slugs was due to the low normalized concentrations of the synthetic surfactant, alkali, and polymer in the core. These low values were obtained due to the adsorption, dispersion, and dilution of these chemical species during the ASP flood. The lack of mobility control at the end of the slug will produce viscous fingering, which also causes dilution and dissipation of the slug. Core flood experiments conducted with the long core gave further evidence that dilution was the dominant factor. When an ASP slug with a small size was followed with a chase polymer having a higher viscosity, oil recovery improved slightly. The viscosity during the chase polymer injection was relatively high; however, oil recovery was low. Obviously, the consumption and dilution of all chemical species, especially the synthetic surfactant, which plays a vital role in attaining low interfacial tension, are very important. In other words, maintaining good mobility control with small ASP slugs does not significantly enhance the efficiency of these slugs.

To assess the effect of sodium carbonate and core geometry on oil recovery, it is important to follow changes in the capillary number ($N_c$) during each flood.

Tables 9 and 10 summarize the apparent viscosity at a shear rate of 5 $s^{-1}$, minimum IFT, and the capillary number for corefloods conducted in radial and short linear cores. The capillary number at the end of the initial brine flood was very low, nearly $10^{-7}$. For slugs having ≤1 wt % sodium carbonate, oil recovery and capillary number was relatively low. This means that the capillary forces are dominant and oil droplets are trapped in the core. A significant increase in oil recovery occurred for slugs having high sodium carbonate concentrations, with capillary numbers greater than $10^{-3}$ (radial core) and $10^{-4}$ (short linear cores). These results indicate that it is necessary to increase the capillary number by several orders of magnitude to successfully recover waterflood residual oil in the system examined in the present study. The effect of capillary number on oil recovery was similar to that observed by other researchers using micellar flooding (2,85). It is worth noting that the variation of oil recovery with the capillary number depended on the core geometry.

## I. Field Application

Several laboratory studies have been reported that examine the application of the ASP flooding process to various reservoir systems (34,36,37,41,71, 86–90). Field trials of this technology have been recently reported (91–95).

One ASP project that has been economically successful and carefully evaluated is the West Kiehl field in Wyoming (49,93,94). The West Kiehl field is a medium permeability (350 mD) sandstone reservoir. The formation brine contains 45,500 ppm total dissolved solids, with about 450 ppm divalent ions. The 24° API crude oil has viscosity of 19 mPa·s at a reservoir temperature of 49°C. The chemical slug used in the project was prepared in a fresh, relatively soft brine (800 ppm TDS, 18 ppm hardness). Based on interfacial tension measurements (phase behavior tests were not reported) a solution of 0.8 wt % $Na_2CO_3$ and 0.1 wt % Petrostep B-100, a petroleum sulfonate, was selected. To this was added 0.1 wt % of a polyacrylamide polymer to provide the necessary mobility control. The project design called for injection of 0.25 pore volume of the ASP slug followed by a similar volume of a chase polymer acting as a mobility control buffer.

Results in the West Kiehl field showed that an additional 20% of original-oil-in-place was recovered above the primary plus waterflood oil recovery estimate of 39.9% OOIP. A total of 29,380 $m^3$ of incremental oil is projected for the project at an incremental cost of $13.40 per $m^3$ ($2.13 per barrel).

A field test of the ASP process in the White Castle field, Louisiana, recovered 38% of the waterflood residual oil in the reservoir (92). In this field test a chemical slug was injected that contained 2.53 wt % sodium carbonate, 0.44 wt % internal olefin sulfonate (IOS), 0.4 wt % sodium chloride, 0.06 wt % Neodol 25-12 nonionic surfactant, and a sodium thiocyanate tracer. Project economics were not discussed. Problems to be addressed include injector scale formation and treatment of produced emulsions.

In Daqing oil field, P.R. China, a successful ASP pilot was started in 1994 (95). Numerical simulation predicts that oil recovery will be increased by 18.1% OOIP. This pilot used a sodium carbonate/petroleum sulfonate/polyacrylamide slug injected for 0.3 PV followed by a 0.28 PV polymer buffer solution.

## IV. CONCLUDING REMARKS

Crude oil becomes trapped in porous media because of capillary forces. The reduction of these forces by surfactants is an important aspect of enhanced oil recovery. In practice, capillary forces are reduced primarily by lowering interfacial tension between oil and water phases, although increasing the viscosity of the water is also important. Lowering interfacial tension with surfactants leads to the formation of emulsions and microemulsions, which are also important in enhanced oil recovery techniques.

Micellar-polymer flooding and alkali-surfactant-polymer flooding both rely on the injection into a crude oil reservoir of surfactants or surfactant-forming materials. Emulsions may be injected into the reservoir, or they may be formed in the reservoir, but their properties will change as they travel through the reservoir to eventually flow from a producing well after weeks or months.

Micellar-polymer flooding is a technically well-developed process. Phase compositional aspects of microemulsion design are relatively well understood, and several technically successful field trials have been carried out. Micellar-polymer floods can be designed and carried out with a good chance of success. However, the process is too expensive. This is due primarily to the high concentrations of synthetic surfactants required. The problem is further compounded because the cost of synthetic surfactants is directly related to the price of crude oil.

Alkali-surfactant-polymer flooding has promise to become economically more attractive than micellar-polymer flooding. The process is inherently less expensive because the concentration of synthetic surfactant is significantly lower. However, the ASP process is more complex and technically less developed than micellar-polymer flooding. In addition, the disappointing history of field trials of classical alkaline flooding has left many researchers skeptical of the process in general. But with the use of high optimal salinity surfactants to lower interfacial tension at realistic reservoir salinities, the use of buffered alkali to reduce alkali–rock interactions, and the addition of water-soluble polymers to the system to increase displacing phase viscosity, many of the problems associated with classical alkaline flooding have been addressed. Areas that require further investigation include the effect of dynamic interfacial tension in the reservoir, conciliation of interfacial tension and phase behavior measurements, computer simulation of the ASP process, and the separation/handling of produced emulsions.

## APPENDIX: NOMENCLATURE

A     alkaline flood
$A_S$     cross-sectional area, m$^2$
AS     alkali-surfactant flood

# Surfactants in Enhanced Oil Recovery

| | |
|---|---|
| ASP | alkali-surfactant-polymer flood |
| $C$ | chemical concentration in the effluent aqueous phase, wt % |
| $C_o$ | chemical concentration in the injected slug, wt % |
| $D$ | dispersion coefficient, m$^2$/s |
| IFT | dynamic interfacial tension, mN/m |
| $k_o$ | effective permeability to oil, mD |
| $k_w$ | effective permeability to brine, mD |
| $L$ | core length, m |
| $m$ | exponent in Equation 3 |
| MP | micellar-polymer flood |
| $n$ | power-law exponent |
| $N_c$ | capillary number, defined in Equation 1 |
| $p$ | pressure, Pa |
| Pe | core Péclet number, defined in Equation 4 |
| $Q$ | flow rate, m$^3$/s |
| $S_{oi}$ | initial oil saturation, volume fraction |
| $S_{or}$ | residual oil saturation at the end of water flood, volume fraction |
| $S_{orc}$ | residual oil saturation at the end of chemical flood, volume fraction |
| $S_{orf}$ | residual oil saturation at the end of experiment, volume fraction |
| $S_{wi}$ | irreducible water saturation, volume fraction |
| SP | surfactant-polymer flood |
| TAN | total acid number, mg KOH/g oil |
| $V$ | frontal advance velocity $(Q/(A_s \epsilon))$, m/s |
| $V$ | Darcy velocity, m/s |
| $V_o$ | oil volume in microemulsion phase, m$^3$ |
| $V_s$ | surfactant volume in microemulsion phase, m$^3$ |
| $V_w$ | water volume in microemulsion phase, m$^3$ |

Greek letters

| | |
|---|---|
| $\gamma$ | shear rate, s$^{-1}$ |
| $\epsilon$ | porosity of the porous medium, volume fraction |
| $\mu$ | apparent viscosity, mPa·s |
| $\mu_n$ | low-shear Newtonian viscosity, mPa·s |
| $\mu_s$ | slug viscosity, mPa·s |
| $\rho_s$ | slug density, kg/m$^3$ |
| $\eta$ | interfacial tension, mN/m |
| $\eta_{mo}$ | interfacial tension between the microemulsion and oil phases, mN/m |
| $\eta_{mw}$ | interfacial tension between the microemulsion and water phases, mN/m |
| $\tau_r$ | rotational relaxation time (inverse of the critical shear rate), s |

## REFERENCES

1. LW Lake, GA Pope. Pet Eng Int 51(13):38–60, 1979.
2. JJ Taber. In: Surface Phenomena in Enhanced Oil Recovery. Shah, DO, ed. New York, Plenum Press, 1981, pp 13–52.
3. I Chatzis, NR Morrow. Soc Pet Eng J 24(5):555–562, 1984.
4. PB Lorenz, JC Trantham, DR Zornes, CG Dodd. In: SPE/DOE Fourth Symposium on Enhanced Oil Recovery. SPE/DOE #12695, Tulsa, OK, April 15–18, 1984.
5. RC Whiteley, JW Ware. J Pet Tech 925–932, August 1977.
6. JR Fanchi, HB Carroll. Soc Pet Eng Res Eng 3(2):609–616, 1988.
7. KT Raterman. Soc Pet Eng Res Eng 5(4):459–466, 1990.
8. K Shinoda, B Lindman. Langmuir 3(2):134–149, 1987.
9. R Aveyard, B Vincent. Prog Surf Sci 8(2-A):59–102, 1977.
10. RC Nelson. In: Surface Phenomena in Enhanced Oil Recovery. Shah, DO, ed. New York, Plenum Press, 1981, pp 73–104.
11. RC Nelson, GA Pope. Soc Pet Eng J 18(5):325–38, 1978.
12. RC Nelson. Chem Eng Prog 50–57, March 1989.
13. RN Healy, RL Reed, DG Stenmark. Soc Pet Eng J 16(3):147–160, 1976.
14. PA Winsor. Solvent Properties of Amphiphillic Compounds. London, Butterworth's Scientific Publication, 1954.
15. A Graciaa, LN Fortney, RS Schechter, WH Wade, S Yiv. In: Second Joint SPE/DOE Symposium on Enhanced Oil Recovery. Paper SPE 9815, Tulsa, OK, 1981.
16. RN Healy, RL Reed. Soc Pet Eng J 14(5):491–501, 1974.
17. CJ Huh. Colloid Interface Sci 71(2):408–426, 1979.
18. GR Glinsmann. In: 54th Annual Fall Technical Conference and Exhibition of the Society of Petroleum Engineers of AIME. Paper SPE 8326, Las Vegas, NV, September 23–26, 1979.
19. JM Maerker, WW Gale. In: SPE/DOE Seventh Symposium on Enhanced Oil Recovery, SPE/DOE #20218, Tulsa, OK, April 22–25, 1990.
20. TR Reppert, JR Bragg, JR Wilkinson, TM Snow, NK Maer, Jr., WW Gale. In: SPE/DOE Seventh Symposium on Enhanced Oil Recovery. SPE/DOE #20219, Tulsa, OK, April 22–25, 1990.
21. WW Gale, MC Puerto, TL Ashcraft, RK Saunders, RL Reed. U.S. Patent No. 4,293,428 (October 6, 1981).
22. CE Johnson, Jr. J Pet Technol 14:85–92, 1976.
23. TP Castor, WH Somerton, JF Kelly. In: Surface Phenomena in Enhanced Oil Recovery. Shah, DO, ed. New York, Plenum Press, 1981, pp 249–291.
24. EH Mayer, RL Berg, JD Carmichael, RM Weinbrandt. J Pet Tech 35(2):209–221, 1983.
25. VA Kuuskraa. In: SPE/DOE Fifth Symposium on Enhanced Oil Recovery. SPE/DOE #14951, Tulsa, OK, April 20–23, 1986.
26. DL Dauben, RA Easterly, MM Western. U.S. Department of Energy, Report DOE/BC/10830-5, 1987.

27. RF Burdyn, HL Chang, EL Cook. U.S. Patent No. 4,004,638 (January 25, 1975).
28. RC Nelson, JB Lawson, DR Thigpen, GL Stegemeier. Presented at the SPE/DOE Fourth Symposium on Enhanced Oil Recovery. SPE/DOE 12672, Tulsa, OK, April 15-18, 1984.
29. KC Taylor. In Situ 16(3):229-250, 1992.
30. PJ Shuler, DL Kuehne, RM Lerner. J Pet Tech 41(1):80-88, 1986.
31. FFJ Lin, GJ Besserer, MJ Pitts. J Can Pet Technol 26:54-65, 1987.
32. PH Krumrine, JS Falcone, Jr., TC Campbell. Soc Pet Eng J 22:503-513, 1982a.
33. PH Krumrine, JS Falcone, Jr., TC Campbell. Soc Pet Eng J 22:983-992, 1982b.
34. FD Martin, JC Oxley, H Lim. Presented at the 60th Annual Tech. Conf. and Exhibition of the SPE. SPE 14293, Las Vegas, NV, September 22-25, 1985.
35. J Ball, H Surkalo. Am Oil Gas Rep 31:46-48, 1988.
36. SM Saleem, MJ Faber. Rev Tec Intevep 6:133-142, 1986.
37. SM Saleem, A Hernandez. J Surf Sci Tech 3:1-10, 1987.
38. A Al-Ghamdi, HA Nasr-El-Din. Colloids Surfaces, 125:5-18, 1997.
39. CJ Radke. DOE Contract No. 7405-ENG-48, 1982.
40. B Sloat, D Zlomke. Presented at the SPE/DOE Third Symposium on Enhanced Oil Recovery. SPE 10719, Tulsa, OK, April 4-7, 1982.
41. JT Ball, MJ Pitts. Presented at the International Symposium on Oilfield and Geothermal Chemistry. SPE 11790, Denver, CO, June 1-3, 1983.
42. JT Ball, MJ Pitts. Presented at the SPE/DOE Fourth Symposium on Enhanced Oil Recovery. SPE/DOE 12650, Tulsa, OK, April 15-18, 1984.
43. IM Mihcakan, CW van Kirk. Presented at the SPE Rocky Mountain Regional Meeting. SPE 15158, Billings, MT, May 19-21, 1986.
44. JH Burk. Soc Pet Eng Res Eng 2:9-16, 1987.
45. DE Potts, DL Kuehne. Soc Pet Eng Res Eng 3:1143-1152, 1988.
46. MW Alam, D Tiab. Energy Resources 10:1-19, 1988.
47. PH Krumrine, JS Falcone, Jr. Presented at the International Symposium on Oilfield and Geothermal Chemistry. SPE 11778, Denver, CO, June 1-3, 1983.
48. KH Manji, BW Stasiuk. J Can Pet Tech 27(3):49-54, 1988.
49. SR Clark, MJ Pitts, SM Smith. Presented at the SPE Rocky Mountain Regional Meeting. SPE 17538, Casper, WY, May 11-13, 1988.
50. J Labrid. J Can Pet Technol 3:67-74, 1991.
51. B Bazin, J Labrid. Soc Pet Eng Res Eng 6:233-238, 1991.
52. J Labrid, JP Duquerroix. Revue de l'Institut Français du Pétrole 46:199-219, 1991.
53. PH Krumrine, JS Falcone, Jr. Soc Pet Eng Res Eng 3:62-68, 1988.
54. SD Thornton, CJ Radke. Soc Pet Eng Res Eng 3:743-52, 1988.
55. PB Lorenz, DA Peru. Oil Gas J 87:53-57, 1989.
56. RC Nelson. Soc Pet Eng J 22:259-270, 1982.
57. KC Taylor, BF Hawkins, MR Islam. J Can Pet Tech 29(1):50-55, 1990.
58. E Rubin, CJ Radke. Chem Eng Sci 35:1129-38, 1980.

59. HA Nasr-El-Din, KC Taylor. Colloids Surfaces 66:23-37, 1992.
60. HA Nasr-El-Din, KC Taylor. Colloids Surfaces A: Physicochemical and Engineering Aspects 75:169-183, 1993.
61. KC Taylor, HA Nasr-El-Din. Colloids Surfaces A: Physicochemical and Engineering Aspects 108(1):49-72, 1996.
62. JS Ward, FD Martin. Soc Pet Eng J 21:623-631, 1981.
63. A Ait-Kadi, PJ Carreau, G Chauveteau. J Rheology 31:537-561, 1987.
64. H Nasr-El-Din, BF Hawkins, KA Green. Presented at the International Symposium on Oilfield and Geothermal Chemistry. SPE 21028, Anaheim, CA, Feb. 20-22, 1991.
65. PG Carreau. Trans Soc Rheol 16:99-127, 1972.
66. KC Tam, C Tiu. J Rheology 33:257-280, 1989.
67. RG Ryles. Soc Pet Eng Res Eng 3:23-34, 1988.
68. HA Nasr-El-Din, J Noy. Revue de l'Institut Français du Pétrole 47:771-791, 1992.
69. H Nasr-El-Din, BF Hawkins, KA Green. J Pet Sci Eng 6:381-401, 1992.
70. HA Nasr-El-Din, KA Green, LL Schramm. Revue de l'Institut Français du Pétrole 49:359-377, 1994.
71. BF Hawkins, KC Taylor, HA Nasr-El-Din. J Can Pet Tech 33:52-63, 1994.
72. HA Nasr-El-Din, BF Hawkins. Revue de l'Institut Français du Pétrole 46:199-219, 1991.
73. JC Shaw, PL Churcher, BF Hawkins. Presented at the International Symposium on Oilfield and Geothermal Chemistry. SPE 18463, Houston, TX, February 8-10, 1989.
74. S Ma, NR Morrow. Presented at the International Symposium on Oilfield and Geothermal Chemistry. SPE 21045, Anaheim, CA, Feb. 20-22, 1991.
75. VW Reid, GF Longman, E Heinerth. Tenside 4:292-304, 1967.
76. KC Taylor. Presented at the International Symposium on Oilfield and Geothermal Chemistry. SPE 21007, Anaheim, CA, Feb. 20-22, 1991.
77. JA Jensen, CJ Radke. Soc Pet Eng Res Eng 3:849-856, 1988.
78. MR Islam, A Chakma. J Pet Sci Eng 5:105-126, 1991.
79. W Kwok, RE Hayes, HA Nasr-El-Din. Chem Eng Sci 50:769-783, 1995.
80. MS Bidner, VC Vampa. J Pet Sci Eng 3:267-281, 1989.
81. RC Nelson. Chem Eng Prog 85:50-57, 1989.
82. GJ Hirasaki, HR van Domselaar, RC Nelson. Soc Pet Eng J 23:486-500, 1983.
83. W Kwok, HA Nasr-El-Din, RE Hayes. J Can Pet Tech 32:39-48, 1993.
84. CE Cooke, Jr., RE Williams, PA Kolodzie. J Pet Technol 12:1365-74, 1974.
85. K Patel, M Greaves. Can J Chem Eng 65:676-678, 1987.
86. B Bazin, CZ Yang, DC Wang, XY Su. In: SPE International Meeting on Petroleum Engineering. SPE 22363, Beijing, China, March 24-27, 1992.
87. J Rudin, DT Wasan. SPE Res Eng 8(4):275-280, 1993.
88. G Shutang, L Huabin, L Hongfu. SPE Res Eng 194-197, August 1995.
89. M Bavière, P Glénat, V Plazanet, J Labrid. SPE Res Eng 187-193, August 1995.
90. PB Lorenz. In: Research Needs to Maximize Economic Producibility of the

Domestic Oil Resource. Tham, MK, Burchfield, T, Chung, T, Lorez, P, Bryant, R, Sarathi, P, Chang, MM, Jackson, S, Tomutsa, L, Dauben, DL, eds. National Institute of Petroleum and Energy Research-527, October 1991.
91. TR French, CB Josephson. Surfactant-Enhanced Alkaline Flooding Field Project, Annual Report. National Institute of Petroleum and Energy Research-714, December 1993.
92. AH Falls, DR Thigpen, RC Nelson, JW Ciaston, JB Lawson, PA Good, RC Ueber, GT Shahin. SPE Res Eng 217-223, August 1994.
93. JJ Meyers, MJ Pitts, K Wyatt. In: SPE/DOE Eighth Symposium on Enhanced Oil Recovery. Paper #24144, Tulsa, OK, April 22-24, 1992.
94. SR Clark, MJ Pitts, SM Smith. SPE Adv Tech Series 1(1):172-179, 1993.
95. G Shutang, L Huabin, Y Zhenyu, MJ Pitts, H Surkalo, K Wyatt. In: SPE/DOE Tenth Symposium on Improved Oil Recovery. Paper #35383, Tulsa, OK, April 21-24, 1996.

# 12
# Nanosized Particles: Self-Assemblies, Control of Size and Shape

**M. P. Pileni, I. Lisiecki, L. Motte, C. Petit, and J. Tanori**
*Pierre and Marie Curie University, Paris, and C.E.N. Saclay, Gif sur Yvette, France*

**N. Moumen**
*Pierre and Marie Curie University, Paris, France*

## I. INTRODUCTION

The use of dispersed media to synthesize microparticles *in situ* has made considerable progress in the last few years, including reverse micelle (1), Langmuir–Blodgett film (2), zeolite (3), vesicle (4), glass matrice, and sol-gel methods (5).

In material physics and chemistry preparation of nanosized particles, an artificially engineered new structure with new properties is a major goal. Due to their small size, the nanoparticles exhibit novel material properties that largely differ from the bulk solid state (6). Many reports on quantum size effect on photochemistry of semiconductor (7,8) or the emergence of metallic properties with the size of the particles (9–12) have appeared during the past few years. In this new emerging field, finely divided magnetic nanoparticles are widely desired due to their large domains of application, especially for high density magnetic recording media or magnetically responsive fluids.

The electronic and optical properties of "quantum dots" or semiconductor crystallites that are small in comparison to bulk electron delocalization length (1 to 10 nm) are the subject of investigation (13). The ability to assemble molecules into well-defined two- and three-dimensional spatial configurations is a major goal in the field of self-assembled monolayers. Such assemblies can then be used to build up more complex structures in three dimensions (14), enabling chemists to engineer complex organic structures on top of macroscopic surfaces. Organized assemblies of crystallites should show interesting physical properties. Difficulties in producing and manipulating nearly monodisperse nanometer-size crystallites of arbitrary diameter have prevented the fabrication of such well-defined two- or three-dimensional structures. A recent exception is the special case of single crystals of identical 1.5 nm CdS clusters (15).

Because they have an extremely large surface area and are quite active, it is well known that fine metal structures constitute a wide class of catalysts. On the other hand, these systems are being investigated by many researchers due to optical properties that are affected not only by the size of particles (16,17) but also by their aggregation (18) and oxidation state (19).

Relatively little work exists for magnetic materials having sizes smaller than 10 nm. Nanocrystallites of $\gamma Fe_2O_3$ having an average size of 8 nm have been synthesized using a polymer matrix (20). Recently, reverse micelles have been used to synthesize metallic or boride cobalt nanoparticles having an average size equal to 3 and 4 nm (21–23). Magnetic particles, $Fe_3O_4$, have been obtained on bilayer lipid membranes, wherein magnetic domains made of small particles have been obtained (24).

In this chapter, we will discuss formation of monolayers made of silver sulfide particles organized in a hexagonal network on a long-distance range without any external forces. These monolayers are obtained for various particle sizes having a diameter equal to 3, 4, and 6 nm. The self-assembly takes place on various solid supports. Layers on three dimensions can be obtained when the support is left for some time in the presence of the solution containing these nanocrystals. Hence, a three-dimensional crystal is obtained.

By conducting syntheses in various colloidal assemblies, it is possible to control the size and the shape of metallic copper particles. In reverse micellar solution (25), the size of spherical particles is controlled by the size of the droplet. Syntheses performed in bicontinuous phase induce formation of 35% of cylindrical particles, whereas in lamellar phase large rods are formed. Syntheses performed in oil-in-water micellar solutions induce a control in the shape and the size of copper metallic particles. At the critical micellar concentration (cmc), a large network made of metallic copper is observed. Above the cmc, fragmentation of the network is observed with an increase in the sphericity of the particles and a decrease in the particle size. Below the cmc, a large network made of copper oxide is observed.

Magnetic particles can be obtained in water-in-oil and oil-in-water micelles. In reverse micelles, the size of cobalt boride, $Co_2B$, is controlled by the water content and the magnetic properties are characterized by a superparamagnetic behavior. In oil-in-water micelles, cobalt ferrite, $CoFe_2O_4$, magnetic fluid is obtained. The size of $CoFe_2O_4$ is controlled by the reactant concentration and varies from 2 to 5 nm. The magnetic properties of particles differ by their sizes.

## II. SELF-ASSEMBLED MONOLAYER OF NANOSIZED SILVER SULFIDE PARTICLES DIFFERING BY SIZE

Aqueous solution containing sodium sulfide is mixed to reverse micelles formed by Na(AOT) and Ag(AOT), sodium and silver bis(2-ethylhexyl) sulfosuccinate, respectively. The water content $w = [H_2O]/[AOT]$, is fixed by the amount of aqueous solution containing sodium sulfide added. The size of the $Ag_2S$ particles, observed by Transmission Electron Microscopy (TEM) (Figure 1A), increases linearly from 2 to 10 nm with the water content, $w$. This indicates a control of the crystallite size by the water content. At low water content (below $w = 10$), the size distribution (30%) is large compared to that observed at higher water content (13%) (26).

The silver sulfide nanosize particles synthesized in reverse micelles are coated by dodecanethiol and dispersed in heptane (27). The solution made of coated particles is optically clear. The sizes of coated particles are determined by TEM (Figure 1B) and compared to those observed by Small Angle x-ray Scattering (SAXS). Figure 1C shows the simulated and experimental data obtained on the Porod plot representation of particles made in reverse micelles having various water contents. A well-defined maximum and minimum indicating a low polydispersity in size is observed (Figure 1C). The particle diameter is deduced from the maximum and minimum of the Porod plot representation. A good agreement between the average size determined by SAXS and TEM is obtained.

The silver sulfide particles are well crystallized. Figure 2A shows the interreticular distance of the lattices obtained from high electron microscopic resolution. The interreticular distance is found equal to 0.26 nm and is attributed to ($-1\ 2\ 1$) lattice planes. Figure 1B shows a monolayer made of nanosized particles organized in a hexagonal network. This process takes place for various particle sizes. Contrary to what has been already observed (26–31), no external force has to be applied to induce the self-assembly. The formation of monolayer is observed by using various solid supports. This indicates that monolayer formation is not due to specific interactions with the solid surface.

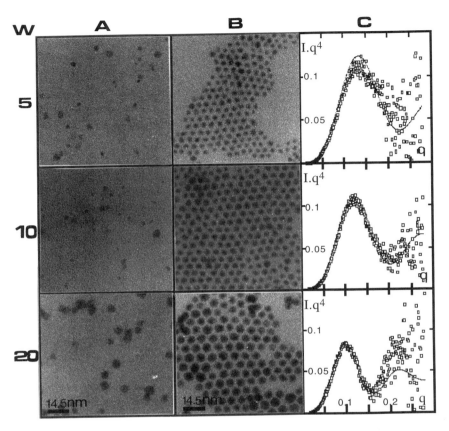

**Figure 1** (A) Electron microscopy patterns of silver sulfide particles in reverse micelles at $w = 5$, $w = 10$, and $w = 20$. (B) Coated silver sulfide particles obtained by extraction from reverse micelles having water content equal to $w = 5$, 10, and 20. (C) Porod plot obtained by SAXS of coated particles dispersed in heptane.

The interparticle spacing is unchanged with the particle size. It is close to the length of hydrocarbon tail of the dodecanethiol derivative. As a matter of fact, from TEM, the average distance between particles is found close to 1.8 nm. The length of dodecanethiol tail, $l$, is evaluated from the empirical equation given by Bain and colleagues (32): $l = 0.25 + 0.127 \cdot n$ (nm) where $n$ is the number of $CH_2$ groups. The value calculated is found equal to 1.77 nm, which is very close to the average distance determined by TEM (Figure 2A).

By leaving a solid support such as various TEM grids or silicium

# Self-Assembly of Nanosized Particles

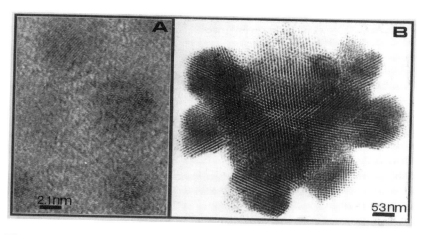

**Figure 2** High resolution of $Ag_2S$ particles (A), large domain of observation by TEM of self assembly of particles (B).

wafers in heptane solutions containing the particles characterized by a given size, large domains of self-assemblies are observed (Figure 2B). The surface of the network is larger than $10^5$ nm$^2$ and does not drastically change with the solid support used. The self-assembly is observed for particles differing by their size. The formation of such self-assemblies made of nanosized particles could be due to several factors acting simultaneously, such as the following:

1. The low polydispersity in the particle size favors the self-organization of particles.
2. The interactions between particles can be strong enough to induce such organization. The best defined network is obtained for particles having the largest sizes (Figure 2B). This can be attributed to an increase in van der Waals interactions with size. Vrij et al. (33) show, by SAXS experiments performed on silica spheres coated by octadecyl chains suspended in benzene, a strong increase in the attraction between the solute particles when the particle volume fraction increases. Similar attractive interactions can be taken into account for silver sulfide particles coated by dodecanethiol. By evaporation of the solvent containing the particles coated on the surface, the particle volume fraction increases, inducing an increase in the interaction between particles. The formation of multilayers confirms that such self-assembly is mainly due to van der Waals and dispersion forces.

## III. CONTROL OF THE SIZE AND THE SHAPE OF COPPER METALLIC PARTICLES

The change in size and in the shape has been performed either in water-in-oil or oil-in-water colloidal solution. From absorption spectroscopy it has been demonstrated that colloidal copper particles are characterized by a plasmon peak. The intensity and the position of this peak depend on the particle size (9,10) and shape (34,35). The particles, observed by high-resolution electron microscopy, show an interreticular distance of 0.2 nm, which is in good agreement with the lattice of bulk metallic copper. Electron diffraction showed concentric circles characteristic of a face-centered cubic phase with a lattice dimension equal to 0.361 nm. The presence of oxide is detected by electron diffraction.

### A. Oil Used as the Bulk Phase

Hydrazine, $N_2H_4$, used as a reducing agent is added to micellar solution formed either by mixed $Na(AOT)$–$Cu(AOT)_2$ or $Cu(AOT)_2$ reverse micelles. The water content of the micelle, $w$, is defined as $w = [H_2O]/[AOT]$. The total AOT concentration is kept equal to 0.1 M.

In reverse micelles, the size of copper metallic particles is controlled by the water content. As matter of fact, Figure 3 shows the data obtained by TEM at various water contents. Below $w = 10$, an increase in the particle size from 2 to 12 nm with the water content is clearly observed. At water content up to 10, the size of the particles remains unchanged with an increase in the polydispersity.

Previous structural study of $Cu(AOT)_2$–water–isooctane reverse micelles (36) demonstrates a change in the shape of the water droplets with the water content, $w = [H_2O]/[Cu(AOT)_2]$. Below $w = 3$, spherical droplets are formed, whereas between $w = 3$ to 5, the droplets are elongated with a cylindrical structure. Syntheses performed in cylindrical droplets are shown on Figure 4A. We observe the formation of spherical copper metallic particles with an average radius equal to 10 nm. However, the appearance of a small amount of cylindrical particles is also observed.

At $w = 5.5$, the water in oil droplets ($L_2$ phase) are destabilized with the appearance of a liquid–gas phase transition. The upper phase is pure isooctane and the surfactant is located in the lower phase. Synthesis performed at $w = 5.6$ shows the formation of 35% of cylindrical copper metallic particles with more than 65% of spherical particles. The average cylinders are 19 nm long with a diameter of 7 nm, whereas the average diameter of the spheres remains equal to 10 nm (Figure 4B).

# Self-Assembly of Nanosized Particles

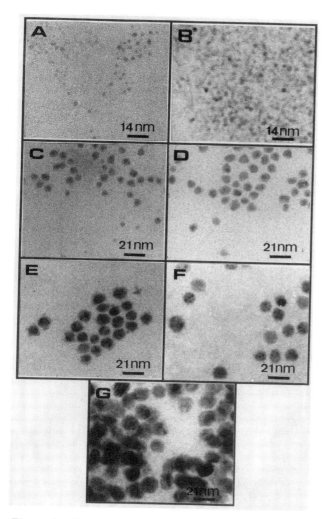

**Figure 3** Electron microscopy patterns of copper metallic particles obtained at various $w$ values: (A) $w = 1$; (B) $w = 2$; (C) $w = 3$; (D) $w = 4$; (E) $w = 5$; (F) $w = 10$; (G) $w = 15$.

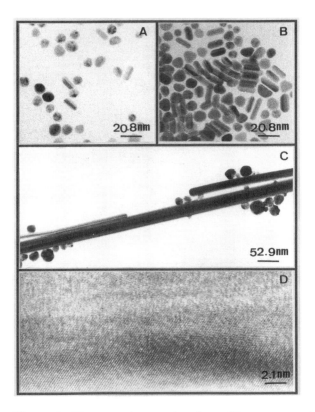

**Figure 4** Electron micrographs of metallic copper particles prepared in the $L_2(I)$ (A), $L_2(III)$ (B) phases, in lamellar phases $L_\alpha$(C), and high resolution of (C) in (D).

At $w = 9.8$, a third turbid and birefringent phase appears. Synthesis performed at $w = 9.8$ shows the formation of rods of metallic copper particles (Figure 4C). The diameter of the rods varies from 10 to 30 nm and the length from 300 to 1500 nm. Similar average size of copper metallic particle is obtained at $w = 14$. High resolution electron microscopy indicates a high crystallinity with a very low amount of defects (Figure 4D). The formation of such large rods having a high crystallinity indicates that the phase structure governs a slow nucleation.

Above $w = 16$, a new isotropic phase appears in coexistence with $L(\alpha)$ and isooctane phases. As above, the increase in water content induces a decrease in the $L(\alpha)$ and isooctane phases and an increase of the isotropic phase called $L_2(II)$ phase. Copper metallic particles made at $w = 20$ are spherical nanosized crystallites with an average diameter equal to 10 nm. Again, some amount of cylindrical particles is also obtained.

## B. Water Used as the Bulk Phase

In oil-in-water micelles, copper dodecyl sulfate, $Cu(DS)_2$ is used as the reactant. Synthesis performed below the cmc ($[Cu(DS)_2] < 1.2 \times 10^{-3}$ M) does not induce flocculation, and the solution remains optically clear. By TEM, isolated aggregates forming an interconnected network are observed (Figure 5). Electron diffraction of the network indicates the formation of copper oxide material, $Cu_2O$. Syntheses performed at higher surfactant concentration but still below the cmc induce a decrease in the size of the

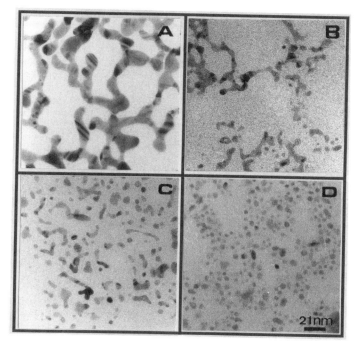

**Figure 5** TEM patterns of copper metallic particles synthetized at various $Cu(DS)_2$ concentrations. (A) $[Cu(DS)_2] = 1.2 \times 10^{-3}$ M; (B) $[Cu(DS)_2] = 2 \times 10^{-3}$ M; (C) $[Cu(DS)_2] = 3 \times 10^{-3}$ M; (D) $[Cu(DS)_2] = 1 \times 10^{-2}$ M.

interconnected network with formation of a mixture of metallic and oxide copper material. At the cmc, the network is made of pure metallic copper material. The network can be observed in a very large domain (Figure 6). Above the cmc ($1.2 \times 10^{-3}$ M), a progressive fragmentation of the network takes place with formation of isolated particles (Figure 5). Furthermore, a change in the shape of the particles from elongated to spherical one is observed.

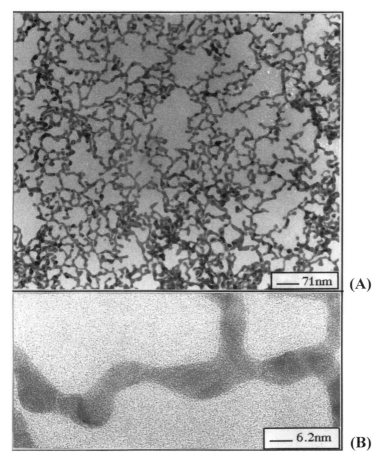

**Figure 6** Interconnected network made of pure metallic copper particles obtained from synthesis performed at the cmc (A) and expanded picture (B).

## IV. CONTROL OF THE SIZE OF COBALT BORIDE PARTICLES

By using mixed reverse micelles $Co(AOT)_2$–NaAOT, cobalt boride nanoparticles are synthesized. By increasing the water content, Figure 7 shows an increase in the $Co_2B$ particle size from 4 to 7.5 nm (23).

Cobalt boride particles are extracted from the micellar solution, as described previously (22). A superparamagnetic behavior is observed at room temperature. Magnetization properties of cobalt boride nanoparticles is closed to the bulk phase (Table 1). The particles show a low coercivity (between 50 and 70 Oe) which does not drastically change with the particle size (Table 1). Because the particle size and the coercivity is very low in comparison to bulk phase ($H_c \approx 2000$ Oe [37]), in first approximation, it is assumed to be superparamagnetic behavior. The average magnetic size of the particle can be deduced from simulation of the first magnetization curve by using Langevin relationship and assuming a log normal size distribution (38). A good agreement between the size determined by TEM after extraction and from magnetization curve is observed (Table 1). The magnetic susceptibility, $\chi$ determined from the initial slope of the first magnetization curve increases with the magnetic size of the particles (Table 1). This is consistent with the theoretical models (39) and with the data obtained previously for metallic cobalt particles (22).

## V. CONTROL OF THE SIZE OF COBALT FERRITE FLUIDS

A synthesis is performed in oil in water micellar solution by mixing $Co(DS)_2$ and $Fe(DS)_2$ micellar solution with methylamine, $CH_3NH_3OH$. The change in the particle size is obtained by changing $Fe(DS)_2$ concentration, keeping the ratio $[Co(DS)_2]/[Fe(DS)_2]$ and $[Fe(DS)_2]/[CH_3NH_3OH]$ constant (40).

By changing the $Fe(DS)_2$ concentration, the average diameter of cobalt ferrite changes equal to 5 nm to 2 nm (Figure 8). The size distribution remains unchanged and is found close to 30%. The electron diffractogram pattern indicates the formation of particles having a spinel crystalline structure as in the bulk phase. The average diameter of the particles determined by SAXS is confirmed by TEM.

Magnetic particles are dispersed in 50% of ethylene glycol in water. For particles having 5 nm as an average diameter, the variation of magnetization, $M$, with applied field, $H$, is given, at room temperature, in Figure 9A. As expected, no hysteresis is observed, indicating that the particles are in a superparamagnetic regime. The saturation magnetization is found

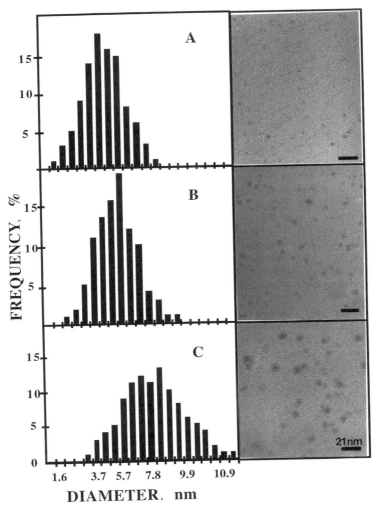

**Figure 7** (A–C) Electron microscopy patterns and histograms of cobalt boride nanoparticles.

## Self-Assembly of Nanosized Particles

**Table 1** Diameters Determined by TEM, $D_{TEM}$ (nm), Langevin Curve Simulation, $D_M$ (nm), and Magnetic Properties of Cobalt Boride Nanoparticles

| $D_{TEM}$ (nm) | $M_S$ (emu g$^{-1}$) | $H_C$(Oe) | $D_M$ (nm) | $\chi$ (emu g$^{-1}$Oe$^{-1}$) |
|---|---|---|---|---|
| 4.4 | 70 ± 10 | 35 | 2.6 | 4.8 × 10$^{-3}$ |
| 5.5 | 50 ± 5 | 75 | 4.4 | 10.5 × 10$^{-3}$ |
| 7.5 | 55 ± 5 | 35 | 5.4 | 15.5 × 10$^{-3}$ |

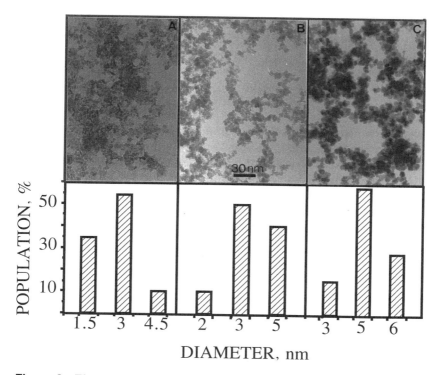

**Figure 8** Electron microscopy pattern and histogram of magnetic fluid made at various surfactant concentrations keeping [Co(DS)$_2$]/[Fe(DS)$_2$] = 0.325, [Co(DS)$_2$]/[NH$_2$CH$_3$] = 1.3 × 10$^{-2}$, [Fe(DS)$_2$] = 6.5 × 10$^{-3}$ M (A); [Fe(DS)$_2$] = 1.3 × 10$^{-2}$ M (B); [Fe(DS)$_2$] = 2.6 × 10$^{-2}$ M (C).

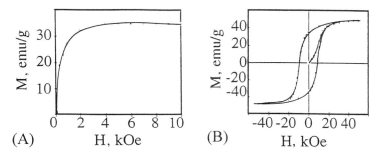

**Figure 9** Magnetization as function of applied field at room temperature (A) and at 20 K (B) for particles having a 5 nm average diameter.

equal to 35 emu · $g^{-1}$, whereas for bulk phase it is equal to 78 emu · $g^{-1}$ (at 300 K) (41). By decreasing the diameter of the particles to 2 nm, the saturation magnetization is not reached for an applied field equal to 30 kOe. This decrease in the saturation magnetization with the particle size can be attributed to a surface effect with formation of inert layer(s).

The field needed to reach the saturation magnetization of bulk Co-$Fe_2O_4$ depends on the crystallographic direction. As matter of fact, for (100) and (111) atomic density orientations, the saturation magnetization is equal to 1 and 9 kOe, respectively (42). For particles having an average diameter equal to 5 nm, the saturation magnetization is reached at 4 kOe. This value is in good agreement with the data obtained in bulk phase.

The magnetic particles having an average diameter equal to 5 nm are frozen in zero field at 10 K. Figure 9B shows the presence of a hysteresis. For random distribution of easy axes of particles with cubic magnetocrystalline anisotropy, the ratio of the remanence to saturation magnetizations is expected to be equal to 0.83 at 0 K. At 20 K it is found equal to 0.74. If the anisotropy had been primarily due to shape, then the ratio would have been 0.5 at 0 K. The coercivity at 10 K is found 8.8 kOe for particles having a diameter equal to 5 nm. By decreasing the diameter of particle from 5 to 2 nm, the coercivity decreases from 8.8 kOe to 4.8 kOe. The large remanence and coercivity values and the remance to saturation magnetization of 0.83 indicate that the system consists of randomly oriented equiaxial particles with cubic magnetocrystalline anisotropy (43).

Figure 9 shows an increase in the saturation magnetization with decreasing temperature from 300 K (Figure 9A) to 10 K (Figure 9B). This is similar to the data obtained in bulk phase. As matter of fact, the ratio of saturation magnetizations at 300 K and at 20 K is found equal to 0.83 and 0.88 for particles having 5 nm diameter and for the bulk phase, respectively.

## REFERENCES

1. MP Pileni. J Phys Chem 97:6961, 1993.
2. XK Zhao, J Yang, LD McCormick, JH Fendler. J Phys Chem 96:9933, 1992.
3. N Herron, Y Wang, MM Eddy, GD Stucky, DE Cox, K Moller, T Bein. J Am Chem Soc 111:530, 1989.
4. HC Youn, S Baral, JH Fendler. J Phys Chem 92:6320, 1988.
5. K Osakada, A Taniguchi, E Kubota. Chem Mater 4:562, 1992.
6. GA Ozin. Adv Mater 4:612, 1992.
7. MP Pileni, L Motte, C Petit. Chem Mater 4:338, 1992.
8. MG Bawendi, ML Steigerwald, LE Brus. Ann Rev Phys Chem 41:477, 1990, and references therein.
9. I Lisiecki, MP Pileni. J Am Chem Soc 115:3887, 1993.
10. I Lisiecki, MP Pileni. J Phys Chem 99:5077, 1995.
11. C Petit, MP Lixon, MP Pileni. J Phys Chem 97:12974, 1993.
12. A Henglein. Chem Rev 89:1861, 1989.
13. DAB Miller, DS Chemla, S Schmitt-Rink. In: Optical Nonlinearities and Instability in Semiconductors. Haug, H ed. Orlando, FL, Academic Press, 1988.
14. A Ulman, N Tillman. Langmuir 5:1418, 1989.
15. N Herron, JC Calabrese, WE Farneth, Y Wang. Science 259:1369, 1993.
16. A Creighton, J Desmond, G Eadon. J Chem Soc Faraday Trans 87:3881, 1991.
17. N Satoh, H Hasegawa, K Tsujii. J Phys Chem 98:2143, 1994.
18. BV Enüstün, J Turkevich. J Am Chem Soc 21:3317, 1963.
19. A Yanase, H Komiyama. Surf Sci 248:20, 1991.
20. RF Ziolo, EP Giannelis, BA Weinstein, MP O'Horo, BN Ganguly, V Mehrotra, MW Russell, DR Huffman. Science 257:219, 1992.
21. JP Chen, KM Lee, CM Sorensen, KJ Klabunde, GC Hadjipanayis. J Appl Phys 75:5876, 1994.
22. JP Chen, CM Sorensen, KJ Klabunde, GC Hadjipanayis. J Appl Phys 76:6316, 1994.
23. C Petit, MP Pileni. J Mag Mag Mat 166:82, 1997.
24. XK Zhao, PJ Herve, JH Fendler. J Phys Chem 93:908, 1989.
25. MP Pileni, ed. Structure and Reactivity in Reverse Micelles. Elsevier Amsterdam, 1989.
26. L Motte, F Billoudet, MP Pileni. J Mater Sci 31:38, 1996.
27. L Motte, F Billoudet, MP Pileni. J Phys Chem 99:16425, 1995.
28. M Giersig, P Mulvaney. Langmuir 9:3408, 1993.
29. BO Dabbousi, CB Murray, MF Rubner, MG Bawendi. Chem Mater 6:216, 1994.
30. VL Colvin, AN Goldstein, AP Alivisatos. J Am Chem Soc 114:5221, 1992.
31. D Heitman, JP Kotthaus. Physics Today 46:56, 1993.
32. CD Bain, J Evall, GM Whitesides. J Am Chem Soc 111:7155, 1989.
33. A Vrij, JW Jansen, JKG Dhont, C Pathmamanoharan, MM Kops-Werkhoven, HM Fijnaut. Faraday Discuss Chem Soc 76:19, 1983.
34. J Tanori, MP Pileni. Adv Mater 10:7, 1995.

35. I Lisiecki, F Billoudet, MP Pileni. J Phys Chem 100:4160, 1996.
36. C Petit, P Lixon, MP Pileni. J Phys Chem 97:1274, 1993.
37. Y Yiping, GC Hadjipanayis, CM Sorensen, KJ Klabunde. J Mag Mag Mater 79:321, 1989.
38. SW Charles, J Popplewell. Ferromagnetic Materials. Vol. 2. Wohlfarth EP, ed. Amsterdam, Northholland Publishing, 1982.
39. JL Dorman. Revue Phys Appl 16:275, 1981.
40. N Moumen, P Veillet, MP Pileni. J Mag Mag Mat 149:67, 1995.
41. EP Wohlfarth, ed. Ferromagnetic Materials. Amsterdam, Northholland Publishing, 1982.
42. KH Hellwege, AM Hellwege, eds. Landolt-Bornstein Numerical Data and Functional Relationships in Science and Technology. Vol. 12. Berlin, Springer-Verlag, 1980.
43. AE Berkowitz, E Kneller, eds. Magnetism and Metallugry. Vol. 1, Chap. 8. Academic Press, New York, 1969.

# 13
# Microemulsions as Tunable Media for Diverse Applications

**Syed Qutubuddin**
*Case Western Reserve University, Cleveland, Ohio*

## I. INTRODUCTION

A microemulsion may be defined as a thermodynamically stable isotropic solution of two immiscible fluids, generally oil and water, containing one or more surface-active species. The surface-active species or surfactant molecules are mostly located at the interface between the domains of polar and nonpolar fluids. Historically, microemulsions were first described by Schulman and coworkers as transparent or translucent systems formed spontaneously by mixing oil and water with relatively large amounts of an ionic surfactant together with a cosurfactant (1,2). Microemulsions need not be transparent and are not required to contain cosurfactants or cosolvents and electrolytes (3). Many systems of practical interest do contain a cosolvent such as a medium chain alcohol and also electrolytes (4). The above definition is a rather broad one that includes among microemulsions the classical ones with cosolvents, reverse micellar solutions, and "swollen" or "solubilized" micellar solutions. This definition also does not identify a specific structure for microemulsion aggregates.

The utilization of microemulsions preceded their introduction into the scientific community by Schulman and coworkers (1,2). For instance, microemulsions formulated with eucalyptus oil, water, soap, and white spirit were used more than a century ago to wash wool efficiently. Several industrial products such as lubricating oils, liquid waxes, and detergent formulations were patented during the 1930s (5).

Microemulsions have received great attention during the last quarter century because of their interesting thermodynamics, intricate physicochemical behavior, and increasing applications. The importance of microemulsions for enhanced oil recovery (6,7) has played a key role in the fostering of extensive theoretical and experimental research programs worldwide. The salient features of microemulsions include (1) spontaneous formation and thermodynamic stability, (2) intricate phase behavior, (3) large interfacial area per unit volume, (4) a wide range of interfacial curvature, (5) ultralow interfacial tensions and critical phenomena, and (6) large solubilization of both organic and aqueous phases. The special features of microemulsions make them tunable for a wide variety of applications. Commercial applications are increasing and the scope is tremendous, as evident from the increasing number of papers and patents on microemulsions. Some of the important applications pursued at Case Western Reserve University are summarized in this article.

## II. PARAMETERS AFFECTING MICROEMULSION PHASE BEHAVIOR

Physicochemical aspects of the phase behavior of microemulsions continue to be extensively studied, as evident from several chapters on this topic in this book. The phase behavior exhibited by microemulsions is very rich. Besides being single phases, microemulsions can exist in equilibrium with excess oil, water, or both. Winsor referred to these equilibria as Types I, II, and III, respectively (8). Such transitions occur when an appropriate variable such as salinity is changed. Microemulsions containing a small percentage of an anionic surfactant (e.g., petroleum sulfonates), a cosurfactant such as a short-chain alcohol, and equal volumes of oil and NaCl brine have been extensively studied (7). The phase behavior is termed "simple" (9) when the microemulsions behave as though composed of three pure components having ternary diagrams, as shown in Figure 1. Figure 1a at low salinities illustrates a two-phase region in which microemulsions along the binodal curve exist in equilibrium with oil containing molecularly dispersed surfactant. The surfactant is partitioned predominantly into the aqueous phase. This microemulsion system, which corresponds to Winsor's Type I, is often called a "lower" phase and is water-continuous. Figure 1 at high salinities indicates a two-phase region in which microemulsions exist in equilibrium with excess brine, i.e., Winsor's Type II. Such microemulsions are commonly named "upper" phase and are oil continuous. Figure 1b at an intermediate salinity illustrates Winsor's Type III microemulsion in the lower triangle. Three phases exist in equilibrium: a microemulsion corresponding to composition M, excess brine, and excess oil. Type III micro-

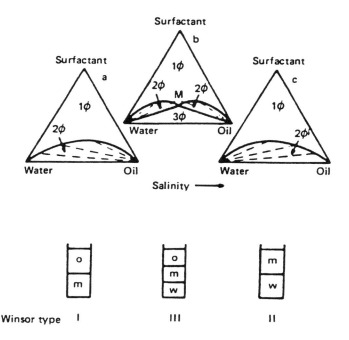

**Figure 1** An illustration of "simple" phase behavior of microemulsions and Winsor I, III, and II systems.

emulsions have been designated as "middle" phase (9) or "surfactant" phase (10).

Although not all microemulsions conform qualitatively to Figure 1, "simple" behavior appears to be a good first approximation for many systems of practical importance. A notable difference between real and simple systems is the locus of middle phase compositions rather than a single point. It is also possible to have four coexisting phases, two of which may be microemulsions in equilibrium with one another, or a microemulsion in equilibrium with a liquid crystalline phase. A liquid crystalline phase may also exist in equilibrium with only brine and oil. The pseudoternary diagrams are slices at constant salinity of the tetrahedron necessary to display the phase boundaries of a four-component system.

Parameters that may be used to tune the phase behavior of microemulsions include salinity, surfactant type and concentration, cosolvent type and concentration, pH, oil composition, temperature, and pressure. As salinity increases, there is a steady progression from lower-phase to middle-phase to upper-phase microemulsions. This reflects a continuous evolution of the preferred curvature of the surfactant film, and corresponds to an increased hydrophobicity with added salts like NaCl.

At low salinities the droplet size in the water-continuous lower phase increases with increasing salinity. This corresponds to an increase in solubilization of oil and is reflected in increased scattering. As salinity increases further, the middle phase appears and is initially water-continuous. The microemulsion is apparently bicontinuous over an intermediate salinity range. The middle phase becomes oil-continuous at higher salinities. After the transition from three to two phases, the microemulsion remains oil-continuous with drop size decreasing with increasing salinity.

Table 1 summarizes the qualitative changes in the phase behavior of microemulsions containing ionic surfactants. Some details of the effects of different variables are available in Reference 7. The phase transitions are generally understood in terms of relative strengths of hydrophilic and hydrophobic properties of the surfactant film in the microemulsions. The phase behavior depends strongly on the type and structure of the surfactant. For example, microemulsion containing nonionic surfactants are less sensitive to salinity, but are more sensitive to temperature. The partitioning of cosolvents such as alcohols between the surfactant film, the organic phase, and the aqueous phase also affects the phase behavior. This chapter illustrates how to tailor microemulsions for specific applications by changing an appropriate variable. For example, as indicated in Table 1 and described later in this chapter, the effect of salinity on the phase behavior can be counterbalanced by an increase in the pH of an appropriate microemulsion.

There are two bulk interfaces in middle-phase microemulsions and one in lower- or upper-phase microemulsions. Thus, one or three values of

**Table 1** Variables for Tuning the Phase Behavior of Ionic Microemulsions

| Variable | Phase transitions |
| --- | --- |
| Salinity ↑ | lower → middle → upper |
| Oil: alkane carbon number ↑ | upper → middle → lower |
| Oil: aromaticity ↑ | lower → middle → upper |
| Alcohol concentration ↑ (low MW) | upper → middle → lower |
| Alcohol concentration ↑ (high MW) | lower → middle → upper |
| Surfactant: hydrophobic chain length ↑ | lower → middle → upper |
| Temperature ↑ | upper → middle → lower |
| pH ↑ (for carboxylic acids, amines, and other pH-sensitive surfactants) | upper → middle → lower |

# Microemulsions as Tunable Media

interfacial tension (IFT) may be measured depending on system composition: (1) $\gamma_{mo}$ between microemulsion and excess oil phase, (2) $\gamma_{mw}$ between microemulsion and excess brine phase, and (3) $\gamma_{ow}$ between excess oil and brine phases. Phase volumes and, consequently, the volumes of oil ($V_o$) and brine ($V_w$) solubilized in the microemulsion depend on the variables that control the phase behavior. The solubilization parameters are defined as $V_o/V_s$ and $V_w/V_s$, where $V_s$ is the volume of the surfactant in the microemulsion phase. These parameters are easily determined from phase volume measurements if all the surfactant is assumed to be in the microemulsion phase.

The magnitude of $\gamma_{mo}$ decreases as $V_o/V_s$ increases, i.e., as more oil is solubilized. Similarly, the magnitude of $\gamma_{mw}$ decreases as $V_o/V_s$ increases. The salinity at which the values of $\gamma_{mo}$ and $\gamma_{mw}$ are equal is known as the optimal salinity based on IFT. Similarly, the intersection of $V_o/V_s$ and $V_w/V_s$ defines the optimal salinity based on phase behavior. The optimal salinity concept is very important for enhanced oil recovery, as discussed below.

## III. MICROEMULSIONS FOR ENHANCED OIL RECOVERY

Conventional production methods for crude oil do not recover the total amount of oil present in the reservoirs. During the primary recovery stage immediately following drilling, oil is driven to individual wells by natural forces in the reservoir. Water flooding is a common practice after primary recovery. This secondary recovery process involves injecting water into some wells to drive oil through the reservoir to other wells. Even after secondary recovery, about one-half of the original oil remains underground as immobile, discrete globules or "ganglia." The oil ganglia are trapped in the small pores by capillary forces. Enhanced or tertiary recovery methods are aimed at recovering this oil efficiently. The capillary forces have to be reduced drastically in order to recover the residual oil. This requires the interfacial tension to be reduced by several orders of magnitude, typically from about 30 dyn/cm between crude oil and brine to "ultralow" values on the order of 0.001 to 0.01 dyn/cm. Such lowering of interfacial tension is possible with microemulsions. The potential of enhanced oil recovery (EOR) using microemulsions is enormous. Considerable research has been done and several field tests conducted (4).

A typical surfactant flooding process involves sequential injections of (1) a preflush of low salinity brine, (2) a surfactant slug, (3) a dilute aqueous polymer solution, and (4) brine. The surfactant slug may take the form of a "soluble oil" containing the surfactant and an alcohol cosolvent but little or no brine, a microemulsion having substantial contents of both oil and brine, or an aqueous solution of the surfactant and alcohol containing little or no added oil. Economic factors dictate the use of as little surfactant as possible and hence a relatively small slug. The use of water-continuous slugs with little oil has the advantage that it minimizes injection of expensive oil along with the expensive surfactant (typically 2-10% by weight).

One of the challenges of surfactant flooding is to develop processes that will be effective under diverse conditions. Large composition differences exist among crude oils and reservoir brines. Reservoir temperatures may exceed 100°C in deep reservoirs. Porosity, permeability, and chemical composition of the rocks vary widely. Hence, it is important to understand and thereby manipulate the phase behavior of microemulsions. The optimization of surfactant, cosolvent, oil, temperature, pressure, and other variables for effective microemulsion flooding is well documented (4,6,7).

The optimal salinity corresponds to the composition where middle phase microemulsions exist and both $\gamma_{mo}$ and $\gamma_{mw}$ are ultralow. The displacement of oil by the microemulsion depends on $\gamma_{mo}$, while the displacement of the microemulsion by the water phase is governed by $\gamma_{mw}$. Both of these tensions should be ultralow to prevent plugging and achieve effective displacement of oil. Thus the optimal salinity is a useful guide in developing surfactant formulations. Besides phase behavior studies, core tests are necessary to optimize microemulsion formulations for specific field conditions. Methods to maintain ultralow tensions during surfactant flooding are reviewed in Reference 7 along with the present problems (e.g., adsorption, polymer-surfactant interactions) and future prospects.

The use of pH as a tool to tune the phase behavior of microemulsions was described by Qutubuddin et al. (11). Oleic acid was studied as a model pH-sensitive surfactant to illustrate how salt-tolerant microemulsions may be developed for enhanced oil recovery. In general, an increase in pH by addition of a base (e.g., NaOH) at constant salinity makes the surfactant more hydrophilic by ionizing the carboxylic acid. Therefore, under certain conditions, the effect of increasing salinity, which makes the surfactant hydrophobic, is counterbalanced by an appropriate change in the pH. The amount of NaOH, or equivalently, the pH needed for an upper phase (oil-continuous) microemulsion to shift to a middle phase increases with increasing salinity. Upper to middle to lower phase transitions were observed in the oleic acid system for salinities less than 5 g/dl NaCl. For higher salinities, the oleic acid microemulsion remained as a middle phase even with an excess of NaOH. All of the carboxylic groups are ionized in such a situation, and the salinity is too high to be counterbalanced by pH adjustments only.

Instead of using only a carboxylic acid, it is possible to manipulate the salt tolerance of microemulsions using a mixture of surfactants. The phase behavior of microemulsions containing a mixture of carboxylic acid and synthetic sulfonate has been studied (12). Middle-phase microemulsions are reported over a wide range of salinity (2 to 13 g/dl NaCl) with IFT on the order of 0.01 dyn/cm near the optimal salinity. By comparison, the synthetic sulfonate (TRS 10-410) system without any added carboxylic acid has a middle phase range of only 1.2 g/dl. Hence, the pH manipulation has extended the middle phase region by about an order of magnitude, and yet maintained low IFT values.

Yet another approach to optimize microemulsions is the use of zwitterionic surfactants. In recent years, zwitterionic surfactants have received less attention than other systems although they are used in a wide variety of applications including cosmetics, health care products, and pharmaceuticals. Light scattering results are reported on a series of alkyl dimethyl betaines $C_m$DMB ($m$ = 12, 16, 18) in aqueous solutions as functions of surfactant concentration, salt concentration, and temperature (13). The characteristics of the micelles change dramatically as the alkyl chain length is increased from 12 to 18. In the case of $C_{18}$DMB, large micelles form at very low surfactant concentrations (about 0.2 wt %) and grow rapidly to become entangled at slightly higher concentrations (about 1 wt %). Upon addition of salt, $C_{18}$DMB micelles become very long and semiflexible. In contrast, $C_{12}$DMB micelles remain small and spherical over a wide range of surfactant concentration. The phase behavior of zwitterionic microemulsions has been studied in the light of EOR application (14). Salt-tolerant microemulsions were formulated using $C_{18}$DMB surfactant, $n$-butanol as cosolvent, NaCl as electrolyte, and $n$-decane as oil. The dependence of phase behavior and solubilization parameters on electrolyte concentration is illustrated in Figure 2. The surfactant concentration is 1 wt %. The cosolvent to surfactant ratio is 4 and the water-to-oil ratio (WOR) is unity. The typical lower to middle to upper phase transitions are observed on increasing the salinity. The intriguing aspect of this zwitterionic system is the optimal salinity which is around 11 wt %. This is a very high value relative to the optimal salinities achievable with conventional anionic and nonionic microemulsions. The solubilization parameters at the optimal salinity are also high, indicating that the IFT values are ultralow as desirable for EOR. The effects of other variables such as cosolvent concentration are discussed in Reference 14.

Apart from salt tolerance, an important issue in EOR is the compatibility of the polymer and surfactant slugs. Surfactant–polymer flooding involves successive injections into the reservoir of an aqueous surfactant-cosurfactant solution and a dilute aqueous solution of the surfactant slug (6,7). As previously described, the role of the surfactant slug is to reduce the interfacial tension. The polymer solution is injected to control the vis-

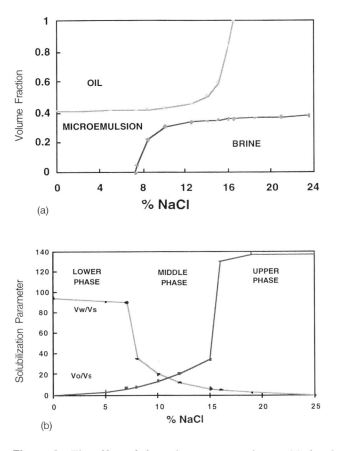

**Figure 2** The effect of electrolyte concentration on (a) the phase behavior and (b) solubilization parameters of $C_{18}$DMB microemulsions.

cosity of the water solution and thereby to avoid viscous fingering. Viscous fingering occurs when the viscosity of the aqueous drive is less than the surfactant solution. The mixing of the high molecular weight polymer (e.g., polyacrylamide and xanthan gum) with the surfactant solution may cause phase separation (12). Such polymer addition also causes phase separation in an otherwise single phase oil-in-water microemulsion. One approach to avoid such polymer–surfactant incompatibility is to use new types of polymeric surfactant that will reduce interfacial tension and also provide high viscosity for mobility control. The phase behavior of amphiphilic polymers such as Dapral (Figure 3) is being investigated in this context. A micro-

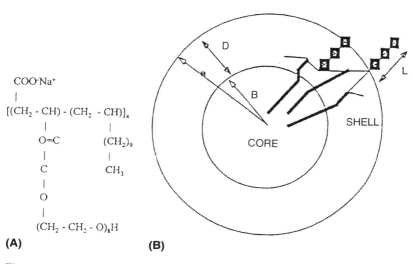

**Figure 3** (A) The chemical structure of a model amphiphilic polymer, Dapral GE 202. (B) Geometry of a spherical micelle of an amphiphilic polymer. The squares represent the hydrophilic groups, and the thick lines represent the hydrophobic side chains. The core contains hydrophobic side chains only. The spherical shell contains the polymer backbone part of the side chains, and solvent. (From Ref. 15, used with permission.)

scopic model for the formation of micelles by amphiphilic side-chain polymers predicts the free energy, micelle size, and micellar phase composition (15). A self-consistent field method was used to obtain the distribution of the backbone around the hydrophobic core (16). Some solubilization and interfacial tension studies have been conducted. Preliminary results indicate that the solubilization of hydrocarbons is relatively low. This opens up the challenge to develop polymeric surfactants with high solubilization parameters, low IFT, salt and temperature tolerance, etc. Perhaps the most important factor determining the future of microemulsions for EOR is the price of oil!

## IV. MICROEMULSIONS AS LIQUID MEMBRANES FOR SEPARATIONS

Separation processes are widespread in the chemical industry, accounting for a major portion of capital and operating costs. Typically, the market price of a chemical varies inversely with the concentration at which it is present in the original mixture. The need for economical and energy-efficient recovery of chemical from dilute solutions is obvious. The effective separation of toxic and hazardous components from dilute streams is also critical for environmental factors. One type of separation that attempts to fulfill this need is based on emulsion liquid membranes as first developed by Li (17). The method uses an immiscible liquid phase that separates two miscible phases. The desired chemical species (or solute) is selectively transported from the feed phase across a thin film of the immiscible phase (membrane) and enriched in the receiving phase. A coarse emulsion of the membrane and receiving phase is obtained using a surfactant as stabilizer. The receiving phase inside the emulsion may contain complexing reagents to help facilitate mass transport across the membrane. The main advantage of emulsion liquid membranes over other separation techniques is the large surface area available for mass transfer that results in a fast rate of separation. Two disadvantages of the emulsion technique are (1) lack of stability of the emulsion that allows leakage of the solute and unreacted internal reagent back into the feed phase, and (2) swelling of the internal macrodrops with water from the feed phase.

Microemulsions offer certain advantages over coarse emulsions as media for separations based on the liquid membrane concept (18,19). The low interfacial tensions typical of microemulsions lead to smaller drops that imply faster mass transfer rates due to increased surface area per unit volume. Microemulsions are thermodynamically stable, unlike coarse emulsions, and hence offer a more stable liquid membrane. Another key advantage is the ease of emulsification and demulsification steps in the separations scheme. A simple adjustment of temperature can cause spontaneous emulsification or demulsification of the microemulsion. Also, formation of the microemulsion does not require any special mixing. However, gentle mixing is usually provided to minimize the time required for the microemulsion to form spontaneously.

A microemulsion suitable for use as a liquid membrane must fulfill several constraints. The microemulsion must be oil-continuous if utilized to separate components from an aqueous feed phase. The receiving phase should constitute a significant volume fraction of the microemulsion to achieve high separation capacity. The microemulsion needs to be tolerant to large pH changes because pH is often used as the driving force for separation. Thus, nonionic surfactants are promising candidates due to their low sensitivity to pH compared to ionic or zwitterionic surfactants. Finally, the microemulsion should not contain components that easily partition into the aqueous feed phase. This requirement limits the use of cosurfactants or cosolvents to those which are water immiscible. Developing a microemulsion system that fulfills all the above constraints is nontrivial.

## Microemulsions as Tunable Media

Coarse emulsions are easily formulated to contain a wide range of concentration of the various constituents. In contrast, the thermodynamic equilibrium in a microemulsion puts an upper limit on the amount of receiving phase that may be solubilized in it.

Considering the thermodynamic constraints on solubilization, phase behavior studies are necessary for the optimization of microemulsion formulations for specific separations (18-20). Oil-continuous microemulsions containing nonionic surfactants and free of any cosolvent or cosurfactant were studied. Surfactants examined include an ethoxylated dinonyl phenol (DNP-8) and ethoxylated alcohol (Neodol 91-2.5). Solubilization was studied as a function of electrolyte type and concentration, temperature, hydrocarbon chain length, etc. Depending on the type of electrolyte, i.e., whether it is hydration-enhancing or hydration-depleting, the solubilization may increase or decrease with electrolyte concentration at a given temperature (19). Increasing NaOH or NaCl concentration leads to a decrease in solubilization, whereas KSCN or KI increases solubilization. Increasing temperature results in a reduced solubilization, while increasing the alkane chain length leads to higher solubilization. A semiempirical model adequately accounts for the dependence of solubilization on key experimental parameters (18-20). The aqueous phase solubilization parameter previously defined as $V_w/V_s$ is depicted for simplicity as $SP_w$. The pertinent correlations are

Electrolyte effect: $$SP_w = SP_{wo} \exp(k_e C) \quad (1)$$
Hydrocarbon effect: $$SP_{wo} = mV_0 + b \quad (2)$$
Temperature effect: $$SP_w = SP_{w,T_{ref}} \exp[-k_T(T - T_{ref})] \quad (3)$$

where $SP_{wo}$ is the value of $SP_w$ for pure water in the absence of an electrolyte, $k_e$ is an experimental decay constant for the effect of electrolyte concentration, $C$. The molecular volume of the hydrocarbon is denoted by $V_0$ while $m$ and $b$ are empirical parameters for a given nonionic surfactant. $SP_{w,T_{ref}}$ is the solubilization parameter at a specified reference temperature, $T_{ref}$, and $k_T$ is the decay constant for the effect of temperature. The major advantage of this simple model is the limited number of parameters required to predict the solubilization behavior over a wide range of variables. Expressions are also suggested for the dependence of interfacial tension on electrolyte concentration in nonionic microemulsions (20).

The above solubilization model was used to tune microemulsions for the separation of acetic acid (21), copper (22), and proteins (23). The mechanism of separating an organic like acetic acid is illustrated in Figure 4. Acetic acid (HAc) is soluble in the organic phase of the microemulsion membrane in its undissociated form. Thus, HAc partitions from the aque-

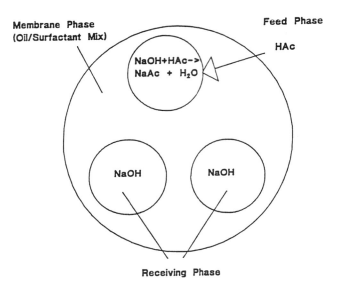

**Figure 4** Mechanism for acetic acid (HAc) removal from an aqueous feed solution using a microemulsion membrane. Sodium hydroxide present in the microemulsion droplets "binds" the acid by reacting to form sodium acetate that has negligible solubility in the oil phase of the microemulsion.

ous phase into the dispersed microemulsion globule and diffuses inward. Upon encountering the aqueous droplets (receiving phase) which contain NaOH, the HAc reacts to form sodium acetate (NaAc). NaAc is insoluble in the organic phase and is thus isolated from the original feed phase. Nonionic microemulsions were successfully used to separate acetic acid (21). The effects of mixing intensity, feed concentration, treat ratio, and microemulsion viscosity on the kinetics of separation were evaluated and analyzed in terms of the advancing reaction front model. The reversible phase behavior of the microemulsion was utilized to demulsify the liquid membrane phase and recover the acetate ion via a temperature increase.

The separation of copper ions from buffered aqueous feed phases was also accomplished using microemulsions (22). A complexing reagent (such as benzoyl acetone) is necessary in the membrane to transport the metal ion from the feed phase to the water droplets (receiving phase). The water droplets in the microemulsion contain an exchangeable ion such as $H^+$. Figure 5 compares the kinetics of separation of copper from a buffered feed phase using a microemulsion versus a coarse emulsion (22). Using a nonionic microemulsion, the separation was completed within 2 min and no

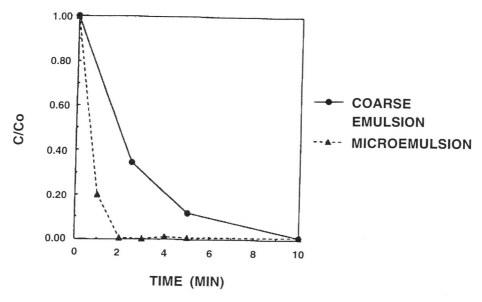

**Figure 5** A comparison of copper separation from buffered feed via microemulsion and coarse emulsion liquid membranes. Experimental conditions: treat ratio (volume of liquid membrane to volume of feed solution) of 1 : 5 and a mixing speed of 300 rpm.

leakage occurred up to 1 h. The same separation took about 10 min using a coarse emulsion. The faster kinetics in the microemulsion system is explained by the fact that the average diameter of the dispersed globules was on the order of 0.1 mm as compared to 1–2 mm in the emulsion case. Some undesirable partitioning of the surfactant into the feed phase was observed in the microemulsion systems. This problem may be eliminated or minimized by fine-tuning the surfactant characteristics. A less water soluble surfactant (lower HLB) is desirable. If no buffer is added to the feed phase, the separation capacity and rates decrease with benzoyl acetone as the complexing reagent. Using a commercial reagent (LIX 860), it was demonstrated that the interfacial complexation plays a major role (22). LIX 860 is effective in removing copper from unbuffered media if its concentration equals or exceeds that of the stabilizing surfactant in the microemulsion.

The concept of using microemulsions as liquid membranes has also been applied to separate proteins. Unlike the use of reverse micelles (23), this new method employs a preformulated cosurfactant-free nonionic microemulsion of high water content (24). The main advantage of using microemulsions as media for separations is their relatively high water content, which promotes solubilization and maintains the structural integrity of proteins. The liquid membrane process involves four steps as in the case of organic acid and metal ion separation. Figure 6 depicts the separation scheme for the case of proteins. A microemulsion is formulated in the first step by mixing oil and surfactant with an aqueous solution (e.g., NaCl solution). The saturated microemulsion is contacted with the protein feed solution. The protein partitions into the microemulsion droplets, and separation is achieved. Next, the immiscible phases are allowed to disengage via gravity or centrifugation. Finally, the temperature is raised to separate most of the water containing the protein. The oil and surfactant mixture may be recycled to Step 1. Unlike a reverse micellar solution, the preformulated microemulsion is saturated with aqueous solution. A moderate temperature increase may be used to strip the protein because the surfactant employed is nonionic with phase behavior that is temperature-sensitive. Of course, the final temperature should not be excessive to avoid protein denaturation. An ethoxylated dinonyl phenol was used as the surfactant in cosurfactant-free microemulsions for the extraction of human hemoglobin (24). Two ion exchangers were investigated, benzoyl acetone and Aliquat 336. The extent of extraction depends strongly on the pH of the aqueous solution and the microemulsion composition. The type of liquid ion exchanger used to facilitate the extraction affects the dynamics of phase separation. The

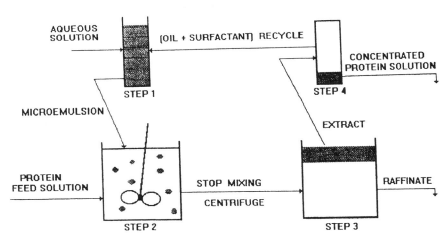

**Figure 6** A schematic of protein extraction and recovery using microemulsion liquid membrane.

extraction results are not explainable solely in terms of ionic interactions. Hydrophobic interactions are also important. Further research is necessary to understand and manipulate the mechanisms involved in the separation of various species using microemulsion media.

## V. MICROEMULSIONS AS REACTION MEDIA: GENERAL FEATURES

A rapidly growing field of microemulsion application is as media for a variety of chemical reactions, including electrochemical, photochemical, enzymatic, and polymerization reactions. The existence of microdomains or droplets with large interfacial area per unit volume allows the possibility of controlling the reaction rates, pathways, and stereochemistry, as well as the morphology of the precipitation products. Catalysis of various chemical reactions has great potential for industrial applications (25,26). One factor in the rate enhancement is the tremendous increase of contact surface between the reactant molecules that may be adsorbed at the interfacial film or solubilized inside the dispersed phase. The rate enhancement may far exceed the value predicted by simple partitioning of substrates between the dispersed phase and the continuous phase. The pathway of a reaction system and the stereochemistry may be controlled by the local environment of the microdomains. For instance, in the case of enzymatic reactions, the microenvironment may mimic the active sites, and certain configurations or orientations could be favorable for the bound substrate.

Some factors to be considered in utilizing microemulsions as reaction media include the dynamics of the surfactant molecules, the collisions between the droplets, the flexibility of the interfacial film, and the partitioning of the reactants. The average lifetime of a surfactant molecule in the interfacial film is on the order of microseconds while the exchange time between droplets is on the order of milliseconds (27). The number of "sticky" collisions which allow exchange between droplets decreases with increasing rigidity of the interfacial film. When the reaction rates are faster than the exchange times as in electron and proton transfers, the droplets may be treated as virtually isolated or "frozen." On the other extreme, when the rates are very slow, the dispersed phase may be regarded as "continuous" from the reaction kinetics point of view, allowing regular rate expressions to apply. For the intermediate case, statistical distributions and interdroplet exchange will affect the reaction rate (28). Applications of microemulsions as media for electrochemical and polymerization reactions are illustrated in the following sections.

## A. Electrochemical Characterization of Microemulsions

Research on electrochemistry in microemulsions may be viewed from two different perspectives, (1) electrochemistry as a characterization tool and (2) microemulsions as media to conduct electrochemical reactions. A battery of techniques is available to characterize the microstructure and also understand the interactions in microemulsions. Tools widely used include dynamic light scattering, NMR, neutron scattering, small-angle X-ray scattering, and fluorescence spectroscopy. Some of these topics appear elsewhere in this book. Electrochemical techniques provide a complementary tool that is simple, fast, and inexpensive for characterization of microemulsions (29).

Any electrochemical technique that allows the determination of the diffusion coefficient of an electroactive substance can, in principle, be used for measuring diffusion in micellar and microemulsion systems (29–32). A predetermined concentration of an electroactive probe is dissolved in the surfactant system, and an apparent diffusion coefficient of the aggregate (micelle or microemulsion droplet) is measured. The information obtained depends on the nature of the electroactive probe, its relative partitioning between the continuous and discontinuous pseudophases, and interactions. Electrochemical techniques used for such studies include polarography, cyclic voltammetry (CV), rotating disk voltammetry (RDV), chronoamperometry/chronocoulometry, and chronopotentiometry. The current–diffusion coefficient relationships that are applicable for each of these techniques are as follows (33):

Polarography (Ilkovic equation)

$$i_d = 708nCD^{1/2}m^{2/3}t^{1/6} \tag{4}$$

Cyclic voltammetry (Randles-Sevcik equation)

$$i_p = 0.4463 \frac{n^{3/2}F^{3/2}}{R^{1/2}T^{1/2}} ACD^{1/2}\nu^{1/2} \tag{5}$$

Rotating disk voltammetry (Levich equation)

$$i_l = 0.62nFACD^{2/3}\nu^{-1/6}\omega^{1/2} \tag{6}$$

Chronoamperometry (Cottrell equation)

$$i(t) = nFACD^{1/2}\pi^{-1/2}t^{-1/2} \tag{7}$$

Chronopotentiometry (Sand equation)

$$i(\tau) = 1/2 \, nFACD^{1/2}\pi^{1/2}\tau^{-1/2} \tag{8}$$

The symbols in the above mean the following: $i_d$ is the polarographic diffusion-limited current (A), n is the number of electrons transferred, C is the concentration of the electroactive probe (mol cm$^{-3}$), D is the diffusion coefficient of the electroactive probe (cm$^2$ s$^{-1}$), m is the mass flow rate of mercury at the dropping electrode (mg s$^{-1}$), t is the drop time (s), $i_p$ is the peak current (A), F is Faraday's constant (C mol$^{-1}$), A is the area of the electrode (cm$^2$), R is the gas constant (J mol$^{-1}$), T is the temperature (K), $\nu$ is voltage scan rate (V s$^{-1}$), $i_l$ is the limiting current (A), $\nu$ is the kinematic viscosity of the solution (cm$^2$ s$^{-1}$), $\omega$ is the angular velocity of the rotating disk electrode (rad s$^{-1}$), i(t) is the diffusion current (A) at time t (s), and $\tau$ is the transition time (s).

One of the shortcomings of EC measurements in surfactant systems used to be that the measurements are only possible in systems that are fairly conducting. Generally speaking, results are not reliable due to charging current and iR drop contributions in systems where sufficient indifferent electrolyte is absent. The added electrolyte very often changes the microstructure of the surfactant system compared to the extant microstructure without the electrolyte. EC measurements were usually restricted to surfactant systems containing added electrolytes. Recent developments have made it possible to extend voltammetry to highly resistive systems using ultramicroelectrodes (34). The limiting current ($i_l$) is given by

$$i_l = 4nFDCr \tag{9}$$

where r is the radius of the ultramicroelectrode (cm) and the other terms are as defined above. A few investigators have applied this technique for diffusion measurements in surfactant systems (35–38).

Mackay and coworkers used polarography to measure diffusion coefficients in microemulsions (30,32). Chokshi et al. described the use of two techniques, cyclic voltammetry and rotating disk voltammetry, to characterize oil-in-water microemulsions (29). Diffusion coefficients of microemulsion droplets were determined using ferrocene as the hydrophobic electroactive probe. The diffusion coefficients are comparable to values obtained from quasielastic light scattering measurements. The electrochemical techniques yield values of the self-diffusion coefficients, whereas light scattering techniques yield mutual-diffusion coefficients. The diffusion coefficients are strongly affected by interactions between the microemulsion droplets. Compared to light scattering, the electrochemical approach provides a faster and less expensive tool for characterizing microemulsions. Electrochemical techniques do not require any prior information of physical properties except viscosity, and are also applicable to opaque systems.

The application of various electrochemical techniques for measuring diffusion coefficients in micellar and microemulsion systems is reviewed by Dayalan et al. (39). Results are presented for anionic, cationic, nonionic, and zwitterionic surfactant systems. Along with the microemulsion structure, the probe partitioning equilibria (40) and kinetics (41,42) are critically important in interpreting the electrochemical measurements of apparent diffusion coefficients. The in situ analysis of the distribution of electroactive solutes between the aqueous and organic domains in microemulsions has been demonstrated. Dayalan et al. determined the apparent diffusion coefficients for substituted para-phenylenediamines (PPDs) in oil-in-water microemulsions (40). The partitioning of the electroactive solutes into the organic domains depends on hydrophobic interactions as well as electrostatic interactions for ionic surfactants.

In summary of this section, electrochemical techniques are particularly useful for studying microemulsion droplet self-diffusion coefficients. Explicit accounting of partitioning is necessary in interpreting the measured apparent diffusion coefficient when the probe is soluble in both the continuous and discontinuous pseudophases. However, when more than 95% of the probe is present in the dispersed phase, both slow and fast kinetics limiting expressions yield diffusion coefficients of comparable magnitude (39). In conjunction with light scattering, NMR, and other measurements, electrochemical experiments provide qualitative insight into interactions such as electrostatic and hydrophobic effects.

## B. Microemulsions as Media for Electrochemistry

Methylviologen (MV) was chosen as a model system to illustrate the usefulness of microemulsions as media for fundamental electrochemical studies (43). This reaction involves a water-soluble reactant, a sparingly water-soluble intermediate, and a water-insoluble product. Microemulsions are very suitable for conducting electrochemical investigations of such systems since the three different types of species can be solubilized in the same medium. Methylviologen has been widely investigated to understand the electron-transfer mechanism between the various redox forms, namely, the dication $MV^{2+}$, the cation radical, $MV^{\cdot +}$, and the neutral form MV. An understanding of the electrochemistry of methylviologen is important because of its frequent use as an electron acceptor in photochemical energy conversion and electrochemical display devices, as a mediator in electron transfer in biological studies, and in reducing electrochemically inactive compounds.

The electrochemical behavior of MV in aqueous media (Figure 7) differs significantly from the behavior in microemulsions (Figure 8). The cyclic voltammogram (CV) for 1 mM $MV^{2+}$ in NaBr solution at 100 mV/s

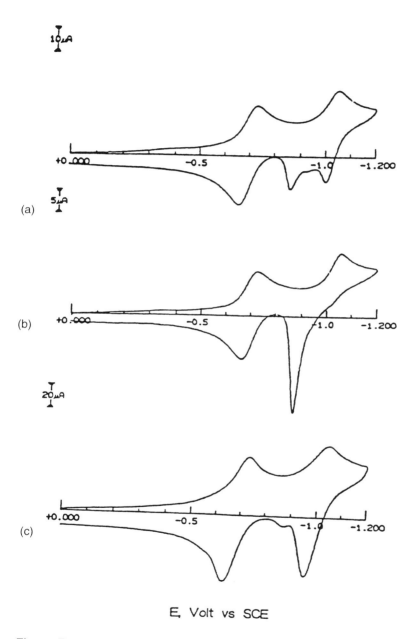

**Figure 7** Cyclic voltammograms of 1 mM methylviologen in aqueous 100 mM NaBr. Sweep rate (mV/s): (a) 100; (b) 20; (c) 500.

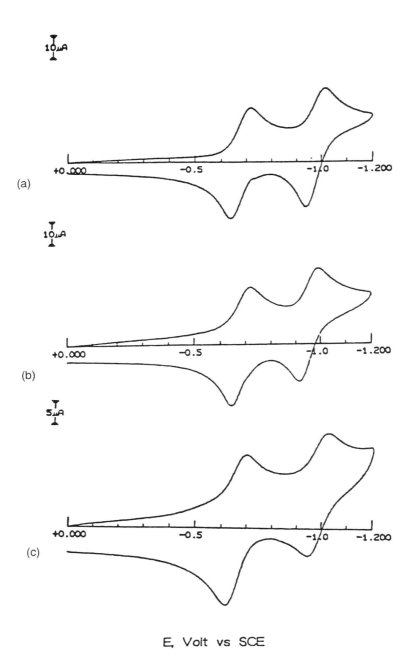

**Figure 8** Cyclic voltammograms of 1 mM methylviologen in microemulsions: (a) cationic CTAB; (b) nonionic Triton X-100; (c) anionic SDS. Sweep rate 100 mV/s.

scan rate (Figure 7A) shows two peaks for the anodic reaction corresponding to the second step of MV reduction. One peak is at $-1.0$ mV and the other at $-0.875$ mV. This implies that two forms of the product of the second reduction step (MV) are present, one easily oxidizable compared to the other. The peak positions change with scan rate. At 20 mV/s only the peak at $-0.875$ V remains (Figure 7B). However, at 500 mV/s (Figure 7C) the peak at $-1.0$ V is predominant, while a very small shoulder is left at 0.875 V. The hydrophobic nature of MV leads to the adsorption peak. The reduction of $MV^{2+}$ in three different types of microemulsions (Figure 8) takes place in two reversible steps (42). The complications observed in aqueous solutions due to the adsorption and rearrangement of MV on the electrode surface are eliminated in microemulsion media. Microemulsions provide a nonpolar environment for the preferential solubilization of the hydrophobic product, thereby preventing its adsorption on the electrode surface. The half-wave potentials for the reduction steps and the diffusion currents depend on the surfactant type and the composition of the microemulsion.

The electrochemistry of 1-amino-9,10-anthraquinone (1AAQ) in microemulsion media has been investigated (44). Reduction of 1AAQ to the hydroquinone form is quasi-reversible with a net transfer of two electrons. Rapid disproportionation of the anion radical (the product of the first electron transfer) to the hydroquinone and the original quinone is apparent. As with MV, the adsorption problem does not exist in the aminoanthraquinone system due to the presence of compartmentalized hydrophilic and hydrophobic domains in microemulsion media. Research is in progress on other electrochemical reactions such as the reduction of nucleic acids (cytosine and adenine) in various types of microemulsions.

## C. Microemulsions as Media for Polymerization

Microemulsions provide particular advantages over aqueous or organic solutions as media for polymerization to obtain nanoparticles or microlatexes and specialty polymers including porous solids and composites. Microemulsions provide tunable media where the solubilization of either hydrophilic or hydrophobic monomer is possible in the dispersed or continuous phase. A low viscosity, optically transparent and stable dispersion of microlatex particles may be obtained under appropriate conditions. Coagulum formation is also avoidable. The dispersions are easy to store and handle due to inherent thermodynamic stability.

Interest in microemulsion polymerization started in the early 1980s with the primary objective of producing stable and uniform latexes or nanoparticles (10–50 nm) not obtainable using classical emulsion polymerization. The early studies dealt with the polymerization of a hydrophobic monomer (primarily styrene) in oil-in-water microemulsions (45a–49). Surfactants used were anionic (e.g., SDS), cationic (e.g., CTAB) and to a less extent mixtures of anionic and nonionic systems. The typical formulation contained a short-chain alcohol as a cosurfactant. Atik and Thomas reported the polymerization of styrene using gamma radiation to obtain monodisperse latices (45). Jayakrishnan and Shah studied the polymerization of styrene and methylmethacrylate using a nonionic surfactant (46). Phase separation or polymer precipitation often occurs when the styrene content exceeds the solubilization limit of a few percent (46,47). Gulari and coworkers also faced limitations on solubilization of the monomer and diluted the microemulsion during polymerization to avoid turbidity (48,49). The instability is due to entropic effects and/or the incompatibility between polymer and cosurfactant (50). Alcohols also adversely affect the molecular weight due to chain transfer reactions. Polymerization of cosurfactant-free oil-in-water microemulsions has been investigated using cationic (51) and nonionic (52,53) surfactants. El-Aasser and coworkers have discussed the mechanism of polymerization based on observed polymerization rate and particle size (54,55).

Candau and coworkers pioneered the research on the polymerization of a hydrophilic monomer (primarily acrylamide) in water-in-oil microemulsions to obtain inverse latices of small size ($<50$ nm) (56–60). Aerosol OT was the surfactant and toluene the oil in most investigations. In general, clear or slightly bluish microlatexes were obtained after polymerization. The inverse microlatexes contain polymer particles highly swollen by water and dispersed in the continuous oil phase. Details of the kinetics and mechanism of polymerization in microemulsion media are discussed by Candau in an excellent review (61) and in Chapter 8 of this book.

Microemulsions may also be used as media to obtain solid materials rather than latex or inverse latex dispersions. Stoffer and Bone observed phase separation during polymerization of the continuous phase containing hydrophobic monomer (62,63). Rabagliati et al. studied the polymerization of styrene in a three-phase system (64). Qutubuddin and coworkers demonstrated the preparation of porous polymeric solids using microemulsions (65,66). The morphology and porosity of the polymeric solid depend on the initial microstructure of the microemulsion as well as the type and concentration of surfactant and cosurfactant. The middle phase or bicontinuous microemulsions yield maximum porosity in the polymerized solid. The porous morphology is illustrated in Figure 9 for the case of styrene polymerization in a SDS microemulsion (66). Two ranges of pore size are visible. The large pores are on the order of a few microns. Numerous smaller pores of submicron size are present throughout the solid. Thus, the

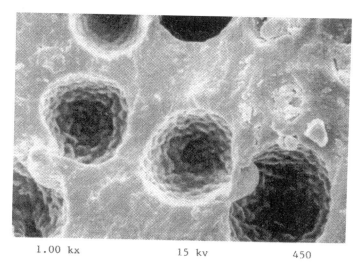

**Figure 9** Electron micrograph of porous polymeric solid obtained by polymerizing a bicontinuous microemulsion containing styrene (magnification X1000). (From Ref. 65, used with permission.)

original microstructure of the bicontinuous microemulsions is not retained during polymerization. Besides the morphology, the thermal behavior of the polymeric solid depends on the nature of the surfactant that becomes trapped during polymerization. The glass transition of polystyrene is enhanced by the ionic interactions with SDS. The microemulsion microstructure and composition also affect the permeability of the porous solid to various gases (66).

Microemulsions are also useful media for the synthesis of polymer composites. Qutubuddin and coworkers have developed a novel approach to prepare composites of different porosity and morphology from microemulsions (67,68). Instead of blending two polymers, appropriate hydrophilic and hydrophobic monomers are solubilized in a microemulsion medium and then polymerized to obtain composites with desirable properties (67). The term "composite" is preferable over "copolymer" or "blend" since the product obtained is a solid mass rather than a dispersion or powder. This technique is exemplified by the polymerization of microemulsions containing styrene and acrylamide as the hydrophobic and hydrophilic monomer, respectively. Different surfactants were used in the microemulsion formulation, anionic SDS, cationic CTAB, and zwitterionic $C_{18}DMB$. No macroscopic phase separation was apparent during polymerization. The morphology and thermal behavior of the composites resulting from microemulsion polymerization are strongly influenced by the surfactant embedded inside the porous solid (68). The resulting morphology is sensitive to the sample-drying and preparation protocol used for the scanning electron microscopy (SEM) study. SEM micrographs of composites prepared from microemulsions are illustrated in Figure 10. $C_{18}$DMB microemulsions produced the largest pore size and most rigid composite. While random copolymers are obtained with SDS or CTAB microemulsions, the $C_{18}$DMB system is more heterogeneous, with homopolymer domains present in addition to the copolymer. The zwitterionic surfactant also acts as a plasticizer in the composite, reducing the glass transition temperature of the homopolymers. Therefore, the surfactant type and concentration are key parameters for tuning the polymerizable microemulsion to specific applications of the porous composites.

Novel conductive composite films have been developed using a two-step process: microemulsion polymerization to form a porous conductive coating on an electrode followed by electropolymerization of an electroactive monomer, such as pyrrole. The porous matrix was prepared by polymerizing a SDS microemulsion containing two monomers, acrylamide and styrene, as described in the previous section. The electropolymerization

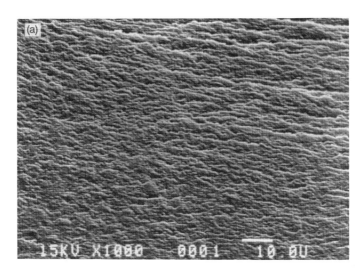

**Figure 10** SEM micrographs (magnification X1000) of dried composites prepared by polymerizing microemulsions containing (a) SDS, (b) CTAB, and (c) $C_{18}$DMB. (From Ref. 68, used with permission.)

# Microemulsions as Tunable Media 329

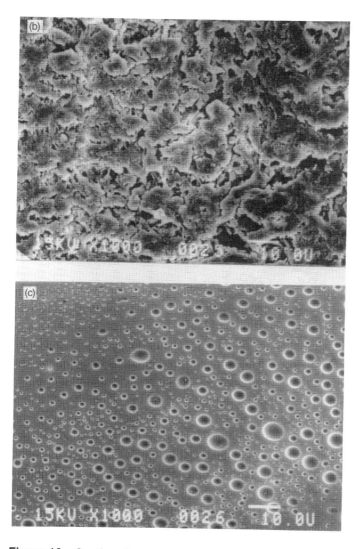

**Figure 10** Continued.

of pyrrole was performed in an aqueous perchlorate or toluenesulfonate solution. The effects of polymerization potential on the electropolymerization, morphology and electrochemical properties have been reported (69). The copolymer matrix improves the mechanical behavior of the polypyrrole composite film.

## VI. CONCLUDING REMARKS

Microemulsions possess very interesting properties that continue to be investigated in terms of fundamentals. Improved understanding of the phase behavior and detailed characterization of microemulsions by various techniques have paved the way for exciting applications. Besides enhanced oil recovery, microemulsions have been demonstrated to be useful as tunable media for separations and a variety of reactions including electrochemistry and polymerization to obtain nanoparticles, porous solids and composites. Microemulsions are no longer of primarily academic interest, as commercial applications are expanding fast. For instance, carboxylic acid microemulsions have found intriguing application for microencapsulation of nutrients for bioremediation. It is heartening to note that such microemulsions were successfully employed in the environmental cleanup of Prudhoe Bay following the Exxon oil spill. Many more success stories are expected in the next few years as the potential of microemulsions continues to be developed around the globe.

## REFERENCES

1. JP Hoar, JH Schulman. Nature 152:102, 1943.
2. JH Schulman, EG Cockbain. Trans Faraday Soc 36:651, 1940.
3. SE Friberg. J Dispersion Sci Technol 6:317, 1985.
4. S Qutubuddin. In: Encyclopedia of Chemical Processing and Design. Vol. 30. JJ McKetta, ed. New York, Dekker, 1989.
5. LM Prince. Microemulsions. New York, Academic Press, 1977.
6. DO Shah, ed. Surface Phenomena in Enhanced Oil Recovery. New York, Plenum Press, 1984.
7. CA Miller, S Qutubuddin. In: Interfacial Phenomena in Non-Aqueous Media. Eicke, HF, ed. New York, Marcel Dekker, 1986.
8. PA Winsor. Solvent Properties of Amphiphilic Compounds. London, Butterworths, 1954.
9. RL Reed, RM Healy. In: Improved Oil Recovery by Surfactant and Polymer Flooding. Shah, DO, Schecter, RS, eds. New York, Academic Press, 1977.
10. K Shinoda, S Friberg. Adv Colloid Interface Sci 91:223, 1983.
11. S Qutubuddin, CA Miller, T Fort, Jr. J Colloid Interface Sci 101:46, 1984.

## Microemulsions as Tunable Media

12. S Qutubuddin, CA Miller, WJ Benton, T Fort, Jr. In: Macro- and Microemulsions: Theory and Applications. Shah, DO, ed. ACS Symposium Series, Vol. 272. Washington, DC, American Chemical Society, 1985.
13. S Qutubuddin, E Fendler, A Nabi. manuscript in preparation.
14. A Bhatia, S Qutubuddin. Colloids Surfaces 69:277, 1993.
15. E Hamad, S Qutubuddin. Macromolecules 23:4185, 1990.
16. E Hamad, S Qutubuddin. J Chem Physics 96:6222, 1992.
17. NN Li. U.S. Patent 3,410,794, (1968).
18. J Wiencek, S Qutubuddin. Colloids Surfaces 29:119, 1988.
19. S Qutubuddin, J Wiencek. In: Surfactants in Solution. Mittal, KL, ed. Vol. 10. New York, Plenum Press, 1989, p. 1882.
20. J Wiencek, S Qutubuddin. Colloids Surfaces 54:1, 1991.
21. J Wiencek, S Qutubuddin. Sep Sci Technol 27:1211, 1992.
22. J Wiencek, S Qutubuddin. Sep Sci Technol 27:1407, 1992.
23. KE Goklen, TA Hatton. Sep Sci Technol 22:831, 1987.
24. S Qutubuddin, J Wiencek, A Nabi, J Boo. Sep Sci Technol 29:923, 1994.
25. JH Fendler, EJ Fendler. Catalysis in Micellar and Macromolecular Systems. New York, Academic Press, 1975.
26. JH Fendler. Membrane Mimetic Chemistry. New York, Wiley, 1982.
27. M Zulauf, HF Eicke. J Phys Chem 84:1503, 1980.
28. PE Luisi, BE Straub, eds. Reverse Micelles. New York, Plenum, 1984.
29. K Chokshi, S Qutubuddin, A Hussam. J Colloid Interface Sci 129:315, 1989.
30. RA Mackay, NS Dixit, R Agarwall, RP Seiders. J Dispersion Sci Technol 4:397, 1983.
31. J Georges, A Berthod. Electrochim Acta 28:735, 1983.
32. R Zana, RA Mackay. Langmuir 2:109, 1986.
33. AJ Bard, LR Faulkner, eds. Electrochemical Methods: Fundamentals and Applications. New York, Wiley, 1980.
34. RM Wightman, DO Wipf. In: Electroanalitical Chemistry. AJ Bard, ed. Vol. 16. 1988, p. 267.
35. J-W Chen, J Georges. J Electroanal Chem 210:205, 1986.
36. Z Wang, A Owlia, JF Rusling. J Electroanal Chem 270:407, 1989.
37. A Owlia, Z Wang, JF Rusling. J Am Chem Soc 111:5091, 1989.
38. JF Rusling, Z Wang, A Owlia. Colloids Surfaces 54:1, 1990.
39. E Dayalan, S Qutubuddin, J Texter. In: Electrochemistry in Colloids and Dispersions. Mackay, RA, Texter, J, eds. New York, VCH Publishers, 1992.
40. E Dayalan, S Qutubuddin, J Texter. J Colloid Interface Sci 143:423, 1991.
41. J Texter, FR Horch, S Qutubuddin, E Dayalan. J Colloid Interface Sci 135:263, 1990.
42. J Texter. J Electroanal Chem 304:257, 1991.
43. E Dayalan, S Qutubuddin, A Hussam. Langmuir 6:715, 1990.
44. S Qutubuddin, Y Shalaby, E Dayalan. "Electrochemical Investigations in Microemulsion Media: 2. Aminoanthraquinone reduction," to be published.
45a. SS Atik, KJ Thomas. J Am Chem Soc 103:4279, 1981.
45b. SS Atik, KJ Thomas. J Am Chem Soc 104:5868, 1982.
45c. SS Atik, KJ Thomas. J Am Chem Soc 105:4515, 1983.

46. A Jayakrishnan, DO Shah. J Polm Sci Polym Lett Ed 22:31, 1984.
47. CK Gratzel, M Jirousek, M Gratzel. Langmuir 2:292, 1986.
48. PL Johnson, E Gulari. J Polm Sci Polym Chem Ed 22:3967, 1984.
49. HI Tang, PL Johnson, E Gulari. Polymer 25:1357, 1984.
50. LM Gan, CH Chew, SE Friberg. J Macromol Sci Chem A 19:739, 1983.
51. VH Perez-Puna, JE Puig, VM Castano, BE Rodriguez, AK Murthy, E Kaler. Langmuir 6:1040, 1990.
52. TD Leman. M.S. Thesis. Case Western Reserve University, 1988.
53. S Qutubuddin, TD Leman, to be published.
54. PL Kuo, NJ Turro, CM Tseng, MS El-Aasser, JM Vanderhoff. Macromolecules 20:1216, 1987.
55. JS Guo, MS El-Aasser, JM Vanderhoff. J Polym Sci Polym Chem 27:691, 1989.
56. YS Leong, G Reiss, F Candau. J Chim Phys 78:279, 1981.
57. YS Leong, F Candau. J Phys Chim 86:2269, 1982.
58. F Candau, YS Leong, G Pouyet, SJ Candau. J Colloid Interface Sci 101:167, 1984.
59. F Candau, YS Leong, RM Fitch. J Polym Sci Polym Chem Ed 23:193, 1985.
60. MT Carver, U Dreyer, R Knoesel, F Candau, RM Fitch. J Polym Sci Polym Chem Ed 27:2161, 1989.
61. F Candau. In: Polymerization in Organized Media. Paleos, C, ed. Reading, England, Gordon and Breach, 1992. Chapter 4, pp 215–282.
62. JO Stoffer, T Bone. J Polym Sci Polym Chem Ed 18:2641, 1980.
63. JO Stoffer, T Bone. J Dispersion Sci Technol 1:393, 1980.
64. FM Rabagliati, AC Falcon, DA Gonzales, C Martin, RE Anton, JL Salager. J Dispersion Sci Technol 1:393, 1980.
65. E Haque, S Qutubuddin. J Polymer Sci Poly Letters Ed 26:429, 1988.
66. S Qutubuddin, E Haque, WJ Benton, EJ Fendler. In: Polymer Association Structures: Microemulsions and Liquid Crystals. El-Nokaly, M, ed. ACS Symposium Series, Vol. 384. Washington, DC, American Chemical Society, 1989. Chap. 5, pp 65–83.
67. S Qutubuddin. U.S. Patent 5238992, (1993). Assigned to Edison Polymer Innovation Corporation, Ohio.
68. S Qutubuddin, CS Lin, Y Tajuddin. Polymer 35:4605, 1994.
69. DA Kaplin, S Qutubuddin. Synthetic Metals 63:187, 1994.

# 14
# Double Emulsions Stabilized by Macromolecular Surfactants

**Nissim Garti and Abraham Aserin**
*Casali Institute of Applied Chemistry, The Hebrew University of Jerusalem, Jerusalem, Israel*

## I. DEFINITIONS, FORMATION, APPLICATIONS

Emulsions have been described as heterogeneous systems of one immiscible liquid dispersed in another, in the form of droplets that have thermodynamic instability (1). Multiple emulsions are more complex systems, termed "emulsions of emulsions"; the droplets of the dispersed phase themselves contain even smaller dispersed droplets. Each dispersed globule in the double emulsion forms a vesicular structure with single or multiple aqueous compartments separated from the aqueous phase by a layer of oil phase compartments (2–6).

Multiple emulsions were first described by Seifriz (5) in 1925, but only in the past 20 years have they been studied in more detail. The two major types of multiple emulsions are the water–oil–water (w/o/w) and oil–water–oil (o/w/o) double emulsions.

A schematic representation of a w/o/w double emulsion droplet is shown in Figure 1.

Multiple emulsions have been prepared in two main modes: one-step emulsification and two-step emulsification. There have been several reports on one-step emulsification for the preparation of w/o/w double emulsion that included strong mechanical agitation of the water phase containing an hydrophilic emulsifier and an oil phase containing large amounts of hydrophobic surfactant. A w/o emulsion is formed, but it tends to invert and form a w/o/w double emulsion (4a,4b) (Figure 2). In addition, double

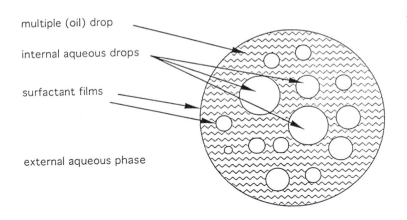

**Figure 1** Schematic representation of a w/o/w double emulsion droplet.

emulsions can be prepared by forming w/o emulsion with a large excess of relatively hydrophobic emulsifier and a small amount of hydrophilic emulsifier followed by heat-treating the emulsion until, at least in part, it will invert. At a proper temperature, and with the right hydrophilic–lipophilic balance (HLB) of the emulsifiers w/o/w emulsion can be found in the system. However, there is usually little chance of reproducing these "accidental" preparations.

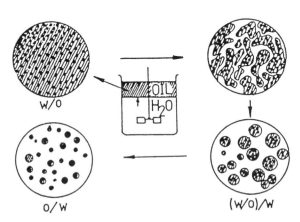

**Figure 2** Possible sequence of events leading to the final formation of an o/w emulsion via a transient (w/o/w) emulsion, when hydrophilic surfactant is initially located in the oil phase.

In most of the recent studies, double emulsions are prepared in a two-step emulsification process by two sets of emulsifiers: a hydrophobic "Emulsifier I" (for the water-in-oil emulsion) and a hydrophilic "Emulsifier II" (for the oil-in-water emulsion) (Figure 3).

The primary w/o emulsion is prepared under high shear conditions (ultrasonication, homogenization), while the secondary emulsification step is carried out without any severe mixing (an excess of mixing can rupture the drops resulting in a simple oil-in-water emulsion).

The composition of the multiple emulsion is of significant importance, since the different surfactants along with the nature and concentration of the oil phase will affect the stability of the double emulsion (2,6–8). Much work was done on the nature of the oils and their influence on the manufacturing conditions, as well as on the stability of the double emulsion (7).

Ionic and nonionic surfactants have been used for different applica-

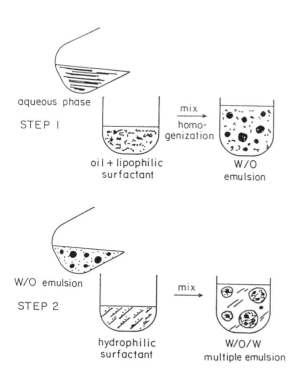

**Figure 3** Schematic illustration of a two-step process in formation of double emulsion.

tions, in accordance with health restrictions. It was, however, well established that combinations of emulsifiers, at the outer phase, have a beneficial effect on stability, and that the inner hydrophobic Emulsifier I must be used in great excess (10-30 wt % of the inner emulsion) while the hydrophilic Emulsifier II must be used in low concentration (0.5-5 wt %) (Figure 4). The inner emulsifier was found to migrate, in part, to the outer interface and influence the outer emulsifiers (8-10). The HLB of the outer emulsion was found to be a weighted HLB of the contribution of the two types of emulsifiers (11a,11b).

In addition, parameters such as the oil phase volume and the nature of the entrapped materials in the inner phase have been discussed and optimized (9-11).

Multiple emulsions have shown significant promise in many technologies, particularly in food, pharmacology, and separation science. Their potential biopharmaceutical applications, as a consequence of the dispersal of one phase inside droplets of another, include uses such as adjuvant vaccines (12,13), prolonged drug delivery systems (14-19), sorbent reservoirs of drug overdose treatments (20,21), taste masking (11a,11b), and

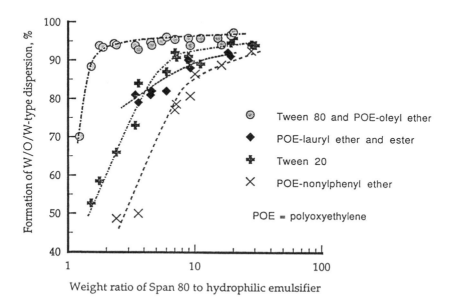

**Figure 4** Weight ratio of Span 80 in the oil phase to hydrophilic emulsifiers in the aqueous suspending fluid affecting the formation of water-liquid-paraffin-water emulsions due to the two separated steps of emulsification.

immobilization of enzymes (18,19). Multiple emulsions have also been formulated as cosmetics and as household products such as wax polish (25) and foods (26-29).

In some disciplines, certain multiple emulsions have been termed "liquid membranes," as the liquid film which separates one liquid phase from the other liquid phases acts as a thin semipermeable film through which solute must diffuse as it moves from one phase to another. The use of multiple emulsions in the separation field has included, for example, separation of hydrocarbons, and the removal of toxic materials from wastewater (20-25).

## II. STABILITY AND TRANSPORT MECHANISMS FOR DOUBLE EMULSIONS STABILIZED BY MONOMERIC EMULSIFIER

Many review articles have been written on the potential practical applications of the multiple emulsions and on the main problems associated with this technology—their inherent thermodynamic instability (20).

It was concluded that the classical double emulsions prepared with two sets of monomeric emulsifiers—one hydrophobic in nature (to stabilize the inner w/o interface) and the other hydrophilic in nature (to stabilize the outer o/w interface)—cannot provide long-term stability to the double emulsion. As a result, relatively large w/o/w droplets with short-term stability are obtained, which cannot be used in practice (21) (Figure 5).

The complex nature of double emulsions has caused significant difficulties in the assessment of the stability and in detecting rupture and coalescence phenomena. The main technique is based on measurement of number and size of the double emulsion drops over a period of time. Such measurements produce only limited information on double emulsion stability, since no information on the coalescence of the inner droplets can be deduced. Similar information is obtained from photomicrography.

It was and still is very difficult to determine if the internal droplets coalesce, aggregate, or tend to rupture. Several modern techniques have been applied, including freeze-etching techniques, viscosity measurements, and quantitative estimation of addenda transported from the inner phase to the outer phase and vice versa. In addition, engulfment or shrinkage of the double emulsion in the presence of water migration in or out of the droplets has been studied and interpreted in terms of stability (Figure 6).

Several possible mechanisms by which materials may be transported across the oil layer in the w/o/w systems have been proposed and discussed (2,5,8). The most common is the "diffusion-controlled mechanism for ion-

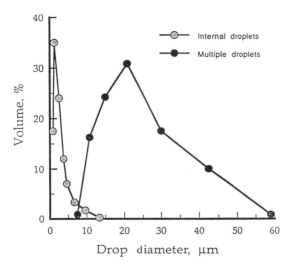

**Figure 5** Size distribution of the internal aqueous and multiple drops of a multiple emulsion at the preparation stage.

ized lipid-soluble materials." The transport will be dependent on the nature of the entrapped material (including its dissociation constant), the oil, as well as on the pH of the aqueous phase. Such systems could be used, for example, in the treatment of overdose of acidic drugs such as barbiturates (30). At low pH values the barbiturate would exist almost exclusively as the unionized form and so would be readily soluble in the oil phase. The drug could therefore easily pass across the oil layer to the internal aqueous phase containing a basic buffer that would ionize the addenda that is now insoluble in the oil phase and would become trapped within the internal aqueous phase. The drug would then be carried out with the emulsion and be voided from the gastrointestinal tract (Figure 7). The drug transport was found to follow first-order kinetics according to Fick's Law (30).

Ionized compounds are not the only materials to be transported across the oil membrane. It has been demonstrated that both water molecules, electrolytes, and nonelectrolyte water-soluble substances can easily migrate through the oil membrane without affecting the double emulsion stability (31–33). It was demonstrated that one can alter the diffusion rates through the oil membrane. This suggests that the diffusion of the water-soluble substances through the oil is the rate-determining step, and that the inner water phase does not have any effect on the determination of the release rates. Kita, Matsumoto, and Yonezawa (34) have suggested two possible

# Surfactant-Stabilized Double Emulsions

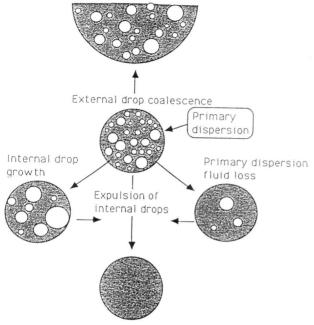

**Figure 6** Schematic illustration of several possible breakdown pathways that may occur in w/o/w or o/w/o systems.

mechanisms for the permeation of water and water-soluble materials through the oil phase, the first being via the "reverse micellar transport" (Figure 8) and the second by "diffusion across a very thin lamellae of surfactant" formed, where the oil layer is very thin (Figure 9).

Our detailed studies on the release of electrolytes from the inner aqueous phase to the outer aqueous phase in double emulsions stabilized by monomeric nonionic emulsifiers (Span and Tween) (31–33) have indicated that the osmotic pressure gradient between the two aqueous phases is a strong driving force for mutual migration of electrolytes and nonelectrolytes, and water from one phase through the other by both mechanisms. However, it was clearly demonstrated that even if the osmotic pressure of the two phases was equilibrated and no visual coalescence took place (neither of the inner phase droplets nor of the outer phase droplets), electrolytes tend to be transported out mostly through a "reverse micellar mechanism" controlled by the viscosity of the oil phase and the nature of the oil mem-

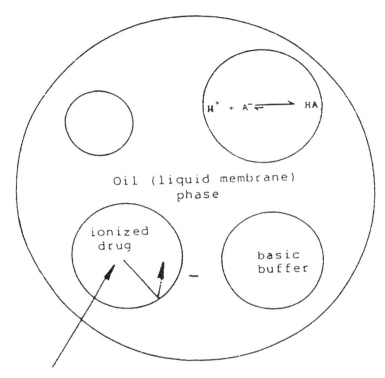

**Figure 7** A w/o/w *liquid membrane* system for removal of acidic drugs from an aqueous system.

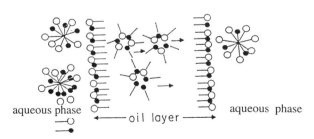

**Figure 8** Schematic illustration of a model for *micellar transport* of water from the outer aqueous phase to the inner aqueous phase through the oil layer in w/o/w emulsion.

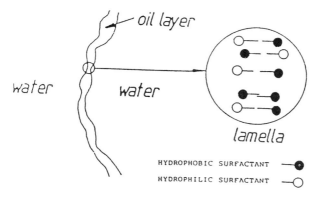

**Figure 9** Schematic illustration of a model for water transport through the *lamellae of surfactant*, due to fluctuation in the thickness of oil.

brane. The mechanism is similar to the one described by Higuchi for release from polymeric matrices — "slab into the sink" of the oil phase (Figure 10) (34–36).

A modified diffusion-controlled release equation was adapted to explain the release results. It has been demonstrated that the release factor $B$ —

$$B = 3/2[1 - (1 - F)^{2/3}] - F; \qquad B = \frac{3\mathrm{De}\,t}{r_o^2 C_o}$$

where $F$ is the release fraction of the marker soluble in the inner phase, De is the effective diffusion coefficient, $r_o$ is the radius of the droplets, $t$ is time of release, $C_o$ is initial concentration of marker — can be plotted against $1/C_o$ and $t$, with excellent correlation coefficients suggesting diffusion-controlled release mechanism of reverse micellar transport (32,33). The "effective" as well as the "real" diffusion coefficient could therefore be calculated.

Although the release mechanism was clarified to some extent, it is still very difficult to control the rate of release (to slow it down) mainly due to the fact that the monomeric emulsifiers that have been used serve both as stabilizing moieties and as transport species. In addition, the fast exchange that those emulsifiers will undergo between the two interfaces and the fact that shear should be avoided in the second emulsification stage lead to the formation of relatively large double emulsion droplets with very limited thermodynamic stability.

The main progress that was made in recent years on in vitro and in

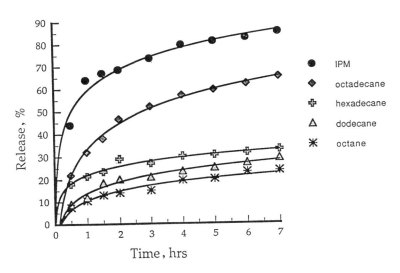

**Figure 10** Release of MTX from multiple W/O/W emulsions. The emulsions contained 2.5% Span 80 and 0.2% BSA as primary emulsifiers, with MTX (1 mg/ml) in the internal phase and the following oil phases: * octane: △ dodecane; + hexadecane; ◆ octadecane; and ● isopropylmyristate.

vivo studies was related to the release patterns of some specific suggested drugs (2,37-42) from the double emulsions.

## III. NEW APPROACHES TO IMPROVED STABILITY AND CONTROLLED RELEASE IN DOUBLE EMULSIONS

One can think of several approaches to overcoming stability and release problems in double emulsions.

1. Stabilizing the inner w/o emulsion by (1) reducing its droplet size, (2) forming $L_2$-microemulsions, (3) forming microspheres, or (4) increasing the viscosity of the inner water.
2. Modifying the nature of the oil phase by (1) increasing its viscosity, (2) adding carriers, or (3) adding complexants.
3. Stabilizing the inner and/or the outer emulsion by (1) using polymeric emulsifiers or (2) adding colloidal solid particles to form stronger and more rigid film at the interface.

The following review evaluates the suggestions related to the use of polymeric emulsifiers to stabilize both the inner and the outer interfaces.

One can consider both naturally occurring macromolecules (gums, proteins) or synthetic, amphiphilic grafted block copolymers. It should be noted, although it is beyond the scope of this text, that polymerizable nonionic surfactants can form *in-situ* cross-linked membranes (after adsorption and carrying out a polymerization process). This concept was well documented and tested (43–45). The polymeric complex that was formed was able to withstand extensive thinning (caused by osmotic driven influx of water) and the resulting swelling of the internal water droplets (46). Our main discussion will be related to recent achievements in stabilizing double emulsions using polymeric (not polymerizable) emulsifiers to enhance steric stabilization.

## IV. NATURALLY OCCURRING POLYMERIC EMULSIFIERS

The stability of double emulsions can be improved (as explained above) by forming a polymeric film or macromolecular complex across the oil–water interfaces. Omotosho and Florence have suggested using macromolecules and nonionic surfactants to form such stabilizing complexes. The film is formed through an interfacial interaction between macromolecules such as albumin and nonionic surfactants (47,48).

Release rates of methotrexate (MTX) encapsulated in the internal phase of w/o/w emulsions stabilized by a film, formed as a result of an interfacial interaction between albumin and sorbitan monooleate (Span 80), were measured as functions of two formulation variables: the oil phase and the secondary emulsifier composition. The release rate was significantly affected by the nature of the oil phase and decreased in the order of:

isopropylmyristate > octadecane > hexadecane >
dodecane > octane

which was a reflection of the increasing internal droplet size of the emulsion (Figure 10).

The release rate data conforms with first-order kinetics. Comparison of the effective permeability coefficients—calculated from the experimental apparent first-order rate constants—with the effective permeability coefficient of water in planar oil layers containing nonionic surfactants (determined by a microgravimetric method) supported the hypothesis of diffusion of MTX via loaded inverse micelles. Surfactants with high HLB values, used as the secondary hydrophilic emulsifier, increased the release rates, primarily by increasing the rate of diffusion of MTX through the nonaqueous liquid membrane (Figure 11). Omotosho and colleagues have also reported the use of other macromolecule complexes (47,48).

Multiple emulsions containing chloroquine phosphate (one of the most effective antimalarial drugs) in the internal phase that had been stored for 2 weeks surprisingly showed a reduced rate of release of chloroquine phosphate as compared with freshly prepared emulsions, suggesting the release from these systems occurs by the process of diffusion as opposed to the physical breakdown of emulsions (48) (Figure 12). The intramuscular administration of chloroquine in the form of w/o/w emulsions has a significant advantage, since it could reduce frequency of administration, improve patient compliance, and increase the therapeutic efficacy of chloroquine. The drug can be formulated as a combination of two types of doses: starting dose, which is incorporated into the external phase, and a maintenance dose, which is encapsulated in the internal phase of the double emulsion.

Wasan and coworkers (49,50), has published a novel method for forming stable hemoglobin–oil-in-water (Hb/o/w) multiple emulsion for use as an artificial red cell substitute. A concentrated Hb solution was

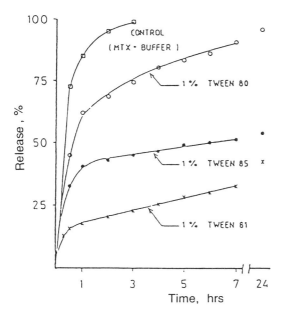

**Figure 11** Effect of the secondary emulsifier (Emulsifier II) on release of MTX from multiple w/o/w emulsions prepared with 2.5% Span 80 and 0.2% BSA as primary emulsifiers with isopropylmyristate as the oil phase.

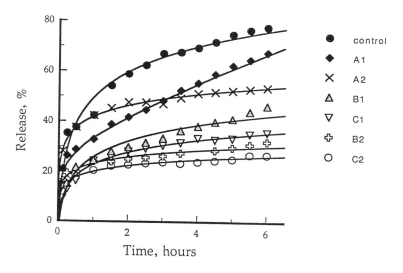

**Figure 12** Profile of chloroquine phosphate release from w/o/w multiple emulsions. $A_1$, freshly prepared PVP emulsions; $A_2$, PVP emulsions stored for 2 weeks; $B_1$, freshly prepared gelatin emulsions; $B_2$, gelatin emulsions stored for 2 weeks; $C_1$, emulsion prepared with gum acacia; $C_2$, acacia emulsion stored for 2 weeks.

emulsified in oil to form microdroplets (with Pluronic F101 and Span 80), followed by dispersion of the primary emulsion into an outer aqueous phase containing hydrophilic surfactants (Pluronic F68 or Tween 80). Addition of human serum albumin into both the inner and the outer phase along with added dextran into the outer phase, seems to be stabilizing the emulsion. The average diameter of the prepared multiple emulsions after homogenization and filtration was 2–3 $\mu$m with good hydrodynamic stability (sensitivity to shear). The formulation showed very small release of Hb from the primary emulsion to the outer aqueous, and good stability of the multiple emulsion during short-term storage (Figure 13).

Oza and Frank (51,52) have suggested the use of colloidal microcrystalline cellulose (CMCC) and various monomeric surfactants to stabilize double emulsions via a mechanical stabilization mechanism. It was shown that the double emulsions were stable over a period of 1 month (monitored by microscopy). Slow release of certain drugs was achieved by gelling the oil phase (Table 1).

Recently (53), we have used BSA (bovine serum albumin) as a polymeric emulsifier added both to the inner and the outer interfaces in the absence and in the presence of conventional monomeric emulsifiers (such

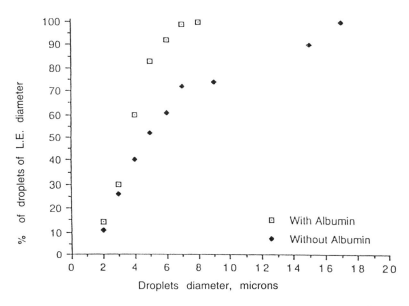

**Figure 13** Effect of albumin on size distribution of multiple emulsion.

**Table 1** Comparison of Release of Lidocaine Base from Different Formulations

| Formulation | Percentage of lidocaine base incorporated within formulation | | Effective diffusion coefficient $D \times 10^6$ $(cm^2/s)$ | Time for 30% release (h) | Release after 5 h (%) |
| --- | --- | --- | --- | --- | --- |
| 2% CMCC[a] dispersion | 1 | — | 0.8<br>0.9 | 6.9<br>6.2 | 25.4<br>27.0 |
| w/o/w[b] (4:6/1:1), 2% CMCC | 1 | Innermost aqueous phase | 0.03<br>0.04 | 186<br>141 | 4.9<br>5.7 |

[a]Release of lidocaine from a 2 wt % CMCC dispersion containing 1 wt % lidocaine base.
[b]Release of lidocaine from a w/o/w (4:6/1:1) emulsion containing 1 wt % lidocaine base 25 ± 0.1°C.

# Surfactant-Stabilized Double Emulsions

as Span 80 and Tween 80). We have tried to evaluate in more detail the mechanism of release of an electrolyte, NaCl, from the inner phase to the outer phase. Since it was believed that the release mechanism is associated with a micellar transport that is diffusion controlled, attempts were made to obtain stable double emulsions with relatively small droplets and a minimum oil micellization capacity.

Double emulsions were prepared with 10 wt % Span 80, 0–0.5 wt % BSA, and 2 wt % NaCl in the inner aqueous phase. The outer interface was stabilized with 5 wt % of Span 80–Tween 80.

The percentage release plot of NaCl versus weight percentage of the inner BSA reveals that BSA retards the release of NaCl (Figure 14). Its maximum effect was obtained at 0.2 wt % BSA. In addition, no emulsions remained stable without any droplet size change for over 25 h, which indicates improved stability in the presence of BSA.

In accordance with the previous studies, it was suggested that the polymeric surfactant forms a "complex" with the monomeric lipophilic surfactant. The complex is probably a thick, strong gelled film that imparts elasticity and resistance to rupture of the inner droplets. The film improves

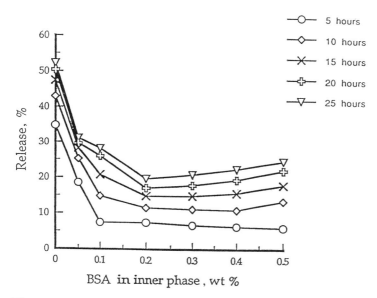

**Figure 14** Percentage release with time of NaCl from double emulsion prepared with 10 wt % Span 80 and various BSA concentrations in the inner phase, and 5 wt % Span + Tween (1:9) in the outer aqueous phase.

the mechanical and steric stability of the double emulsions and slows the coalescence rates. In addition, it appears that it depresses the formation of reverse micelles in the oil phase and slows the transport of the electrolyte via the reverse micelle, or mechanism.

However, the monomeric emulsifiers covering the outer interface do not prevent the coalescence of the outer droplets in the double emulsions. In fact, after 1 week of aging, a significant increase in droplet size distribution, as well as strong flocculation, was detected. Therefore, BSA was also added to the outer aqueous phase (during the second step of the emulsification) in addition to the monomeric Span 80–Tween 80. The Coulter counter measurements (Figure 15) clearly indicate that BSA when present in the outer phase contributes effectively to the stability of the double emulsions. Double emulsion prepared with BSA (in the outer interface) consisted of droplets significantly smaller than those of any other emulsions prepared without the BSA. The improved droplet size distribution was found for any BSA-emulsion prepared with any level of monomeric emulsifiers.

The photomicrographs and the Coulter counter measurements of the

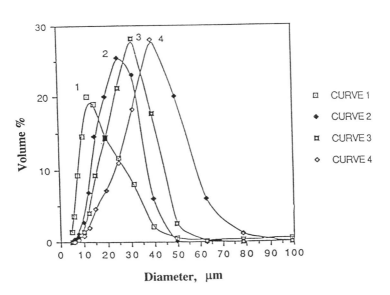

**Figure 15** Droplet size distribution of four emulsions prepared with (1) Span 80 + Tween 80 (9:1) in the inner phase and 0.2 wt % BSA in the outer aqueous phase; (2) Span 80 in the inner phase and without BSA in the outer phase; (3) Span 80 in the inner phase and without BSA in the outer phase; (4) Span 80 in the inner phase and without BSA in the outer phase.

droplet size distribution after 6 weeks of aging show practically no change in droplets size distribution.

## V. TRANSPORT MECHANISM IN DOUBLE EMULSIONS STABILIZED BY NATURALLY OCCURRING EMULSIFIERS

The release rates as a function of the BSA concentration in the outer phase (Emulsifier II) show a minimum at 0.2 wt % BSA and a slight increase in the release rates at higher BSA concentration (Figure 16).

The Higuchi model for the release of drug from solid polymeric matrix (35) as well as other models such as Garti's modification (32,33) for multiple emulsions (based on Fick's diffusion) have been used for testing the double emulsions prepared with BSA in the inner phase. The $B$ parameter (as described previously) was plotted against time ($t$) in order to determine the effective diffusion coefficient De. Figures 17 and 18 are typical plots of the parameter $B$ versus $1/C_o$ of entrapped NaCl, and $B$ versus the time of aging of the emulsion.

The De values for each BSA concentration calculated from plot $B$ against $1/C_o$ and against $t^n$ ($n = 1$) are linear and quite similar, indicating that the Higuchi model dominates the release mechanism. However, better correlation coefficients ($r^2 = 0.998$–$1.000$) will be obtained if the $B$ param-

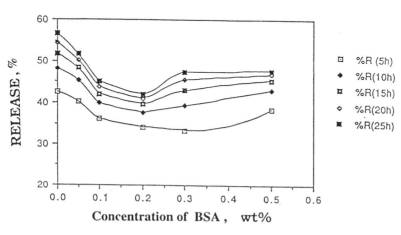

**Figure 16** Percentage release of NaCl from double emulsion prepared with 10 wt % Span 80 in the inner phase versus BSA concentration in the outer phase.

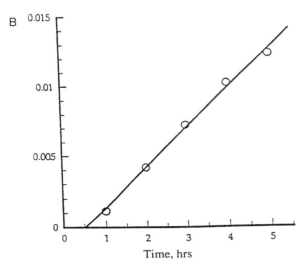

**Figure 17** Factor $B$ plotted against time of release for double emulsion stabilized with 10 wt % Span 80 + 0.2 wt % BSA in the inner phase and 5 wt % of Span 80/Tween 80 (9:1) in the outer phase.

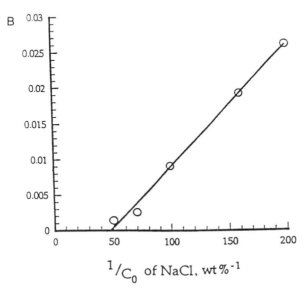

**Figure 18** Factor B plotted against $1/C_o$ of the NaCl concentration in the inner aqueous phase for double emulsion described in Figure 17.

## Surfactant-Stabilized Double Emulsions

eter is plotted against $t^n$ where $n$ (termed arbitrarily as the "diffusion order") varies from 0.5 to 3.0 as a function of the BSA concentration (see Table 2).

Plots of parameter $B$ versus $t^n$ for double emulsions stabilized with BSA in the outer phase showed similar trends of linearity. Best correlation coefficients for linearity of the curves were obtained for $t^n$ in which $n$ was in the range of 0.5 to 1 for 0–0.5 wt % BSA in $W_2$ (see Table 2).

The differences between the functionality and performance of the BSA present in the inner ($W_1$) or the outer ($W_2$) interface can be seen from the plot of the time exponent $n$ (the "diffusion order") versus the BSA concentration (Figure 19).

The effective diffusion coefficients (De) were calculated from the corrected Higuchi equation and plotted versus the BSA concentrations in the inner and outer phases (Figure 20).

Significant differences in the performance of BSA have been found. The outer BSA has only a limited retarding effect on the electrolytes transport, limited to concentration at 0–0.1 wt %, while the inner BSA (at 0.02 wt % levels) has a strong slowing effect on the release of NaCl.

It has been assumed that the parameter $n$ is a reflection of the nature of the film that is formed on the interface. When $n \leq 1$ the film is rather thin, and no significant viscoelastic gel-film (complex between the Span and

**Table 2** Values of $n$ ("Diffusion Order") Obtained from Plots of $B$ Factor Corresponding to the Percentage of Release of NaCl versus $t^n$ from Double Emulsions Prepared with BSA in the Inner Phase ($W_1$) and in the Outer Phase ($W_2$)

| BSA concentrations in $W_1$ or $W_2$ (wt %) | $n_{w1}$[a] | $n_{w2}$[b] |
|---|---|---|
| 0 | 0.5 | 0.5 |
| 0.05 | 1.25 | 0.5 |
| 0.1 | 2.0 | 0.5 |
| 0.2 | 2.0 | 0.5 |
| 0.3 | 2.0 | 1.0 |
| 0.5 | 3.0 | 1.0 |

[a]$n_{w1}$ is calculated from the best fit for linearity of the plot of $B$ against $t^n$ for double emulsions stabilized with BSA added to the inner water phase ($W_1$).
[b]$n_{w2}$ is calculated from the best for linearity of the plot of factor $B$ against $t^n$ of double emulsions stabilized with BSA added to the outer water phase ($W_2$).

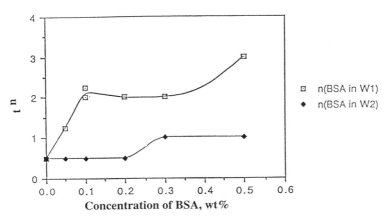

**Figure 19** Plot of exponent $n$ of time ($t^n$) against BSA concentration in both inner and outer interfaces.

BSA) is formed. When $n \geq 1$ the film is viscous due to the formation of a strong Span–BSA complex.

From Figure 19 it can be concluded that the internal w/o film ($W_1$-interface) is more pronounced and stronger than the o/w ($W_2$-interface) film. The $W_1$ film develops as the BSA concentration increases and is well

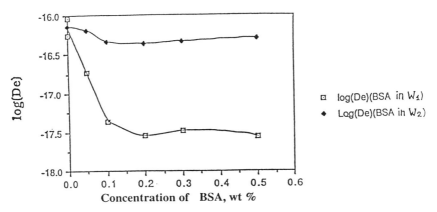

**Figure 20** Plot of log De (effective diffusion coefficient) of NaCl versus BSA concentration in the inner and the outer interfaces.

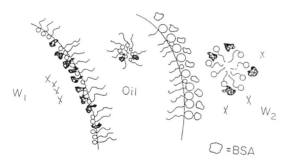

**Figure 21** Schematic illustration of the two interfaces of the double emulsion stabilized by combination of monomeric and polymeric (BSA) emulsifiers.

defined at 0.1–0.2 wt % BSA. On the other hand, the BSA in the outer phase ($w_2$) does not contribute to the formation of a viscoelastic network with the hydrophilic combination of Span–Tween, and has only limited effect on the diffusion coefficient.

Figure 21 is a schematic illustration of the organization of the monomeric and polymeric emulsifiers onto the inner ($W_1$) and the outer ($W_2$) interfaces. Note that the BSA is coadsorbed together with the monomeric emulsifier at the inner interface and serves only as protective colloid at the outer interface.

From a careful evaluation of the release and stability results, it is possible to formulate an optimal double emulsion consisting of Span 80/BSA in the outer interface and Span 80/BSA in the inner interface.

## VI. SYNTHETIC POLYMERIC SURFACTANTS

Florence and Whitehill (54,55) reported the gelation of the aqueous solutions of poloxamer surfactants by gamma-irradiation. The poloxamers are relatively nontoxic poly(oxyethylene)–poly(oxypropylene) block copolymers, with the general formula $HO[C_2H_4O]_a[C_3H_6O]_b[C_2H_4O]_aH$.

Cross-linking of the surfactant molecules may be induced by simultaneous activation of two neighboring molecules with the net resulting in an increase of the molecular weight of the polymer until a three-dimensional network is formed (gel formation). Poloxamers with an ethylene oxide content of less than 70% were shown to degrade on irradiation. This was explained on the basis that a certain minimum chain length was required for gel formation. After emulsification, the surfactant molecules can be cross-linked at the oil-in-water interface and in the continuous phase by gamma-irradiation, forming a network of surfactant molecules that link the dispersed oil globules. Similarly, the poloxamer compounds can be used in the second emulsification step in the preparation of the w/o/w emulsion.

Gamma-irradiation of poloxamer gels in the aqueous phases of isopropylmyristate (IPM) emulsions leads to systems that have a greater intrinsic stability than untreated multiple emulsions. If the internal aqueous phase is gelled, coalescence is prevented. When the continuous outer phase is gelled an opaque emulsion is produced in which the dispersed w/o droplets are held in a hydrophilic polymer network. These w/o droplets are released upon contact with water. Several attempts have been made to improve this technology, but not much has been achieved so far.

To the best of our knowledge no other attempts have been made to use synthetic, well-designed polymeric surfactants to stabilize w/o/w emulsion. Lack of availability of such polymeric emulsifiers and the fact that they are toxic and prohibited systems for food, cosmetic, or pharmaceutical applications are the main reasons for work discontinuity.

The use of double emulsions is not limited to food or pharmaceuticals. Various agricultural and industrial applications can be considered for which the use of synthetic emulsifiers would prove beneficial.

In our recent studies (56–58) attempts have been made to produce and use synthetic polymeric surfactants based on polysiloxane-graft-poly(oxyethylene) for the stabilization of water-in-oil-in-water multiple emulsions. Hydrophobic comb grafted copolymers, polyhydrogen methyl siloxanes grafted to polyethylene glycols (PHMS-PEG) have been used at the inner interface to obtain stable small (1 $\mu$m droplet size) water-in-oil emulsions. Hydrophilic comb grafted copolymers with similar structures but with high density grafting and long polyethylene glycol chain length hooked to the backbone through a hydrophobic spacer of undecanoic acid (UPEG) have been designed, prepared, characterized, and used to stabilize the outer interface of the w/o/w emulsion. The hydrophilic emulsifiers were marked and named, in short, PHMS-PDMS-UPEG.

Two-step emulsification was employed in all multiple emulsions. In the first step, three types of water-in-oil emulsions were prepared with three families of surfactants:

1. Small molecular weight classical emulsifiers such as sorbitan monooleate (Span 80).
2. Graft copolymer polyglycerol polyricinoleate (PGPR, also known commercially as ETD).
3. Grafted PDMS with PEG (commercially known as Abil EM-90).

The above W/O emulsions were further emulsified with the PHMS-PDMS-UPEG emulsifiers. The outer phase consisted of hydrophilic emulsifier prepared in our labs; basically PHMS-PDMS-UPEG with a molecular weight of PHMS-PDMS backbone of 3300, 55% of UPEG substitution, and 45 EO units at hydrophilic moiety (for detailed structure see Reference 45). In the second step, six different polymeric, grafted polymers were used. The two most interesting polymers are the following:

1. The most hydrophilic polymer consisted of PHMS-PDMS of MW 3300 with 52% substitution of UPEG with 45 EO units.
2. The most hydrophobic, grafted polymer (Abil EM-90, ETD, and Span 80) consisted of PHMS-PDMS of MW 3300 with 5% substitution and UPEG with 7 EO units.

The double emulsions were segregated into three sets. In set I, Span 80 w/o emulsion was further emulsified with six different polymeric PHMS-PDMS-UPEG. In set II were the emulsions prepared with ETD in the inner phase and the six different grafted polymers with outer phase, and in set III were the emulsions prepared from w/o siliconic Abil EM-90 emulsions further emulsified with the newly designed PHMS-PDMS-UPEG emulsifier.

All three sets of emulsions were very stable. The droplets were evenly distributed and uniformed. The droplet size distribution was narrow and the particles in general were smaller in size than double emulsions prepared with monomeric emulsifiers (see Figure 22).

The outer aqueous phase electrical conductivity as a function of time (results from migration of NaCl from the inner to the outer phase) can be plotted and correlated to the percentage of release of the electrolyte to the outer phase (Figure 23).

The differences between emulsions stabilized with the three types of Emulsifier I are clearly seen in the release rates of NaCl from the inner to the outer phase. The siliconic hydrophobic emulsifier (Abil EM-90) showed the best trapping capacity and the slowest release rates, followed by ETD and Span 80. Span 80 was unable to stabilize the multiple emulsion, probably due to its fast migration to the outer interface (34). Most emulsions prepared with Span 80 showed low trapping capacity (low yield of preparation) and fast release.

For any given Emulsifier I (Span 80, ETD, and siliconic polymer), the stability of the double emulsion and the fate of release is directly affected by the hydrophilicity of the Emulsifier II. The more hydrophilic surfactant will entrap more electrolytes and release them slower than the more hydrophobic surfactant. The surfactant consisting of the longer EO units attached to the backbone will best retain the NaCl within the inner phase.

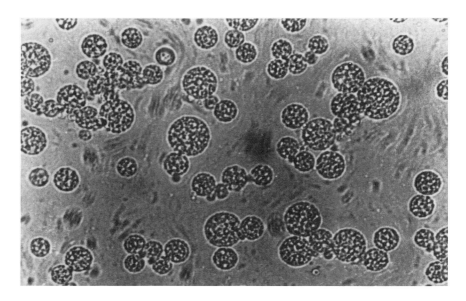

**Figure 22** Photomicrograph of the best double emulsion obtained from the use of Abil-EM 90 in the inner interface and PHMS-PDMS-52% UPEG-45 EO at the outer interface.

It should also be noted that when the inner interface is covered with polymeric surfactant (Abil EM-90) the need for polymeric surfactant for the outer interface is less pronounced, and good stability and slow release rates can be obtained even with less efficient hydrophilic (Emulsifier II) surfactant.

## VII. RESISTANCE TO SHEAR

It has been well documented that shear should be avoided in the second emulsification stage in order to prevent possible rupture of the double emulsion droplets that can lead to the inversion of the w/o/w emulsion into simple o/w emulsion and low yields of preparation. This rupture of the double emulsion will lead to premature release of the entrapped matter during the homogenization. On the other hand, slow or gentle stirring will form somewhat large multiple emulsion droplets, with limited thermodynamic stability and a strong tendency to coalescence.

Polymeric surfactants are expected to form a thick, rigid film at the oil phase, resistant to shear.

# Surfactant-Stabilized Double Emulsions

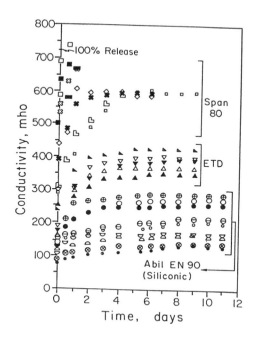

**Figure 23** Plot of conductivity of the outer aqueous phase reflecting the concentration of NaCl in the outer phase (% release) versus time (days) in three sets of double emulsions. *Set I* (the circles): Abil EM-90 in the internal interface and polymeric silicones in the outer phase. All circles indicate use of Abil EM-90 as hydrophobic Emulsifier I. The lower curves (●, ⊗, ◠, ◡, ⊖, ·, ⊠) indicate the most hydrophilic PHMS-PDMS-UPEG Emulsifier II and the upper curves (⊕, ○, ●) indicate the most hydrophobic PHMS-PDMS-UPEG with 52% substitution and 45 EO units. Each circle symbol represents a different polymeric emulsifier. *Set II* (the triangles): All triangles represent use of polyglycerol polyricinoleate (ETD) as Emulsifier I, and the curves are arranged again with increasing hydrophobicity of Emulsifier II. The lower curve (▲) represents the most hydrophilic emulsifier and the upper curve in the set represents the most hydrophobic one (▲). *Set III* (the squares): All squares represent the use of Span 80 as Emulsifier I and the curves are arranged with increasing hydrophobicity of Emulsifier II.

Multiple emulsions consisting of siliconic lipophilic Emulsifier II (Abil EM-90) and siliconic grafted copolymer Emulsifier II (PHMS-PDMS-52%UPEG-45 EO) have been sheared for up to 60 min and evaluated under light microscope for the release rates of NaCl.

Figure 24 compares the release rates of reference multiple emulsions (stabilized with Span–Tween combinations) to the siliconic stabilized multi-

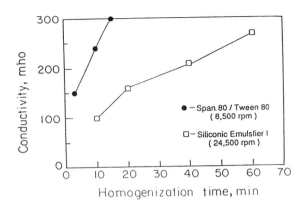

**Figure 24** Conductivity of the outer aqueous phase plotted against homogenization time and speed (minutes) for two sets of double emulsions: control double emulsions stabilized with (●) 10 wt % Span 80 + Tween 80 (5:1) as Emulsifier I, (□) 10 wt % silicone-based double emulsion consisting of Abil-EM-90 as Emulsifier I. In both emulsions, PHMS-PDMS-52% UPEG-45 EO served as Emulsifier II (2%). Conductivity correlates to percentage release.

ple emulsions. The Span–Tween double emulsions were inverted almost immediately after preparation into simple o/w emulsions, and practically no significant entrapment was achieved. On the other hand, the silicone-based emulsions remained double, and slow release of NaCl was detected.

The siliconic emulsions were shear-resistant, with high yields of preparation. After 60 minutes of homogenization, 50% of the NaCl was released. Furthermore, the average droplet size in the sheared emulsion was reduced and narrowed (Figure 25).

The mechanical strength of double emulsions stabilized with polymeric-silicone-based surfactant was surprising and encouraging. Centrifugation usually ends in rupture of most of the conventional double emulsions and in fast creaming. The reference Span–Tween double emulsions separated into two phases after moderate centrifugation action. However, the silicone-based double emulsions remained stable (double emulsions) even after 45 minutes of high-speed centrifugation. Even ultracentrifugation did not cause severe damage to the double emulsions.

As a consequence, w/o/w emulsions consisting of the lipophilic silicone (Abil EM-90) Emulsifier I and PHMS-PDMS-52%UPEG-45 EO as Emulsifier II were prepared with the microfluidizer under very high shear rates, and were found to be small in diameter with excellent stability to

# Surfactant-Stabilized Double Emulsions

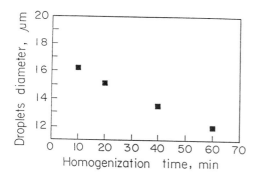

**Figure 25** Average droplet size of double emulsion immediately after preparation as a function of homogenization time (double emulsion of 10 wt % Abil-EM-90 as Emulsifier I and 2 wt % PHMS-PDMS-52% UPEG-45 EO as Emulsifier II).

coalescence (Figure 26). The average droplet size is 4.5 µm and grew to only 7.0 µm after 30 days of aging.

## VIII. CONCLUSIONS

Double emulsions of w/o/w have many potential applications, but no real commercial product exists yet in the market place. The main reason is the

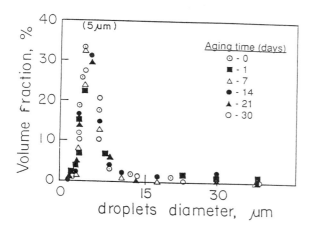

**Figure 26** Droplet size distribution of double emulsions stabilized with siliconic Emulsifier I (Abil EM-90) and siliconic Emulsifier II (PHMS-PDMS-52% UPEG-45 EO) immediately after preparation and after 30 days of aging.

inherent instability of the preparation and the uncontrolled release of the entrapped matter (both on the self and within the time of the application). The use of macromolecules to serve as steric stabilizers for both the inner and the outer interfaces has opened new options and possibilities.

Naturally occurring macromolecules such as selected proteins (BSA, HSA, gelatin) and hydrocolloids (gum arabic) have been used with great success to improve the film formation over the water and the oil phase (better anchoring, full coverage, thick layer, low desorption, no interfacial migration) together with monomeric classical hydrophobic and hydrophilic emulsifiers. The polymer–surfactant complex is an ideal interfacial barrier for diffusion controlled transport of both hydrophobic (nonionized, lipid-like molecules) and hydrophilic (ionized molecules, hydrophilic organic molecules, and electrolytes) substances. It has been shown that both thermodynamic stability and entrapment were significantly improved. In addition, the micellar transport via reverse-micelles was reduced and therefore migration was very limited.

The use of synthetic, well-designed and characterized polymeric surfactants was very helpful in reducing leakage of addenda, improving shear resistance, and obtaining small double emulsions with excellent shelf-life and stability. For agricultural and industrial applications it seems to be an ideal formulation.

Polymeric surfactants, in combination with the conventional small molecular weight emulsifiers, are suggested as the future emulsifiers for double emulsions.

## REFERENCES

1. P Becher. Emulsion Theory and Practice. 2nd edition. New York, Reinhold, 1965.
2. SS Davis, J Hadgraft, KJ Palin. In: Encyclopedia of Emulsion Technology. Vol. 2. Becher, P, ed. New York, Marcel Dekker, 1985, p 159.
3. AT Florence, D Whitehill. Int J Pharm 11:277, 1982.
4a. P Sherman, C Parkinson. Prog Coll Polym Sci 63:10, 1978.
4b. P Docic, P Sherman. Coll Polym Sci 258:1159, 1980.
5. W Seifriz. J Phys Chem 29:738, 1980.
6. TJ Lin, H Kurihara, H Ohta. J Soc Cosmet Chem 26:121, 1975.
7. AT Florence, D Whitehill. J Colloid Interface Sci 79:243, 1981.
8. S Matsumoto, WW Kang. J Dispersion Sci Technol 10:455 1989.
9. S Matsumoto, Y Kida, D.Yonezawa. J Colloid Interface Sci 57:353, 1976.
10. AT Florence, D Whitehill. J Colloid Interface Sci 79:243, 1981.
11a. M Frenkel, R Shwartz, N Garti. J Colloid Interface Sci 94:174 1983.
11b. N Garti, M Frenkel, R Schwartz. J Dispersion Sci Technol 4:237, 1983.
12. WJ Herbert. Lancet 2:771, 1965.

13. RH Engel, SJ Riggi, MJ Fahrenbach. Nature 219:856, 1968.
14. JA Omotosho. Int J Pharm 62:81, 1990.
15. Y Morimoto, K Sugibayashi, Y Yamaguchi, Y Kato. Chem Pharm Bull 27:3188, 1979.
16. PJ Taylor, CL Miller, TM Pollock, FT Perkins, MA Westwook. Hygiene Cambridge 67:485, 1969.
17. LA Elson, BCV Mitchley, AJ Collings, R Schneider. 15:87, 1970.
18. AJ Collings. U.K. Patent 1,235,667 (1971).
19. CH Benoy, LA Elson, R Schneider. Rev Europ Etudes Clin Biol 45:135, 1972.
20. AF Brodin, SG Frank. Acta Pharm Suec 15:1, 1978.
21. AF Brodin, SG Frank. Acta Pharm Suec 15:111, 1978.
22. NN Li. U.S. Patent 3,410,794 (1968).
23. NN Li. AI Ch EJ 17:459, 1971.
24. NN Li, AL Shrier. Recent Developments in Separation Science 1. Cleveland, Chem Rubber Co., 1972, p. 163.
25. L Mackles. U.S. Patent 3,395,028 (1968).
26. E Dickinson, G Stainsby. In: Advances in Food Emulsions and Foams. Dickinson, E, Stainsby, G, eds. London, Elsevier Applied Science, 1988.
27. RK Owusu, Z Qinhong, E Dickinson, Food Hydrocolloids 6:443, 1992.
28. E Dickinson, J Evison, JW Gramshaw, D Schwope. Food Hydrocolloids 8:63, 1994.
29. E Dickinson, J Evison, RK Owusu. Food Hydrocolloids 5:481, 1991.
30. C Chiang, GC Fuller, JW Frankenfeld, CT Rhodes. J Pharm Sci 67:63, 1978.
31. S Magdassi, M Frenkel, N Garti. J Dispersion Sci Technol 5:49, 1984.
32. S Magdassi, M Frenkel, N Garti. Drug Ind Pharm 11:791, 1985.
33. S Magdassi, N Garti. J Controlled Release 3:273, 1986.
34. Y Kita, S Matsumoto, D Yonezawa. J Colloid Interface Sci 62:87, 1977.
35. T Higuchi. J Pharm Sci 52:1145, 1963.
36. TK Law, AT Florence, TL Whateley. J Pharm Pharmacol 36:50, 1984.
37. MS Mohamed, FS Ghazy, MA Mahdy, MA Gad. J Pharm Sci 5:56, 1989.
38. KP Oza, SG Frank. J Dispersion Sci Technol 10:163, 1989.
39. JA Omotosho, AT Florence, TL Whateley. Int J Pharm 61:51, 1990.
40. JA Omotosho, TL Whateley, AT Florence. Biopharm Drug Dispos 10:257, 1989.
41. B Mishra, JK Pandit. J Controlled Release 14:53, 1990.
42. M Mathew, CS Thampi. East Pharm 33:385, 1990.
43. TK Law, TL Whateley, AT Florence. Int J Pharm 21:277, 1984.
44. TK Law, TL Whateley, AT Florence. J Controlled Release 3:279, 1986.
45. AT Florence, TK Law, TL Whateley. J Colloid Interface Sci 107:584, 1985.
46. JA Omotosho, TK Law, TL Whateley, AT Florence. Colloid Surfaces 20:133, 1986.
47. JA Omotosho, TK Law, TL Whateley, AT Florence. J Pharm Pharmacol 38:865, 1986.
48. JA Omotosho, TK Law, TL Whateley, AT Florence. J Microencapsulation 6:183, 1989.

49. RL Beissinger, DT Wasan, LR Sehgal, AL Rosen. U.K. Patent Appl GB 2,221,912 (1990).
50. S Zheng, RL Beissinger, DT Wasan. J Colloid Interface Sci 144:72, 1991.
51. KP Oza, SG.Frank. J Dispersion Sci Technol 7:543, 1986.
52. KP Oza, SG.Frank. J Dispersion Sci Technol 10:187, 1989.
53. N Garti, A Aserin, Y Cohen. J Controlled Release 29:41, 1994.
54. AT Florence, D Whitehill. J Pharm Pharmacol 32:641, 1980.
55. AT Florence, D Whitehill. J Pharm Pharmacol 34:687, 1982.
56. Y Sela, S Magdassi, N Garti. Colloids Surfaces A 83:143, 1994.
57. Y Sela, S Magdassi, N Garti. Colloid Polymer Sci 272:684, 1994.
58. Y Sela, S Magdassi, N Garti. J Controlled Release 29:41, 1994.

# 15
# Interparticle Forces from SANS Measurements of Frozen Dispersions

**David C. Steytler and Brian H. Robinson**
*University of East Anglia, Norwich, England*

**Julian Eastoe**
*University of Bristol, Bristol, England*

**Isabel MacDonald**
*Exxon Chemicals Ltd., Abingdon, England*

## I. BACKGROUND

Steric stabilization by surfactants and polymers of liquid and solid dispersions in oil media plays a crucial role in applied colloid science and is responsible for the stability of a range of commercially important products including oil additives, paints, inks, etc. The theoretical treatment of the interparticle forces characterizing the stability of these dispersions is now well developed. This includes both the attractive potential determined mainly by van der Waals forces and described by the Lifshitz approach, and the steric stabilizing (repulsive) forces that arise when adsorbed monolayers of adjacent particles overlap (1). For polymer stabilizers, interactions between overlapping monolayers on adjacent particles have been treated by the Flory-Kringbaum Theory and more recently by de Gennes (2a,2b). However, until recent years experimental investigation of the steric interactions between surfactant monolayers has not been extensive.

In this paper we describe a new technique (3,4) utilizing the osmotic pressure of concentrated particle dispersions in a frozen solvent. In the freezing process two microdomains are formed within the frozen dispersion: one, the pure oil solvent that is selectively solidified (I), the other, a concentrated "liquid" dispersion of particles in oil (II) which separates when the solvent freezes. These two domains are intimately mixed within the frozen colloid and exist in a state of equilibrium at fixed pressure and temperature. The position of equilibrium, which can be represented by the proportion of the solvent that is solidified, and thereby the concentration of particles within the flluid microdomains (II), depends on both temperature and pressure. With increasing pressure (and/or decreasing temperature) the particles become more concentrated within the fluid domains. Combined with SANS measurements to determine the interparticle separation, the osmotic pressure provides a measure of the interparticle repulsion forces between the adsorbed monolayers on a 3-D configuration of particles. Owing to the $Q$-range of available small-angle neutron scattering (SANS) instruments, our technique is ideally suited to the study of small, spherical (nano)particles and water-oil (w/o) microemulsions with radius $r = 15 - 40$ Å.

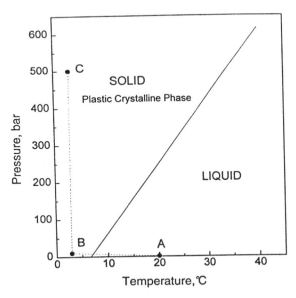

**Figure 1** Pressure-temperature phase diagram for the l-s transition of cyclohexane. Points A, B, and C refer to SANS measurements of Figure 7 (see text for details).

## II. HIGH MELTING POINT OILS (PLASTIC-CRYSTALLINE SOLIDS)

Over recent years the structure and properties of dilute dispersions of colloidal particles and self-assembly systems in oil media (reversed micelles, microemulsions) have been extensively investigated (5–10). In these studies the dispersion is usually in the liquid state and there have been virtually no structural studies of colloidal dispersions in a solidified or "frozen" condition. A partial explanation for this is that the liquid alkanes that are often used as the dispersion medium, for example $n$-heptane or isooctane have freezing points of $\sim -100\,°C$, at which temperature dispersions such as w/o microemulsions are unstable. However for cyclic ("globular") alkanes, for example cyclohexane ($C_6H_{12}$), the freezing point is near ambient temperature. Dispersions can therefore be studied in such oils in a frozen state without destabilizing the colloid. Owing to a relatively high plasticity (11), the solid state of these alkanes is often referred to as a "plastic crystalline" phase.

The dispersion medium used in this study is cyclohexane with a normal freezing point, $T_f^*$, of 6.6°C (Figure 1). Owing to the higher degree of rotational freedom in the solid phase the entropy of freezing, $\Delta S_f$, of cyclic alkanes forming plastic crystalline phases is much lower (Table 1) than for the straight-chain alkanes. This explains both the high freezing point and lower gradient, $dP/dT$, of the freezing line for cyclohexane relative to $n$-hexane.

## III. DISPERSED SYSTEMS EXAMINED IN THE SOLID PHASE

A schematic representation of the surfactant-stabilized particles and droplets examined, defining dimensions relevant to our SANS measurements, is shown in Figure 2. The overall radius, $r_p$, is given by the sum of the core radius ($r_c$) and the thickness of the curved monolayer of surfactant ($t_s$).

**Table 1** Comparison of Freezing Points and Volume and Entropy Changes on Freezing for the Oils Cyclohexane and $n$-Hexane

| Alkane | m.p. ($T^*$) (°C) | $\Delta V_f$ (cm$^3$ mol$^{-1}$) | $\Delta S_f$ (J K$^{-1}$ mol$^{-1}$) |
|---|---|---|---|
| Cyclohexane | 6.6 | $-5.10$ | $-9.40$ |
| $n$-Hexane | $-95.3$ | $-8.90$ | $-73.3$ |

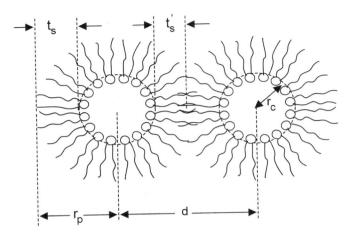

**Figure 2** Schematic representation of surfactant-stabilized particles (or microemulsion droplets); $r_c$ is the particle core radius, $t_s$ is the surfactant layer thickness, $t_s'$ is the "effective surfactant layer thickness, and $r_p$ the overall particle radius.

## A. Water-in-Oil Microemulsions

Dilute sodium bis-(2-ethylhexyl) sulphosuccinate (AOT)-stabilized w/o microemulsions are thermodynamically stable and may be considered to be a dispersion of essentially spherical water droplets with mean water core radius $r_c$ given by

$$r_c = \frac{3V_w}{N_{Av}A_S} \omega \qquad (1)$$

where $V_w$ = molar volume of water, $\omega = [H_2O]/[\text{surfactant}]_i$, $A_S$ = surface area of surfactant at the water/surfactant interface, and $[\text{surfactant}]_i$ = concentration of surfactant in the system located at the interface (6,9,12).

For small droplets, $\omega < 25$, the extent of polydispersity is small (6,9). Numerous studies have shown that Equation 1 is valid over a range of (1) temperature (15–100°C [10,13,14]), (2) pressure (1–1000 bar [13,16]), (3) alkane chain length (ethane to dodecane [11,17]), and (4) droplet volume fraction, $\phi$ up to 0.7 [6,18]).

At constant composition, AOT-stabilized w/o microemulsions have an upper (UTPB) and lower (LTPB) temperature phase boundary where the one phase (1$\phi$) isotropic $L_2$ microemulsion becomes unstable, as shown in Figure 3 for microemulsions formed in cyclohexane. The conditions that induce phase separation of the microemulsion and the processes that occur

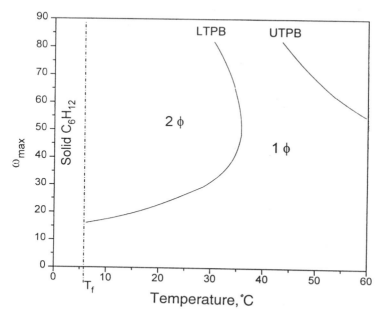

**Figure 3** Solubilization–temperature ($\omega_{max} - T$) phase diagram for AOT-stabilized w/o microemulsions in cyclohexane.

have been studied previously, and both the thermodynamic and kinetic driving forces for phase separation have been discussed in detail (11–14). At the LTPB, the instability is believed to be due to curvature effects such that Winsor II systems (15) are formed with the droplet radius governed by a minimum energy condition for curvature of the surfactant layer often referred to as "natural interfacial curvature." Instability at the UTPB is more complex in nature with increasing contributions from (1) attractive droplet interactions and (2) a propensity to positive interfacial curvature, i.e., formation of o/w microemulsions (Winsor I systems). It is clear from Figure 3 that a $\omega = 10$ AOT-stabilised w/o microemulsion is well removed from the UTPB when $T < 20°C$, so attractive interactions between droplets are not expected to be significant for the system examined in this paper.

### B. Calcium Carbonate Nanoparticles

We have used $CaCO_3$ particles similar to those described as "V-series" in a previous SANS investigation by Markovic and Ottewill (5,7,8) with a core radius $r_c = 26$ Å. The surfactant stabilizing the $CaCO_3$ core particles is a mixture of calcium alkylbenzene sulfonates with a distribution of alkyl chain lengths centered on carbon number $C_{24}$ (19) giving a layer thickness, $t_s$, of approximately 16 Å (8). Under ambient conditions, dispersions of the particles are stable and discrete in the liquid state of the solvent (cyclohexane) over the temperature range employed (3–20°C).

Details of the particle radii and surfactant layer dimensions, obtained from SANS data for dilute systems, are given in Table 2.

## C. Freezing Point Depression of the Dispersions

For the systems examined in this work the concentration of particles, and of microemulsion droplets, in the solvent is approximately $1 \times 10^{-3}$ mol $dm^{-3}$, so that the mole fraction of particles $x_B$ is $\sim 1 \times 10^{-4}$. The freezing point depression ($T_f - T_f^*$) for an ideal cyclohexane solution with $x_B = 10^{-4}$ can be estimated using Equation 2:

$$\Delta T = T_f - T_f^* = \frac{RT_f^{*2}}{\Delta H_f} x_B \qquad (2)$$

where $T_f^*$ is the freezing temperature (at 1 atm) and $\Delta H_f$ is the molar enthalpy of freezing of the pure solvent. For cyclohexane $T_f^* = 6.6°C$ and $\Delta H_f = -2.63$ kJ $mol^{-1}$ which gives $\Delta T = -0.03°C$. The effect of the particles on the freezing point of the continuous phase should therefore be negligible.

It is important to note that for the systems examined the transition is entirely reversible. With the microemulsion samples, no water separation is observed for the liquid–solid–liquid transition, i.e., the microemulsion is thermodynamically stable in both states of the system. Similarly, no destabilization of the $Ca(ABS)_2$-stabilized $CaCO_3$ particles is observed on inspection of the samples on melting the frozen dispersion.

**Table 2** Dimensions of the $Ca(ABS)_2$-Stabilized Carbonate Particles and AOT without Microemulsion in Cyclohexane Obtained from SANS Measurements

| Dispersion | $r_c$ (Å) | $t_s$ (Å) | Surfactant |
|---|---|---|---|
| w/o Microemulsion ($\omega = 10$) | 19 | 8 | AOT |
| V-series $CaCO_3$ particles | 26 | 16 | $Ca(ABS)_2$ |

## IV. LARGE-SCALE MICROSTRUCTURE IN THE FROZEN DISPERSIONS

The structural relationship between the microdomains I and II was examined by both microscopy and SANS measurements.

### A. Light Microscopy

The freezing process was examined by viewing the movement of the freezing "front" through AOT-stabilized w/o microemulsions and $CaCO_3$ particle dispersions using an optical microscope (Wild Heerburg M8, magnification ×60). The sample was frozen by passing a coolant from right to left through a small sample cell such that a temperature gradient was always present, as shown in Figure 4a. The liquid sample was slowly cooled below the freezing point, and photographs were taken of the freezing process induced by the temperature gradient. A typical photograph is shown in Figure 4b, which clearly illustrates the dendritic growth pattern observed for all samples. The "tree-like" structure of the dendrites represents the pure

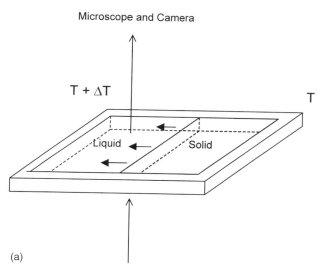

**Figure 4** (a) Schematic representation of the experimental configuration used in optical microscopy measurements. (b) Photograph showing the dendritic growth pattern of the freezing front in an AOT-stabilized w/o microemulsion in cyclohexane. ($\omega$ = 10, [AOT] = 0.2 mol dm$^{-3}$).

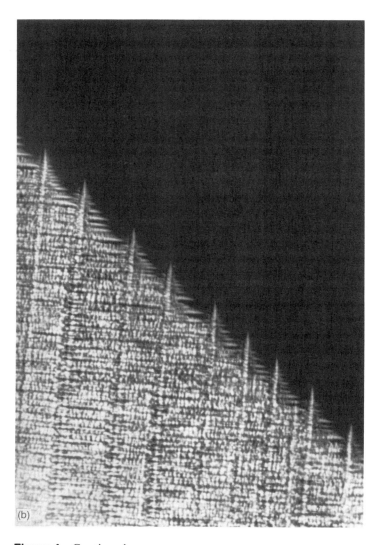

**Figure 4** Continued.

solid phase (I) between which is trapped the "liquid" domains (II) containing the droplets ejected in the freezing process. It should be emphasized that the process is dynamic and Figure 4b shows a "snapshot" in time of the dendritic growth process within the sample.

## B. Electron Microscopy

Scanning electron microscopy (SEM) was also employed to examine the configuration adopted by the microdomains (II) within a frozen dispersion of $CaCO_3$ particles. A liquid dispersion of the V-series colloid in cyclohexane ($\phi = 0.10$) was initially frozen by cooling to $-40°C$. The sample was then removed from the cooling bath, connected to a vacuum line, and allowed to warm. When the melting temperature was reached, liquid cyclohexane was continuously removed from the sample by a process of freeze-drying, leaving behind a "skeleton" of the $CaCO_3$ colloid. A small sample of this highly porous material was carefully transferred to an SEM stub and coated with a layer of gold approximately 20 nm thick. SEM measurements were made using a Phillips 501B scanning electron microscope with an accelerating voltage of 15 kV.

Electron micrographs of the freeze-dried dispersion are shown in Figure 5. Although the method of sample preparation may well remove some of the finer features, the micrographs provide a clear illustration of the "coarse" structure of the microdomains containing the particles. The photographs show a highly porous, connected structure that is adopted by the particles as they are progressively rejected by the pure solvent in the freezing process. In the "frozen" state the system appears to be bicontinuous with respect to both the microdomains of pure frozen oil (I) and ejected particles (II). It may be inferred that this sponge-like structure, with a pore size distribution in the range 5–50 $\mu m$, represents the space between the dendrites (Figure 4b) in the solidified dispersion. Indeed, evidence of an "imprint" of dendritic growth, in support of this hypothesis, is clearly visible in Figure 5. The spatial configuration of the particle clusters is therefore determined by the specific manner in which the dendrites grow and interpenetrate in forming the frozen solid phase (20).

## C. SANS Measurements at Low Q: Microstructure of the Concentrated Phase

The available wavelength range (4–20 Å) and low cross section for absorption of neutrons makes SANS a powerful technique for structural studies of colloidal systems with size correlations in the range 10–1000 Å. Moreover, the appreciable difference in coherent scattering lengths between hy-

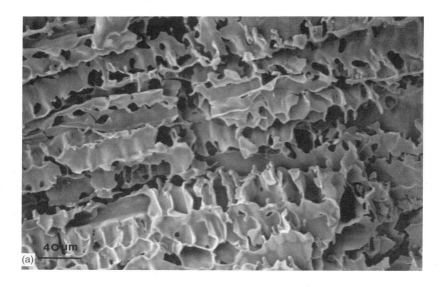

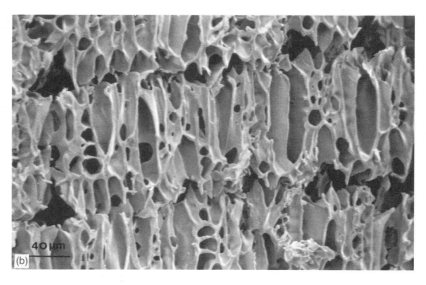

**Figure 5** (a–d) Electron micrographs of a freeze-dried solution of the V-series CaCO$_3$ particles in cyclohexane ($\phi = 0.1$) showing microstructure of the particle cluster in the solid phase. The pore space represents the region originally occupied by the pure solvent (I) that was selectively frozen in the freezing process.

# SANS Measurements of Frozen Dispersions

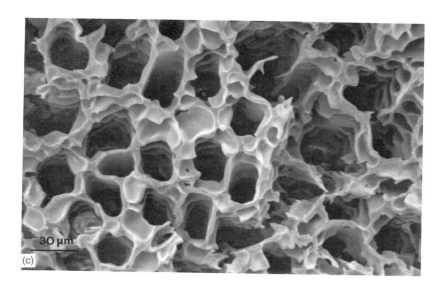

(c)

(d)

drogen ($-3.74 \times 10^{-11}$ cm) and deuterium ($6.67 \times 10^{-11}$ cm) introduces the possibility of controlled contrast variation whereby different regions of a hydrocarbon- or water-containing system can be contrasted and measured by selective deuteration of the components.

The scattering intensity $I(Q)$ is shown in log $I(Q)$ versus log $Q$ form in Figure 6 for a 5.0% mass/volume dispersion of $CaCO_3$ particles in H-cyclohexane at 1 bar and 20°C (liquid phase) and 500 bar and 3°C (solid phase). The data sets were measured on two different instruments (D16 and D11) and have been superimposed to cover a wide range of $Q$. Above $Q = 1.6 \times 10^{-1}$ Å$^{-1}$, no significant scattering is observed, and data for this $Q$ regime are not shown.

At 1 bar, cyclohexane is liquid and the $I(Q)$ versus $Q$ profile shows the characteristic form of a dilute system of noninteracting spherical particles in a liquid dispersion. The SANS pattern from the same system in the solid phase is very different. First, there is a significant $S(Q)$ contribution, which implies that the particles are in a highly concentrated state (this is discussed in detail in the following section). Second, there is evidence for a scattering contribution at very low $Q$ resulting from large structures, i.e., large-scale correlations from the microdomains formed between I and II.

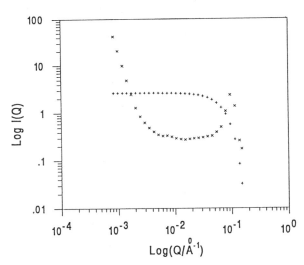

**Figure 6** SANS profiles, plotted as $\log(I(Q))$ versus $\log(Q)$ for Ca(ABS)$_2$-stabilized $CaCO_3$ particles (5.0% m/v) in the liquid phase at 1 bar, 20°C (+), and solid phase at 500 bar, 3°C (×). The Q-range covered in the plot is $6.3 \times 10^{-3} - 1.6 \times 10^{-1}$ Å$^{-1}$.

Under the same conditions of temperature and pressure, $I(Q)$ profiles for the microemulsion droplets exhibit similar features.

According to Porod (21), for a sharp (stepwise) interface the asymptotic value of $I(Q)$ should be

$$I(Q) = 2\pi\Delta\rho^2 \frac{S}{V} Q^{-4} \qquad (3)$$

where $S/V$ is the total interfacial area per unit volume of the sample, and $\Delta\rho$ is the difference in the scattering length densities of the scattering entity (particle aggregate) and the medium (cyclohexane).

Equation 3 can be used to estimate the value of the parameter S/V of the microdomains (II) in the frozen colloid. At high $Q$, plots of $I(Q)Q^4$ versus $Q$ (not shown) exhibit a plateau, characteristic of scattering from the interface. Since the aggregate contains $CaCO_3$ particles, including H-$Ca(ABS)_2$ surfactant, $\rho_{agg}$ can be estimated using the $\rho$ values for $CaCO_3$ and $Ca(ABS)_2$ determined by Markovic and Ottewill (7,8). The calculated value of $S/V$ at 3°C, 1 bar is approximately 600 cm$^{-1}$, which is in broad agreement with the pore size revealed by the SEM measurements.

## V. SANS MEASUREMENTS AT HIGH Q: PARTICLE SIZE, SHAPE AND INTERACTIONS

The total SANS scattering from a system of monodisperse particles can be expressed as

$$I(Q) = n_p V_p^2 \Delta\rho^2 P(Q) S(Q) \qquad (4)$$

The measured scattering cross section, $I(Q)$, contains a dimensionless intraparticle function, $P(Q)$, characterizing the size and shape of the individual particles, and an interparticle structure factor, $S(Q)$, which includes spatial correlations arising from interactions between the particles. The overall scattering intensity is also proportional to the number density of particles, $n_p$.

For a dilute system of noninteracting particles at low volume fraction (typically $\phi < 0.02$), $S(Q) \sim 1.0$ and $I(Q)$ is then uniquely determined by the particle size and shape. In the case of monodisperse spherical particles, the radius is easily obtained by fitting the observed $I(Q)$ to a form-factor for a single sphere. However, the microemulsion droplets and $CaCO_3$ particles are known to exhibit a small degree of polydispersity that can be accounted for (22) using a modified Schultz distribution model with a width parameter ($\sigma/r_c = 0.2$) defining the extent of polydispersity in the system.

The structure factor $S(Q)$ is the Fourier transform of the particle center–center correlation function $g(r)$. For noninteracting (i.e., hard-sphere) particles, excluded volume effects introduce specific structural features into $g(r)$ and therefore $S(Q)$ as $\phi$ increases. The predominant form of the $S(Q)$ function for noninteracting hard spheres is a broad peak with a position of maximum intensity ($Q_{max}$) given by (23)

$$Q_{max} = \frac{c}{d} \tag{5}$$

where $d$ is the mean interparticle separation, $c$ is a constant dependent on the interparticle configuration, which is equal to $2\pi$ for a random distribution of spherical particles.

A typical set of $I(Q)$ data is shown in Figure 7 for $\omega = 10$ microemulsion droplets in cyclohexane. At 20°C and 1 bar, cyclohexane is in the liquid state, and the scattering closely resembles that of a dilute system of spheres. Since the scattering-length density of H-cyclohexane closely matches that of the surfactant hydrocarbon chains, the coherent small-angle scattering arises from the region of high contrast between the surfac-

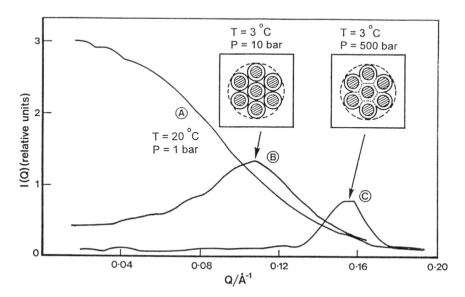

**Figure 7** $I(Q)$ SANS profiles for AOT-stabilized w/o microemulsion droplets ($\omega = 10$, [AOT] = 0.10 mol dm$^{-3}$) in cyclohexane at state points A, B, and C shown in Figure 1.

tant layer and $D_2O$ core. By fitting a form-factor for polydisperse spheres (Schultz distribution, $\sigma/r_c = 0.2$) the mean radius of the water core ($r_c$) was obtained as 19.0 ± 0.5 Å. Similarly, using an $H_2O/H$-$AOT/C_6D_{12}$ contrast profile, the total radius ($r_p$) of the droplet, including the surfactant layer, was found to be 27.0 ± 0.5 Å. The length of the AOT surfactant molecule is then obtained as approximately 8.0 ± 1.0 Å, in good agreement with previous measurements (12,13).

On cooling to 3 °C, a large proportion of the cyclohexane freezes to a pure solid phase microdomain (I) that is in equilibrium with a fluid phase microdomain (II) containing a high concentration of droplets. The scattering is then dominated by the strong $S(Q)$ contribution arising from interparticle correlations in the concentrated configuration of droplets in II. On gradually increasing the pressure to 500 bar, the position of the $S(Q)$ peak, $Q_{max}$, moves consistently to higher $Q$ values, indicating that the interparticle separation is decreasing according to Equation 5. This systematic response is both reversible and reproducible for all the systems studied. The reversibility of the changes suggests that the system is always thermodynamically stable with respect to irreversible association/coagulation of droplets or separation of the dispersed water phase. The SANS behavior of the corresponding $CaCO_3$ particle system shows a similar response to pressure and temperature on solidification of the dispersion.

The effect of pressure and temperature on the microemulsion droplet separation in the concentrated microdomains is predictable. In order for more solvent to be "frozen out" from these domains (transferred from II to I), as is required when the temperature is further decreased (or pressure increased), the liquid domains must shrink, thereby concentrating and ultimately "compressing" the droplets contained within them.

In our experiment the response of the system to pressure and temperature variation is measured by the droplet separation in the "liquid" microdomains within the frozen dispersion. $S(Q)$ profiles for the 3 °C data are shown in Figure 8 from which the mean interparticle separation, $d$, is obtained using Equation 5. A hexagonal close-packed arrangement of droplets (for which $c = \pi\sqrt{6}$) has been assumed in this calculation, which is in accord with results of previous SANS studies on this system at high droplet volume fractions (6,18). To represent the state of "compression" of the surfactant monolayers between the points of contact of the droplets, we define an "effective surfactant layer thickness" ($t_s'$) representing the thickness of the compressed layer as shown schematically in Figure 2 for interdigitated layers.

$$t_s' = \frac{d}{2} - r_c \qquad (6)$$

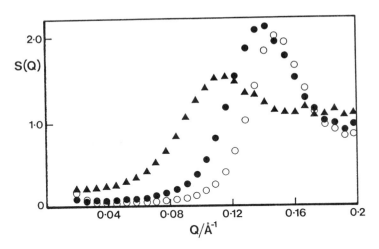

**Figure 8** Structure factors $S(Q)$ obtained at 20°C and 500 bar (●), 3°C and 10 bar (▲), and 3°C and 500 bar (○) for the AOT $\omega = 10$ w/o microemulsion system.

Isotherms showing the dependence of $t_s'$ on pressure and temperature are presented in Figure 9. The plots follow a consistent trend of behavior, with the droplet separation decreasing with increasing pressure and/or decreasing temperature. This response is consistent with a balance of the opposing forces operating in the system between expansion and contraction of the fluid domains, which may be visualized by considering (1) the balance of energies to concentrate the droplets against an osmotic pressure gradient and (2) the energy released ($T\Delta S$ and $P\Delta V$) on freezing out the pure solvent.

## VI. OSMOTIC PRESSURE AND MEAN INTERPARTICLE POTENTIAL IN PARTICLE CLUSTERS

Representations of the data using pressure and temperature as variables (Figure 9) can be rationalized qualitatively but do not provide a quantitative description. An analysis based on the osmotic pressure of the droplets was therefore developed that describes the simultaneous dependence of $t_s'$ on both $P$ and $T$. The approach also provides a measure of the steric forces operating between the stabilizing surfactant layers on adjacent particles/droplets.

At each $P, T$ state point, the osmotic pressure, $\pi$, of the droplets in the liquid microdomains (II) is given by Equation 7 (3):

$$\pi = \frac{\Delta S_f \Delta T - \Delta V_f \Delta P}{V_m^L} \tag{7}$$

where $\Delta S_f$ = the entropy of freezing of the pure solvent (cyclohexane), $\Delta V_f$ = the volume change on freezing of the pure solvent, $V_m^L$ = the molar volume of the solvent, $\Delta T = T - T^*$, and $\Delta P = P - P^*$.

The dependence of $\pi$ on $t_s'$ is shown in Figure 10 for the $\omega = 10$ microemulsion in cyclohexane. The data represent a unification of the three different experimental isotherms of Figure 9, which demonstrates that the interaction energy is not significantly temperature and pressure dependent and confirms the general applicability of osmotic pressure as a parameter for correlation of the effects of temperature and pressure on the droplet configuration in the frozen dispersion.

It is clear that some energy needs to be expended before the droplets are brought into contact ($t_s' = t_s$) which arises from the increased osmotic pressure ($\pi'$) of the system solely due to concentration effects. The magnitude of these colligative properties was found to be greater than could be explained in terms of the "molecular weight" of the droplets as determined by SANS measurements. In order to explain the larger magnitude of $\pi'$ it is

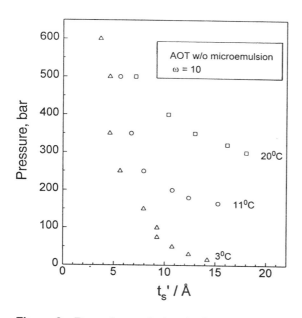

**Figure 9** Dependence of $t_s'$ on hydrostatic pressure and temperature for the AOT $\omega = 10$ w/o microemulsion.

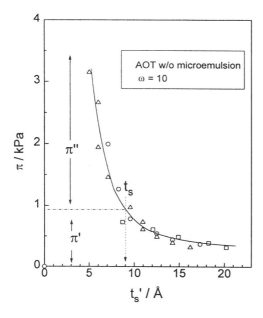

**Figure 10** Correlation of effective surfactant thickness ($t_s'$) with osmotic pressure ($\pi$) for the AOT $\omega = 10$ w/o microemulsion at 20°C (□), 11°C (○), and 3°C (△).

necessary to invoke the presence of low molecular weight components dissolved in the solvent. This could arise either through an impurity in the surfactant, or by partitioning of a small proportion of the surfactant from the droplet surface to the surrounding solvent medium.

The interdigitation of the surfactant layers is represented in Figures 9 and 10 by the region in which $t_s' < t_s$ where the curve appears to be approaching a limiting value of $t_s'$ of ~5.0 Å. The general form of the curve, and absence of a minimum, is in accord with the behavior of sterically stabilized dispersions in better than "theta" solvents for which attractive interactions are negligible compared to thermal energy.

A previous analysis of the osmotic pressure in the "overlap" regime of concentration represented by $\pi''$ has been used to obtain the interparticle pair potential, $E(t_s')$, in the cluster of particles (1).

$$E(t_S') = \int_{t_S}^{t_S'} \left\{ \frac{\Delta S_f \Delta T - \Delta V_f \Delta P}{V_m^L} \right\} (r_c + t_S')^2 dt_S' \tag{8}$$

It is, however, important to recognize that $E(t_s')$ represents the "pair" potential function for particles in a hexagonal close-packed configuration and is not that between an isolated pair of particles. Figure 11 shows $E(t_s')$ for the AOT-stabilized w/o system as calculated from Equation 8. The energies probed by the range of pressure and temperature employed are low and are considerably less than that required to fuse the droplets ($\sim 100$ kJ mol$^{-1}$) (9). It is also clear from Figure 11 that droplets moving with thermal energies of order $kT$ will undergo rather soft collisions with significant interpenetration (or compression) of the surfactant layers.

The interparticle pair-potential for the V-series carbonate dispersion in cyclohexane at 3°C is also shown in Figure 11. As for the microemulsion system, it was observed that a significant energy is initially required to overcome osmotic effects and bring the particles into contact. This is not surprising since these colloids are known to contain a minor proportion (<2% by weight) of alkane residues that are closely associated with the particles and are difficult to remove. The surfactant is also known to partition into the oil phase to a small extent; it might be expected that this will also contribute to the osmotic pressure of the system. Direct force measurements of interparticle interactions for similar, but larger ($r_c = 66$ Å), Ca(ABS)$_2$-stabilized carbonate particles have recently been reported

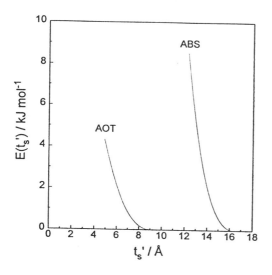

**Figure 11** Interparticle potential, $E(t_s')$, for the $\omega = 10$ AOT w/o microemulsion and Ca(ABS)$_2$-stabilized CaCO$_3$ particles in cyclohexane.

using a surface balance technique (24). A fundamental difference between this study and our SANS technique is the dimensionality of the particle configuration, which is 2-D and 3-D, respectively. The results of the surface balance measurements have been presented in terms of a normalized force ($F/r_c$). Data for our $CaCO_3$ particle system in cyclohexane give a remarkably similar profile to that of Reference 24 when displayed in this form.

## VII. QUARTER CENTURY PROGRESS AND NEW HORIZONS

A pioneering method for measuring interactions between adsorbed polymer layers on particles dispersed in solution has been developed by Ottewill and colleagues (25). The technique involves "classical" measurement of osmotic pressure in a piston/cylinder arrangement incorporating a membrane at one end such that the osmotic pressure can be measured as a function of particle concentration by application of pressure on the piston. It has been applied to measure interactions between polymer-stabilized latices in both water and oil solvents, e.g., latices of methyl methacrylate stabilized by a mixed stabilizer of poly-12-hydroxy stearic acid and glycidyl methacrylate in dodecane. This system showed an initial soft repulsive interaction at $\phi = 0.55$ followed by a much steeper increase in pressure at $\phi = 0.565$ characteristic of a more hard-sphere like interaction. A Flory-Kringbaum approach incorporating the cell model of van Megen adequately represented the initial rise in pressure with concentration but underestimated the extent of the steeper rise in the overlap region at high volume fractions. The inability of the theory to account for the observed behavior was ascribed to a nonuniform density within the stabilizing layer due to a distribution in the length of the stabilizer chains. One drawback with the method as applied to polymer latices in oil appears to be the long equilibration time required, which can be up to several weeks. This complication is likely to become of greater significance in the study of smaller nanoparticles, which would necessitate the use of narrower-pore membranes.

The work of Israelachvili and coworkers (26) in developing instrumentation for direct measurement of surface forces (Surface Force Apparatus — SFA) has had enormous impact recently in experimental surface science. This elegant and delicate technique, with a resolution of a few angstroms, was first applied to studies of forces operating within liquids confined between mica plates. On close approach, the force distance profile was found to follow an oscillatory pattern (27) representing the progressive exclusion of "layers" of molecules from the liquid sample between the plates with the "period" of the oscillation approximately equal to the molecular diameter of the molecules. This behavior was observed for spherically symmetric molecules, such as cyclohexane and octamethylcyclotetrasilane (OMCTS), and anisotropic straight chain alkanes such as tetradecane and hexadecane. However, for branched alkanes such as isononadecane, the force distance profiles were initially monotonically attractive but then monotonically repulsive at close approach. The SFA has also been used to examine interactions between monolayers adsorbed on the mica plates in oil media. Early studies included physisorbed polymers such as polystyrene (28) in cyclohexane. Force-distance profiles for these polymer systems with molecular weight $1 \times 10^5$ and $6 \times 10^5$ g mol$^{-1}$, showed an initial attraction followed by a steep repulsion giving a minimum at a plate separation close to that of the radius of gyration of the free polymers in solution (85 Å and 210 Å, respectively).

For shorter chain surfactant stabilizers, the measured force-distance profiles showed a significant contribution arising from the packing of the oil molecules between the adsorbed layers (29). The form of the interactions in this regime was found to depend strongly on the regularity of the adsorbed surfactant monolayers. For tightly packed monolayers with a high degree of order (e.g., didodecylammonium bromide, DDAB), an oscillatory force-distance profile was again observed for oils such as $n$-tetradecane. However, for surfactants forming less ordered monolayers (cetyltrimethyl ammonium bromide, CTAB and Ca(ABS)$_2$) the profile for $n$-tetradecane was monotonic. In the presence of moisture, the behavior of these surfactant monolayers was complex, with pronounced interactions occurring at large plate separation. This behavior is difficult to reconcile but may be due to induced rod-shaped micelle or liquid crystal formation.

The SFA technique has also been used to examine the effects of water adsorption in adsorbed surfactant layers in the absence of solvent following equilibration with a controlled-humidity atmosphere (30). A range of surfactants were examined including CTAB, dihexadecyldimethylammonium acetate (DHDAA), and the phospholipid $L$-$\alpha$- dimyristoyl-phosphatidylethanolamine (DMPE). For all the surfactants studied, the thickness of the monolayers formed increased with increasing humidity in order of the surfactant headgroup polarity, i.e., CTAB > DHDAA > DMPE. Parallel studies of the extent of solvation of the adsorbed surfactants with oils (cyclohexane and $n$-alkanes $n$-C$_6$-$n$-C$_{16}$) were also undertaken. In agreement with conventional wisdom, the extent of solvation was found to decrease with (1) increasing oil molecular weight and (2) decreasing packing density of the monolayer. The compressibility of the monolayers was also found to increase with oil adsorption.

Both water- and oil-swollen surfactant lamellar ($L_\alpha$) phases have been examined using the SFA method with the liquid crystal adsorbed such that the sheet structure is aligned parallel with the mica plate surfaces (31). The force-distance profiles showed many oscillations characteristic of the spacing within the liquid crystals, which were 280 Å for the oil-swollen phase and 88 Å for the water-swollen phase. The spacing observed for oscillations near close-contact of the plates was found to be lower than expected, due possibly to dislocation effects. The two types of $L_\alpha$ phases are distinctly different in their stabilizing forces which are electrostatic for the water-swollen phase but originate from "undulation" forces in the oil-swollen lyotropic. Analysis of the force-distance relationship gave the compression modulus for the two phases to as $2.8 \times 10^4$ J m$^{-3}$ (water-swollen $L_\alpha$) and $2.8 \times 10^4$ J m$^{-3}$ (oil-swollen $L_\alpha$). The latter figure is significantly higher than the predicted value of $\sim 3 \times 10^2$ J m$^{-3}$.

Recently, Cosgrove et al. have advanced the SFA technique by interfacing to the CRISP neutron reflectometer (32). They are then able to measure F(r) profiles and structural changes within the adsorbed layer simultaneously for high molecular weight polymers adsorbed on a macroscopically flat quartz block. Block copolymers of polystyrene and poly(ethylene oxide) adsorbed on quartz plates were studied under compression. Adsorption from a good solvent (toluene) led to preferential adsorption of the ethylene oxide chains, but because of the very high asymmetry of the polymers and the favorable adsorption energy, the polystyrene was also adsorbed. The extent of the polystyrene-adsorbed layer was found to be less than the radius of gyration for both samples studied. On the approach of a second coated plate, the polystyrene volume fraction was found to increase near the interface, indicative of a strong interlayer repulsion. On changing the solvent to octane, the profiles indicated collapse of the polymer layer with a dramatic decrease in the pressure required to hold the quartz plates at the same separation.

Future measurements using the pressure cell technique are planned to provide a systematic examination of surfactant layer interactions as a function of [1] surfactant structure—tail length, branching, and aromaticity; [2] surfactant type—hydrocarbon, silicone, and fluorocarbon; and [3] monolayer curvature—particle/droplet size and shape. These studies will include both surfactant/polymer-stabilized particles and surfactant self-assembly systems such as w/o microemulsions and liquid crystal phases.

## REFERENCES

1. DH Napper. In: Polymeric Stabilisation of Colloidal Dispersions. New York, Academic Press, 1983.
2a. PG de Gennes. C R Acad Sci (Paris) 300:839–843, 1985.
2b. PG de Gennes. Adv Coll Int Sci 27:189–209, 1987.

3. DC Steytler, BH Robinson, J Eastoe, K Ibel, JC Dore, IP MacDonald. Langmuir 9:903–911, 1993.
4. J Eastoe, BH Robinson, DC Steytler, JC Dore. Chem Phys Letts 166:153–158, 1990.
5. RH Ottewill. Langmuir 5:4–9, 1989.
6. S-H Chen. Ann Rev Phys Chem 37:351–399, 1986.
7. I Markovic, RH Ottewill. Coll Pol Sci 264:454–462, 1986.
8. I Markovic, RH Ottewill. Coll Pol Sci 264:65–76, 1986.
9. PDI Fletcher, AM Howe, BH Robinson. J Chem Soc Faraday Trans 1 83:985–1006, 1987.
10. PDI Fletcher, S Clarke, X Ye. Langmuir 6:1301–1309, 1990.
11. J Timmermans. J Phys Chem Solids 18:1–8, 1961.
12. M Kotlarchyk, JS Huang, S-H Chen. J Phys Chem 89:4382, 1985.
13a. J Eastoe, BH Robinson, DC Steytler. J Chem Soc Faraday Trans 86:511–517, 1990.
13b. J Eastoe, BH Robinson, DC Steytler, WK Young. J Chem Soc Faraday Trans 86:2883–2889, 1990.
14. AM Howe. Ph.D. Thesis, University of Kent at Canterbury, 1986.
15. PA Winsor. Solvent Properties of Amphiphilic Compounds. London, Butterworths, 1954.
16. JL Fulton, RD Smith. J Phys Chem 92:2903–2907, 1988.
17. J Eastoe, BH Robinson, DC Steytler, D Thorn-Leeson. Adv Coll Int Sci 36:1–31, 1991.
18. M Kotlarchyk, S-H Chen, JS Huang, MW Kim. Phys Rev Letts 53:941–944, 1984.
19. JF Marsh. Chem Ind 14:470–473, 1987.
20. KA Jackson, JD Hunt, DR Uhlmann, TP Steward. Trans Met Soc AIME 236:149–158, 1966.
21. G Porod. Koll Z 124:82, 1951.
22. M Kotlarchyk, S-H Chen. J Chem Phys 79:2461, 1983.
23. NW Ashcroft, J Lekner. Phys Rev 83:145, 1966.
24. JH Clint, SE Taylor. Coll Surf 65:61–67, 1992.
25. RJR Cairns, RH Ottewill, DWJ Osmond, I Wagstaff. J Coll Int Sci 54:45, 1976.
26. JN Israelachvili, GE Adams. J Chem Soc Faraday Trans 74:975, 1978.
27. JN Israelachvili. Acc Chem Res 20:415, 1987.
28. JN Israelachvili, M Tirrell, J Klein, Y Almog. Macromolecules 17:204–209, 1984.
29. Y-L Chen, Z Xu, JN Israelachvili. Langmuir 8:2966, 1992.
30. YLE Chen, ML Gee, CA Helm, JN Israelachvili, PM McGuiggan. J Phys Chem 93:7057–7059, 1989.
31. P Kekicheff, P Richetti, HK Christenson. Langmuir 7:1874–1879, 1991.
32. T Cosgrove, A Zarbakhsh, PF Luckhman, ML Hair, JRPW Webster. Faraday Discuss 98:189–201, 1994.

# 16
# Phase Transitions in Lipid Monolayers at the Air–Water Interface

**Harden M. McConnell**
*Stanford University, Stanford, California*

## I. INTRODUCTION

Over the past decade much progress has been made in understanding the physical chemistry of monomolecular films at the air–water interface (1–4). This is especially true for films composed of amphiphilic lipids, such as phosphatidylcholines and cholesterol. Such lipid films generally exhibit coexisting phases such as liquid and gas, or liquid and solid. Coexisting liquid phases (5,6) or coexisting liquid and gas phases have been of particular interest from a theoretical point of view, as noted below.

In monolayers of amphiphilic lipids, coexisting phases tend to break up into domains that are observable using epifluorescence microscopy (1–3). A major theoretical challenge has been to understand the physical and chemical properties of these domains, their shapes and sizes as well as the kinetics of change of shape and size. The theoretical problems posed by domain properties are simplest for liquid domains surrounded by a second liquid, or by gas (liquid) domains surrounded by a liquid (gas). This is because the gas and liquid phases are isotropic in two dimensions. Further, the dipole–dipole interaction between molecules has a particularly simple mathematical form, since these dipoles are on average perpendicular to the plane of the monolayer (2,4). Moreover, the line tension at the interface between domains can be assumed to be constant for specified chemical compositions of the neighboring phases. In the following discussion, I outline some of the progress that has been made in studies of the properties of domains formed in coexisting liquid phases composed of binary mixtures of phosphatidylcholines and cholesterol.

## II. PHASE TRANSITIONS AND SHAPE TRANSITIONS

Many experimental observations on monolayers exhibiting liquid–liquid immiscibility can be summarized using Figures 1 to 3. Figure 1 depicts schematically the phase diagrams of a number of binary mixtures of cholesterol and phosphatidylcholines (2,5,6). At monolayer pressures above the phase boundary the monolayer is homogeneous; at pressures below the phase boundary the monolayer consists of two phases, a "white" liquid phase and a "black" liquid phase. The white phase is richer in the phosphatidylcholine, and the black phase is richer in cholesterol. The white phase is the more

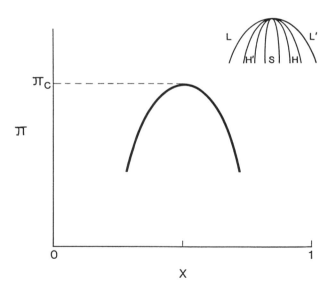

**Figure 1** Schematic representation of the phase diagram for a binary mixture of lipids showing a liquid–liquid phase separation, and a critical composition at mole fraction of one component $X = 0.5$. At monolayer pressures $\pi$ above the critical pressure $\pi_c$ the monolayer is homogeneous, whereas at lower pressures within the two-phase boundary region the monolayer separates into two immiscible liquid phases. The inset at the upper right is a theoretical representation of how a monolayer with two liquid phases gives rise to a stripe (S) or hexagonal (H, H′) phases. See References 1–4, 10–12.

# Phase Transitions in Lipid Monolayers

fluorescent as seen in the epifluorescence microscope, due to the fact that fluorescent lipid probes (typically present to the extent of 1%) preferentially dissolve in this phase. The critical composition in Figure 1 is $X = 1/2$. In recent unpublished studies with John Hagen, it has been found that a number of binary mixtures of cholesterol with unsaturated phosphatidylcholines have quite symmetrical phase diagrams with critical compositions near 1/2.

Figure 2 shows a fluorescence microscope photograph of a monolayer composed of roughly equal amounts of dimyristoyl phosphatidylcholine (DMPC) and cholesterol. This photo is taken from unpublished work of Rice and McConnell under conditions essentially identical to those described earlier (7). This particular monolayer was prepared by reducing the

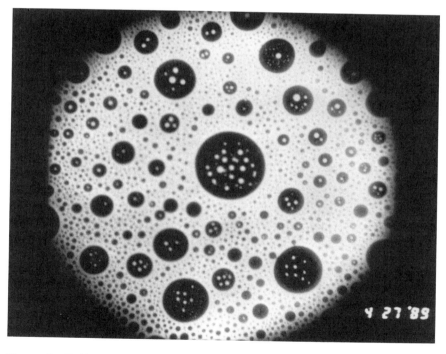

**Figure 2** Epifluorescence microscope photograph of a lipid monolayer composed of dimyristoyl phosphatidylcholine and cholesterol under conditions identical to those described in Reference 7. The dark phase is rich in cholesterol and the light phase is rich the phosphatidylcholine. The contrast between the two phases is achieved by means of a low concentration of a fluorescent lipid probe that partitions preferentially into the phosphatidylcholine phase.

applied pressure through the critical pressure, thus giving rise to the two liquid phases. Note the wide diversity of domain sizes, which doubtless represent a nonequilibrium state, or perhaps a state of metastable equilibrium (8). We believe that this diversity of domain sizes reflects the history of the sample as it passed through the critical point, with its large associated composition fluctuations.

Monolayers can exhibit highly ordered domain patterns, or "superstructure phases," as illustrated in Figure 3 (9-13). For example, decompressing a monolayer of DMPC and cholesterol having a composition removed from the critical composition generally gives rise to circular domains as sketched along path (1') in Figure 3. With further decompression, these domains grow in size until they undergo "shape transitions" (14-17). For example, as a circular domain grows in size it can undergo a (weakly first-order) transition to a dogbone shape, according to path (2') (18). A more rapid growth in domain size can give rise to the formation of transient shapes such as those sketched along paths (2"), (2"'), etc. These simple, symmetrical transient shapes ultimately develop complex labyrinthine patterns (15). However, a slow increase in size of domains with dogbone shapes generally gives rise to the formation of a stripe phase, along path (3').

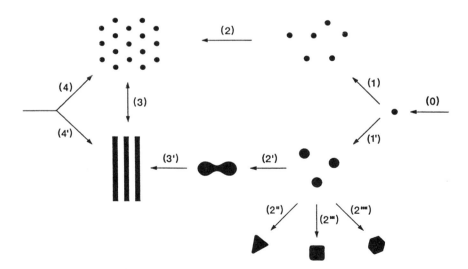

**Figure 3** Schematic representation of the sizes and shapes of lipid domains that are formed from two coexisting liquid monolayer phases at the air-water interface. See text for details.

# Phase Transitions in Lipid Monolayers 391

These various domain shape transitions have been observed experimentally, and have been analyzed theoretically (14–18).

Under idealized equilibrium conditions, the initial formation of circular domains would not proceed according to path (**1'**) but according to path (**1**), in which all the circular domains have their equilibrium radius (**2**). At larger domain concentrations, where the domain–domain repulsions become comparable to $kT$, the domains are expected to form an hexagonal lattice, following path (**2**). This situation may have been observed experimentally in one case (13). In principle there can be transitions between an hexagonal array of circular domains and the stripe phase, according to path (**3**) (10,12). Both hexagonal and stripe phases have been observed in monolayers. There may also be transitions between ordered and disordered stripe phases (9,12).

The shapes and sizes of lipid domains at the air–water interface are determined by two competing forces (1–4). One force is the line tension at the interface along domain boundaries. For liquid domains this line tension is expected to be constant along a given domain boundary. The net effect of the line tension is to favor large, circular domains. The dipolar electrostatic repulsion between molecules at the air–water interface has a component that acts at long range. This force keeps domains apart from one another, as seen in Figure 2, and also favors domains with elongated shapes and domains that are small.

## III. THE KINETICS OF SIZE AND SHAPE CHANGES

One of the major challenges in this field has been to account theoretically for the kinetics of shape and size changes of lipid domains. Experimentally, domain shape changes for isolated domains are found to take place quite rapidly, within seconds or a few minutes. Examples are the circle-to-dogbone transition (**2'**) or the transitions to harmonic shapes with higher rotational symmetry, such as (**2"**), (**2'''**), etc. (15). The kinetics of these transitions are determined by the interplay of line tension, electrostatic dipole–dipole repulsions, and hydrodynamic drag in the subphase (14). (Viscous resistance within the monolayer is negligible for liquid monolayers.) The translation of lipid domains under the influence of their own electric fields, or applied electric fields, is also quite rapid, and has been analyzed theoretically (19,20).

One of the more difficult problems has been to understand the rate of size equilibration of lipid domains. This problem is clearly illustrated by the photograph in Figure 2. Here the two lipid phases do not form a stripe phase or an hexagonal phase, nor do the individual domains have circular shapes all with the same radius. It is likely that such domain patterns do not represent a state of thermodynamic equilibrium, or even states of metastable equilibrium. Preliminary theoretical calculations for liquid–liquid domain systems such as that illustrated in Figure 2 indicate that this rate of equilibration is intrinsically slow due to its dependence on the diffusion of lipid molecules in a gradient of the electrochemical potential (21). This process of domain size change is to be contrasted with the much more rapid process of domain shape change which involves macroscopic hydrodynamic flow in the monolayer and subphase (14).

## IV. CHEMICAL AND BIOLOGICAL SIGNIFICANCE

The quantitative understanding of the various microscopic domain structures observable in lipid monolayers should facilitate the use of these systems for chemical and biological purposes. As one example, the dependence of domain shapes on line tension and dipole density provides a sensitive means to detect photochemical reactions in these systems (7). Monolayers have been used for many years to model various aspects of biological membranes. The ability to visualize phase separations in monolayers should be an impetus for the investigation of such phase separations in bilayers and biological membranes.

## ACKNOWLEDGMENT

This research was supported by the National Science Foundation, grant MCB9316256.

## REFERENCES

1. H Möhwald. Annu Rev Phys Chem 41:441, 1990.
2. H McConnell. Annu Rev Phys Chem 42:171, 1991.
3. CM Knobler. Science 249:870, 1990.
4. D Andelman, F Broachard, CM Knobler, F Rondelez. In: Micelles, Membranes, Microemulsions and Monolayers. Gelbard, WM, Ben-Shaul, A, Roux, DA, eds. New York, Springer, 1994, pp 559–602.
5. S Subramanian, HM McConnell. J Phys Chem 91:1715, 1987.
6. CL Hirshfeld, M Seul. J Phys 52:1537, 1990.
7. P Rice, HM McConnell. Proc Nat Acad Sci 86:6445, 1989.
8. HM McConnell, R De Koker. Langmuir 12:4897, 1996.
9. M Seul, VS Chen. Phys Rev Lett 70:1658, 1993.

10. D Andelman, F Broachard, J Joanny. J Chem Phys 86:3673, 1987.
11. HM McConnell, VT Moy. J Phys Chem 20:2311, 1986.
12. MM Hurley, SJ Singer. Phys Rev B 46:5873, 1992.
13. M Löesche, H Möhwald. Eur Biophys J 11:35, 1984.
14. HA Stone, HM McConnell. Proc Roy Soc Lond 448:97, 1995.
15. KYC Lee, HM McConnell. J Phys Chem 97:9532, 1993.
16. SA Langer, RE Goldstein, DP Jackson. Phys Rev A 46:4897, 1992.
17. TK Vanderlick, H Möhwald. J Phys Chem 94:886, 1990.
18. R De Koker, HM McConnell. J Phys Chem 97:13419, 1993.
19. JK Klingler, HM McConnell. J Phys Chem 97:2962, 1992.
20. KYC Lee, JF Klingler, HM McConnell. Science 263:655, 1994.

# 17
# Surfactant Monolayers in Relation to Foam Breaking by Particles

**R. Aveyard, B. P. Binks, and P. D. I. Fletcher**
University of Hull, Hull, England

## I. INTRODUCTION

It is well known that to be an effective aqueous foam breaking entity, a particle—liquid or solid—must be capable of entering the air–water (aw) surface of a foam lamella (1–4). The feasibility of entry depends ultimately on the free energy of the various interfaces involved, and for the most part the interfaces will be coated with surfactant monolayers.

A liquid droplet, after entry into a lamella surface, may rupture the film by spreading along the surface, which results in thinning and rupture by a Marangoni mechanism (Figure 1a). Spreading is usually discussed in terms of the spreading coefficient, $S_{o,aw}$, defined in terms of interfacial tensions $\gamma$ by

$$S_{o,aw} = \gamma_{aw} - (\gamma_{ow} + \gamma_{oa}) \tag{1}$$

where subscripts ow and oa refer to the oil–water and oil–air (vapor) interfaces, respectively. For a system at adsorption equilibrium, for spreading to occur $S_{o,aw} = 0$; otherwise $S_{o,aw} < 0$ and spreading does not occur (2). In a nonequilibrated system, the spreading coefficient can be positive (spreading) or negative (nonspreading) (2). As will be discussed, surfactant monolayers are often capable of solubilizing hydrocarbons and other oils, and adsorption into the monolayer reduces the value of $\gamma_{aw}$ and hence of $S_{o,aw}$ and the tendency of the oil to spread.

If an oil drop is to enter the aw surface, the aqueous phase must be incapable of spreading spontaneously on the oil. For this reason it is conve-

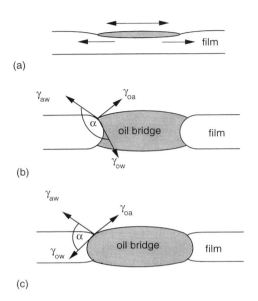

**Figure 1** Oil drops in aqueous films. (a) Oil drop spreading along surface; arrows show direction of flow of drop and underlying liquid. (b) Unstable drop; angle $\alpha >  90°$. (c) Stable drop; angle $\alpha < 90°$.

nient to define an entry coefficient for the oil droplet at the aw surface, $E_{o,aw}$, by

$$E_{o,aw} = (\gamma_{aw} + \gamma_{ow}) - \gamma_{oa} = -S_{w,oa} \qquad (2)$$

The entry coefficient is the negative of the spreading coefficient, $S_{w,oa}$, of the aqueous phase on the surface of oil. In a system at adsorption equilibrium, entry is feasible if $E_{o,aw} > 0$. The minimum value of $E_{o,aw}$ is 0, when entry is not thermodynamically feasible. Clearly, the value of the entry coefficient like that of the spreading coefficient is dependent on the capability of a surfactant monolayer to solubilize hydrocarbon.

If an oil drop enters the aw surface of a soap film and does not spread, then if the film does not rupture for other reasons, the drop will ultimately bridge the film. The stability or otherwise of this oil bridge depends on its shape, which in turn depends on the relative values of the interfacial tensions in the system (1,4,5). Both a stable and an unstable oil bridge in a soap film are depicted in Figure 1. The bridge is unstable when the angle $\alpha$ (between ow and aw interfaces) exceeds 90° (Figure 1b), and stable if $\alpha < 90°$ (Figure 1c). It can be appreciated that when $\alpha < 90°$ the

## Foam Breaking by Particles

ow interface is convex toward the oil bridge and the Laplace pressure causes the aqueous phase in the film to flow away from the oil bridge, accelerating the thinning of the film and its eventual rupture. The angle is related to the tensions by (6)

$$\alpha = \cos^{-1}[\gamma_{oa}^2 - (\gamma_{aw}^2 + \gamma_{ow}^2)]/2\gamma_{aw}\gamma_{ow} \tag{3}$$

so that for $\alpha > 90°$ (unstable bridge), the quantity

$$[(\gamma_{aw}^2 + \gamma_{ow}^2) - \gamma_{oa}^2] = B_{o,aw}$$

termed the bridging coefficient by Garrett (5), must be greater than 0. As with spreading and entry coefficients, it is clear that the bridging coefficient will be influenced by uptake of oil by the surfactant monolayer at the aw surface of the film.

Like a liquid droplet, a solid particle (assumed here to be composed of non-surface-active material) must be able to penetrate into the aw surface of a film if it is to cause rupture. Assuming entry can occur, then the particle will ultimately bridge the film, and if the contact angle of the aw surface with the particle, $\theta_{aw}$, has an appropriate value (which is dependent on particle geometry), the particle will de-wet and film rupture ensues (1,7,8). In systems where oil and particles are present in combination, as in commercial antifoams comprised of hydrophobic particles dispersed in mineral oil, the oil can adsorb onto exposed solid–vapor (sa) interfaces as well as into surfactant monolayers at the aw surface, and modify $\theta_{aw}$. Further, the contact angle which the ow interface makes with the solid ($\theta_{ow}$) is also an important factor in determining the efficacy of a foam-breaking formulation.

Although for a given system, droplet (or particle) entry into an interface may be thermodynamically feasible (entry coefficient exceeds 0 for a liquid drop), entry may be prevented by the existence of a metastable thin aqueous film ("pseudoemulsion" film [9]) between droplet or solid particle and the aw surface (9,10). Thus the interaction between two surfaces (particle/aqueous phase interface and aw surface) can play a determining role in the entry process.

A contact line existing at a three-phase boundary has associated with it a line tension, $\tau$, which is a one-dimensional analogue of interfacial tension. It is known that the existence of the line tension can prevent entry of small particles into a fluid–fluid interface (11) even though the contact angle as normally measured (by placing a relatively large drop of liquid on a plane solid surface) is quite large. The effects of line tension depend upon the size of the particles involved as well as on the magnitude of $\tau$. Values of $\tau$ reported in the literature have usually been in the region of $10^{-12}$ to $10^{-8}$ N, but recently Neumann and coworkers (12,13) have claimed values of

$10^{-6}$ N and higher. Such values, if they exist in systems of interest here, should give rise to significant effects in foam breaking by small solid particles (14).

In what follows we discuss, in relation to effects that liquid and solid particles have on foam stability, some of our results on the solubilization of hydrocarbons (mainly alkanes) in close-packed surfactant monolayers at the aw surface; we contrast these results with those obtained for the adsorption of alkanes on hydrophobic solid surfaces similar to those of particles which are effective foam breakers. We describe also some of our recent work on particle wettability in relation to line tension. In our investigation of the effects of oil and particles in combination on foam stability we have sought to bring about large changes in $\theta_{ow}$, for reasons that will become apparent later. This we have done by selecting systems containing ionic surfactant in which $\gamma_{ow}$ is very low and then varying $\theta_{ow}$ by addition of low concentrations of inorganic electrolyte. We explore the physical significance of these large changes in contact angle which are related to the properties of surfactant monolayers at the ow and sw interfaces.

## II. SOLUBILIZATION OF OIL IN SURFACTANT MONOLAYERS

Although the two-dimensional solubilization of hydrocarbons in surfactant monolayers has a wide relevance in the surface chemistry of oil–water–surfactant systems, our original work in this area was carried out in connection with an investigation of the effects of oil droplets on film and foam stability (15). As mentioned, uptake of oil by surfactant monolayers at the aw surface lowers the surface tension and affects the entry, spreading, and bridging coefficients and hence the potential for foam breaking. Using a homologous series of alkyltrimethylammonium bromides ($C_n$TAB) in aqueous solution together with liquid $n$-alkanes chain length $N_a$, we found that a (nonspreading) lens of alkane placed on the surface of a surfactant solution above its critical micelle concentration (cmc) lowers the surface tension by an amount ($\Delta\gamma$) which depends on the surfactant and alkane chain lengths. For a given alkane, the larger the surfactant chain length the larger $\Delta\gamma$, whereas for a given $n$, $\Delta\gamma$ is smaller the larger $N_a$. The magnitude of $\Delta\gamma$ can be large, as can be seen from Figure 2 where $\Delta\gamma$ is plotted against the difference in chain lengths of surfactant and alkane.

More recently we have determined complete adsorption isotherms for the uptake of alkanes by close-packed monolayers of surfactants of the type $C_nH_{2n+1}(OCH_2CH_2)_mOH$, ($C_nE_m$), at the aw surface (16). This has been achieved by placing (nonspreading) lenses of the required alkane mixed

# Foam Breaking by Particles

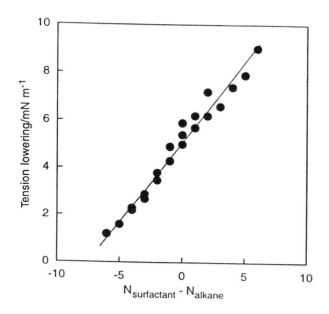

**Figure 2** Lowering of the surface tension, $\Delta\gamma$, of aqueous solutions of alkyltrimethylammonium bromides caused by adsorption of alkanes in the surfactant monolayers at 298 K. The abscissa is the difference between surfactant and alkane chain lengths (see text).

with a nonadsorbing diluent oil (usually squalane) over a range of known activities, on the surface of the surfactant solution, and determining $\Delta\gamma$. The experiment is equivalent to contacting the surfactant solution with alkane vapor over a range of relative pressures. Indeed we have performed such experiments directly using equipment of the kind constructed by Hauxwell and Ottewill, who studied the uptake of alkanes by the air-water interface (17). From a knowledge of the surface tensions of the surfactant solutions in the presence of alkane at various mole fraction activities ($a_o$) (or relative pressures $p/p^o$, where $p^o$ is the vapor pressure of pure alkane) the surface excesses of the alkane ($\Gamma_o$) can be obtained in the usual way using the appropriate form of the Gibbs adsorption isotherm

$$d\Delta\gamma/d \ln a_o = kT\Gamma_o \qquad (4)$$

A representative isotherm for the adsorption of dodecane into monolayers of $C_{12}E_5$ is shown in Figure 3. Results obtained using both methods described above are shown, and the agreement is seen to be excellent (16,18).

We have sought to correlate adsorption of hydrocarbons by surfac-

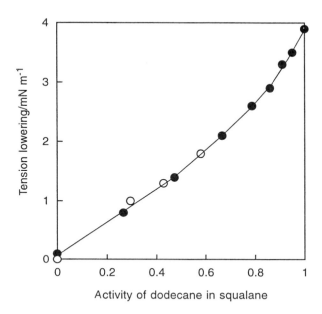

**Figure 3** Isotherm for the adsorption of dodecane into monolayers of $C_{12}E_5$ on water at 298 K. Filled symbols refer to adsorption from nonspreading lenses of dodecane in squalane and open symbols to adsorption directly from the vapor phase.

tant monolayers with droplet entry behavior, and consequent effects on film and foam stability, using solutions of sodium bis(2-ethylhexyl) sulfosuccinate (AOT) at a concentration above its cmc in 0.03 M aqueous NaCl (19). We show in Figure 4 the initial and equilibrium (final) entry coefficients for a homologous series of alkanes with the AOT solutions. The initial entry coefficients, $E_i$, are those calculated using the surface tension of the surfactant solution ($\gamma_{aw}$ in Equation 2) in the absence of alkane, and the equilibrium coefficients, $E_e$, were obtained using the surface tension recorded in the presence of pure alkane. Consideration of these data illustrate some difficulties in the use of interfacial tensions (and entry coefficients) to predict entry behavior and concomitant film and foam breaking. Initial entry coefficients change considerably with alkane chain length, and pass through 0 for tridecane. If it is appropriate to use $E_i$, e.g., if the vapor space over a film is open to the atmosphere so that alkane evaporates from the surface of an aqueous film, then it is feasible that all alkanes with chain length lower than 13 can cause film rupture, whereas those from tetradecane up should be incapable of giving film rupture. If, however, the vapor

# Foam Breaking by Particles

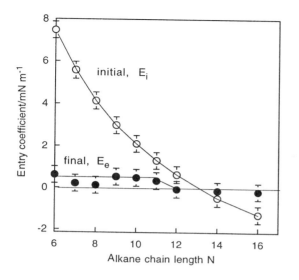

**Figure 4** Initial and final entry coefficients for alkanes at the surface of 3.8 mM AOT in 0.03 M NaCl at 298 K.

space is enclosed and saturated with alkane vapor, the (equilibrium) entry coefficients are all very close to 0 (which is the lowest possible value for a system at adsorption equilibrium). The values for hexane to undecane are all slightly positive, but it is not possible to predict with confidence the effects of the alkanes on film stability. In fact, it is found that in equilibrated systems, single droplets of dodecane and higher alkanes do not enter the aw surface from AOT solution (i.e., $E_e = 0$), whereas drops of the shorter alkanes do ($E_e > 0$). Thus, although $E_e$ values cannot be used confidently in a predictive sense, the findings concerning drops at interfaces are entirely consistent with the equilibrium entry coefficients shown in Figure 4. Further, the observations on single drop entry are also consistent with the way in which dispersed droplets affect the lifetimes of single soap films and foams (19).

As seen then, it is difficult on the basis of values of entry coefficients (i.e., interfacial tensions) alone to distinguish between lack of entry due to the presence of a metastable thin film on the one hand and absence of thermodynamic feasibility on the other (4). The problem is compounded, since reported values of (supposedly equilibrium) entry coefficients are often erroneous. However, we note that if entry is feasible but prevented by a metastable pseudoemulsion film, the aqueous phase will not spread spontaneously on the oil to give a duplex film, a test which is readily made.

## III. ENTRY OF SOLID PARTICLES INTO FLUID INTERFACES

For simplicity, we consider here a spherical solid particle and for the moment we denote the contact angle which the aw surface makes with the particle, $\theta$ (Figure 5a). If the particle is completely wetted, i.e., $\theta = 0°$, then it will not remain in or enter the aw surface. But there are circumstances where in principle although $\theta$ for, say, a drop of aqueous phase on a plane solid surface (Figure 5b) is nonzero, a particle with a similar surface constitution is kept out of the surface by the existence of the line tension $\tau$, which would act in the three-phase contact line around a particle if it were to be in the surface (Figure 5a). It is assumed in this discussion that the contact line is tending to contract, i.e., that $\tau$ is positive (by analogy with surface tension), although we note that negative line tensions are possible.

For a spherical particle, radius $R$ resting in a planar aw surface that has a contact angle $\theta$ with the particle (Figure 5a), it can be shown (11,14) that

$$\bar{\tau} = \frac{\tau}{\gamma_{aw} R} = \sin\theta \left[ 1 - \frac{\cos\theta_o}{\cos\theta} \right] \quad (5)$$

where $\theta_o$ is the contact angle in the absence of line tension effects (e.g., for a large liquid drop resting on a plane solid surface). Equation 5 shows that for a given $\theta_o$, $\bar{\tau}$ (the reduced line tension) passes through a maximum ($\bar{\tau}_m$) with respect to $\theta$ at a value of $\theta_m$, as illustrated schematically in Figure 6 for $\theta_o < 90°$. The angle $\theta_m$ can be shown to be related to $\theta_o$ by (14)

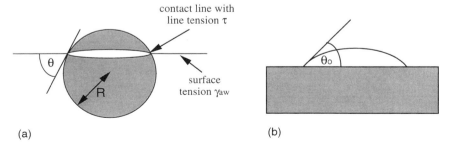

**Figure 5** (a) Spherical particle resting at an air–solution interface. The contact angle $\theta$ depends on the value of the line tension $\tau$ acting in the contact line where the liquid–vapor interface meets the particle. (b) Large liquid drop resting on a plane solid surface, with contact angle $\theta_o$.

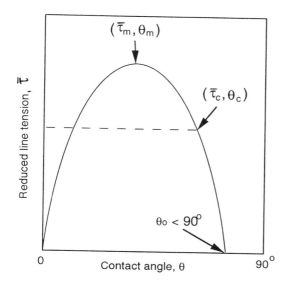

**Figure 6** Relationship between reduced line tension and contact angle for a spherical particle at a fluid–fluid interface with $\theta_o < 90°$. The thermodynamic wetting transition occurs at the condition $(\bar{\tau}_c, \theta_c)$. The maximum possible reduced line tension for a spherical particle to remain in the interface is $\bar{\tau}_m$.

$$\cos \theta_m = (\cos \theta_o)^{1/3} \quad (6)$$

The significance of the maximum is that stable or metastable equilibrium for the particle in the interface is only possible for $\theta_m < \theta < \theta_o$. In this regime (for $\theta_o < 90°$), an increase in $\bar{\tau}$ pushes the particle further into the liquid, giving a reduction in $\theta$. For $\theta < \theta_m$, however, only unstable equilibrium is possible; $d\bar{\tau}/d\theta$ is positive, which means that an increase in line tension would pull the particle further out of the liquid phase into the interface, which is not physically possible for a positive line tension. When $\theta$ falls below $\theta_m$, therefore, the particle leaves the interface and becomes completely wetted.

The maximum in the $(\bar{\tau}, \theta)$ curve does not however represent a wetting transition in a thermodynamic sense (14). This is because the energy of the particle in the interface becomes equal to that of a completely wetted particle at a critical reduced line tension of $\bar{\tau}_c$ corresponding to a contact angle $\theta_c > \theta_m$. Local equilibrium for the particle at the interface is possible between the conditions corresponding to $(\bar{\tau}_c, \theta_c)$ and $(\bar{\tau}_m, \theta_m)$, but a particle at the interface in this regime is only in a metastable state. Thus an "activation" energy is required to remove the particle from the interface. It follows that, for reduced line tensions in excess of $\bar{\tau}_c$, if a particle is already in bulk liquid adjacent to the surface, it cannot spontaneously enter the interface.

The possible effects of line tension are of particular interest, since recently very high values have been reported (12,13). Theoretical estimates of line tension have been in the region of $10^{-11}$ N (6), but results of Neumann and coworkers (12,13) are of the order $10^{-6}$ N. In previous work we have studied effects on foam stability of spherical hydrophobic particles with radii centered on 22 $\mu$m (3). We found that for contact angles of the foaming solution with the particles greater than around 93°, foams became very unstable in the presence of the particles, clearly indicating that the particles are able to enter and remain in the aw surfaces of the foam lamellae. It can be shown that (14), for $\theta_o = 93°$ and a surface tension of the solution = 40 mN m$^{-1}$, assuming the effective wetting transition occurs at ($\bar{\tau}_m$, $\theta_m$), particles with radius 0.3 nm and less cannot remain in the interface for $\tau = 10^{-11}$ N. For $\tau = 10^{-6}$ N, however, particles with radii $\leq$ 30 $\mu$m cannot remain in the interface, and would become completely de-wetted. Thus, in the systems used by us, if the line tension were to be around $10^{-6}$ N, we would not expect the particles to act as effective foam breakers since they could not remain in the aw surfaces and bridge the lamellae.

## IV. ADSORPTION OF HYDROCARBON VAPOR ON HYDROPHOBIC SOLIDS

In a number of commercial systems of interest, particles and oil are present together and act synergistically in destabilizing foams. A possible reason for the synergy is that the solid particles at the surfaces of the oil droplets immersed in surfactant solution (Figure 7) accelerate the entry of the droplets into the aw surface (4,5,20). On the basis of static contact angles, the air-water-solid and oil-water-solid contact lines will meet, and the thin oil-water-air film will rupture, if ($\theta_{aw} + \theta_{ow}$) exceeds 180°. The contact angle $\theta_{aw}$ depends on the surface tension $\gamma_{aw}$ and hence on the solubilization of hydrocarbon into the surfactant monolayers at the aw surface, as already discussed. Hydrocarbon can however also adsorb at the solid-vapor (sa) interface and modify $\theta_{aw}$.

Consider a system consisting of a planar horizontal solid with a drop of aqueous solution resting on it in the presence of saturated hydrocarbon vapor (Figure 8a). Hydrocarbon can adsorb at the aw and sa interfaces. It is also possible that hydrocarbon can diffuse along the sw interface beneath the drop and exert a surface pressure (tension lowering). Suppose for simplicity that the aqueous phase is pure water. Alongside the system depicted in Figure 8a we also consider (b) a drop of water on the solid immersed in

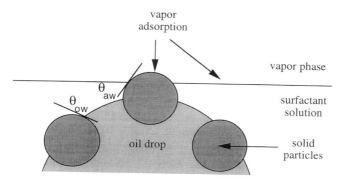

**Figure 7** Oil drop, coated with spherical solid particles, at an air–water interface. The particles bridge the thin aqueous "pseudoemulsion" film between oil and air.

liquid hydrocarbon and (c) a drop of hydrocarbon on the solid in contact with a vapor phase saturated with hydrocarbon. We may write Young's equation for each of these systems:

$$\gamma_{sa} - \gamma_{sw} = \gamma_{aw} \cos \theta_{aw} \tag{7}$$

$$\gamma_{sa} - \gamma_{so} = \gamma_{oa} \cos \theta_{oa} \tag{8}$$

$$\gamma_{so} - \gamma_{sw} = \gamma_{ow} \cos \theta_{ow} \tag{9}$$

Suitable combination of any two of Equations 7, 8, and 9 will yield the third equation. We have found that it is difficult to obtain reproducible values for $\theta_{aw}$, but $\theta_{ow}$ and $\theta_{oa}$ are easier to measure than $\theta_{aw}$. Combination of Equations 7, 8, and 9 gives for $\theta_{aw}$

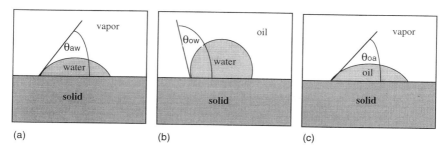

**Figure 8** Liquid drops resting on plane hydrophobic solid surfaces. (a) Aqueous drop in hydrocarbon vapor. (b) Aqueous drop immersed in liquid hydrocarbon. (c) Liquid hydrocarbon drop in its equilibrium vapor.

$$\theta_{aw} = \cos^{-1}(\gamma_{oa}\cos\theta_{oa} + \gamma_{ow}\cos\theta_{ow})/\gamma_{aw} \qquad (10)$$

In Figure 9 we show values of $\theta_{aw}$, for the case where the aqueous phase is pure water, obtained from $\theta_{oa}$ and $\theta_{ow}$ and the use of Equation 10 for a series of alkanes; the solid surface is glass coated with octadecyltrichlorosilane (OTS). Interestingly, the values of the contact angles of water on the solid in the presence of the vapor of the alkanes are very similar to those (hypothetical) angles of water on alkanes calculated from interfacial tensions using Young's equation. It is clear from Equation 7 that a change in $\theta_{aw}$ brought about by introducing hydrocarbon to the system depicted in Figure 8a reflects changes in the wetting tension ($\gamma_{sa} - \gamma_{sw}$) as well as in $\gamma_{aw}$. If we know the changes in $\gamma_{aw}$ that accompany hydrocarbon adsorption at the aw interface (see earlier), then we can obtain changes in the wetting tension caused by adsorption, but we cannot obtain the lowering of $\gamma_{sa}$ alone.

We are currently measuring directly the adsorption of hydrocarbons (alkanes) onto solid surfaces coated with close-packed hydrocarbon chains

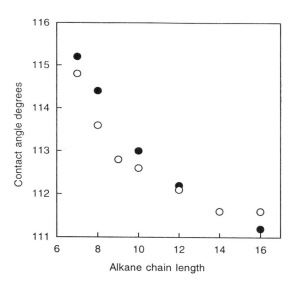

**Figure 9** Contact angle of water resting on hydrocarbon surfaces in the presence of saturated alkane vapor. Open symbols refer to water on hydrophobized (OTS-coated) plane glass; the contact angles have been calculated using Equation 10 and values of $\theta_{oa}$ and $\theta_{ow}$ (see Figure 8). Filled symbols are for water resting on liquid alkane; these angles are calculated from interfacial tensions using Young's equation (see text).

# Foam Breaking by Particles

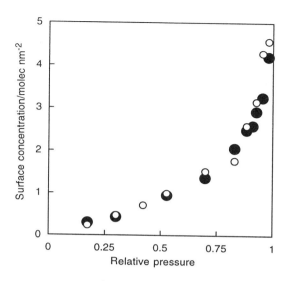

**Figure 10** Isotherms for the adsorption of decane from vapor at 298 K on gold surfaces coated with decyl (open circles) and octadecyl (closed circles) chains.

using a quartz crystal microbalance (21). The resonant frequency of a quartz crystal electrode coated with gold and grafted hydrocarbon chains is being determined for a range of relative pressures of alkane vapor in contact with the electrode. Adsorption changes the frequency, and the change can be converted to a surface concentration of alkane. Adsorption isotherms for decane on gold coated with decyl and octadecyl chains are shown in Figure 10. It is seen that the two isotherms are virtually coincident, suggesting that the decane does not penetrate into the chain region on the surface but rather adsorbs on top of the grafted layers that exhibit methyl groups for both the chain lengths studied. This observation is in marked contrast to the findings for adsorption of alkanes into close-packed surfactant layers on water discussed earlier. In the latter case, alkanes are known to mix with the surfactant chains (15,22).

## V. SOLID PARTICLES AT OIL/WATER INTERFACES

The prevention or acceleration of entry of an (entering) drop of oil at the aw surface caused by solid particles adsorbed at the ow interface (Figure 7) depends on $\theta_{aw}$ and $\theta_{ow}$, which in turn depends upon the extent of surfactant adsorption at the various interfaces. Contact angles can be influenced by the concentration of surfactant in the solution. This is seen in Figure 11 for $\theta_{aw}$ for aqueous solutions of hexadecyltrimethylammonium bromide (CTAB) in contact with a smooth plate of paraffin wax in air (23). For changes in contact angle brought about by changes in surfactant concentration it is readily shown, by combination of Young's equation and the appropriate form of the Gibbs adsorption isotherm, that (24)

$$\frac{d(\gamma_{aw} \cos \theta_{aw})}{d\gamma_{aw}} = \frac{\Gamma_{sa} - \Gamma_{sw}}{\Gamma_{aw}} \tag{11}$$

Clearly then, changes in contact angle with surfactant concentration are related to surface concentrations of surfactant ($\Gamma$) at the various interfaces. If it is (reasonably) assumed that no surfactant adsorption occurs at the sa interface, the right-hand side of Equation 11 becomes $-\Gamma_{sw}/\Gamma_{aw}$. An equation equivalent to Equation 11 holds of course for systems where the air is replaced by oil. For the latter systems, the contact angle $\theta_{ow}$ remains high and independent of surfactant concentration, as illustrated in Figure 11 for systems containing CTAB with dodecane as oil.

We show plots in Figure 12 of $\gamma \cos \theta$ against $\gamma$ for the same systems as those represented in Figure 11. On the assumption that $\Gamma_{sa} = \Gamma_{so} = 0$,

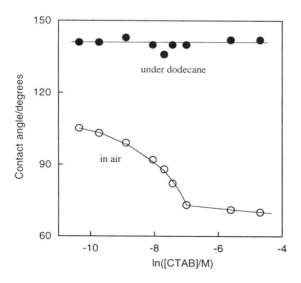

**Figure 11** Contact angles of aqueous solutions of CTAB with planar surfaces of paraffin wax in air and under dodecane at room temperature.

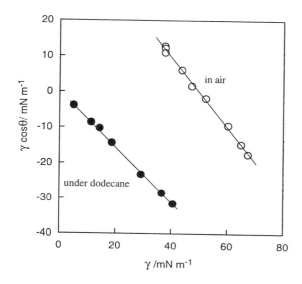

**Figure 12** Plots according to Equation 11 for the systems represented in Figure 11.

we find from the slopes of the straight line plots in Figure 12 that $\Gamma_{sw}/\Gamma_{aw} = 0.97$ and $\Gamma_{sw}/\Gamma_{ow} = 0.78$. This means that $\Gamma_{ow}/\Gamma_{aw} = 1.24$, i.e., the adsorption of CTAB is, over range of surfactant concentration, greater at the ow than at the aw interface by about 24%. This appears reasonable in that the adsorption free energy per methylene group at the ow interface ($-3.2$ kJ mol$^{-1}$) is more negative than that at the aw surface ($-2.7$ kJ mol$^{-1}$) (25).

The sum ($\theta_{aw} + \theta_{ow}$) for systems depicted in Figure 11 is always in excess of 180°. This is a common situation for many surfactants and as mentioned this could be the cause of the synergy between oil and solid hydrophobic particles in foam and film breaking. In order to test such a proposition we intend to determine the effects of spherical hydrophobic particles together with those alkanes which enter the aw surface (in the absence of the particles) on foam stability. On the basis of our earlier discussion in connection with Figure 7, one might suppose that if systems could be found in which ($\theta_{aw} + \theta_{ow}$) is *less* than 180°, the solid particles could have a *stabilizing* effect in the presence of oil, relative to the effect of oil alone. That is, the particles could stabilize the (otherwise unstable) pseudoemulsion oil-water-air film between oil drop and air (Figure 7) by bridging it.

In order to obtain values of ($\theta_{aw} + \theta_{ow}$) < 180°, we need to reduce

$\theta_{ow}$ substantially from around the commonly observed values of between 140° and 160°. This reduction can be achieved by using systems that exhibit very low oil–water interfacial tensions (26). Equation 9 can be rearranged to give

$$\cos \theta_{ow} = \frac{(\gamma_{so} - \gamma_{sw})}{\gamma_{ow}} \tag{12}$$

If $\gamma_{ow}$ can be changed fractionally by a large amount, it is likely that $\theta_{ow}$ will also be changed substantially. We show in Figure 13 $\theta_{aw}$ and $\theta_{ow}$ for systems containing the twin-tail anionic surfactant sodium diethylhexyl sulfosuccinate (AOT) above its cmc in dilute aqueous NaCl; the oil employed is heptane and the solid surface is glass coated with octadecyltrichlorosilane (OTS). It is known that systems with alkane and aqueous NaCl containing AOT above the cmc can give ultralow oil–water tensions by adjusting the concentration of the salt (27). The sums of the oil–water and air–water contact angles are ≤180°, as can be seen from the inset in Figure 13.

The significance of the changes in contact angle, brought about by addition of salt in systems at the cmc in terms of (anionic) surfactant

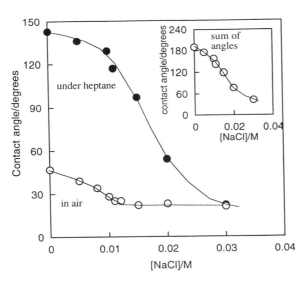

**Figure 13** Effect of NaCl concentration on the contact angles of 3.8 mM aqueous solutions of AOT with OTS-coated plane glass in air and under liquid heptane at room temperature. The inset shows the sum of the aw and ow contact angles as a function of salt concentration.

## Foam Breaking by Particles

monolayer behavior, can be demonstrated as follows. We have shown elsewhere (28) that the change in interfacial tension with surfactant counter-ion (Na$^+$) concentration, $m_{Na}$, in a system with a 1:1 ionic surfactant is given by

$$-\frac{d\gamma}{dm_{Na}} = \frac{RT\Gamma}{m_{Na}}\left[\left(1 + \frac{\partial \ln f_\pm}{\partial \ln m_{Na}}\right)(\alpha_m - \alpha_p)\right] \quad (13)$$

in which $\Gamma$ is the surface concentration of surfactant at the appropriate interface whose tension is $\gamma$. The counter-ion concentration $m_{Na}$ arises from the surfactant and the added salt, which has a common cation, Na$^+$; $f_\pm$ is the mean ionic activity coefficient in the electrolyte solution. The $\alpha$ are degrees of dissociation in the micelles ($\alpha_m$) and the plane interface ($\alpha_p$), defined as (29)

$$\alpha = -2\Gamma_{co}/\Gamma \quad (14)$$

Here, $\Gamma_{co}$ is the (negative) surface excess of the co-ion (Cl$^-$ in the case of systems with NaCl). The physical significance of $\alpha$ is not simple in systems of present interest where the surface charge density can be quite high. But in a general way, it can be seen that the more the surfactant "monolayer" in micelle or plane surface is dissociated, the more the co-ions will be repelled and the greater will be $\alpha$.

In order to make use of Equation 13, we need to have an expression for $d\gamma/dm_{Na}$. Young's equation for (say) a drop of aqueous surfactant resting on a plane solid surface immersed in oil in the absence of added electrolyte is

$$\gamma^o_{so} = \gamma^o_{sw} + \gamma^o_{ow} \cos\theta^o \quad (15)$$

In the presence of added electrolyte which is insoluble in the oil, we have

$$\gamma_{so} = \gamma_{sw} + \gamma_{ow} \cos\theta \quad (16)$$

Thus, the change in tension $\Delta\gamma_{sw}$ of the sw interface on addition of electrolyte is (assuming that $\gamma_{so} = \gamma^o_{so}$)

$$\Delta\gamma_{sw} = \gamma_{sw} - \gamma^o_{sw} = \gamma^o_{ow}\cos\theta^o_{ow} - \gamma_{ow}\cos\theta_{ow} \quad (17)$$

We also note that $d\Delta\gamma_{sw} = d\gamma_{sw}$. Thus, from a knowledge of changes in $\theta_{ow}$ and $\gamma_{ow}$ with electrolyte concentration, it is possible to obtain changes in $\gamma_{sw}$ with electrolyte concentration. Then from Equation 13 we may write

$$\frac{d\Delta\gamma_{sw}}{d\Delta\gamma_{ow}} = \frac{d\gamma_{sw}}{d\gamma_{ow}} = \frac{\Gamma_{sw}(\alpha_m - \alpha_{p,sw})}{\Gamma_{ow}(\alpha_m - \alpha_{p,ow})} \quad (18)$$

The value of $\alpha_m$ can be obtained from the variation of the surfactant cmc with electrolyte concentration using the expression given by Hall (29)

$$-\frac{d \ln cmc}{d \ln m_{Na}} = (1 - \alpha_m) + (2 - \alpha_m) \frac{\partial \ln f_{\pm}}{\partial \ln m_{Na}} \tag{19}$$

For a given salt concentration $\Gamma_{sw}/\Gamma_{ow}$ can be calculated from a knowledge of changes in $\theta_{ow}$ with surfactant concentration at that salt concentration by use of Equation 11. Thus it is possible in principle to calculate the difference in degree of dissociation of surfactant at the sw and ow interfaces. If however we choose to assume for simplicity that $\alpha_{p,sw} = \alpha_{p,ow}$, then we obtain $\Gamma_{sw}/\Gamma_{ow}$ from the slope of a plot of $\Delta\gamma_{sw}$ versus $\Delta\gamma_{ow}$. Such a plot is shown in Figure 14 for the same systems as those represented in Figure 13. The plot is linear with a slope of 0.87. We have observed elsewhere (23) for aqueous AOT solutions (in the absence of salt) on paraffin wax under dodecane, that $\Gamma_{sw}/\Gamma_{ow}$ is very similar, being 0.83 over a wide range of surfactant concentration.

Having shown how low values of $\theta_{ow}$ and hence of $(\theta_{ow} + \theta_{aw})$ can be attained, and discussed the significance of the changes in contact angle with electrolyte concentration in terms of monolayer properties, we now consider very briefly some preliminary findings on how the contact angles influence film and foam stability. Before considering effects on foam stability, we investigated the way in which close-packed layers of spherical hydro-

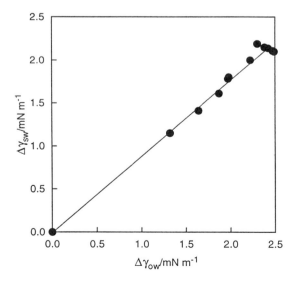

**Figure 14** Plot of $\Delta\gamma_{sw}$, obtained from contact angles and tensions according to Equation 17, against $\Delta\gamma_{ow}$. The systems are the same as those represented in Figure 13.

phobic particles at the oil–water interface of single alkane drops affect the drop entry into the air–surfactant solution interface.

For a given system, the rest time prior to entering (from the aqueous surfactant solution) into the aw surface of between 20 and 40 single oil droplets (volume 0.2 μl) containing dispersed spherical hydrophobic particles (diameter 3 μm) was determined visually using an apparatus described in Reference 19. The particles adhere to the ow surface around a drop giving an essentially close-packed particle monolayer. From the measurements the drop half-lifetimes have been calculated. The results obtained have been discussed in part before (4) and will be presented in full elsewhere, but it emerges from our study that the presence of particles at the surface of an oil drop can either facilitate or prevent drop entry into the aw interface. It is found that entry is prevented when ($\theta_{ow} + \theta_{aw}$) is low, considerably less than the 180° anticipated on the basis that static contact angles alone determine the stability of the pseudoemulsion film between oil drop and aw surface. Equally, for a sum of angles that is higher but still much less than 180°, drop entry can be greatly accelerated. This is illustrated in Figure 15 where half-lives of heptane drops initially in 3.8 mM

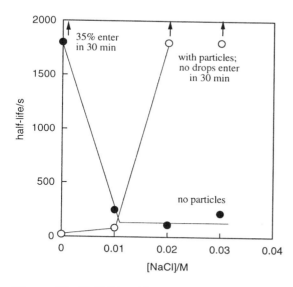

**Figure 15** Half-life of heptane drops (0.2 μl) at the surface of 3.8 mM aqueous solutions of AOT in contact with saturated heptane vapor as a function of the concentration of NaCl in the aqueous phase. Open symbols are for systems containing spherical hydrophobic particles (diameter 3 μm), and the filled symbols are for systems without particles.

AOT solution are shown for a range of NaCl concentrations both in the presence and absence of spherical hydrophobic particles. At low salt concentrations, where $(\theta_{ow} + \theta_{aw}) > 100°$, the particles aid drop entry. For [NaCl] ~ 0.02 M, however, where $(\theta_{ow} + \theta_{aw}) \sim 60°$, the particles confer considerable stability on the drops that do not enter the aw surface within 30 min.

Only very limited measurements on the effects of particles on the stability of foams containing dispersed oil drops have been carried out to date. However, early indications are that the effects of particles on single oil drop behavior at the aw surface are mirrored by the effects observed for foams. That is, where particles enhance the rate of drop entry, oil and particles act synergistically in breaking foams. On the other hand, in cases where drop entry is prevented by particles, oil and particles act antagonistically so that foam half-life is longer in the presence of oil and particles than with just dispersed oil drop alone.

## VI. QUARTER CENTURY PROGRESS AND NEW HORIZONS

Any view on what the significant advances have been over the last 25 years must inevitably be subjective. Further, the seeds of much of what is now accepted knowledge concerning the mechanisms of foam and film breaking were sown much earlier. Various of the approaches have however been refined and tested more recently.

Direct visualization of what happens when particles approach and are held in surfaces and rupture films is very valuable if it can be realized. Dippenaar (8) has produced high-speed photographic images of particles of well-defined geometry (spheres, cubes) that support the ideas on the effects of contact angle in foam breaking by bridging and subsequent de-wetting of particles by films. Johansson and Pugh (30) have also recently observed the movement and location of (irregular) solid particles in single horizontal films using the microinterferometric method of Scheludko (31). Notable work has been reported by Frye and Berg (7) and by Garrett (see Reference 1) relating contact angles to foam breakdown. Included in the work of Frye and Berg is a study of the effects that surface-treated glass rods inserted through single vertical films have on the film stability.

A deeper understanding of the way in which insoluble oil drops rupture aqueous films has begun to emerge in recent years. There is a clear distinction between effects arising from the presence of metastable pseudo-emulsion films between drop and the air-solution interface on the one hand (9,10), and effects related to thermodynamic feasibility of drop entry into the interface on the other (19). Bergeron et al. (10), in an elegant piece of work, determined disjoining pressure isotherms (disjoining pressure versus film thickness) for thin aqueous surfactant films between oil and vapor phase, demonstrating the importance of metastable films in foam breaking by oils.

For the most part, work on foam breaking has been carried out using rather ill-defined systems, i.e., impure materials and solid particles of irregular shape and poorly characterized surfaces. Future investigations can be anticipated in which monodisperse (say spherical) particles of known surface constitution are employed. For such systems it is now possible to measure directly the contact angles that the particles make with foaming solutions, using a Langmuir trough technique (32,33).

The effects of particle size in systems with monodisperse (spherical) particles is still not clearly understood. Insofar as small particles are not effective foam breakers, one possible contributory factor could be that as a result of the existence of positive line tension, such particles are unable to enter the interface. Since very large values of line tension have been reported recently (12,13), it is possible that even relatively large particles could be kept out of the liquid surface and hence be unable to break films and foams.

The effects of the magnitude of bridging coefficients on foam breaking by oil droplets warrants investigation. With reference to Figure 1 and Equation 3, if the angle $\alpha$ is large, a lens of oil resting on the liquid–vapor interface will be thin. This means that the lens will not bridge the film until the film itself is very thin. On the other hand, the larger $\alpha$ the greater the driving force (Laplace pressure) tending to force liquid in the film away from the bridging oil drop.

The synergy between oils and solid particles in foam breaking has not been thoroughly investigated to date. It is clear from our own work that explanations based on static contact angles alone will not be adequate, and that an understanding of dynamic effects associated with the entry of a solid particle into a liquid surface will be necessary. This will include the rate of thinning of pseudoemulsion films and also the dynamics of dewetting of particles as they emerge into a liquid surface. It is also clear that hydrocarbons can be quite strongly solubilized in surfactant monolayers. This could conceivably affect film stability, say by reducing the surface elasticity, and could usefully be investigated in the future.

## REFERENCES

1. PR Garrett. In: Defoaming: Theory and Industrial Applications. Garrett, PR, ed. New York, Marcel Dekker, 1993, p 1.

2. R Aveyard, BP Binks, PDI Fletcher, TG Peck, and CE Rutherford. Adv Colloid Interface Sci 48:93, 1994.
3. R Aveyard, BP Binks, PDI Fletcher, CE Rutherford. J Dispersion Sci Technol 15:251, 1994.
4. R Aveyard, JH Clint. JCS Faraday Trans 91:2681, 1995.
5. PR Garrett. J Colloid Interface Sci 76:587, 1980.
6. JS Rowlinson, B Widom. Molecular Theory of Capillarity. Oxford, Oxford University Press, 1989, chap. 8.
7. GC Frye, JC Berg. J Colloid Interface Sci 127:222, 1989.
8. A Dippenaar. Int J Miner Process 9:1, 1982.
9. L Lobo, DT Wasan. Langmuir 9:1668, 1993.
10. V Bergeron, ME Fagan, CJ Radke. Langmuir 9:1704, 1993.
11. A Scheludko, BV Toshev, D Platikanov. In: The Modern Theory of Capillarity. Goodrich, RC, Rusanov, AI, eds. Berlin, Akademie-Verlag, 1981, p 163 et seq.
12. D Duncan, D Li, J Gaydos, AW Neumann. J Colloid Interface Sci 169:256, 1995.
13. D Li, AW Neumann. Colloids Surf. 43:195, 1990.
14. R Aveyard, JH Clint. JCS Faraday Trans 92:85, 1996.
15. R Aveyard, P Cooper, PDI Fletcher, JCS Faraday Trans 86:3623, 1990.
16. R Aveyard, BP Binks, PDI Fletcher, JR MacNab. Langmuir 11:2515, 1995.
17. F Hauxwell, RH Ottewill. J Colloid Interface Sci 34:473, 1970.
18. R Aveyard, BP Binks, D Crichton, PDI Fletcher. Work in progress.
19. R Aveyard, BP Binks, PDI Fletcher, PR Garrett, TG Peck. JCS Faraday Trans 89:4313, 1993.
20. K Koczo, JK Koczone, DT Wasan. J Colloid Interface Sci 166:225, 1994.
21. R Aveyard, BD Beake, JH Clint. JCS Faraday Trans 92:4271, 1996.
22. JR Lu, RK Thomas, R Aveyard, BP Binks, P Cooper, PDI Fletcher, A Sokolowski, J Penfold. J Phys Chem 96:10971, 1992.
23. R Aveyard, P Cooper, PDI Fletcher, CE Rutherford. Langmuir 9:604, 1993.
24. EH Lucassen-Reynders. J Phys Chem 67:969, 1963.
25. R Aveyard, DA Haydon. Introduction to the Principles of Surface Chemistry. Cambridge, Cambridge University Press, 1973, chap. 3.
26. R Aveyard, BP Binks, PDI Fletcher, TG Peck. Unpublished work.
27. R Aveyard, BP Binks, J Mead. JCS Faraday Trans I 81:2169, 1985.
28. R Aveyard, BP Binks, S Clark, J Mead. JCS Faraday Trans I 82:125, 1986.
29. DG Hall. In: Aggregation Processes in Solution. Wyn-Jones, E, Gormally, J, eds. Amsterdam, Elsevier, 1983, chap. 2.
30. G Johansson, RJ Pugh. Int J Miner Process 34:1, 1992.
31. A Scheludko. Adv Colloid Interface Sci 1:397, 1967.
32. JH Clint, SE Taylor. Colloids Surf 65:61, 1992.
33. R Aveyard, BP Binks, PDI Fletcher, CE Rutherford. Colloids Surf A 83:89, 1994.

# 18
## Dynamic Adsorption and Tension of Spread or Adsorbed Monolayers at the Air–Water Interface

**Elias I. Franses, Chien-Hsiang Chang,* Judy B. Chung,[†] Karen Coltharp McGinnis,[‡] and Sun Young Park[§]**
*Purdue University, West Lafayette, Indiana*

**Dong June Ahn**
*Korea University, Seoul, Korea*

## I. INTRODUCTION

The dynamic behavior of air-water interfaces is important in the behavior of detergents, in foaming, in coating flows, in biological membranes, and in lung surfactants (1-5). The adsorption dynamics depends on diffusion-adsorption-desorption processes of molecules, monomer-micelle exchange rates, and (when applicable) exchange rates of monomers with dispersed surfactant particles, such as liposomes, vesicles, or microcrystallites (6-14). The dynamic adsorption density determines the dynamic surface tension and surface rheology. The latter involves the shear and dilatational surface viscosities, which may depend on the rate of area deformation and time, and are important in foam stability.

This article focuses on (1) monolayer surface pressure hysteresis,

---
*Current affiliations:*
*National Cheng Kung University, Tainan, Taiwan.
[†]Clorox Co., Pleasanton, California.
[‡]Bayer Corporation, Addyston, Ohio.
[§]International Paper, Tuxedo, New York.

which depends on surface density and rate of change of surface area; (2) monolayer compositional variations caused by surface compression and selective collapse; and (3) monolayer density and composition changes occurring during pulsating area changes and resulting from exchanges between the monolayer and extra layers or dispersed particles. The latter two topics are reviewed in more detail in relation to the dynamic adsorption hysteresis mechanism and the appearance of nonequilibrium tensions, which are lower than the equilibrium tensions. The implications to superlow tensions for lung surfactants (3,4) are discussed. Finally, emerging issues in the field of monolayers are delineated.

## II. DYNAMIC PROPERTIES OF SPREAD MONOLAYERS

It is well known that the surface pressure ($\Pi$)-surface area ($\overline{A}$) isotherm of a spread monolayer depends on the rate of area compression (15–17). After compression to monolayer collapse, and often after compression to lower pressures, the $\Pi$-$\overline{A}$ curve during area expansion (descending curve) rarely overlaps with the area expansion (ascending curve). This is observed for single-component surfactants and lipids, as well as for mixed lipids. Examples for such behavior are shown in Figure 1. Experimental details are given in Section VIII of this chapter and elsewhere (12–14). As a monolayer is compressed, it undergoes several phase changes, from liquid-expanded (LE) to liquid-condensed (LC) to solid, with two-phase regions observed sometimes. The transition thresholds may depend on the compression rate. Ideally, the monolayer stability at constant area or constant pressure, and the time dependence (of $\Pi$ or $\overline{A}$) must be examined before one assigns thermodynamic significance to $\Pi$-$\overline{A}$ curves (16,18).

If the monolayer loses material—via desorption, partial dissolution (for sparingly soluble surfactants), evaporation, monolayer folding and collapse, or monolayer leakage around the trough surface barriers—then there could be significant errors in the areas per molecule or the "apparent" mean molecular areas. Substantial errors in the measurement of the surface pressure are also possible if the aqueous system does not completely wet the Wilhelmy plate normally used for measurements of surface tension $\gamma$ (19) (or surface pressure $\Pi \equiv \gamma_o - \gamma$, with $\gamma_o$ being the tension of the aqueous solution). Nonzero dynamic contact angle ($\theta$) effects of the Wilhelmy plate may result in tension measurements lower than the actual tension, since the measured force decreases by a factor $\cos \theta$ and the decrease cannot be accounted accurately.

The collapse surface pressures observed for dipalmitoylphosphatidylcholine (DPPC), hexadecanol, and their mixtures can be quite high (60–72

# Adsorption and Tension of Monolayers

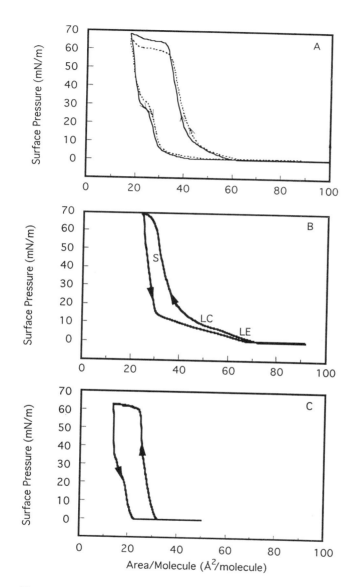

**Figure 1** Isotherms of surface pressure versus mean molecular area (for the initially spread materials) for spread monolayers of DPPC–hexadecanol (h) at the air–saline interface. Arrows indicate possible monolayer phase changes (LE: liquid expanded; LC: liquid condensed; S: solid monolayer). (A) 3:1 mol/mol, 23°C, at 15 cm$^2$/min, solid line, or 1.5 cm$^2$/min, broken line; (B) 3:1 mol/mol, 37°C, 15 cm$^2$/min; (C) 1:1 mol/mol, $d$-hexadecanol, 23°C, 15 cm$^2$/min.

mN/m), and higher than the equilibrium surface pressures of the respective solid particles. The corresponding surface tensions can be quite low, or "superlow," from 0.1 to about 10 mN/m, and lower than those of any equilibrium hydrocarbon surfactant solution or any hydrocarbon liquid at around room temperature. For the effective performance of natural or exogenous lung surfactants in stabilizing the lung alveoli, and for better fundamental understanding of compressed monolayers, superlow tensions have received much attention (4,8-14,20-22). Results of such low surface tensions have been verified by a strong body of evidence from other tension measuring techniques. Using the spinning bubble method, Chung et al. determined for compressed surfaces of DPPC tensions as low as 0.6 mN/m for fluid axisymmetric interfaces with no potential artefacts from contact angle or other effects (8,9). Park et al. and others obtained low tensions for DPPC, hexadecanol, and certain lung surfactants (12-14), using the pulsating bubble (PBS) method developed by Enhorning (22). In the PBS method one uses small, nearly spherical bubbles and determines $\gamma$ from the Laplace-Young pressure difference and the bubble diameter (22-25). Schürch et al. have developed the captive bubble method, in which the bubble is larger and deforms under gravity. The surface tension is determined from the bubble shape using standard methods of axisymmetric drop shape analysis (20,21). Schürch's method is not susceptible to contact angle effects or monolayer leakage effects. In our opinion, in the pulsating bubble method any potential monolayer leakage artefacts could affect the first or second measurement but would not affect measurements of low tensions at later cycles or for subsequent bubbles. Moreover, superlow nonequilibrium surface tensions can exist even when only spread monolayers are present or when the monolayers have formed by adsorption from the aqueous phase (usually a dispersion of lipid particles) (4,8-14,20-25).

## III. SPECIAL EFFECTS FOR MIXED MONOLAYERS

Mixtures of lipids with two saturated hydrocarbon chains (e.g., DPPC), together with lipids with unsaturated chains (e.g., PG or phosphatidylglycerol) or with single-chain surfactants such as hexadecanol (10,15,16,26-28), may demix (form coexisting immiscible domains rich in one component or another). The first, albeit indirect, evidence of such monolayer demixing comes from the monolayer isotherm, when the surface pressure increases beyond the collapse point. This is usually interpreted by the surface refining or "squeeze-out" hypothesis. As the monolayer density increases, different molecules cannot pack effectively in liquid-condensed or solid monolayer phases and segregate into regions. As the monolayer collapses, some regions capable of sustaining the lower surface pressures collapse to a larger extent than the "stronger" regions. The remaining monolayer may then increase its surface pressure by virtue of being enriched in the component with the higher collapse pressure.

Supporting evidence of the squeeze-out hypothesis has come from FT-IR spectroscopy, either directly by external reflection techniques (FT-IR-ER) or less directly by Langmuir–Blodgett (LB) sampling of monolayer, followed by FT-IR-ATR (Attenuated Total Reflection) or other analysis methods. Chung and colleagues found significant evidence of DPPC enrichment of compressed monolayers of DPPC-PG, or of sheep lung surfactant. In one example, DPPC increased from 50 to about 95 mol % at the highest surface pressures (10). Qualitatively consistent results were obtained by thin-layer chromatography, which involves another step of dissolving the LB monolayer as by FT-IR-ATR analysis. For DPPC-PA (palmitic acid) monolayers, at 7 : 3 molar ratio, no significant squeeze-out effect was detected. Using the FT-IR-ATR technique, Rana and colleagues reproted negligible squeeze-out effects for DPPC/DPPG (dipalmitoylphosphatidylglycerol) monolayers at an initial molar ratio of 7 : 1 (27). Their DPPG was synthetic, single-component, and had two saturated chains, unlike PG which was natural, multicomponent, and had at least one unsaturated chain (10). No results were reported at molar ratios of 1 : 1 or so, which have higher probability of being in the biphasic region. Rana and colleagues used smaller compression rates and had smaller LB film transfer ratios than those by Chung and colleagues. The differences in the second lipid used and the molar ratios preclude definitive comparisons of the two sets of results.

Squeeze-out effects were probed by direct in situ FT-IR-ER techniques (28), which showed up to 90% selective removal of DOPG (dioleylphosphatidylglycerol) from DPPC–DOPG mixtures at surface pressures of 51 to 68 mN/m. For DPPC–DPPG there was up to 20% selective removal of DPPG. These authors postulated that lipids with two saturated chains may mix fully, whereas the presence of unsaturated chains may promote demixing (28).

In Figures 2 through 4 and Tables 1 through 4, we present new data for mixed monolayers of DPPC with $d$-hexadecanol, which support some limited squeeze-out effect for an initial molar ratio 1 : 1 at the higher surface pressures (13). The maximum observed surface pressure is about 63 mN/m (Figure 1c). The actual surface pressure may be slightly higher than this value, because of small calibration errors or monolayer leakage effects, as judged from the pure lipids collapse pressures, which have been found to be about 60 mN/m for hexadecanol (12) and 70–72 for DPPC (4,14,20). The collapse pressure for DPPC–hexadecanol is between those of the pure components. The mean area/molecule ($\text{Å}^2$/molecule) at the collapse point

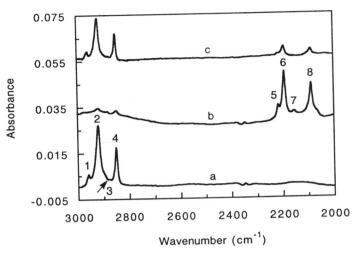

**Figure 2** FTIR-ATR spectra of pure DPPC (a), $d$-H (b), and 1/1 mol/mol DPPC/$d$-H (c) deposited at $\Pi = 10$ mN/m at 23°C. The incident beam was unpolarized. Band assignments and peak positions are given in Table 1. The spectra for other surface pressures and for polarized incident light look similar and are not shown. Dichroic ratios are listed in Table 2.

is about 27 Å$^2$, intermediate between those, 45 and 20 Å$^2$, for DPPC and hexadecanol, respectively. Little difference was observed between hexadecanol and $d$-hexadecanol.

The $\Pi$–$\overline{A}$ isotherm for DPPC–$d$-hexadecanol shows some evidence of the collapse pressure increasing with increasing compression (Figure 1c). This suggests a mild squeeze-out effect. For probing this, DPPC–$d$-hexadecanol monolayers at $\Pi$ from 10 to 63 mN/m were deposited via the LB method on Ge ATR crystals, and were then examined by FTIR. Typical spectra of DPPC, $d$-hexadecanol, and mixtures (Figure 2, $\Pi = 10$ mN/m) show absorbances adequate for sensitive monolayer composition analysis. Deuterated hexadecanol was used for distinguishing its chain bands from those of DPPC ([10]; Figure 2 spectrum b). The results on band assignments and peak positions are given in Table 1. Some dichroic ratios and average tilt angles of the methylene transient dipole moments or of the chains are also given in Table 2. The results indicate that the LB films have ordered chains (mostly all-trans), consistently with the monolayers being in well-ordered states (29,30). Thus, the absorbance intensities should be proportional to the concentrations in the monolayer. The absorbances of the CH$_2$ and CD$_2$ peaks increase with increasing surface pressure, as ex-

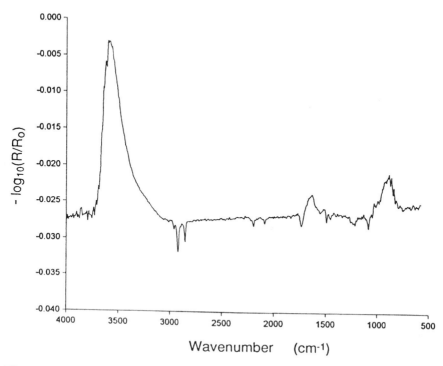

**Figure 3** External reflection–absorption spectrum of a mixed monolayer of DPPC and $d$-hexadecanol at an initial molar ratio of 1:1, at ~25°C and a mean molecular area of 37.5 Å$^2$.

pected, and those of DPPC increase more than those of hexadecanol (Table 3). The ratios of $CH_2$ to $CD_2$ symmetric stretch absorbances increase from 3.7 to 5.0, indicating an increase of the DPPC mole fraction to 0.61 from the value of 0.50 of the initial deposited monolayer.

The results of these experiments, especially those of the collapsing monolayer ($\Pi$ = 50–63 mN/m), should be interpreted with some caution, because at these conditions the transfer ratio was as high as 1.8 (13). This was because the film area at fixed $\Pi$ changed even without moving the LB substrate, indicating the film instability during the collapse stage. Thus, the deposited film may be more than a monolayer, and the absolute surface densities are not the ones indicated from the isotherm. Although the low $\Pi$ data, for which there is no monolayer collapse and the LB transfer ratio is 1.0, could be used for absolute surface density calibration, mainly the rela-

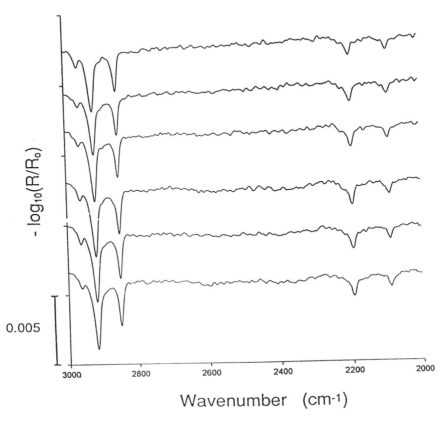

**Figure 4** Effect of mean molecular area (MMA) on absorbance intensities for the system of Figure 3. Only the hydrocarbon stretching bands are shown. From top to bottom, MMA = 37.5, 35.0, 32.5, 30.0, 27.5, and 25.0 Å$^2$; see also Table 4.

tive intensities were used for the film composition analysis, since they should not depend on the LB transfer ratio or other monolayer losses.

To obtain additional information on the state of the mixed layer, we obtained some preliminary in situ FT-IR-ER (external reflection) results. In these results, the mean molecular area (MMA) is reported as the independent variable, since Π was not controlled. Figure 3 shows the spectrum after dividing by the reflectance of a pure water surface. Despite the residual water lines (at ca. 3500 and 1700 cm$^{-1}$), the spectra of both DPPC and $d$-hexadecanol can be clearly seen. Figure 4 shows the hydrocarbon stretching region. The absolute and relative absorbances of the methylene symmet-

**Table 1** Band Assignments and Peak Positions (±2 cm$^{-1}$) for FTIR-ATR Spectra of LB Monolayers

| Peak vibration mode | DPPC | | | | | d – H | | | DPPC/d – H (1/1 mol/mol) | | | | |
|---|---|---|---|---|---|---|---|---|---|---|---|---|---|
| | Π = 5 | Π = 10 | Π = 20 | Π = 50 | | Π = 10 | Π = 20 | Π = 50 | Π = 10 | Π = 20 | Π = 50 | Π = 60 | Π = 63 |
| 1 CH$_3$-$\nu_a$ | 2961 | 2960 | 2959 | 2959 | | a | a | a | 2959 | 2959 | 2959 | 2959 | 2959 |
| 2 CH$_2$-$\nu_a$ | 2922 | 2921 | 2919 | 2918 | | a | a | a | 2920 | 2919 | 2919 | 2919 | 2918 |
| 3 CH$_3$-$\nu_s$ | 2881 | 2880 | 2880 | 2881 | | a | a | a | 2879 | 2881 | 2881 | 2879 | 2880 |
| 4 CH$_2$-$\nu_s$ | 2852 | 2852 | 2850 | 2851 | | a | a | a | 2851 | 2851 | 2850 | 2581 | 2851 |
| 5 CD$_3$-$\nu_a$ | a | a | a | a | | 2217 | 2217 | 2217 | 2218 | 2216 | 2216 | 2216 | 2217 |
| 6 CD$_2$-$\nu_a$ | a | a | a | a | | 2193 | 2193 | 2193 | 2195 | 2195 | 2194 | 2194 | 2195 |
| 7 CD$_3$-$\nu_s$ | a | a | a | a | | 2156 | 2156 | 2158 | 2156 | 2156 | a | 2158 | a |
| 8 CD$_2$-$\nu_s$ | a | a | a | a | | 2089 | 2089 | 2089 | 2089 | 2089 | 2090 | 2088 | 2089 |

*Note*: $\nu_a$: Asymmetric stretching vibration; $\nu_s$: symmetric stretching vibration.
[a] Peak was small or not observable.

**Table 2** Dichroic Ratios (±0.04) and Average Tilt Angles (±5°) for LB Monolayers of DPPC and $d$-Hexadecanol at 23°C

|  |  |  | $\Pi = 5$ | $\Pi = 10$ | $\Pi = 20$ | $\Pi = 50$ | $\Pi = 60$ | $\Pi = 63$[a] |
|---|---|---|---|---|---|---|---|---|
| DPPC | Peak #2[b] | DR | 0.90 | 0.97 | 0.97 | 0.99 | [d] | [d] |
|  | Peak #4[b] | DR | 0.98 | 0.99 | 0.98 | 1.03 | [d] | [d] |
|  |  | $\overline{\gamma_o}$[e] | [f] | 33 | 38 | 32 | [d] | [d] |
|  |  | $\overline{\gamma_o}$[e] | [f] | 67 | 65 | 68 | [d] | [d] |
| $d$-H | Peak #6[c] | DR | [d] | 0.94 | 1.02 | (0.87[g]) | [d] | [d] |
|  | Peak #8[c] | DR | [d] | 0.77 | 1.03 | (0.87[g]) | [d] | [d] |
|  |  | $\overline{\gamma_o}$[e] | [d] | [f] | 32 | (55[g]) | [d] | [d] |
|  |  | $\overline{\gamma_o^M}$ | [d] | [f] | 68 | (55[g]) | [d] | [d] |
| DPPC/$d$-H | Peak #2[b] | DR | [d] | 0.87 | 1.0 | 1.0 | 1.0 | 1.0 |
|  | Peak #4[b] | DR | [d] | 1.0 | 1.0 | 1.0 | 1.0 | 1.0 |
|  |  | $\overline{\gamma_o}$ | [d] | [f] | 32 | 32 | 32 | 32 |
|  |  | $\overline{\gamma_o^M}$ | [d] | [f] | 68 | 68 | 68 | 68 |

[a] The molecules were deposited soon after monolayer collapse was observed.
[b] Peaks #2 and #4 are for the asymmetric and symmetric stretching vibrations, $\nu_a(CH_2)$ and $\nu_s(CH_2)$; see Figure 2 and Table 1.
[c] Peaks #6 and #8 are for the asymmetric and symmetric stretching vibrations, $\nu_a(CD_2)$ and $\nu_s(CD_2)$; see Figure 2 and Table 1.
[d] No deposition of molecules was made at this surface pressure.
[e] The average tilt angle $\overline{\gamma_o}$ of the chains is calculated, if the chains are all trans and in a uniaxial orientation distribution (29).
[f] Uniaxial model was not applied, because the dichroic ratios of asymmetric and symmetric vibrations were quite different.
[g] This monolayer may have been disordered due to collapse.

**Table 3** Absorbances and Number of Moles Deposited from Monolayers of DPPC/$d$-Hexadecanol (Originally 1/1 by Mole) at the Indicated Surface Pressures

|         | Π = 10 | Π = 20 | Π = 50 | Π = 60 | Π = 63[a] |
|---------|--------|--------|--------|--------|-----------|
| Peak #2 | 0.018  | 0.035  | 0.051  | 0.051  | 0.057     |
| Peak #4 | 0.011  | 0.024  | 0.035  | 0.036  | 0.040     |
| Peak #6 | 0.005  | 0.008  | 0.011  | 0.011  | 0.011     |
| Peak #8 | 0.003  | 0.005  | 0.008  | 0.007  | 0.008     |
| x (DPPC)| 0.52   | 0.57   | 0.57   | 0.59   | 0.61      |

*Note:* Absorbance intensities (±0.0005) of FTIR-ATR spectra for unpolarized incident light. Number of moles deposited was calculated based on measured absorbances and absorbances of monolayers at small Π, for which no collapse or squeeze-out are expected (13).
[a]The molecules were deposited soon after a monolayer collapse was observed.

ric stretching bands (Table 4) indicate that (1) even at the low MMAs (high Π), the film remains substantially a monolayer; however, the MMA decrease clearly indicates a substantial loss of material from the probed surface layer; and (2) the observed surface layer becomes enriched in DPPC, although by less than indicated in the LB films results. If the mixed monolayer is partly "collapsed" but remains attached to the observed surface film, this could explain the observed small intensities in the ER spectra compared to those in the ATR spectra, and perhaps accounts for the differ-

**Table 4** Results of E-IR for DPPC/$d$-Hexadecanol Mixed Monolayer

| MMA (Å²/molecule) | Π (mN/m) | $R$[a] | x(DPPC) |
|-------------------|----------|--------|---------|
| 37.5              | ~2       | 2.77   | 0.50    |
| 35.0              | ~5       | 2.81   | 0.51    |
| 32.5              | ~16      | 2.83   | 0.51    |
| 30.0              | ~32      | 3.15   | 0.54    |
| 27.5              | ~50      | 3.24   | 0.55    |
| 25.0              | ~61      | 3.20   | 0.55    |

[a]Ratio of measured intensities of $CH_2$ $\nu_s$ bands to $CD_2$ $\nu_s$ bands from Figure 4; asymmetric stretching bands showed less clear trends, possibly due to overlapping with adjacent bands.

ences in the mole fractions of DPPC (Tables 3 and 4). Although more research is needed, both techniques indicate a small but significant squeeze-out effect for the 1 : 1 molar ratio of DPPC–hexadecanol. No results are available for the 3 : 1 ratio, for which squeeze-out effects are expected to be less pronounced. This molar ratio is perhaps more relevant to the commercial lung surfactant Exosurf (13,500 ppm DPPC, 1500 ppm hexadecanol, molar ratio 3 : 1 [31]), whose surface composition may differ of course from its bulk composition.

## IV. MONOLAYERS OF DISPERSIONS OF SPARINGLY SOLUBLE SURFACTANTS

When a spread or adsorbed monolayer of hexadecanol is compressed beyond its equilibrium spreading pressure, it may remain an intact monolayer, retaining the same number of molecules, or it may collapse losing some material. In the first case, the monolayer becomes supersaturated and may remain in a metastable state for some time, before it collapses, and produce lower surface tension (higher surface pressure) than its equilibrium value and lower tension than in the second case. If there are hexadecanol particles on or near the surface, they can act as nucleation sites, or as a depository or the extra molecules in the nonequilibrium layer. Because hexadecanol has a finite solubility ($\sim$ 40 ppb), there is a possible mechanism for receiving the extra monolayer molecules, as has been detailed elsewhere (12). Conversely, hexadecanol dispersed particles are a substantial source of molecules for the surface. Thus, the nature of the dispersed particles is a key variable of the adsorption dynamics, as has been demonstrated for many systems including DPPC, hexadecanol, and salts of fatty acids (12–14,32). The results for DPPC and hexadecanol dispersions with constant or pulsating area conditions have been reported in detail elsewhere (33). In the next section, we will report some examples for salts of fatty acids.

## V. MONOLAYERS OF SOLUBLE SURFACTANTS

The behavior of monolayers of single or mixed soluble surfactants at constant area conditions has been known for a long time (6,7). At pulsating area conditions, the dynamic tension oscillates between a maximum and minimum value, $\gamma_{max}$ and $\gamma_{min}$. In all cases examined, $\gamma_{min} \leq \gamma_e$ (the equilibrium surface tension). This phenomenon also occurs with insoluble or sparingly soluble surfactants (3,4). Certain fundamental aspects of the dynamic monolayer behavior under pulsating area, and of the technique for measuring $\gamma(t)$ under conditions that bulk and surface viscosity effects are unimportant, have been published (23,24). Nonequilibrium tensions lower than the equilibrium tensions can arise even with simple monolayers, because of a dynamic adsorption hysteresis mechanism that is distinct from the classical monolayer hysteresis mechanism described in Section II of this chapter (12,24). As the surface monolayer is stretched, starting from the equilibrium surface density (and tension), the monolayer is diluted. The resulting concentration gradient leads to mass transfer (by diffusion and adsorption) from the bulk phase. The mass transfer rate depends on the timescale of adsorption relative to the period of oscillation. As the surface area returns to its initial value, only some of the material adsorbed in each first half cycle has the time to return (desorb and diffuse) to the bulk solution in the second half cycle. The net result is the transient generation of nonequilibrium surface densities that are higher than the equilibrium surface densities (24), of course if monolayer collapse is not substantial in the timescale of the area pulsation. This yields values of $\gamma_{min} < \gamma_e$.

Results for octanol, Triton X-100, sodium dodecylsulfate, and other surfactants have been published (23–25). Here, we present some results for sodium laurate in saline/buffer. The effect of surfactant concentration on $\gamma_e$ and the amplitude of tension oscillation $\Delta\gamma \equiv \gamma_{max} - \gamma_{min}$ are shown in Figure 5. The values of $\gamma_{min}$ are consistently lower than $\gamma_e$. As the concentration increases, adsorption becomes faster and $\Delta\gamma$ decreases. For the higher concentration, this results from the timescale $\tau$ for tension equilibration being lower than the period of oscillation $T$ (32). If, now $\tau \gg T$ (Figure 5, low concentrations), then the adsorbed monolayer would behave as an effectively insoluble monolayer in the timescale of the pulsation, with $\gamma_{min} \approx \gamma_e$ and $\gamma_{max} \gg \gamma_e$. At the higher concentrations, the system is a dispersion, with the particles playing important roles both on the equilibrium and dynamic tension behavior (see Tables 5 and 6).

## VI. CONCLUSIONS

The dynamic behavior of single-component spread monolayers—compression rate dependence, hysteresis, collapse—has to be considered carefully in both fundamental thermodynamic monolayer studies and in practical applications. In addition, in mixed monolayers, one should consider the possibilities of squeeze-out (surface refining) effects even below the collapse pressure region. Supersaturated monolayers can have higher surface densities, and lower tensions, than those of the saturated monolayer at the equilibrium surface pressure. Supersaturation may be reduced more rapidly when dispersed particles are available to "receive" the extra molecules. The

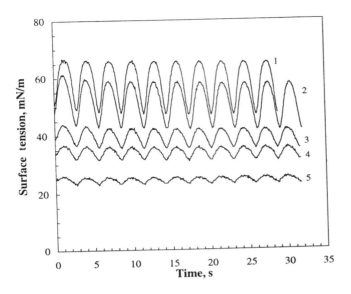

**Figure 5** Dynamic surface tension of sodium laurate in saline/phosphate buffer solution at 37°C and 20 cycles/min. Data were obtained with the PBS method for (1) 0.010, (2) 0.50, (3) 2, (4) 4.5, and (5) 10 mM. For constant area data see Table 5 and Reference 32.

presence and nature of dispersed surfactant microparticles at or near the surface monolayer are key factors in adsorption dynamics of sparingly soluble surfactants.

Even soluble surfactants can exhibit nonequilibrium surface densities higher than their equilibrium values, in a general dynamic adsorption hysteresis mechanism, which depends on the area pulsation frequency, timescale of adsorption, and possible monolayer collapse effects. The results have important implications for low surface tensions, which at present are primarily relevant to lung surfactants.

## VII. KEY ISSUES AND NEW HORIZONS

Despite significant advances during the past 25 years, several important questions remain:

1. What are the causes and events which trigger monolayer collapse?
2. What happens to the collapsed structures, especially those of the mixed monolayers?

# Adsorption and Tension of Monolayers

**Table 5** Summary of Tension Data for Aqueous Sodium Laurate with the Bubble Method

| Concentration (mM) | Solution (S) or dispersion (D)[a] | $\gamma_{ss}^{b}$ (mN/m) | $t_{95}^{c}$ (s) |
|---|---|---|---|
| 0.010 | S | 51 | 1200 |
| 0.50 | S | 54 | 20 |
| 2 | D | 46 | 20 |
| 4.5 | D | 26 | 11 |
| 10 | D | 35 | 7 |
| (<10)[d] | S[d] | 47 | 10 |

*Note:* All solutions at 37°C in saline/phosphate buffer.
[a] Dispersions were cloudy and contained dispersed microcrystallites of sodium laurate.
[b] Steady-state (close to equilibrium) tension (32).
[c] Time for the surface tension to drop to 95% of its total decrease.
[d] The 10 mM dispersion was filtered to yield a clear solution, whose concentration is not precisely known.

**Table 6** Pulsating Area Results for the Aqueous Laurate Systems of Table 5 and Figure 5

| Concentration (mM) | $\nu$ (cycles/min) | $\gamma_i^a$ (mN/m) | $\gamma_{max}^b$ (mN/m) | $\gamma_{min}^b$ (mN/m) | $\gamma_{max} - \gamma_{min}$ (mN/m) |
|---|---|---|---|---|---|
| 0.010 | 20 | 51 | 65 | 47 | 18 |
| 0.010 | 80 | 51 | 66 | 48 | 18 |
| 0.50 | 20 | 48 | 58 | 41 | 17 |
| 0.50 | 80 | 51 | 64 | 42 | 22 |
| 2.0 | 20 | 39 | 42 | 35 | 7 |
| 2.0 | 80 | 42 | 48 | 37 | 11 |
| 4.5 | 20 | 34 | 35 | 31 | 4 |
| 4.5 | 80 | 34 | 37 | 30 | 7 |
| 10 | 20 | 25 | 26 | 22 | 4 |
| 10 | 80 | 25 | 25 | 22 | 3 |
| (<10)[c] | 20 | 49 | 56 | 42 | 14 |
| (<10)[c] | 80 | 48 | 58 | 42 | 16 |

[a] The initial tension at the start of the pulsation. This tension was usually, but not always, equal to the steady tension of Table 5.
[b] Values at the tenth cycle, at which the maxima and minima were usually repeatable.
[c] Sample was the filtrate from the 10 mM dispersion.

3. What thermodynamic and molecular factors determine whether these will be a surface refining or squeeze-out of mixed monolayers?
4. What thermodynamic, dynamic, and molecular factors determine the lowest tensions achievable by a monolayer?
5. What are the monolayer and overall interface microstructures at the low surface tension states?

To lead to new applications of spread and adsorbed monolayers in biological and biochemical systems, and in foaming, film coating, and LB film technologies, several research groups are working actively to address many important issues in this field. The emergence of new quantitative optical and spectroscopic techniques (X ray and neutron reflection, Brewster angle microscopy, sum-frequency and other nonlinear optical techniques, etc.) will add data and insights for better understanding of the dynamic properties of monolayers. This bodes well for the science and future applications of spread and adsorbed monolayers.

## VIII. EXPERIMENTAL DETAILS

### A. Materials and Sample Preparation

Synthetic $L$-$\alpha$-dipalmitoylphosphatidylcholine (DPPC) (+99%) and $n$-hexadecanol (H) (~99%) were purchased from Sigma Chemical Co (St. Louis, MS), and were used without further purification. Perdeuterated $n$-hexadecanol (d-H), $C_{16}D_{33}OH$, with over 96% atomic purity was obtained from Isotech, Inc. (Miamisburg, Ohio). Sodium laurate (over 99% pure) was purchased from Fluka Chemical Co. (Ronconkoma, NY). The saline/phosphate buffer had pH = 7 and a total ionic strength of 0.467 M (32). Sodium chloride, analytical reagent (AR®) grade was purchased from Mallinckrodt, Inc. (Paris, KY). HPLC grade hexane from Aldrich (Milwaukee, WI), and absolute alcohol from Midwest Grain Products Co. (Pekin, IL), were used as solvents for making spread monolayers. All the experiments were done with ultrapure Millipore water which was prepared by passing distilled water through a Milli-Q four-bowl system. The water had an initial resistivity of 18 M$\Omega \cdot$ cm. All dispersions were prepared in saline solution (0.9 wt % aqueous sodium chloride). A saline solution was also used as the subphase in the Langmuir trough experiments.

### B. Spread Monolayer Experiments

For surface pressure ($\Pi$)–surface area per molecule ($\overline{A}$) monolayer isotherm studies and for fabrication of Langmuir–Blodgett (LB) monolayers, solutions of DPPC in 9/1 v/v hexane/ethanol (0.5 mg/ml), hexadecanol (H),

# Adsorption and Tension of Monolayers

and perdeuterated hexadecanol (d-H) in hexane (0.5 mg/ml) were prepared. For mixed monolayers of DPPC/H and DPPC/d-H, the components were dissolved in 9/1 v/v hexane/ethanol to yield 1 mg/ml solutions. After monolayers were spread, about 7 minutes were allowed for solvent evaporation. The surface area was compressed at 15 cm$^2$/min (from 2 to 7 Å$^2$/molecule/min, depending on the material) until a desired deposition pressure was reached. The monolayer was transferred at room temperature onto an ATR crystal at a deposition rate of 0.5 cm/min. During deposition, the set pressure was maintained to within 1 mN/m.

## C. Apparatus and Procedures

A computer-controlled minitrough from KSV Instruments, Finland, had a roughened platinum Wilhelmy plate. The maximum trough surface area was 250 cm$^2$, and the volume of the subphase was ~220 cm$^3$. Experiments with the trough were done at room temperature or at 37°C by flowing thermostated water through the hollow body of the trough.

A commercial, thermostated, computer-controlled Pulsating Bubble Surfactometer (PBS) from Electronetics Co., Amherst, NY was used for measuring dynamic surface tensions, $\gamma(t)$. The instrument, which is based on Enhorning's design (22), measures the pressure difference across the bubble surface with a pressure transducer. The Laplace–Young equation, $\Delta P(t) = 2 \gamma(t)/R(t)$, where $\Delta P$ is the pressure difference and $R$ is the radius of the bubble, is used for calculating the surface tension (24,25). The sample volume is approximately 25 µl. Under constant area conditions, the radius of the bubble is fixed at $R = 0.40$ mm and the surface tension is measured every 50 ms after a 1 s initial delay or "dead" time. Under pulsating area conditions, the radius varies nearly sinusoidally from $R = 0.40$ and 0.55 mm (area ratio 1.89) by using a liquid volume displacer. The bubble can be pulsated at various pulsating frequencies ranging from 1 to 100 cycles/min. The bubble is quite small and nearly spherical (25). Most of the experiments were done within a day after the dispersions were prepared.

A Nicolet 800 FT-IR Spectrometer (Nicolet Inc., Madison, WI) with a liquid-nitrogen-cooled HgCdTe (MCT) detector was used for the ATR spectra. It was continuously purged with dry air from a Balston air purifier, for reducing the water vapor and carbon dioxide. The horizontal ATR accessory was custom-built by Connecticut Instrument Co., Norwalk, CT. The spectrometer has an internal computer-controlled wire-grid polarizer. Unpolarized or s- and p-polarized incident beams were used (29). Germanium ATR crystals (1 × 10 × 50 mm, 45° trapezoids) were purchased from Wilmad Glass Co., Buena, NJ. The resolution of the FT-IR spectra was about 2 cm$^{-1}$, and 256 scans were taken for each spectrum. The actual frequencies and absorbance intensities were determined without detailed spectral analysis. For the external reflection spectra, a small custom-built minitrough was fitted into an FT-IR sample chamber in a Perkin-Elmer spectrometer (34). The collapse surface pressures of DPPC and $d$-hexadecanol were ~72 and 60 mN/m, respectively. Experiments were done at 25°C. A plane polarized beam was used at 10° from the surface normal.

## IX. ACKNOWLEDGMENTS

This research was supported in part by the National Science Foundation (grant nos. BCS 91-12154 and CTS 93-04328), by a grant from the Showalter Trust, and by a fellowship from the Purdue Research Foundation (to S. Y. Park).

## REFERENCES

1. R Lemlich. Adsorptive Bubble Separation Techniques. New York, Academic Press, 1972.
2. JE Valentini, WR Thomas, P Sevenhuyen, TS Jiang, HO Lee, L Yi, SC Yen. Ind Eng Chem Res 30:453, 1991.
3. RH Notter, PE Morrow. Ann Biomed Eng 3:119, 1975.
4. KMW Keough. In: Pulmonary Surfactant: From Molecular Biology to Clinical Practice. Robertson, B, Van Golde, LMG, Batenburg, JJ, eds. Elsevier, Amsterdam, 1992, pp 109–164.
5. T Fujiwara, B Robertson. In: Pulmonary Surfactant: From Molecular Biology to Clinical Practice. Robertson, B, Van Golde, LMG, Batenburg, JJ, eds. Elsevier, Amsterdam, 1992, pp 561–633.
6. R Miller, P Joos, VB Fainerman. Adv Colloid Interf Sci 49:249, 1994.
7. CH Chang, EI Franses. Colloids Surfaces A 100:1, 1995.
8. JB Chung. Ph.D. Thesis. Purdue University, 1989.
9. JB Chung, PC Shanks, RE Hannemann, EI Franses. Colloids Surfaces 43:223, 1990.
10. JB Chung, RE Hannemann, EI Franses. Langmuir 6:1647, 1990.
11. EC Johannsen, JB Chung, CH Chang, EI Franses. Colloids Surfaces 53:117, 1991.
12. SY Park, CH Chang, DJ Ahn, EI Franses. Langmuir. 9:3640, 1993.
13. SY Park. Ph.D. Thesis. Purdue University, 1995.
14. SY Park, SC Peck, CH Chang, EI Franses. In: Dynamic Properties of Interfaces and Association Structures. Pillai, V, Shah, DO, eds. American Oil Chemists Society, Champaign, IL, 1996, pp 1–22.
15. GL Gaines. Insoluble Monolayers at Liquid-Gas Interfaces. New York, Wiley, 1966.
16. HD Dörfler. Adv Colloid Interf Sci 31:1, 1990.

# Adsorption and Tension of Monolayers

17. AF Mingotaud, C Mingotaud, LK Patterson. Handbook of Monolayers, Vol. 1 and 2. New York, Academic Press, 1994.
18. K Tajima, NL Gershfeld. Biophys J 32:489, 1978.
19. BA Hill. J Appl Physiol 54:420, 1983.
20. S Schürch, H Bachofen, J Goerke, F Possmayer. J Appl Physiol 67:2389, 1989.
21. S Schürch, H Bachofen, J Goerke, F Green. Biochim Biophys Acta 1103:127, 1992.
22. GJ Enhorning. J Appl Physiol 43:198, 1977.
23. CH Chang. Ph.D. Thesis. Purdue University, 1993.
24. CH Chang, EI Franses. Chem Eng Sci 49:313, 1994.
25. CH Chang, EI Franses. J Colloid Interf Sci 164:107, 1994.
26. A Boonman, FHJ Machiels, AFM Snik, J Egberts. J Colloid Interf Sci 120:456, 1987.
27. FR Rana, AJ Mautone, RA Dluhy. Biochemistry 32:3169, 1993.
28. B Pastrana-Rios, CR Flach, JW Brauner, AJ Mautone, R Mendelson. Biochemistry 33:5121, 1994.
29. DJ Ahn, EI Franses. J Phys Chem 96:9952, 1992.
30. SY Park, EI Franses. Langmuir 11:2187, 1995.
31. DJ Durand, RL Clayman, MA Heymann, JA Clements. J Pediatr 107:775, 1985.
32. KA Coltharp. M.S. Thesis. Purdue University, 1995.
33. CH Chang, KA Coltharp, SY Park, EI Franses. Colloids Surfaces A 114:185, 1996.
34. A Berman, DJ Ahn, A Lio, M Salmeron, A Reichert, D Charych. Science 269:515, 1995.

# 19
# What X-rays Tell Us About Langmuir Monolayers

**Pulak Dutta**
*Northwestern University, Evanston, Illinois*

## I. INTRODUCTION

There hasn't been a quarter century of progress in this area—in fact, it's only 10 years since the first (rudimentary) X-ray diffraction data from Langmuir monolayers were published (1,2). Given how commonly words implying knowledge of structure ("expanded liquid," "condensed liquid," "condensed solid," etc.) were and still are used in the field, it may be surprising to realize that no one actually knew anything about Langmuir monolayer lattice structures until a decade ago, and that even today we know about only a handful of systems (most of the work has been done on "simple" saturated fatty acids and alcohols).

As a result of the increased availability of synchrotron radiation, X-ray diffraction studies of monolayers can now be performed routinely, limited only by the availability of beam time. Now that the Advanced Photon Source has been commissioned, there will be more and brighter beam available than ever before, and thus even less justification for just assuming the structure of a monolayer system. In this paper I will first describe the techniques used to scatter off the surface of water, and then discuss some recent experiments. My intent is not to review everything that has been published in the last 10 years, but to give a few examples of the kinds of things that X-rays can tell us about Langmuir monolayers.

## II. TECHNIQUES

The simplest X-ray scattering technique used to study surfaces is specular reflectivity. The incident and reflected beams are in the same plane as and at equal angles to the surface normal (specular condition); the density distribution at the surface determines how the intensity varies as a function of the angle. A simplified picture of reflection from a monolayer on a (solid or liquid) substrate is that X-rays are reflected both from the monolayer-air interface and the monolayer-substrate interface; interference between these two reflections gives rise to intensity maxima and minima from which the monolayer thickness can be determined. The average molecular tilt can be estimated from this thickness if the dimensions of the molecule are known. From detailed analysis of the intensity profiles, much more information can be extracted, loosely analogous to the way diffraction peak positions give lattice spacings, but much more detail can be derived from diffuse scans.

Although X-ray reflectivity is a powerful technique, and has been used to study Langmuir films (3), it has not had much impact in this area simply because it is most useful for monolayers that have no structure within the layer. In such cases reflectivity is the best that can be done. As stated above, the tilt angle can be calculated from the thickness, but at small angles the thickness is insensitive to the tilt. It turns out that most of the Langmuir monolayer phases studied do have some degree of order in the plane, and in such cases X-ray diffraction provides tilt angle and tilt direction information (the latter cannot be determined from the thickness) in addition to lattice spacings.

To diffract from a monolayer it is necessary to distinguish the signal of interest from the scattering taking place in the substrate (where there are many more electrons). This is done by having the incident beam fall on the surface at an angle smaller than the critical angle for total external reflection. In this geometry only an evanescent wave penetrates into the bulk; scattering from the bulk is not eliminated, but it is greatly reduced. This experimental geometry was known and used well before the first Langmuir monolayer studies, but it is particularly important when the substrate is a liquid or glass, because disordered materials scatter X-rays at all angles, while crystalline solids scatter primarily in specific diffraction peak directions. Another difficulty with liquids is that the surface normal cannot be aligned in the beam using a diffractometer; rather, the beam must be brought down to the surface.

Figure 1 shows a schematic of a typical layout. Here the incident X-ray beam is reflected by a mirror down onto the liquid surface, but tilting the diffraction plane of a monochromator crystal also works well. The

# X-rays and Langmuir Monolayers

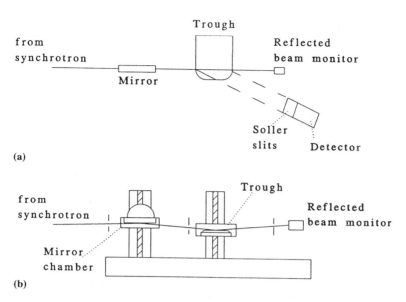

**Figure 1** Schematic diagram showing experimental layout for X-ray diffraction studies from Langmuir monolayers on the surface of water. (a) Top view; (b) side view.

critical angle for total external reflection from water is about 0.1°. The detector looking at the scattered beam moves both in the monolayer plane and above it; we will show data in terms of the difference between incident and scattered wave vectors: $K = K_{in} - K_{out}$. This vector K has an in-plane component $K_{xy}$ and a normal-to-monolayer component $K_z$; all monolayers studied so far have been powders in the plane, and in such cases the two components of $K_{xy}$ cannot be separated.

## III. TYPICAL RESULTS

Figure 2 shows typical diffraction data within the monolayer plane (i.e., with $K_z = 0$). This particular scan is from a monolayer of heneicosanoic acid ($C_{21}$) at low temperature and high pressure (4), in the so-called CS phase. We know the molecules are vertical because of scans as a function of $K_z$ (not shown) where the diffraction peaks are seen to have maximum intensity very close to the horizontal plane ($K_z \sim 0$). The first-order (1A/1B) and second-order (2A/2B) diffraction peaks establish that the lattice is

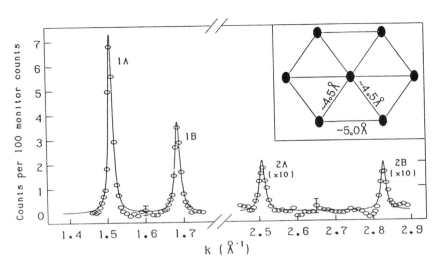

**Figure 2** X-ray diffraction data in the monolayer plane from $C_{21}$ acid at 5°C and 35 dyn/cm. The solid lines are Lorentzian fits. Inset: horizontal-plane lattice. (From Ref. 4, used with permission.)

distorted-hexagonal, i.e., near neighbors form an isosceles triangle. There should in general be six first-order peaks from a distorted-hexagonal structure, but the peaks labeled 1A and 2B are fourfold degenerate, while 1B and 2A are twofold degenerate.

If the molecules tilt, three of these six peaks will be at or below the horizontal plane (peaks below the plane cannot be seen, because the water gets in the way) while the remaining three will be at or above the plane. Figure 3 shows scattered intensity contours in the $K_p$-$K_z$ plane for the first-order peaks of a $C_{20}$ acid monolayer (5) as a function of pressure at 0°C. (The subphase is a pH = 2 solution of HCl, which doesn't freeze at 0°C.) At 5 dyn/cm and 10 dyn/cm (the two panels at left), it can be seen that there are two peaks, one of which is in the plane. The absence of any other peaks means that the off-plane peak is doubly degenerate, and from this we conclude that the molecules are tilted toward a nearest neighbor. This is the so-called $L_2$ phase. Moving next to 20 dyn/cm (bottom right in Figure 3), we see that there are again two peaks, but both are off-plane. $K_z$ at one peak is twice that at the other peak. This can only mean that the molecules are tilted in a direction halfway between near neighbors (normal to the 5 Å bond in the Figure 2 inset), in other words toward a next-nearest neighbor. This is the so-called $L_2'$ phase. These two tilt directions are the symmetry directions of the lattice shown in Figure 2.

# X-rays and Langmuir Monolayers

With data such as these, the structures of saturated fatty acid monolayers have been determined in considerable detail (7–16) Of course, one does not need X-ray data to know that there are more than just solid, expanded liquid, and condensed liquid phases; isotherms will suffice. If you look at a fatty acid isotherm at any particular temperature, there will often be only two features, and tradition requires that the three continuous segments be labeled "solid," "liquid condensed," and "liquid expanded." But families of isotherms recorded as a function of temperature show several triple points, so that there are certainly more than three phases. We "discovered" this in our own isotherm data (8a,8b); in fact it had been reported many years ago (17) but ignored at the time. This is why one should never rely on a single (usually room temperature) isotherm!

Using X-ray diffraction, we now know the structures of most fatty acid phases whose boundaries are visible as isotherm features. Some of these phases, with their conventional if uninformative code names, are listed in Table 1. Notice that there are several pairs of phases with the same symmetry. The CS and S phases, for example, are separated by a small

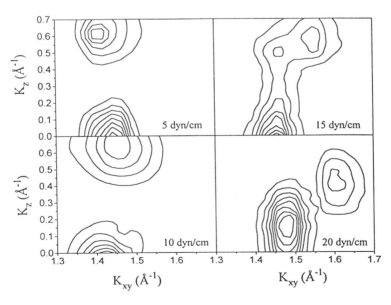

**Figure 3** X-ray diffraction data plotted as intensity contours for a monolayer of $C_{20}$ acid at 0°C and various pressures, as indicated. (From Ref. 5, used with permission.)

**Table 1** Some Phases Seen in Fatty Acid Monolayers

| Name | Pressure | Temperature | Lattice | Tilt direction |
|---|---|---|---|---|
| CS | High | V. low | DH | None |
| S | High | Low | DH | None |
| Rotator I | High | Medium | DH | None |
| Rotator II | High | High | Hexagonal | None |
| $L_2''$ | Medium | Low | DH | NN |
| $L_2'$ | Medium | Medium | DH | NNN |
| $L_2$ | Low | All | DH | NN |
| Ov | Medium | High | DH | NNN |

*Note:* The actual temperatures and pressures depend on chain length (higher for longer chains); for details, see Reference 18. DH means distorted hexagonal; NN means toward a nearest neighbor; NNN means towards a next-nearest neighbor.

first-order structural transition; the CS phase diffraction peaks are resolution limited, but the S phase has broad peaks at higher temperatures and so may be an intermediate-range ordered mesophase. Similarly, the difference between the $L_2''$ and $L_2$ phases appears to be the degree of order: $L_2''$ is long-range-ordered (as far as one can tell given finite experimental resolution) while $L_2$ has uniaxial order (one of the diffraction peaks is sharp while the other is broad).

I said earlier that the phase boundaries were indicated in the isotherms, but the Ov phase is a special case. Nothing in the isotherms marks the $L_2$-Ov boundary; it was identified on the basis of texture changes seen using Brewster angle microscopy (19a,19b). Our studies then showed that it had the same symmetry as the $L_2'$ phase; Fischer and Knobler (20) have shown that in acid–alcohol mixtures the phase boundaries move until the $L_2'$ and Ov phases merge, which would have been impossible if they had different symmetries. The X-ray data also provide a microscopic picture of why there is no isotherm feature: across the $L_2$-Ov phase boundary the tilt direction changes from NN to NNN, but the tilt angle remains the same, and so do the intermolecular spacings measured normal to the alkane chains ($a_p$ and $a_p''$ in Figure 4). Thus the horizontal area/molecule also does not change, even though this is very definitely a first-order transition.

We return now to the top right panel in Figure 3. It shows *three peaks*. We have recently found (5) that in $C_{20}$ acid, unlike $C_{21}$, the transition from NN to NNN tilt takes place not as a first-order jump but continuously via a phase with intermediate tilt direction. Such an intermediate tilt removes the degeneracy, resulting in three distinct peaks. In this region the "flat sections" in the $C_{20}$ acid isotherms actually have small but nonzero slopes. We find that these flat sections are not coexistence regions of a first-order phase transition, but a new phase with a different symmetry from the neighboring phases.

Determining the symmetries of various phases is one of the very simplest things that can be done with X-ray diffraction. As an example of more subtle kinds of information that may be extracted from trends in quantitative X-ray data, Figure 5 shows lattice spacings measured in mixed $C_{21}$ acid–alcohol monolayers (16) as a function of alcohol fraction and pressure at 15°C. We undertook this study of mixtures with different headgroups but the same chain length in order to try to separate the roles of the head- and tailgroups. As with the Ov phase, the spacings normal to the alkane chains (cf. Figure 4) are more informative than the spacings in the horizontal plane. It can be seen in Figure 5 that the $L_2$-$L_2'$ "swiveling transition" moves to lower pressures as the alcohol fraction increases. The lattice spacings measured normal to the molecules do not change either with pressure or with alcohol fraction; in each phase, the chain packing is always the same. Since the tilt angle is a strong function of the pressure, the picture that emerges is that the headgroups form a lattice and the chains then "fall over" until they are at optimum distances from each other, tilting to whatever angle is required to achieve this. In the "chain view" offered by the normal-to-chain spacings, there is no reason why the phase transition should occur at any particular pressure or why this pressure should depend on the alcohol fraction. Therefore, it must be that the headgroups "control"

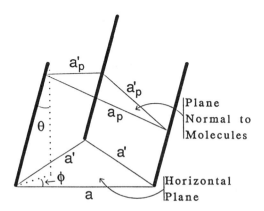

**Figure 4** Schematic diagram defining lattice spacings measured in the horizontal plane versus spacings measured normal to the tilted molecules.

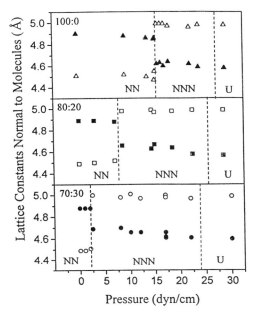

**Figure 5** Intermolecular spacings in the plane normal to the molecules, $a_p$ (open symbols) and $a_p'$ (filled symbols), as a function of pressure along a 15°C isotherm for three different monolayers: pure heneicosanoic acid (triangles), an 80 : 20 mixture of acid and alcohol (squares), and a 70 : 30 mixture of acid and alcohol (circles). Vertical dashed lines mark phase boundaries indicated by isotherms. (From Ref. 16, used with permission.)

the structure and swiveling transition; changing the composition of the mixture changes the interactions among headgroups and shifts the swiveling transition. Notice also that while the normal-to-chain spacings are different in the two phases, if the chains were free to rotate about their long axes as well as to tilt, there would be no reason for these spacings to be different. We conclude that the headgroups are oriented differently in the two phases, and that this then determines the orientations of the chain cross sections. The chains may tilt as needed, but they may not choose the tilt direction or the alignment of their cross sections.

Similar insights may be obtained from quantitative trends in pure acid monolayer spacings as a function of pH (11). It is well known that when there are metal ions in the subphase and the pH is high, the acid isotherms turn into featureless "salt isotherms." We found (with calcium ions) that this is because the features move to lower pressures as the pH is increased, until finally the very dense high-pressure phase occurs even at zero pressure. Looking at the lattice spacings at zero pressure, we see that as the pH increases the spacing in one direction decreases while the other direction remains almost the same. Since calcium is divalent, this suggests that the ions attach to two acid molecules not at random but along a specific direction of the monolayer lattice.

There are some situations where X-ray diffraction isn't useful, such as when the monolayer is completely disordered, although we have found that many supposedly liquid phases are in reality at least partially ordered. Also, diffraction scans take time (e.g., over an hour for each panel in Figure 2), and so it isn't possible as yet to look at dynamic phenomena. The Advanced Photon Source may help a great deal in this regard. Finally, X-rays average over a large region of the monolayer, because at small incident angles even a small synchrotron beam illuminates a large area. Thus, diffraction and microscopy will remain complementary for the foreseeable future; the Ov phase (see above) is one example of how real space and reciprocal space techniques complement each other. In any case X-ray diffraction is one of the core characterization tools for anyone working with three-dimensional materials, and there is no reason why it should not be so for monolayers as well.

## ACKNOWLEDGMENTS

Our work in this area was performed in collaboration with B. Lin, M. C. Shih, M. K. Durbin, T. M. Bohanon, A. Malik, J. B. Ketterson, and P. Zschack, and supported by the U.S. Department of Energy under grant no. DE-FG02-84ER45125. The National Synchrotron Light Source and Beam Line X-6 are also supported by the U.S. Department of Energy.

## REFERENCES

1. K Kjaer, J Als-Nielsen, CA Helm, LA Laxhuber, H Möhwald. Phys Rev Lett 58:2224, 1987.
2. P Dutta, JB Peng, B Lin, M Prakash, JB Ketterson, P Georgopoulos, S Ehrlich. Phys Rev Lett 58:2228, 1987.
3. See e.g. J Als-Nielsen, K Kjær. NATO ASI Ser Ser B 211:113, 1989.
4. T Bohanon, B Lin, M Shih, G Ice, P Dutta. Phys Rev B (Rapid Comm) 41:4846, 1990.
5. MK Durbin, A Malik, A Richter, P Zschack, P Dutta. J Chem Phys 106:8216, 1997.
6. SW Barton, BN Thomas, SA Rice, B Lin, JB Peng, JB Ketterson, P Dutta. J Chem Phys 89:2257, 1988.

7. B Lin, MC Shih, TM Bohanon, GE Ice, P Dutta. Phys Rev Lett 65:191, 1990.
8a. P Dutta. In Phase Transitions in Surface Films 2. Taub, H, Torzo, G, Lauter, HJ, Fain, SC, eds. New York, Plenum, 1991, 183.
8b. B Lin. Ph.D. Thesis. Northwestern University, 1991.
9. RM Kenn, C Böhm, AM Bibo, IR Peterson, H Möhwald, K Kjær, J Als-Nielsen. J Phys Chem 95:2092, 1991.
10. K Kjær, J Als-Nielsen, RM Kenn, C Böhm, P Tippmann-Krayer, IR Peterson, AM Bibo, CA Helm, H Möhwald. Makromol Chem Macromol Symp 46:89, 1991.
11. MC Shih, TM Bohanon, JM Mikrut, P Zschack, P Dutta. J Chem Phys 96:1556, 1992.
12. MC Shih, TM Bohanon, P Zschack, JM Mikrut, P Dutta. J Chem Phys 97:4485, 1992.
13. MC Shih, TM Bohanon, JM Mikrut, P Zschack, P Dutta. Phys Rev A45:5734, 1992.
14. MC Shih, TM Bohanon, JM Mikrut, P Zschack, P Dutta. In Surface X-Ray and Neutron Scattering. Springer Proceedings in Physics Vol. 61. Robinson, IK, Zabel, H, eds. Berlin, Springer-Verlag, 1992, p 151.
15. MK Durbin, A Malik, R Ghaskadvi, MC Shih, P Zschack, P Dutta. J Phys Chem 98:1753, 1994.
16. MC Shih, MK Durbin, A Malik, P Zschack, P Dutta. J Chem Phys 101:9132, 1994.
17. e.g. S Stallberg-Stenhagen, E Stenhagen. Nature 156:239, 1945; M Lundquist. Chem Scripta 1:197, 1971.
18. AM Bibo, IR Peterson. Adv Materials 2:309, 1990.
19a. GA Overbeck, D Möbius. J Phys Chem 97:7999, 1993.
19b. DK Schwartz, CM Knobler. J Phys Chem 97:8849, 1993.
20. B Fisher, CM Knobler. J Chem Phys 103:2365, 1995.

# 20
## Inorganic Extended Solid Langmuir–Blodgett Films

**Daniel R. Talham, Houston Byrd,\* and Candace T. Seip**
University of Florida, Gainesville, Florida

## I. INTRODUCTION

It has long been realized (1) that metal ions added to the subphase under monolayer films of fatty acids can be incorporated into Langmuir–Blodgett (LB) monolayer and multilayer films. Divalent metal ions crosslink the carboxylate groups and are often used to enhance the stability of transferred films (2–4). The difference in solubility between a free acid and its metal salt was used by Blodgett to produce films with a tunable refractive index through a process termed skeletonization (5). The free acid molecules of a partially ionized film were dissolved away with an organic solvent, leaving behind a skeletonized film of metal carboxylate with an enhanced porosity and a reduced refractive index. The influence of metal ions on the film can be detected at the air–water interface by changes in the pressure versus area isotherms (2), and recently the different surface structures that form have been probed using synchrotron radiation (6). Pressure–pH phase diagrams clearly show changes in film structure as metal ions are incorporated (6). Headgroup–headgroup interactions change as the pH is varied and metal ions bind at the interface, leading to changes in the packing of the alkyl tails.

It has also become increasingly clear in recent years that the structure of the film at the air–water interface does not necessarily determine the structure of the transferred film (3,7–9). For monolayer films, the structure may depend on the substrate onto which it is transferred (3,8,10). Interaction of the polar headgroups with the substrate can determine the organiza-

---
\**Current affiliation:* University of Montevallo, Montevallo, Alabama.

tion of the deposited film. For multilayer Y-type films it is the polar interactions, specifically the ionic interactions between the polar headgroup and the metal ion, that determine the structure of the transferred films. Different structures are observed as the identity of the metal ion changes. For example, recent AFM studies on a series of divalent metal carboxylate LB films have revealed a complex assortment of structures that depend on the nature of the metal-ion–headgroup interaction (8). In the transferred films, the alkyl chains vary their local packing, tilt angle, and tilt direction to achieve close-packing in the film, but it is the metal-ion–headgroup lattice energy that dictates the molecular area.

Although a picture is emerging that suggests that polar headgroup interactions can dictate the structure of LB films, historically little attention has been devoted toward using ionic headgroup interactions to purposefully control monolayer structure. Headgroup–metal-ion combinations other than classic fatty acid salts should be considered for organizing LB films. An approach that we are investigating is to utilize known solid-state layered structures to organize LB films (11–13). There are several examples in the solid-state literature of mixed organic–inorganic layered compounds where polar ionic networks are separated by nonpolar organic networks (14). Two examples are shown in Figure 1. Transition metal organophosphonates

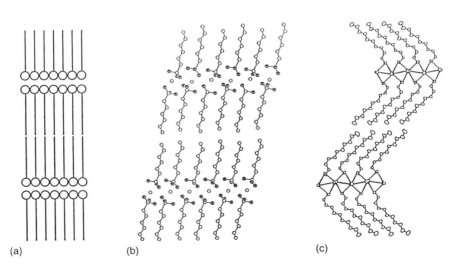

**Figure 1** (a) Conventional schematic of a Y-type Langmuir–Blodgett film. (b) Packing diagram of the layered metal phosphonate $Ca(HO_2PC_6H_{13})_2$ (see Reference 18). (c) The layered perovskite structure of $(C_{10}H_{21}NH_3)_2 CdCl_4$ (see Reference 14). Note the similarity between these solid-state structures and Y-type LB films.

# Extended Solid Langmuir–Blodgett Films

frequently form layered structures where the binding within the metal phosphonate layer is ionic/covalent and the interaction between layers is van der Waals in nature (15–17). The structure (18) of $Ca(O_3PC_6H_{13})_2$ is shown in Figure 1, but layered structures are known for many other organophosphonates with a variety of divalent, trivalent, and tetravalent metal ions (15,16,18–20). The alkylammonium layered perovskites (14), also pictured in Figure 1, are another family of mixed organic–inorganic layered structures based on organic amphiphiles. When comparing structures of this type it becomes clear that it is the inorganic lattice energy that determines these structures since the same ionic lattices are found in examples without organic substituents. For example, the metal halide lattice in the alkylammonium layered perovskites is essentially isostructural with that in the purely inorganic $K_2MnF_4$. Since we can identify inorganic interactions that clearly favor layered structures, it is worthwhile to ask whether or not these inorganic structure types can be used to control in a predictable way the structures of mixed organic–inorganic LB films.

We review here our attempts to employ this strategy for formulating LB films, and describe the preparation of LB films based on transition metal organophosphonate layered structures using octadecylphosphonic acid with tetravalent $Zr^{4+}$ and divalent $Mn^{2+}$ ions. A consequence of incorporating extended lattice layer structures into LB films is that properties normally associated with solid-state inorganic structures can now be incorporated into the organic films. An example is cooperative magnetic phenomena, and the magnetic properties of the manganese phosphonate LB film are compared to the solid-state analogs. Finally, we discuss the magnetic behavior of the manganese phosphonate films in the context of previous work with a brief review of magnetism in LB films.

## II. ZIRCONIUM PHOSPHONATE LB FILMS

Octadecylphosphonic acid forms a stable Langmuir monolayer on a pure water subphase (13). The surface pressure versus area ($\pi - A$) isotherm at pH 7 yields an extrapolated cross-sectional area of 24 $Å^2$/molecule and a collapse pressure of 60 mN/m (13,21). The isotherm is pH-dependent, and the monolayer behavior also depends greatly on the presence of metal ions in the subphase. If octadecylphosphonic acid is compressed on a 5 mM $Zr^{4+}$ subphase, the film becomes rigid (13,21). Accurate pressure versus area measurements using standard Wilhelmy plate techniques are not possible because the rigid monolayer pushes the platinum plate from the vertical. Brewster angle microscopy (BAM) demonstrated that the octadecylphosphonic acid forms monolayer thick aggregates on the surface of the $Zr^{4+}$

subphase and that these domains maintain their shape throughout compression (21). Because of this rigidity, the films do not transfer using conventional vertical deposition methods.

While the strong $Zr^{4+}$-phosphonate binding interaction works against forming a processible Langmuir monolayer, this same feature allowed us to develop a novel stepwise procedure for depositing monolayer and multilayer films of zirconium octadecylphosphonate that involves a combination of LB and "inorganic" self-assembly methods (13). This procedure is outlined in Scheme 1. The first step creates an LB template of octadecylphosphonic acid suitable for binding $Zr^{4+}$ ions by transferring a single LB layer of octadecylphosphonic acid from a pure water subphase onto an octadecyltrichlorosilane (OTS)-covered substrate that is dipped through the film and into a vial sitting in the subphase. The vial containing the substrate, immersed in subphase, is then removed. In step 2, $Zr^{4+}$ ions "self-assemble" at the newly formed organic template upon adding $ZrOCl_2$ to the vial to produce a 5-mM zirconium ion solution. After 30 min in the zirconium solution, the substrate with the zirconated octadecylphosphonic acid layer is removed and placed into another vial containing pure water. Step 3 of the deposition procedure is to cap the zirconated layer with another LB layer to complete the bilayer assembly. To accomplish this, the substrate in pure water is placed back into the LB trough where a new octadecylphosphonic acid film is compressed over the vial, and then transferred to the substrate creating a Y-type zirconium octadecylphosphonate bilayer. Multilayers can be produced by repeating this three-step deposition procedure on the resulting hydrophobic surface.

Each step of the deposition process has been carefully analyzed (13,21). Transfer ratios of unity on both the downstroke and upstroke indicate that complete, close-packed layers are transferred. This is confirmed by FT-IR, where the intensity and position of the C-H stretching modes show that the octadecylphosphonate molecules are close-packed and in an all-trans conformation. X-ray photoelectron spectroscopy (XPS) analysis (Figure 2) is also consistent with the step-wise deposition described in Scheme 1. According to the first spectrum in Figure 2, after zirconium ions bind to the first octadecylphosphonic acid layer, the P to Zr ratio is 1 : 1. This indicates that the molecules in the octadecylphosphonic acid film are spaced to allow binding of one zirconium ion per phosphonate group. Of particular interest is the stability of the zirconated layer. Because of the strong zirconium–phosphonate binding, the layer does not rearrange at this stage to form a bilayer structure. After capping the zirconated surface with a second octadecylphosphonic acid LB layer, the P to Zr ratio is 2 : 1 (Figure 2), which is consistent with the stoichiometry observed in the bulk layered zirconium phosphonates.

# Extended Solid Langmuir–Blodgett Films

Step 1

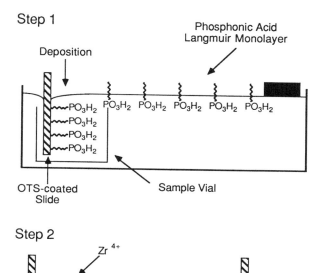

Step 2

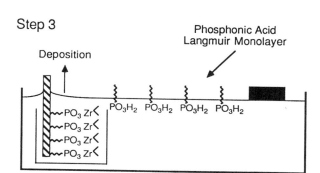

Step 3

**Scheme 1**  Procedure for depositing zirconium octadecylphosphonate LB bilayers.

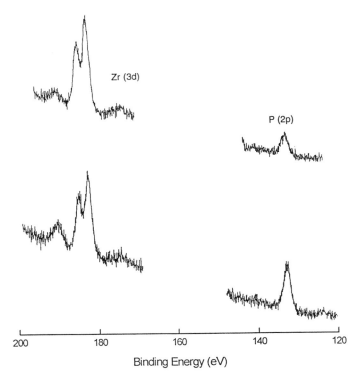

**Figure 2** XPS spectra monitoring the deposition of a zirconium octadecylphosphonate LB film. Top: A single octadecylphosphonic acid layer after binding $Zr^{4+}$ according to Step 2 in Scheme 1. Bottom: A bilayer of zirconium octadecylphosphonate deposited according to Scheme 1. Elemental ratios are obtained from the integrated signal intensities that have been corrected for elemental sensitivity factors and lattice effects. The Zr to P ratios are 1 : 1 in the top spectrum and 1 : 2 in the bottom spectrum.

More information on the structure of the film is learned from X-ray diffraction and polarized attenuated total reflectance (ATR)-FT-IR (13,21, 22). X-ray diffraction from a multilayered sample confirms the layered structure of the film and yields an interlayer d-spacing of 52 Å. Based on molecular modeling, a bilayer thickness of 52 Å requires that the alkyl chains orient nearly perpendicular to the metal ion plane. Polarized ATR-FT-IR was used to establish the tilt angles of the octadecylphosphonate molecules in the zirconium phosphonate LB film (22). The tilt angle of infrared (IR) active vibrational modes within the layered organic film can be determined from the ratio of IR absorbances measured in two polariza-

# Extended Solid Langmuir–Blodgett Films

tion directions (23,24). If the orientation of the transition dipole is known relative to the molecular coordinates, then the orientation of the molecules relative to the surface normal can be obtained. In these experiments, the absorbance of the $\nu_a(CH_2)$ band recorded with s- and p-polarized light was used to determine the tilt angles. Analyses were performed after each step of the deposition procedure so that the template layer could be compared to the capping layer (Scheme 2). Interestingly, we observe that the bottom layer is oriented differently from the capping layer. The tilt angle for the zirconated first layer is 31° (from the normal), while the measured tilt angle for the alkyl chain in the capping layer is 5°, essentially oriented perpendicular to the substrate.

In order to better understand the LB film structure, we used SYBYL (Tripos Associates) molecular modeling to generate a model based on a known zirconium phosphonate layered structure (22). The crystallographic coordinates (25) for $\alpha$-$Zr(HPO_4)_2 \cdot H_2O$ were used to model the $Zr$-$O_3P$ binding, and alkyl chains were grafted on in place of the phosphate OH groups. In the generated structure, the all-trans hydrocarbon chain lies at a tilt angle of 31.3° with respect to the zirconium ion plane, which is a consequence of the phosphonate P–C bond orienting nearly perpendicular to the plane of metal ions, just as the P–OH bond is oriented in the $\alpha$-$Zr(HPO_4)_2 \cdot H_2O$ structure. This is precisely the tilt angle observed for the first LB layer. The 31° tilt angle suggests that this layer relaxes in the pure water subphase after it is transferred. Addition of zirconium ions then cross-links the layer and holds it in place. The low solubility of zirconium phosphonates allows this layer to be used in subsequent assembly steps without loss of organic film or displacement of zirconium ions. The 5° tilt angle of the LB capping layer is consistent with the layer thickness deter-

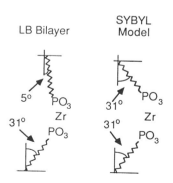

**Scheme 2**  Molecular axis tilt angles in zirconium octadecylphosphonate.

mined from X-ray diffraction data. This nearly perpendicular arrangement of the alkyl chains is also consistent with the 112° water contact angle measured on the complete LB bilayer film (22). The surface pressure used to align the molecules in the LB experiment orients the alkyl chains nearly perpendicular to the surface, and it appears that this packing is preserved when the layer is transferred to the zirconated LB template layer. Once the zirconium-phosphonate bond is formed, this extremely strong binding interaction does not allow the LB capping layer to relax.

The zirconium phosphonate deposition process clearly forms layered LB films, although the in-plane structure has not been easy to discern due to the poor crystallinity of the layers. This is similar to most solid-state zirconium phosphonates, which normally are also poorly crystalline. Spectroscopic analyses suggest that the zirconium phosphonate binding in the LB film is not the same as that observed in the solid-state for $\alpha$-$Zr(HPO_4)_2 \cdot H_2O$. In fact, the tilt angle analyses indicate that the top layer of each bilayer may not even be commensurate with the bottom layer. This is a consequence of the strong zirconium-oxygen binding in the phosphonates, which does not allow the structure to anneal on the time scale of the deposition process. Another consequence of the strong zirconium-phosphonate binding is that solid-state zirconium phosphonates are highly insoluble in both water and organic solvents, and the zirconium phosphonate LB films are similarly insoluble. For example, we have observed a 10-bilayer film of zirconium octadecylphosphonate to remain intact after soaking for several hours immersed in chloroform (13). Although the zirconium phosphonate headgroup binding does not appear to afford much control over LB film structure, these films are extremely stable once deposited.

## III. MANGANESE PHOSPHONATE LB FILMS

The pressure versus area isotherm of octadecylphosphonic acid compressed on a $Mn^{2+}$ subphase at a pH of 5.5 yields an extrapolated cross-sectional area of 27 Å$^2$ (Figure 3), and is expanded relative to the phosphonic acid monolayer compressed on pure water (12,26). As the film is repeatedly compressed and decompressed in a series of hysteresis runs, the cross-sectional area increases (26). The behavior is different than that observed when octadecylphosphonic acid is formed on a pure water subphase, where the cross-sectional area of the film decreases upon repeated compressions due to slow dissolution. The increase in area displayed by the film suggests that the $Mn^{2+}$ ions cross-link the phosphonate groups at the air-water interface and demonstrates that the film does not dissolve into the sub-

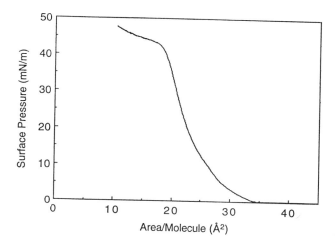

**Figure 3** Pressure versus area isotherm for octadecylphosphonic acid on a water subphase that is $5 \times 10^{-4}$ M $MnCl_2 \cdot 4H_2O$ and pH 5.5.

phase. The film also becomes increasingly rigid with successive compression cycles.

In contrast to the $Zr^{4+}$ case, deposition of the manganese octadecylphosphonate is accomplished using normal vertical deposition methods (12,26). In a typical deposition experiment, octadecylphosphonic acid is compressed over a $Mn^{2+}$ ion subphase to a pressure of 17 mN/m. At this point, an OTS-covered substrate is lowered through the film at the air–water interface, thereby transferring the film in a tail-to-tail fashion. Once the first layer is deposited, the substrate is then raised from the subphase through the film, creating a bilayer. A slow deposition speed on the upstroke, and thus a slow draining of the water from the film, aids in crystallizing the inorganic lattice. We have found that 5 mm/min is the optimum speed for transfer. The pH of the subphase also affects the transfer of the monolayers and the optimum pH range for deposition is 5.2–5.5. XPS analyses of single-bilayer and multilayered films deposited using these conditions show a Mn to P ratio of 1:1, indicating complete ionization of the phosphonic acid.

Figure 4 compares the infrared spectrum of an octadecylphosphonic acid bilayer (top) and a manganese octadecylphosphonate bilayer (bottom). In both spectra the common bands are the asymmetric methyl stretch ($\nu_a(CH_3)$), at 2958 $cm^{-1}$, the asymmetric methylene stretch ($\nu_a(CH_2)$) at 2917 $cm^{-1}$,

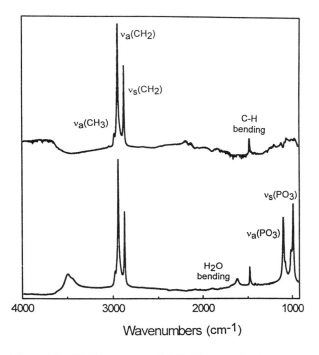

**Figure 4** FT-IR spectra of LB bilayers obtained on germanium ATR crystals. Top: Octadecylphosphonic acid. Bottom: Manganese octadecylphosphonate.

the symmetric methylene stretch ($\nu_s(CH_2)$) at 2850 cm$^{-1}$, and the methylene bending mode at 1467 cm$^{-1}$. The difference between the octadecylphosphonic acid film and manganese octadecylphosphonate film is clearly demonstrated by the P-O stretches. The asymmetric phosphonate stretch ($\nu_a(PO_3^{2-})$) at 1088 cm$^{-1}$ and the symmetric phosphonate stretch ($\nu_s(PO_3^{2-})$) at 978 cm$^{-1}$ are observed only in the film deposited from a Mn$^{2+}$ subphase (Figure 4). The absence of the strong P=O stretch in the 1350-1250 cm$^{-1}$ region or the 1250-1110 cm$^{-1}$ region for free and hydrogen bonded modes, respectively, suggests that the all of the phosphonate groups in the film are ionized. The appearance of the H-O-H bend at 1608 cm$^{-1}$ also indicates that Mn$^{2+}$ ions are present in the deposited films. In the bulk manganese phosphonates, each Mn$^{2+}$ ion is bound by five oxygen atoms from the phosphonate anion and one H$_2$O molecule fills out the coordination sphere. The H$_2$O bend (1608 cm$^{-1}$) along with the PO$_3^{2-}$ phosphonate stretching modes are all observed in the bulk manganese phosphonate materials.

To further compare the LB films to the solid-state analogs, we investigated the magnetic properties of the manganese phosphonate LB films (12, 26–28). We previously prepared a powder sample of $Mn(O_3PC_6H_5) \cdot H_2O$ and observed an antiferromagnetic ordering transition at 12 K by SQUID magnetometry (29). Meanwhile, Carling et al. have shown that the series $Mn(C_nH_{2n+1}PO_3) \cdot H_2O$ ($n = 1$–4) are all canted antiferromagnets with ordering temperatures in the range 14.8–15.1 K (30). Our magnetic studies using electron paramagnetic resonance (EPR) also show evidence for antiferromagnetic exchange and short-range antiferromagnetic order in the manganese phosphonate LB films. The integrated area of the EPR signal, which is proportional to the spin susceptibility, gradually increases with decreasing temperature to a maximum near 30 K, before decreasing at lower temperatures until the signal disappears around 18 K (Figure 5). The shape of the plot is characteristic of antiferromagnetic exchange in a low-dimensional lattice, and has been fit with a numerical expression for the susceptibility of a quadratic-layer Heisenberg antiferromagnet (31). The fit yields a value for antiferromagnetic exchange of $J/k = -2.8$ K, very close to the value observed for the powder sample of manganese phenylphosphonate. In addition, we have now observed magnetic ordering to a canted antiferromagnetic state in the Langmuir-Blodgett film (28). The observed magnetic exchange and magnetic order arises from the crystalline extended-

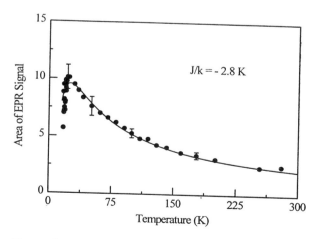

**Figure 5** Temperature dependence of the integrated area of the EPR signal from a manganese phosphonate LB film. The solid line is a fit to the data for a two-dimensional lattice with Heisenberg antiferromagnetic exchange with exchange constant, $J/k = -2.8$ K.

solid structure of the LB layers, and the magnetic behavior is nearly identical to that observed for bulk samples with analogous structures.

In contrast to the tetravalent zirconium phosphonate case, spectroscopic and structural analyses indicate that the manganese phosphonate LB film has the same in-plane structure as the known solid-state analogs. The results show that layered solid-state structures can indeed be prepared using LB methods. Since the lattice energy clearly favors forming this known metal phosphonate layered structure, the divalent metal phosphonates might further be used to control LB film structures. The procedure is not restricted to manganese phosphonates, as we have seen similar results with other divalent metals (28).

## IV. QUARTER CENTURY PROGRESS AND NEW HORIZONS: MAGNETISM IN LB FILMS

The study of the magnetic properties of manganese octadecylphosphonate is one of relatively few investigations of magnetic phenomena in LB films. While EPR is frequently used to study the structure and orientation of organic radicals or metal complexes in transferred films, there are very few examples of magnetic exchange or magnetic phase transitions in LB films. In the late 1970s, Pomerantz and coworkers (32-35) at IBM Yorktown Heights investigated what they called "literally two-dimensional magnets." They formed LB bilayers of Mn stearate which resulted in a layer of $Mn^{II}$ ions held symmetrically between two stearate layers. EPR was used to determine the magnetic properties. The behavior was well described by known models for a two-dimensional magnetic lattice. Pomerantz and coworkers searched for magnetic ordering by measuring the temperature dependence of the EPR signal. At low temperature ($<10$ K), substantial increases in the linewidth and shifts in the resonant field were cited as evidence for magnetic order. The nature of the observed magnetic order is still not certain, although the authors proposed that the system behaves as a "weak ferromagnet." Haseda and coworkers (36) and Martini and coworkers (37) have reported similar results on Mn carboxylate LB films. Divalent copper in fatty acid LB films has also been extensively studied using EPR (38,39). The copper salts normally form exchange coupled dimers and are EPR silent. Exposure to ammonia breaks up the dimers, and the monomeric copper sites can then be studied by EPR, although no cooperative behavior is observed.

Magnetic exchange has also been observed in LB films of organic radicals. Perhaps the best characterized are films based on organic $\pi$-donor and $\pi$-acceptor molecules that have been prepared as part of investigations of conducting LB films. For example, a thorough EPR study of LB films of a series of long-chain pyridinium-(TCNQ)$_2$ salts showed evidence for magnetic exchange in the films, although the systems are best described as quasi-one-dimensional spin systems (40,41). The authors explained the data in terms of a random-exchange Heisenberg antiferromagnetic chain model. There have also been studies of doxyl nitroxide radicals in LB films (42,43). These radicals are commonly used as magnetic probes for investigating structure and dynamics in membranes and membrane-like assemblies. Studies on doxylstearic acid LB films show that EPR linewidths are influenced by Heisenberg spin exchange, and it has been shown that the exchange depends on the position of the doxyl radical along the alkyl chain. In at least one case, the angular dependence of the EPR linewidth is consistent with exchange in a two-dimensional lattice (42).

The interesting phenomenon of spin crossover, where a transition metal ion undergoes a temperature-dependent transition from a high-spin to a low-spin electron configuration, has also been demonstrated in an LB film (44). Amphiphilic analogs of the known spin-crossover complex Fe(phen)$_2$(NCS)$_2$ (where phen is phenanthroline) were prepared by grafting 18-carbon chains onto phenanthroline ligands. The amphiphilic complexes formed good LB films and the spin-crossover transition was observed by monitoring the shift in the frequency of the CN stretching mode of the thiocyanide ligand as the spin-state changes. While spin crossover is not an example of magnetic exchange, the transition can be greatly influenced by lattice effects. The results showed that spin crossover, which is known to occur in 3-dimensional lattices, can also take place in a two-dimensional lattice.

The magnetic data on the manganese phosphonates described here show that the role of the metal–headgroup lattice is not restricted to a structural one. The observed magnetic exchange demonstrates the potential for introducing cooperative phenomena into LB films through the incorporation of extended solid inorganic structures. Future work will be aimed at designing functional organic molecules to complement the extended solid inorganic systems. The possibility exists for using this methodology to prepare "dual-network" assemblies with separate inorganic and organic components resulting in a composite of physical properties that are normally only associated with one type of material or the other.

## ACKNOWLEDGMENT

We would like to acknowledge the U.S. National Science Foundation for financial support.

## REFERENCES

1. KB Blodgett. J Am Chem Soc 57:1007–1022, 1935.
2. GJ Gaines. Insoluble Monolayers at Liquid-Gas Interfaces. New York, Wiley-Interscience, 1966.
3. Roberts, GG ed. Langmuir-Blodgett Films. New York, Plenum Press, 1990.
4. A Ulman. An Introduction to Ultrathin Organic Films: From Langmuir-Blodgett to Self-Assembly. Boston, Academic Press, 1991.
5. KB Blodgett. Phys Rev 55:391–404, 1939.
6. MC Shih, TM Bohanon, JM Mikrut, P Zschack, PJ Dutta. Chem Phys 96:1556–1559, 1992.
7. MC Shih, JB Peng, KC Huang, P Dutta. Langmuir 9:776–778, 1993.
8. JA Zasadzinski, R Viswanathan, L Madsen, J Garnaes, DK Schwartz. Science 263:1726–1733, 1994.
9. DA Outka, J Stöhr, JP Rabe, JD Swalen, HH Rotermund. Phys Rev Lett 59:1321–1324, 1987.
10. R Viswanathan, DK Schwartz, J Garnaes, JAN Zasadzinski. Langmuir 8:1603–1607, 1992.
11. H Byrd, JK Pike, ML Showalter, S Whipps, DR Talham. In: Interfacial Design and Chemical Sensing: ACS Symposium Series 561. Mallouk, TE, Harrison, DJ, ed. Washington, American Chemical Society, 1994, pp 49–59.
12. H Byrd, JK Pike, DR Talham. J Am Chem Soc 116:7903–7904, 1994.
13. H Byrd, JK Pike, DR Talham. Chem Mater 5:709–715, 1993.
14. P Day. Phil Trans R Soc Lond A 314:145–158, 1985.
15. A Clearfield. Comm Inorg Chem 10:89–128, 1990.
16. G Cao, H-G Hong, TE Mallouk. Acc Chem Res 25:420–427, 1992.
17. MB Dines, PM DiGiacomo. Inorg Chem 20:92–97, 1981.
18. G Cao, VM Lynch, JS Swinnea, TE Mallouk. Inorg Chem 29:2112–2117, 1990.
19. G Cao, H Lee, VM Lynch, TE Mallouk. Solid State Ionics 26:63–69, 1988.
20. G Cao, H Lee, VM Lynch, TE Mallouk. Inorg Chem 27:2781–2785, 1988.
21. H Byrd, JK Pike, DR Talham. Thin Solid Films 242:100–105, 1994.
22. H Byrd, S Whipps, JK Pike, J Ma, SE Nagler, DR Talham. J Am Chem Soc 116:295–301, 1994.
23. GL Haller, RW Rice. J Phys Chem 74:4386–4393, 1970.
24. N Tillman, A Ulman, JS Schildkraut, TL Penner. J Am Chem Soc 110:6136–6144, 1988.
25. A Clearfield, GD Smith. Inorg Chem 8:431–436, 1969.
26. H Byrd, JK Pike, DR Talham. Synthetic Metals 71:1977–1980, 1995.
27. CT Seip, H Byrd, DR Talham. Inorg Chem 35:3479–3483, 1996.
28. CT Seip, GE Granroth, MW Meisel, DR Talham. J Am Chem Soc 119, 1996.
29. ML Crews. Master's Thesis, University of Florida, 1992.
30. SG Carling, P Day, D Visser, RK Kremer. J Solid State Chem 106:111–119, 1993.
31. GS Rushbrooke, P Wood. J Molec Phys 1:257, 1958.
32. M Pomerantz. Surf Sci 142:556–570, 1984.

33. F Ferrieu, M Pomerantz. Solid State Commun 39:707–710, 1981.
34. M Pomerantz. In: In NATO ASI Series. Dash, JG, Ruvalds, J, eds. New York, Plenum, 1980, pp 317–346.
35. M Pomerantz, FH Dacol, A Segmüller. Phys Rev Lett 40:246–249, 1978.
36. T Haseda, H Yamakawa, M Ishizuka, Y Okuda, T Kubota, M Hata, K Amaya. Solid State Commun 24:599–602, 1977.
37. F Bonosi, G Gabrielli, G Martini. Colloids Surfaces A 72:105–110, 1993.
38. S Bettarini, F Bonosi, G Gabrielli, G Martini. Langmuir 7:1082–1097, 1991.
39. J Messier, G Marc. J Phys Paris 32:799–804, 1971.
40. K Ikegami, S-I Kuroda, M Sugi, T Nakamura, H Tachibana, M Matsumoto, Y Kawabata. J Phys Soc Jpn 61:3752–3765, 1992.
41. K Ikegami, S-I Kuroda, M Sugi, H Tachibana, T Nakamura, M Matsumoto. Thin Solid Films 242:11–15, 1994.
42. K Suga, Y Iwamoto, M Fujihira. Thin Solid Films 243:634–637, 1994.
43. F Bonosi, G Gabrielli, G Martini, MF Ottaviani. Langmuir 5:1037–1043, 1989.
44. A Ruaudel-Teixier, A Barraud, P Coronel, O Kahn. Thin Solid Films 160:107–115, 1988.

# 21
# Formation and Control of Unit Aggregates of Squaraines and Related Compounds in Langmuir–Blodgett Films

**Huijuan Chen,**[*] **Kangning Liang, Hussein Samha, Xuedong Song,**[†] **and David G. Whitten**[†]
University of Rochester, Rochester, New York

**Kock-Yee Law**
Xerox Corporation, Webster, New York

**Thomas L. Penner**
Eastman Kodak Company, Rochester, New York

## I. INTRODUCTION

Aggregation has been observed as a general phenomenon in a number of different media for a wide variety of aromatic compounds and dyes (1–3). Among the media where aggregate formation has been most commonly observed are Langmuir–Blodgett (LB) films (4), vesicles (5,6), reversed micelles, and other microheterogeneous media (7) which present a hydrophobic–hydrophilic interface. In several cases it has been supposed that the formation of aggregates can be attributed to a combination of very highly effective "local" concentrations of the chromophore in the microphase as well as to the tendency of the medium to "pack" the component molecules into a geometry that favors aggregation of the chromophores (8,9). Two characteristic types of aggregates are frequently encountered; the "J" aggregate characterized by a sharp, intense, and red-shifted absorption compared to the monomer transition (10,11), and the "H" aggregate characterized by

---
*Current affiliations:*
[*]Eastman Kodak Company, Rochester, New York.
[†]National Laboratory, Los Alamos, New Mexico.

a prominent blue-shifted absorption compared to the monomer transition (7,12). An early characterization of these two limiting cases based on a transition dipole–transition dipole coupling model was proposed by Kasha and coworkers (13,14) and attributes the J aggregate spectral shifts to a "head-to-tail" or "brickstone" arrangement of the chromophore transition moments within the aggregate while the H aggregate spectral shifts are attributed to a "card-pack" or "head-to-head" arrangement of the chromophore transition moments within the aggregate. This model was extended by Kuhn and coworkers (15,16), who developed an extended dipole–extended dipole treatment that gives very good agreement between predicted and observed exciton splittings for a variety of different arrays of chromophores and their transition moments within aggregates.

As mentioned above, aggregate formation is very frequently observed in LB assemblies, and the absorption and photophysics of a rather wide variety of aggregates from a diverse array of chromophores has been observed. Typical compounds forming J aggregates in LB assemblies include a variety of bifunctionalized cyanine dyes (17,18), which are anticipated to give "brickstone" arrays of the chomophores as indicated in Scheme 1 (d). In contrast, amphiphiles containing chromophores lying along the long axis

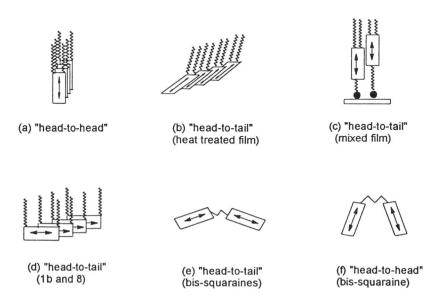

(a) "head-to-head"

(b) "head-to-tail" (heat treated film)

(c) "head-to-tail" (mixed film)

(d) "head-to-tail" (1b and 8)

(e) "head-to-tail" (bis-squaraines)

(f) "head-to-head" (bis-squaraine)

**Scheme 1** Schematic representation of transition dipole arrangements in different monolayers.

of the amphiphile (with their transition moments also lying along the long axis) tend to give blue-shifted or H aggregates; examples include *trans*-stilbenes (19,20), azobenzenes (21), diphenylpolyenes (22), and hemicyanine derivatives (23). A particularly interesting case is that of certain porphyrin derivatives that have two strong transition moments for the chromophore that are mutually perpendicular; for these compounds in LB assemblies aggregates can show both types of spectral shifts simultaneously (24,25).

In very recent work we have been able to show that very similar aggregates of several different aromatic compounds and dyes can be generated in LB assemblies, vesicles, and mixed aqueous–organic solutions (21,24,26,27). These studies have indicated that relatively stable H or blue-shifted aggregates occur for several different chromophores and that a small chiral "unit-aggregate" of remarkably similar structure can be deduced for several different compounds. We have suggested that the extended aggregate may be best described as a mosaic of these unit aggregates. A major question concerns the driving force for forming these and the related J aggregates in the different media. In the present paper we discuss some studies of aggregate formation for a series of amphiphilic squaraine dyes and some related compounds in LB films at the air–water interface and in the corresponding supported LB assemblies.

## II. RESULTS AND DISCUSSION

### A. Amphiphilic Squaraines

The structures of the squaraine amphiphiles and bis-squaraines used in these investigations are shown in Scheme 2. A variety of studies have shown that aggregate formation is very common for substituted squaraine dyes (28–30). In terms of aggregate structures, an X-ray diffraction study of crystals of bis(4-methoxyphenyl)squaraine indicates that the squaraine chromophores are packed into a translation layer structure such that the molecules are in an effective "card-pack" arrangement, resulting in a blue-shifted solid state absorption relative to the monomer absorption in solution (31). In contrast, bis(2-methyl-4-($N,N$-dimethylamino)-phenyl)squaraine forms crystals in which there is a "slipped stack" arrangement of the chromophores with a broad and red-shifted absorption compared to the monomer in solution (32). Several amphiphilic squaraines have also been found to form either red or blue shifted (or both) aggregates in LB assemblies (27,29), and thus it appeared especially appealing to follow the aggregation behavior of squaraines at the air–water interface.

The alkyl-substituted squaraines 1a and 1b in Scheme 2 are isomers differing in the placement of the octadecyl side chains such that 1a, with

(1-3) Mono-squaraines

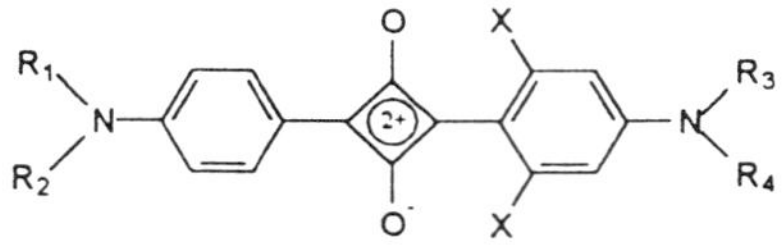

| Squaraine | $R_1$ | $R_2$ | $R_3$ | $R_4$ | X |
|---|---|---|---|---|---|
| 1a | $C_{18}H_{37}$ | $C_{18}H_{37}$ | $CH_3$ | $CH_3$ | H |
| 1b | $C_{18}H_{37}$ | $CH_3$ | $C_{18}H_{37}$ | $CH_3$ | H |
| 2a | $C_4H_9$ | $C_4H_9$ | $CH_3$ | $(CH_2)_3COOH$ | H |
| 2b | $C_8H_{17}$ | $C_8H_{17}$ | $CH_3$ | $(CH_2)_3COOH$ | H |
| 2c | $C_4H_9$ | $C_4H_9$ | $CH_3$ | $(CH_2)_7COOH$ | H |
| 2d | $C_8H_{17}$ | $C_8H_{17}$ | $(CH_2)_3COOH$ | $(CH_2)_3COOH$ | H |
| 3a | $C_4H_9$ | $C_4H_9$ | $CH_3$ | $(CH_2)_3COOH$ | OH |
| 3b | $C_8H_{17}$ | $C_8H_{17}$ | $CH_3$ | $(CH_2)_3COOH$ | OH |

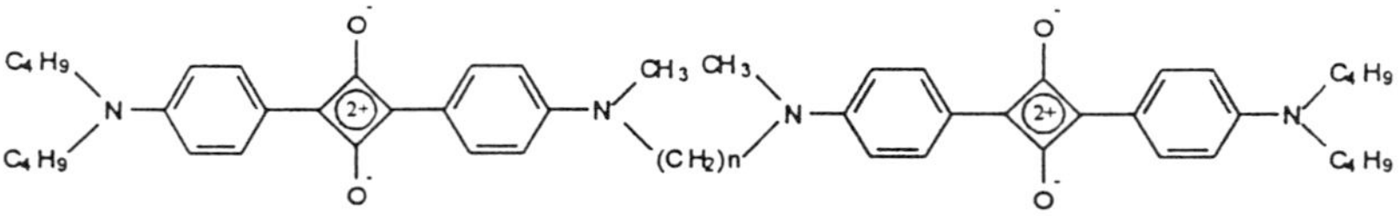

(4) Bis-squaraines (4a-4d: n = 2, 3, 6, 10)

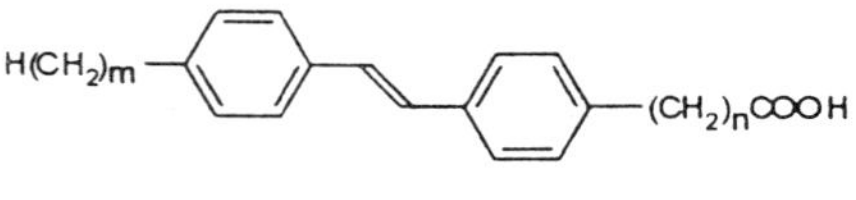

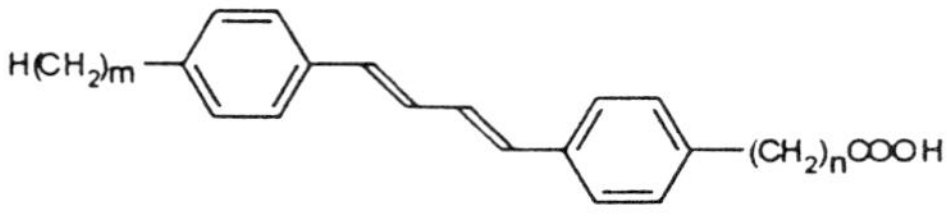

(5)  a: m=4, n=3; b: m=4, n=5; c: m=0, n=9; d: m=0, n=11.

(6)

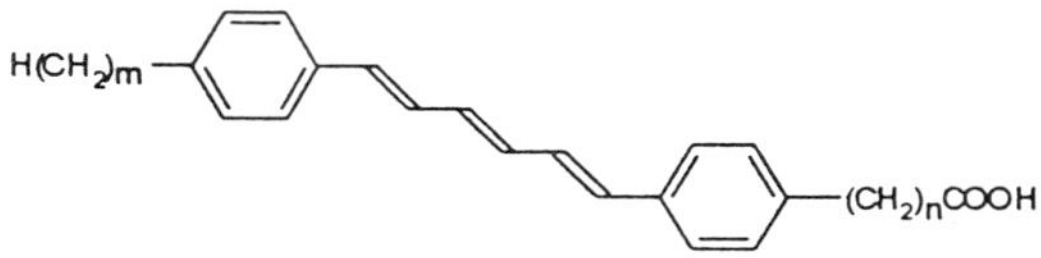

(7)

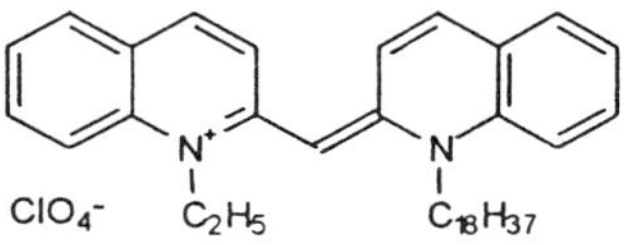

(8)

**Scheme 2** Structures of squaraines and related compounds.

# Aggregate Formation of Squaraines

the two chains on one end of the molecule, should be expected to orient with the squaraine chromophores having their long axis perpendicular to the air–water interface in the compressed films, while 1b, with one chain at each end of the molecule, might be expected to orient with the chromophore having its long axis parallel to the air–water interface. In several respects 1a is similar to the molecules described in the introduction such as the stilbene and azobenzene derivatives, which tend to form H aggregates in LB assemblies, while 1b more closely resembles the cyanine dyes that are typically encountered as J aggregates in assemblies. Previous studies showed that both 1a and 1b form stable monolayer films at the air–water interface with limiting areas of 52 and 60 Å$^2$, respectively (30). The monolayers of both 1a and 1b can be transferred to solid substrates with transfer ratios close to unity. As anticipated, the transferred film of 1a has a blue-shifted absorption with $\gamma_{max}$ at 530 nm relative to 634 nm for the monomer in CHCl$_3$. In contrast, the transferred film of 1b is fairly broad but dominated by a red-shifted peak at 650 nm (30).

When the compression of the two squaraines from dilute spread films at the air–water interface is followed, quite different results are observed for the two isomers. For 1a it is found that the blue-shifted absorption (534 nm) similar to that for the transferred film is observed even at very low compression; no changes are observed as the film is compressed to the collapse point. In contrast, for films of 1b the absorption of the film shows significant change under compression. As indicated by Figures 1a and 1b, which show the pressure-area isotherm and surface absorption of squaraine 1b at the air–water interface, at low surface pressures (<10 mN/m) a broad absorption with a dominant peak at 660 nm is observed. At 10 mN/m, where an apparent phase transition indicated by the isotherm occurs, a very sharp transition at 750 nm emerges and the intensity of this peak increases as the surface pressure is increased beyond the phase transition. These results indicate that in the spread films, squaraine 1a forms a stable H aggregate even before mechanical compression into a "card-pack" array, but that squaraine 1b exists as a weakly aggregated form or a mixture of monomer and aggregate that only forms a J aggregate upon compression. This suggests that the H aggregate may be a rather stable entity that does not depend upon imposed orientation by the LB film but that the J aggregate may be either a less stable species or highly dependent upon the organization of the film for its formation. It is also interesting that the sharp J aggregate observed at the air–water interface is apparently lost or at least significantly rearranged upon transfer of the film to a rigid support.

While the H aggregate is found to be the stable arrangement for squaraine 1a in the spread films at the air–water interface and also in the supported LB assemblies, it has been found that heating of the blue-shifted

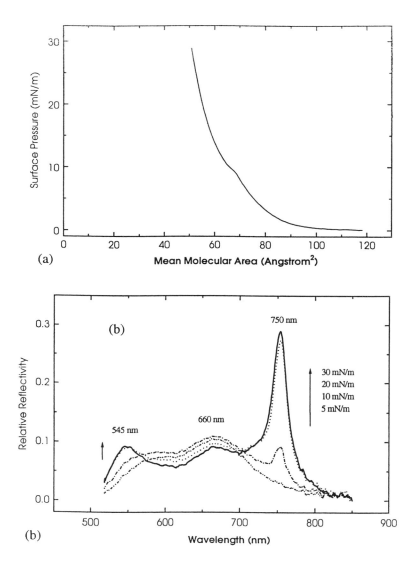

**Figure 1** The surface pressure-area isotherm (a) and surface absorption spectra (b) of the squaraine 1b monolayer.

aggregate results in its conversion to a red-shifted aggregate (660 nm) and J aggregate (690 nm) (27). This process can be thermally reversed by heating the modified film in the presence of steam. We have interpreted this to indicate preferred formation of a card-pack or modified card-pack structure (Scheme 1 (a)) that can be reversibly converted into a slipped stack or head-to-tail structure (Scheme 1 (b)) upon thermal removal of water from the freshly formed transferred film (27).

Since neither squaraine 1a nor 1b is an ideal amphiphile, we thought it would be useful to synthesize and study some amphiphiles containing the squaraine chromophore in the backbone such that better control of orientation might be obtained. Thus we synthesized squaraines containing two different chromophores (2a–2d and 3a–3b) embedded in a fatty-acid-like amphiphile terminated in a carboxylic acid. All of these compounds form stable films with collapse pressures of 40–45 mN/m and limiting areas of 38–45 Å$^2$/molecule. An example of the surface pressure-area isotherms is given in Figure 2 for squaraine 2a. The transferred supported LB assemblies of these compounds show strong absorption near 530–540 nm that is blue shifted from the solution absorption (630–635 nm in chloroform). Figure 3 shows the surface absorption spectrum of squaraine 3a at various surface pressures. As for squaraine 1a, the blue-shifted spectrum is observable even at very low surface pressures (2 mN/m) where little "forced" packing of the chromophore is anticipated. Mixed monolayers of these squaraines such as 3a with saturated fatty acids such as stearic acid can also be prepared. However, dilution of 3a with stearic acid up to 1 : 50 dye to stearic acid has no obvious effect on the absorption of the aggregate at the air–water interface, indicating again that the aggregate is relatively stable and formed largely independent of the compression process. A slight exception to the behavior of most of these squaraines is provided by squaraine 2d. This squaraine, which is anticipated to be somewhat more restricted from aggregating compared to the other structures (due to the four rather bulky chains), shows a sharp characteristic aggregate spectrum at the air–water interface that becomes somewhat more intense and blue shifted as the surface pressure increases. Interestingly, when films of squaraine 2d are transferred to a rigid support (Figure 4), a drastically red-shifted spectrum is observed compared to that seen at the air–water interface. For the rest of these squaraine carboxylic acids transferred films show nearly identical absorption spectra to those obtained at the air–water interface and, as common for the squaraine H aggregates, no fluorescence is detected for the aggregates even though all of the squaraine monomers in solution are strongly fluorescent. Here again, even dilute squaraine mixtures with stearic acid (dye to stearic acid mole ratio 1 : 10 or 1 : 20) show only aggregate absorption.

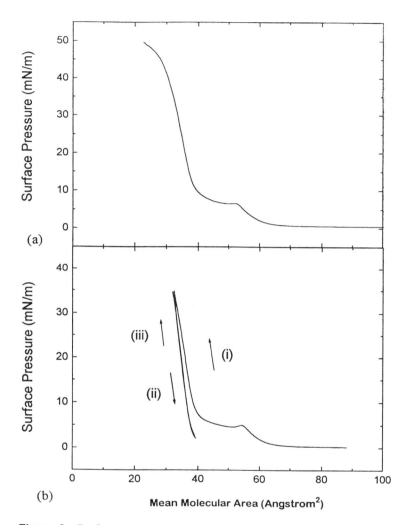

**Figure 2** Surface pressure-area isotherms of squaraine 2a monolayers: (a) normal isotherm, compressed once; (b) hysterisis behavior, three sweeps between 2 and 35 mN/m: (i) first compression; (ii) decompression; (iii) second compression.

# Aggregate Formation of Squaraines

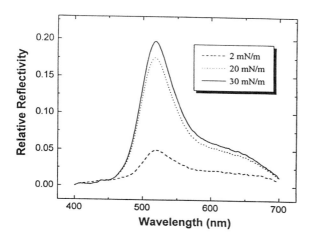

**Figure 3** Surface absorption spectra of the squaraine 3a monolayer at various surface pressures.

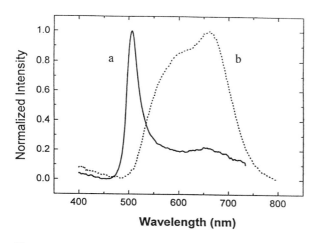

**Figure 4** Absorption spectra of squaraine 2d monolayers: (a) at the air–water interface; (b) on a glass slide.

In contrast to the supported assemblies of 1a, LB films of squaraines 2a–2c and 3a–3b on glass do not undergo appreciable or facile conversion to red-shifted aggregates upon heating, supporting the idea that the card-pack arrangement should be highly favored for these compounds in both films at the air–water interface and in supported assemblies. The difference in behavior for 1a can be attributed, at least in part, to a "free area" effect (27).

As discussed above both squaraines 2b and 2c form blue-shifted aggregates in the pure squaraine monolayer, either at the air–water interface or on solid supports. The two compounds are similar in structure but contain the squaraine chromophore at quite different positions in the amphiphile "backbone." It would be presumed that simple packing of a 1 : 1 mixture of 2b and 2c might lead to an "off-set" arrangement of the squaraine chromophores that would give a J aggregate spectrum such as was observed for the heating of supported films of squaraine 1a as described above. Figure 5a compares the isotherms of pure 2b, 2c and a 1 : 1 (mole ratio) mixture of 2b and 2c. The 1 : 1 mixed monolayer film shows a lower collapse pressure and a somewhat larger mean molecular area than those of the pure squaraine monolayers. The mixture shows a reflectance (surface) spectrum at the air–water interface (Figure 5b) with red-shifted peaks at 670 nm and 612 nm (the latter may result from exciton splitting) that do not change appreciably with compression. The 1 : 1 mixed monolayer of squaraines 2b and 2c can be transferred to a glass support; the absorption spectrum of the mixed film shows both the 670 nm and 530 nm absorptions (Figure 6) in contrast to the supported films of pure 2b and 2c, which show only the 530 nm H aggregate absorption. Interestingly, the blue-shifted band for the mixed aggregate is readily converted to the red-shifted band upon heat treatment (70°C, 5 min) while the aggregates from either "pure" squaraine are only converted slowly under prolonged heating at higher temperatures.

### B. Bis-Squaraines

Several bis-squaraines (4a–4d in Scheme 2) have also been prepared and examined under a variety of conditions (33). These compounds are notable in that the strong oscillator strength of the squaraine transition leads to fairly large exciton coupling even where the indicated separation between the chromophores is fairly large. We have found that in nonpolar solvents, the bis-squaraine shows a J-type spectral shift that is most consistent with the squaraine being in an "extended" configuration such as that shown in Scheme 1 (e); in fact, simulations show that a configuration of this type is a local or global minimum for each of the bis-squaraines and give spectral splittings in an extended dipole-dipole coupling calculation that agree very closely with those observed (33). In contrast, in aqueous acetonitrile, those

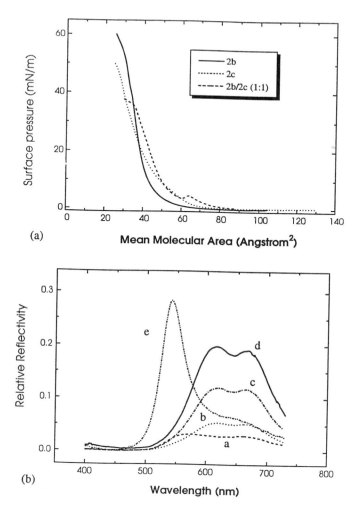

**Figure 5** (a) Surface pressure-area isotherms for the pure and mixed squaraine monolayers. (b) Absorption spectra of squaraines at the air–water interface at different surface pressures: Spectra a to d, squaraine 2b/2c (1 : 1) mixed monolayers (a—upon spread, b—2 mN/m, c—10 mN/m, d—26 mN/m); spectrum e squaraine 2b monolayer at 30 mN/m.

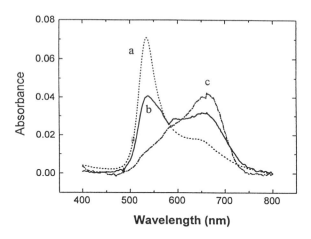

**Figure 6** Absorption spectra of squaraine monolayers on glass slides. a — Squaraine 2b (as prepared); b — squaraine 2b/2c (1 : 1) mixture (as prepared); c — squaraine 2b/2c (1 : 1) mixture, heated at 70°C for 5 min.

bis-squaraines that should be able to fold to the structure shown schematically in Scheme 1 (f) generally give a blue-shifted main transition characteristic of an H dimer. Simulations show that this is a second low energy structure for the bis-squaraines, and extended dipole–dipole calculations give good agreement between the spectral splittings calculated and observed in this solvent (33). Bis-squaraines 4a–4d all form stable monolayers at the air–water interface upon addition from evaporating chloroform solutions. The films of the bis-squaraines show interesting correlations between the isotherms and absorption spectra. The isotherm for squaraine 4a is shown in Figure 7a. A distinct phase transition is observed at a pressure of 10 mN/m, where the limiting molecular area is 150 Å$^2$. The absorption spectra of bis-squarine 4a in the film before and after the phase transition differ significantly, as shown in Figure 7b. Before the phase transition, the main transition is at 795 nm, which is significantly red-shifted relative to its absorption in chloroform (677 nm). This suggests the possibility of intermolecular J aggregation in the case where the film is relatively uncompressed and the bis-squaraine is extended with both chromophores lying on the water surface. Since bis-squaraine 4a cannot be easily arranged into a configuration that permits H dimerization, the absorptions at high compression suggest a staggered array, but with the chromophores sufficiently "offset" to give a red-shifted spectrum as contrasted with the extended J aggregate at low compression. The contrast for bis-squaraine 4b is remark-

# Aggregate Formation of Squaraines

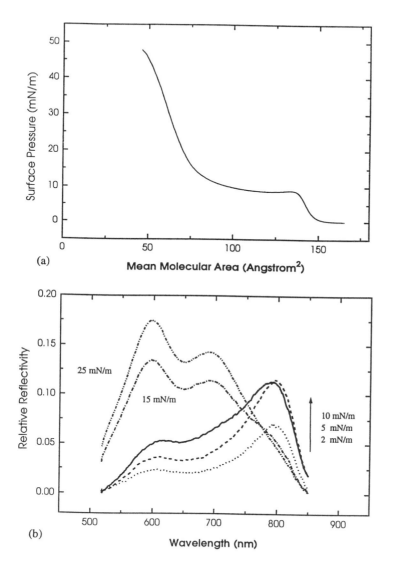

**Figure 7** The surface pressure-area isotherm (a) and surface absorption spectra (b) of the squaraine 4a monolayer.

able. For this compound in solution, a blue-shifted H-dimer-type spectrum is readily obtained in acetonitrile–water; the monolayer at the air–water interface shows a similar spectrum from pressures of 0–30 mN/m, suggesting that "folding" into the H-dimer configuration occurs even without compression and that the H-dimer must be a stable entity at the air–water interface as well as in solution. Similar behavior is observed for the longer chained bis-squaraine 4c, while the compound with ten methylene groups (4d) separating the two squaraine chromophores evidently can form types of aggregates in films, more or less independent of pressure.

## C. Amphiphilic Stilbenes and $\alpha,\omega$-Diphenylpolyenes

We have examined an array of amphiphilic *trans*-stilbene fatty acid derivatives (5 in Scheme 2) and *trans,trans*-1,4-diphenyl-1,3-butadiene (6 in Scheme 2) and *trans,trans,trans*-1,6-diphenyl-1,3,5-hexatriene (7 in Scheme 2) derivatives in supported assemblies (19,22). These compounds exhibit H-type aggregation with a characteristic blue-shifted absorption and a red-shifted fluorescence. In several cases, only H aggregates can be detected, even for LB films of mixtures of the aromatic diluted with a large excess of fatty acid "host." Although it was originally thought that the H aggregate occurred as a consequence of compression of the films, forcing the chromophores into a card-pack array, the persistence of aggregate for compounds 5–7 even up until relatively high dilution suggested that the H aggregate is formed for several of the stilbenes and diphenylpolyene amphiphiles as well. As shown in Figure 8, the H aggregate spectrum is obtained for the uncompressed film of stilbene 5a at the air–water interface and the spectrum changes very little with compression, indicating that the aggregates are preformed even in the expanded liquid or gaseous state. Dilution of the 5a monolayer with stearic acid shows little effect on the H aggregate formation. Similar behavior is observed for the monolayers of stilbene derivatives 5b–5d at the air–water interface. The fact that H aggregate is the only observed aggregation form for stilbene and diphenyl polyene derivatives in supported films and in films at the air–water interface indicates the high stability of the H aggregated form.

## D. Surfactant Cyanine Dyes in LB Films

In related studies we have been examining the photophysics of some cyanine dyes that have been previously shown to form J aggregates in supported LB films (34). We have recently found that these same cyanine dyes such as 8 (shown in Scheme 2) form J aggregates in bilayer vesicles as well as in supported LB films (35), so it became of interest to examine the behavior of

# Aggregate Formation of Squaraines

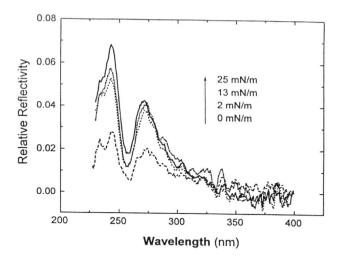

**Figure 8** Absorption spectra of the stilbene fatty acid (5a) monolayer at the air–water interface.

compound 8 in films at the air–water interface. As shown in Figure 9a, compound 8 can be compressed easily with no evident phase transitions observable during the compression. However, as shown in Figure 9b, the absorption spectrum of the spread films of compound 8 shows a dramatic conversion from a mixture of monomer and dimer below a surface pressure of ~10 mN/m to a predominance of J aggregate at pressures of ca. 10 mN/m and above. The J aggregate to monomer/dimer interconversion is reversible and clearly a function of layer compression. In many respects the behavior of compound 8 is very similar to that of squaraine 1b in that the J aggregate can be generated, but only under relatively high compression.

## III. SUMMARY

Although these investigations of aggregation of different types of chromophores in LB films and in films at the air–water interface are far from complete or comprehensive, they indicate that at least some trends can be recognized. While shape of the amphiphile or location of the chromophore in the amphiphile structure plays a clear role in the type of aggregate obtained and its stability, the studies completed thus far underline the difference between the stability and ease of formation of different types of aggre-

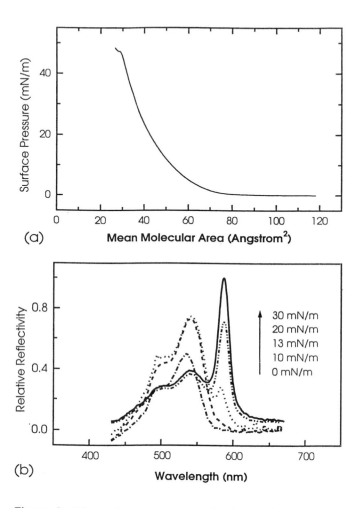

**Figure 9** The surface pressure-area isotherm (a) and surface absorption spectra (b) of the cyanine dye (8) monolayer.

gates. These studies also indicate that there may be great differences in stabilization between H aggregates and J aggregates such that film compression may play very different roles depending both on the chromophore and the type of aggregate it can form. Thus for the squaraines and stilbenes and related diphenylpolyenes it is clear that the H aggregate is a very stable entity that can form in the spread films virtually without compression. This is reinforced by other studies of phospholipid and fatty acid amphiphiles of these chromophores in other media, which indicate these compounds form very stabilized small unit aggregates that may be a supramolecular energy minimum (5,21,26). While the same chromophores may form J aggregates, it is clear that the J aggregate can form only at high compression and that the ordered array giving rise to the J aggregate spectral characteristics is very likely a less stable structure, perhaps in at least some cases consisting of a much larger array of molecules than the relatively small H unit aggregate. The fact that a single chromophore (squaraine) can form both types of aggregates, depending both on specific molecular structure and manipulation of the film, shows the power of the LB technique in assembling arrays of molecules for specific properties of functions.

## ACKNOWLEDGMENTS

We are grateful to the National Science Foundation (grants CHE-9120001 and CHE-9211586) for support of this research. We thank all of the members of the Monolayer Group at the NSF Center for Photoinduced Charge Transfer for lively discussions and suggestions concerning this research.

## REFERENCES

1. VI Yuzhakov. Russ Chem Rev (Engl Transl) 48:1076, 1979.
2. H Sato, M Kawasaki, K Kasatani, Y Kusumoto, N Nakashima, K Yoshihara. Chem Phys Lett 1139, 1980.
3. ES Emerson, MA Conlin, AE Rosenoff, KS Norland, H Rodriquez, D Chin, GR Bird. J Phys Chem 71:2396, 1967.
4. H Kuhn. In: Light-Induced Charge Separation in Biology and Chemistry. Geischer, H, Katz JJ, eds. West Berlin, Dahlem Konferenzen, 1979, p 151.
5. X Song, C Geiger, U Leinhos, J Perlstein, DG Whitten. J Am Chem Soc 116: 10340, 1994.
6. M Shimomura, T Kunitake. J Am Chem Soc 109:5175, 1987.
7. DG Whitten. Acc Chem Res 26:502, 1993.
8. PB Gilman. In: Photographic Sensitivity. Cox, RJ, ed. London, Academic Press, 1973, p 187.
9. GR Bird, KS Norland, AE Rosenoff, HB Michaud. Photographic Sci Eng 12: 196, 1968.
10. S Vaidyanathan, LK Patterson, D Möbius, HR Gruniger. J Phys Chem 89: 491, 1985.
11. H Nakahara, K Fukuda, D Möbius, H Kuhn. J Phys Chem 90:6144, 1986.
12. CE Evans, Q Song, PW Bohn. J Phys Chem 97:12302, 1993.
13. M Kasha, MA El-Bayoumi, W Rhodes. J Chem Phys 58:916, 1961.
14. RM Hochstrasser, M Kasha. Photochem Photobiol 3:317, 1964.

15. V Czikkely, HD Försterling, H Kuhn. Chem Phys Lett 6:207, 1970.
16. V Czikkely, HD Försterling, H Kuhn. Chem Phys Lett 6:11, 1970.
17. H Kuhn, D Möbius. Angew Chem Int Ed Engl 10:620, 1972.
18. D Möbius, H Kuhn. J Appl Phys 64:5138, 1988.
19. I Furman, HC Geiger, DG Whitten, TL Penner, A Ulman. Langmuir 10:837, 1994.
20. X Song, C Geiger, I Furman, DG Whitten. J Am Chem Soc 116:4103, 1994.
21. X Song, J Perlstein, DG Whitten. J Am Chem Soc 117:7816, 1995.
22. SP Spooner, DG Whitten. Proc SPTE-Int Soc Opt Eng 82:1436, 1991.
23. CE Evans, PW Bohn. J Am Chem Soc 115:3306, 1993.
24. H Chen, CW Farahat, M Farahat, HC Geiger, UW Leinhos, K Liang, X Song, TL Penner, A Ulman, J Perlstein, KY Law, DG Whitten. MRS Bull 20(6):39, 1995.
25. GS Cox, Ph. D. Thesis. University of North Carolina, Chapel Hill, 1982.
26. H Chen, KY Law, J Perlstein, DG Whitten. J Am Chem Soc 117:7257, 1995.
27. K Liang, KY Law, DG Whitten. J Phys Chem 98:13379, 1994.
28. E Buncel, A McKerrow, PM Kazmaier. J Am Chem Soc, Chem Commun 1242, 1992.
29. H Chen, WG Herkstroeter, J Perlstein, KY Law, DG Whitten. J Phys Chem 98:5138, 1994.
30. KY Law, CC Chen. J Phys Chem 91:5184, 1987.
31. KY Law. J Phys Chem 92:4226, 1988.
32. RE Wingard. IEEE Ind Appl 1251, 1982.
33. K Liang, J Perlstein, KY Law, DG Whitten. Submitted to JACS.
34. TL Penner, D Möbius. J Am Chem Soc 104:7407, 1982.
35. H Samha, TL Penner, DG Whitten. Unpublished results.

# 22
# Protein and Molecular Assembly Monolayer and Multilayer Film Studies with Scanning Probe Microscopy

**J. A. DeRose and R. M. Leblanc**
*University of Miami, Coral Gables, Florida*

## I. INTRODUCTION

Since the invention of the scanning tunneling microscope (STM) over 10 years ago (1), its use, as well as the use of another related scanning probe microscope (SPM), the atomic force microscope (AFM) [also called the scanning force microscope (SFM)] (2), to study complex molecules and molecular assemblies has grown steadily around the world. The main advantage of these SPMs is their ability to image samples in different environmental conditions, i.e., vacuum, air, and solution, with high resolution, unlike scanning or transmission electron microscopy (SEM or TEM). This versatility enables one to image molecules and molecular assemblies in situ, usually in an aqueous solution.

Both the STM and AFM have achieved atomic resolution while imaging the surface of solid materials in vacuum, air, and solution (3). Both have also been able to resolve features of a few nanometers or less when used to image soft materials adsorbed onto a solid's surface (substrate).

In this report, a representative sample of the current accomplishments in the study of protein and molecular assembly structure and function with the STM and AFM will be discussed.

## II. BASIC PRINCIPLES OF STM AND AFM

Both scanning probe microscopes are able to image a material's surface by scanning a sharp probe very close to it.

For STM, the probe is told to search for a specific value of current that passes between it and the sample. The probe comes very close to the sample's surface (usually less than 1 nm away) but, in theory, should not touch. This means that electrons must pass from the tip to the sample or vice versa by tunneling. The tunneling current, $I$, will be governed by the expression

$$I \propto \frac{V}{z} \exp\{-Cz\sqrt{\bar{\phi}}\}$$

where $V$ is the voltage bias, $z$ is the tip-sample separation distance, $C$ is a constant with the value 10.25 $[nm - (eV)^{1/2}]^{-1}$, and $\bar{\phi}$ is the average of the tip and sample work functions. This expression is called Simmons' formula, which he derived for a plane, parallel tunnel junction (4). Due to the value of $C$ in the argument of the exponential for the tunneling current, the STM is able to achieve atomic resolution (the current will change by nearly a factor of 3 for a change in distance as small as 0.1 nm). Soon after images of atoms on the Si(111) 7 × 7 reconstructed surface were taken by STM (5), theories were developed to explain what the STM was actually imaging (6-7). The theories give a complex expression for the tunneling current, which is proportional to the local density of states (LDOS) at the Fermi level ($E_F$) for the sample surface. The theories imply that the STM image reveals a constant contour of the sample's LDOS in the region being scanned.

The AFM exploits the forces that exist between atoms and molecules. The force exerted upon a tip mounted onto a cantilever with a known spring constant is monitored as the tip passes over the surface. The cantilever deflects during the scan and from the measurement of these deflections an image of the surface topography is obtained. All type of materials experience these forces, so the AFM is not limited to conductors. There are two regimes of force that can be felt by the probing tip. If the tip is scanned extremely close to the surface, the force, $F(r)$, will be expressed by

$$F(r) = \frac{12A}{r^{13}} - \frac{6B}{r^7}$$

where $A$ and $B$ are constants that depend upon the material of both the tip and sample (the charge density for the atoms) and $r$ is the separation distance between atoms of the tip and sample (8). For small separation distances that are usual for contact scanning, the first term (the repulsive force term) will dominate, implying that only the tip atoms nearest the sample surface will contribute most of the information about its topography. Scanning at large separation distances, the second term (the attractive or van der Waals force term) will dominate, and there will be a many-atom interaction between the tip and sample that requires one to integrate the second term of the above equation over the volumes of both the tip and sample in order to quantify the force appropriately (9).

The properties of the probing tip also greatly affect the image obtained with SPM, especially the resolution (7,10a,10b). The general rule is, the sharper the profile of the tip, the better will be the resolution in the image. For AFM, the resolution is best with the tip in contact with the surface, due to the more rapid decay of the repulsive force term (fewer atoms are participating in the imaging).

For more details on the STM and AFM, see References 11–15b.

## III. SUBSTRATE AND SAMPLE PREPARATION FOR SPM

The results obtained when imaging with a SPM depend largely upon the characteristics of the substrate surface (if needed) and sample purity. For molecular films, a substrate will always be needed. The most popular substrates are gold on mica films (Au(111)/mica), graphite (HOPG), muscovite mica, and silicon wafers (Si).

To image molecular films with STM, one requires a substrate with a flat, clean surface that is a good conductor, noncorrosive, and chemically inert. As far as metals are concerned, these criteria limit the choice to the noble metals. Evaporating Au onto the (001) surface of mica results in Au films oriented in the (111) direction (16,17). These films are generally flat and clean and are useful as STM substrates when imaging proteins. HOPG itself can be used as a substrate for the STM. It cleaves easily and is very flat over large areas. It is conductive enough for tunneling, noncorrosive, and relatively clean and inert over flat regions of the surface. However, steps on its surface are known to be reactive, and this has caused HOPG to have a history of artifact problems when imaging molecules such as DNA (18,19). Still, there have been interesting studies of molecular films on HOPG.

To image proteins with the AFM, a substrate with a flat, clean surface that is noncorrosive and inert is preferred. The AFM can scan any type of material, so a conductor is not required. The materials most often used as an AFM substrate are mica, HOPG, and Si wafers. Mica, like HOPG, cleaves easily, is very flat over large areas, and its surface is relatively clean and stable in air or solution. Mica is also hydrophilic, which makes it very useful when studying molecules that have an affinity for an aqueous environment. In some cases, the mica has been chemically modified to increase the strength of the bond between its surface atoms and the molecules of interest (20,21). Silicon wafers are also flat over large areas of the surface, but silicon easily forms an oxide layer when exposed to oxygen and cannot be cleaved easily. Silicon offers the advantage of easily obtaining either a hydrophilic surface (a thin oxide layer present) or a hydrophobic surface (it must be etched with acid to remove the oxide layer).

Sample purity is very important when using SPM. Common sense dictates that one would want the sample to be as pure as possible to minimize ambiguity in the image interpretation. Both the STM and AFM are not able to reliably discern molecules by any means other than physical geometry at present (22). Because many molecules have similar size and dimension, it can be very difficult to differentiate them in an image. This limitation arises for two reasons: first, there is no theory that is generally reliable at the present time for image interpretation when imaging molecules on a substrate and, second, SPMs probe only the electrons near the Fermi level and cannot be used to identify a specific element or molecule by interacting with the deep, inner shell electrons as most spectroscopic techniques do. By using the most pure sample available, one will simplify image interpretation by virtually eliminating the chance of one molecule being mistaken for another.

## IV. RESULTS OF PROTEIN AND MOLECULAR ASSEMBLY FILM STUDIES

The results obtained for some recent protein and molecular assembly studies performed with SPM will now be discussed.

Langmuir–Blodgett (LB) films of the protein cytochrome $f$ (Cyt$f$), which is part of the $b_6/f$ protein complex that resides in the thylakoid membrane and has a well-established role in the electron transfer process of photosynthesis, transferred onto a Au(111)/mica substrate which has been covered with a trilayer of arachidic acid (Ar/Au(111)/mica) have been imaged by Tazi, Boussaad, DeRose, and Leblanc with the STM (23a,23b). From the STM images, they determined the structure of the Cyt$f$ LB films (two monolayers) as a function of surface pressure ($\Pi$) and ethanol to water concentration for the spreading solvent. They observed aggregation of the Cyt$f$ protein in films deposited at 15 and 20 mN/m with ethanol concentrations of 20%, 40%, and 80% for the spreading solvent. The LB films of Cyt$f$ were mainly composed of aggregates a few hundred nanometers in

# Protein and Molecular Assembly Studies

size. Film morphology and aggregate sizes depended upon the ethanol concentration and the surface pressure for transfer onto the solid substrate. The Cyt $f$ LB films were more disordered as the spreading solvent ethanol concentration increased. The images in Figure 1a, b, and c clearly demonstrate the effect on the LB film structure of different ethanol concentrations for the spreading solvent.

The protein complex photosystem II (PSII), which is involved in the oxidation of water during photosynthesis, was studied with STM by our group (24). Biochemical and spectroscopic evidence has helped to build a model for the structure of the PSII complex, but little or no direct evidence, such as microscopy or diffraction data, exists that can verify the model (25–27). Langmuir films of PSII were transferred onto Au(111)/mica (LB films) for imaging with the STM. The PSII particles take a lumen-upright configu-

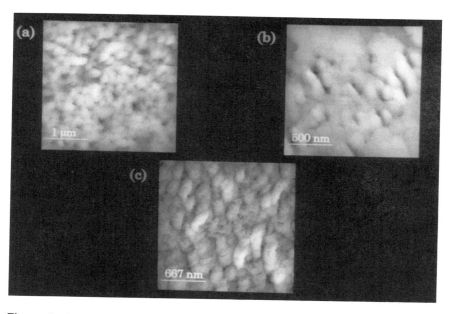

**Figure 1** (a) A 3 × 3 μm STM image of a Cyt $f$, bilayer LB film transferred at 15 mN/m onto Ar/Au(111)/mica. The monolayer at the air–water interface was made with a 20% ethanol spreading solvent. (b) A 1.5 × 1.5 μm STM image of a bilayer, Cyt $f$ LB film transferred at 15 mN/m onto Ar/Au(111)/mica. The monolayer at the air–water interface was made with a 40% ethanol solution. (c) A 2 × 2 μm STM image of a bilayer, Cyt $f$ LB film transferred at 20 mN/m onto Ar/Au(111)/mica. The monolayer at the air–water interface was made with an 80% ethanol solution. Figure adapted from Reference 23.

ration at the air–water interface and, consequently, in the LB film. The STM images showed particles with a shape (curved triangular) and dimensions (approximately 5 × 12 nm) that are in good agreement with the existing biochemical models. The PSII particles were usually found to reside near the Au(111) terrace edges or steps, as one can see in Figure 2. More information on internal structure can be obtained as better methods of sample preparation for imaging with SPM are discovered.

The formation of nanocrystals of the electron acceptor molecule 2-octadecylthio-1,4-benzoquinone, when prepared as LB films on Si, was studied by Garnaes et al. using AFM (28). The film was prepared by spreading a solution of the acceptor in chloroform ($CHCl_3$) on the water surface and compressing it to a surface pressure of 20 mN/m, resulting in a mean

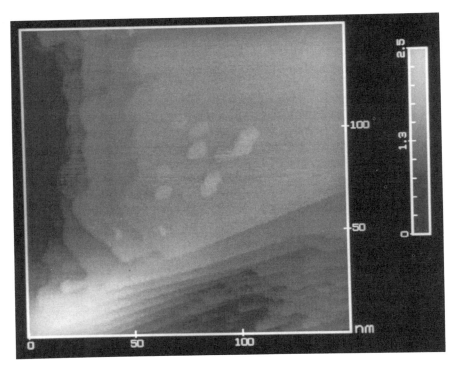

**Figure 2** A 150 × 150 nm STM image of a monolayer, PSII LB film transferred onto Au(111)/mica. The "spots" seen on the Au(111) terrace near the step edges are individual PSII core complexes. They have an average diameter of approximately 10 nm, which is in good agreement with the structural models developed from biochemical and spectroscopic analysis. Figure adapted from Reference 24.

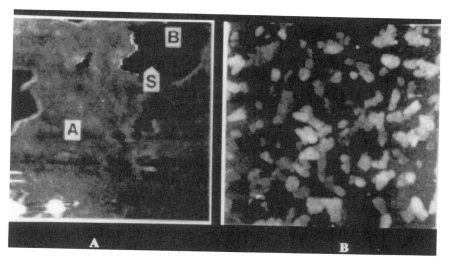

**Figure 3** (A) An $8 \times 8$ $\mu$m AFM image of a bilayer of the electron acceptor 2-octadecylthio-1,4-benzoquinone on a bilayer of cadmium behenate deposited on a polished Si substrate. Letter A marks the acceptor, B the cadmium behenate, and S the Si substrate. (B) Images of ($10 \times 10$ $\mu$m) crystals of the acceptor molecule. The molecules were prepared as four layers of LB film. The maximum $z$ range is approximately 20 nm. Figure adapted from Reference 28.

molecular area of 0.26 nm$^2$. They found the crystals had a layered structure with heights between 3 and 15 nm over an area greater than 1 $\mu$m$^2$ on the substrate. Figure 3A and B show AFM images of the acceptor. The acceptor bilayer lies on a bilayer of cadmium behenate which lies on the Si substrate. All three layered compounds can be seen. The molecules formed a rectangular unit cell with dimensions 0.51 nm and 1.35 nm. The packing density in the plane of a layer, 0.34 nm$^2$, was found to be significantly greater than that on the water subphase during deposition and the LB film bilayer. They concluded that a transition from a two-dimensional crystal phase to a three-dimensional, bulk crystal phase had occurred.

Boussaad, DeRose, and Leblanc used STM to visualize the structure of chlorophyll $\underline{a}$ (Chl $\underline{a}$) films electrodeposited onto Au(111)/mica (29). The photocurrent properties of microcrystalline Chl $\underline{a}$ have been studied (it is the light harvester in photosynthesis) for possible use in photovoltaic cells. The STM results showed that the electrodeposited microcrystalline Chl $\underline{a}$ films had a grain size that ranged from 30 to 100 nm. The electric field effects tremendously the colloidal aggregates and probably modifies their structure upon deposition. Figure 4A and 4B demonstrate this clearly. Figure 4A is an electrodeposited Chl $\underline{a}$ film and 4B is a film created by simply drying a drop of solution containing Chl $\underline{a}$ onto a Au(111)/mica substrate. The electrodeposited Chl $\underline{a}$ films show short range, polycrystalline order with many defects, which correlates with its photoelectric properties. The

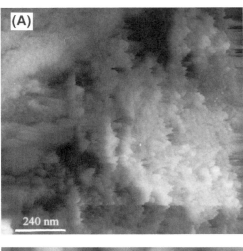

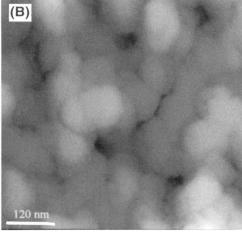

**Figure 4** STM images of chlorophyll $\underline{a}$ deposited on Au(111)/mica. (A) A 1.2 × 1.2 μm image of an electrodeposited Chl $\underline{a}$ film. The mean grain size is between 35 and 50 nm. (B) A 600 × 600 nm image of Chl $\underline{a}$ film deposited from the drop. Figure adapted from Reference 29.

effect of the electric field and the electrode surface is of great importance, and their study shows that the Chl $\underline{a}$ film topography can be greatly affected by them. The grains seen in the Chl $\underline{a}$ films were essentially nanocrystals of the pigment.

Highly ordered adlayers of the porphyrin molecule 5,10,15,20-tetrakis(N-methyl-pyridinium-4-yl)-21H,23H-porphine tetrakis(p-toluenesulfonate) (TMPyP) on an iodine-modified Au(111) [I-Au(111)] substrate was imaged with STM by Kunitake, Batina, and Itaya (30). All imaging was done in a 0.1 M perchloric acid ($HClO_4$) solution with and without the presence of TMPyP (0.5 μM TMPyP) in a three-electrode electrochemical cell. When they raised the electrochemical potential in solution to 0.8 V with respect to the reversible hydrogen electrode (RHE), they found that the TMPyP formed an ordered array on the I-Au(111) surface. The lattice parameters were measured to be $a = 3.4 \pm 0.2$ nm and $b = 1.8 \pm 0.1$ nm with an angle of approximately 60° between the vectors, so the TMPyP molecules formed a hexagonal close-packed structure. Each lattice point seen in the STM images were TMPyP molecules, as seen in Figure 5. Four bright spots were seen around each TMPyP molecule, and these were concluded to be the pyridinium units. The reason for the increased brightness of the pyridinium units was not understood by the authors and, due to the nature of STM, it is possible that it could be for reasons of tilting in the TMPyP molecule at the pyridinium units that caused them to be elevated above the main portion of the molecule or the electronic properties of the TMPyP when it forms an array on the I-Au(111) surface.

Activity of the enzyme lysozyme was monitored with AFM by Radmacher et al. (31). They monitored height fluctuations of the AFM cantilever when positioned over lysozyme molecules adsorbed onto mica from a buffer solution. They first imaged the lysozyme molecules on mica in tapping mode, then measured the fluctuation of the AFM tip/cantilever as it remained over a lysozyme monolayer under various conditions without scanning. In order to make well-founded conclusions from the results, the authors measured the height fluctuations of the tip for: (1) bare mica, (2) lysozyme on mica in a buffer solution, (3) lysozyme in buffer with 10 μM of the substrate 4-methyl-umbelliferyl-$N,N',N''$-triacetyl-chitotriose with which the lysozyme reacts, (4) lysozyme in buffer with 20 μM of $N,N'$-chitobiose which inhibits lysozyme activity, and (5) lysozyme in buffer solution with both the substrate and inhibitor present. When comparing the height fluctuation data from the five cases, they found the AFM tip to have large spike-like height fluctuations on the order of 1 nm when the lysozyme and substrate were both present in solution without the inhibitor (case 3 above). For the other four cases, the fluctuations were much less than 1 nm. They concluded that the AFM is probing the enzyme activity of lysozyme as

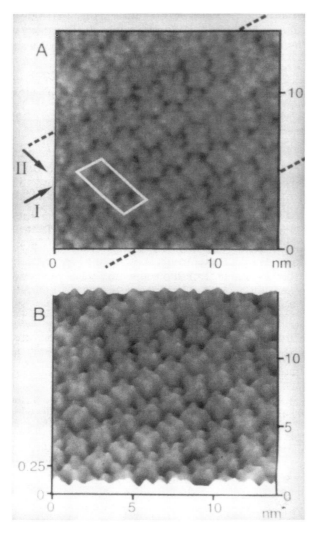

**Figure 5** High-resolution STM images of a TMPyP array in 0.1 M HClO$_4$ solution with 0.5 μM TMPyP. Each image was obtained at 0.82 V versus the RHE, a tip potential of 0.29 V versus the RHE, and 2 nA tunneling current. (A) A 15 × 15 nm top view image showing an outline of the unit cell for the TMPyP lattice on the I-Au(111) surface. (B) A 15 × 15 nm surface plot image showing molecular height information. Figure adapted from Reference 30.

it reacts with the substrate, which leads to a conformational change of the enzyme. The data were analyzed by several methods, and each allowed the same conclusion to be made. The authors do mention that one should be aware that when adsorbed to the mica surface, the lysozyme activity is reduced when compared to a bulk solution, probably due to steric hindrance. It is also possible that not every lysozyme molecule on the mica surface is able to react with a substrate molecule.

Schabert, Henn, and Engel measured the topography of two-dimensional crystals of the native *Escherichia coli* OmpF porin protein, which have been reconstituted in the presence of phospholipids with AFM (32). A lateral resolution of 1 nm and a vertical resolution of 0.1 nm were achieved. The OmpF porin–lipid crystals were adsorbed onto mica from a buffer solution for imaging. They were able to mechanically displace different layers of the crystal with the AFM tip and reveal the extracellular, corrugated extracellular, and periplasmic surfaces of the porin protein. With information received from the AFM images and X-ray diffraction results, the authors constructed an atomic model of the protein–protein and lipid–protein interactions. They found the extracellular surface to exhibit a salt bridge pattern between lysines 25 and carboxyl groups of nearby porin trimers and the periplasmic surface protein–protein interactions to be more hydrophobic. For the extracellular porin surface, two conformations were observed. The difference between the conformations consisted of a 0.5 nm shift of the extracellular domains toward the center of the porin trimer. The displacement was noticed to constrict the channel entrances in the *Escherichia coli* cell membrane, which may explain the two open-channel configurations of the OmpF porins.

The surface of the hexagonally packed intermediate (HPI) layer of *Deinococcus radiodurans* was imaged in 0.1 M phosphate buffer solution on a modified glass substrate with AFM by Karrasch et al. (33). They made comparisons between the AFM images and those obtained by electron microscopy. The Semper image-processing system was employed to calculate correlation averages from AFM images and electron micrographs. Figure 6 shows an AFM image and an electron micrograph. In both, the HPI hexamer units are resolved. The AFM was able to resolve features in the HPI layer larger than 1 nm laterally and 0.1 nm vertically. The authors calculated the modulus of the difference between an average of different AFM topographs and a three-dimensional map constructed from electron micrographs. The results showed that the AFM had more trouble imaging deep trenches and very fine structures of the HPI layer, but most of the data from the two techniques were remarkably similar. They mentioned that high resolution AFM imaging on the HPI layer was obtained by using a sharp tip, an imaging force below 1 nN, and an optimized scan speed (300

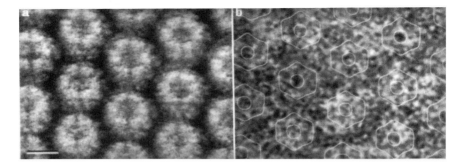

**Figure 6** HPI layer from the cell envelope of *D. radiodurans* (a) imaged with AFM in buffer solution on glass and (b) in cadmium thioglycerol with electron microscopy. Positions of the HPI hexamers have been outlined on the electron micrograph in (b). The gray scale for the AFM image in (a) is 3 nm and the bar represents 20 nm. Figure adapted from Reference 33.

nm/s for this case). The results also indicated the possibility of mapping the elastic properties of the HPI layer, or other biomacromolecules, with the AFM through comparing the images of many molecules.

The structure of the protein membrane obtained from the photosynthetic bacteria *Rhodopseudomonas viridis* was investigated by Yamada et al. with the LB technique and AFM (34). Langmuir films of the photosynthetic protein (chromatophore) membrane were transferred onto glass substrates and then imaged. The authors noticed small domains of hexagonal close-packed (HCP) photoreaction units (PRU), as seen in Figure 7. The mean distance between the PRUs was approximately 12 nm, which agrees with data from TEM. The chromatophore LB films were found, on the mesoscopic scale, to be composed of domains, which are membrane fragments, with diameters between 0.1 and 1 $\mu$m. The domain sizes also show good agreement with TEM data. The authors used force modulation while in contact mode, a new imaging method, to obtain information on the chromatophore elasticity and viscosity while on the glass surface.

Lee et al. used STM and tunneling spectroscopy (TS) to study thylakoids and photosystem I (PSI) reaction centers from spinach (35). The PSI, some having been reacted with Pt and others not, were imaged on Au(111)/mica substrates. The thylakoids were deposited onto mica and then sputtered-coated with a palladium/gold (Pd/Au) alloy. Uncoated thylakoids did not yield stable images by STM due to their poor conductivity (the LDOS near $E_F$ is zero). Thylakoids coated with Pd/Au yielded stable STM images. Both isolated and aggregated thylakoids were seen in the images.

## Protein and Molecular Assembly Studies

The Pd/Au coated thylakoids showed diameters between 150 and 380 nm and heights between 37 and 50 nm. The bare PSI were observed to have dimensions of 6 × 5 nm, and the Pt-reacted PSI were observed to have dimensions of 9 × 7 nm. The height values obtained for the PSI were anomalous, as seen for many complex molecular samples imaged by STM. For the lightly Pt reacted PSI, there was a large change in image contrast when the bias voltage was changed to the opposite value (refer to Figure 8). This change in image contrast was not seen for either the bare PSI or the heavily Pt-reacted PSI. TS was able to distinguish the three types of PSI studied.

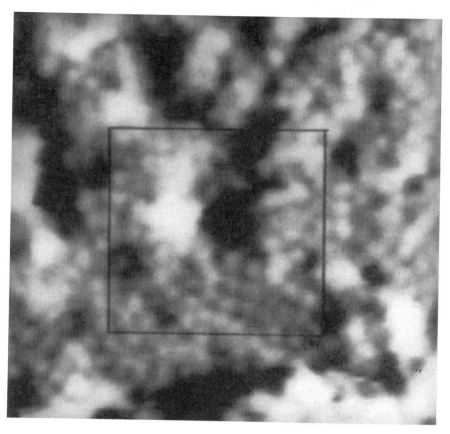

**Figure 7** A 240 × 240 nm AFM image of chromatophore membrane LB film isolated from the photosynthetic bacteria *Rhodopseudomonas viridis*. Each small, bright dot in the image is a photoreaction unit that has adopted a hexagonally packed structure in the LB film. Figure adapted from Reference 34.

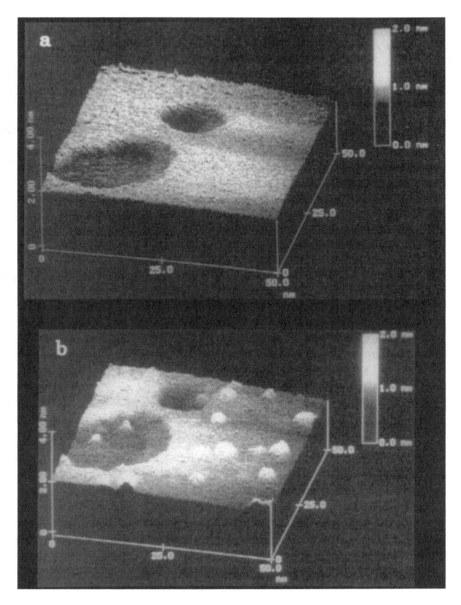

**Figure 8** STM images of lightly Pt-reacted (for 1 h) PSI reaction centers at bias voltages of (a) +0.5 V and (b) −0.5 V. The strong contrast variation with bias seen easily repeatable and reversible. Figure adapted from Reference 35.

# Protein and Molecular Assembly Studies

The bare PSI behaved as an insulator or large band gap semiconductor, the lightly Pt-reacted PSI showed a diode behavior as seen when changing the bias voltage during imaging, and the heavily Pt-reacted PSI showed ohmic behavior characteristic of a metal. In fact, for the bare PSI, the authors were able to measure a band gap of $\sim$1.8 eV from the TS data, which corresponds with spectroscopic results showing a strong optical absorption band at 672 nm (1.84 eV). The structural and electrical properties of the bare and Pt-reacted PSI were found to be stable for periods of at least 4 months.

Adsorption of the proteins immunoglobulin G (IgG) and glucose oxidase (GOx) onto HOPG from an aqueous phosphate-buffered saline (PBS) solution was studied by Cullen and Lowe using AFM (36). The modes of adsorption for the two proteins were seen to be very different over large scan areas (5 $\times$ 5 $\mu$m) with protein concentrations in solution of 50 $\mu$g/ml. No significant distortion of the protein/buffer adlayer was seen during imaging. The IgG displayed uniform adsorption in high density over the HOPG surface for long adsorption times. The mechanism of adsorption was seen to be nucleation or the formation of monolayer IgG islands at many places on the surface that grow and eventually coalesce to form monolayer patches. The HOPG–IgG interaction was found to dominate over the IgG–IgG interaction due to the presence of monolayer patches and no signs of IgG aggregation on the HOPG surface. The IgG did not preferentially adsorb at steps or defects. The GOx also nucleated upon adsorption to the HOPG surface, but preferred to adsorb at steps and defects. Therefore, the GOx did not show uniform adsorption onto the surface. Instead, the GOx islands grew outward from steps and defects on the surface, but were unable to coalesce and form a uniform film as the IgG did. In both cases, the rate of adsorption decayed over time (as the HOPG surface became more passivated), but the rate was greater for the IgG case due to the fact that it was able to find more places at which it could adsorb. From the thickness of the adsorbed films, the authors concluded that IgG adsorbed in a native conformation while GOx denatured upon adsorption. In addition, lateral force microscopy (LFM), which utilizes lateral interaction forces between the tip and sample, was performed upon GOx films and then compared with AFM results.

Yang, Mou, and Shao used AFM to investigate the structure and stability of pertussis toxin protein produced by the bacteria *Bordetella pertussis* (37). The toxin was simply adsorbed onto mica for imaging; no special modification of the mica was needed. They imaged both the complete pertussis toxin and the B-oligomer in situ. All imaging was done in deionized water (see Figure 9). The toxin was stable upon imaging, probably due

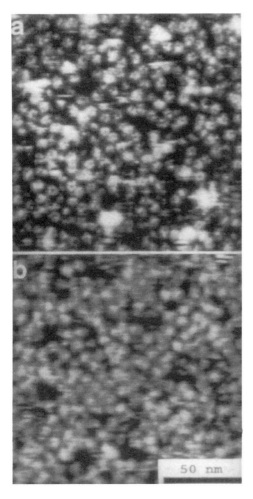

**Figure 9** AFM images of pertussis toxin obtained in deionized water at room temperature. (a) Pertussis toxin B-oligomer showing a pentameric structure with a central pore on most of the molecules. (b) Intact pertussis toxin that does not show a central pore on most of the molecules in contrast to the B-oligomer. The absence of the central pore indicates that the catalytic subunit A (S1) is located at the center on top of the B-oligomer. Figure adapted from Reference 37.

to a strong affinity for the negatively charged mica surface. The B-oligomer was determined from the AFM images to be a flat pentamer with the larger two of its subunits, S2 and S3, located next to each other, and its catalytic A subunit located on top of the B-pentamer in the center. The location of the A subunit was deduced by comparing images of the B-oligomer toxin with that of the intact toxin. The authors also varied the temperature and pH of the protein solution to check its stability under different conditions. For temperatures between room temperature and 60°C and pH values between 4.5 and 9.5, the B-pentamer structure was stable. At 70°C and for pH values below 4.5 or above 9.5, the majority of the subunits dissociated. A spatial distance between the subunits as small as 0.5 nm was resolved in the AFM images.

Clathrin cages and triskelia were attached to $N$-hydroxysuccinimide (NHS) terminated self-assembled monolayers on Au(111)/mica and imaged with AFM by Wagner et al. (38). The monolayers covalently bonded to the Au(111) surface due to the presence of thiols. Proteins such as clathrin cages and triskelia bind to the monolayer covalently and are well immobilized for imaging. Stable images of the clathrin cages and their disassembly in situ producing triskelia were obtained. The authors found the substrate preparation method to be simple, rapid, and reproducible for imaging proteins with AFM.

## V. CONCLUSION

In conclusion, the usefulness of SPM (STM/AFM) for studying the structure, function, and other properties of proteins and molecular assemblies adsorbed onto substrates as monolayer and multilayer films has been demonstrated through the studies discussed above. In the future, it is apparent that SPM will continue to make significant contributions to many diverse fields, but especially molecular biology and surface chemistry. Due to its versatility (ability to image in vacuum, air, and solution), ease of use, and low cost, SPM has already shown and will continue to show that it has many advantages over electron microscopy when imaging proteins, molecular assemblies, and other molecules.

## ACKNOWLEDGMENTS

We would like to acknowledge Drs. Salah Boussaad, Abderrahim Tazi, and Lei Shao, Professor Bruce Mainsbridge, and Dr. Kevi Konka for help with the STM experiments of cytochrome $f$, photosystem II, and chlorophyll.

## REFERENCES

1. G Binnig, H Rohrer, C Gerber, E Weibel. Phys Rev Lett 49:57, 1982.
2. G Binnig, CF Quate, C Gerber. Phys Rev Lett 56:930, 1986.
3. PK Hansma, VB Elings, O Marti, CE Bracker. Science 242:209, 1988.
4. JG Simmons. J Appl Phys 34:1793, 1963.
5. G Binnig, H Rohrer, C Gerber, E Weibel. Phys Rev Lett 50:120, 1983.
6a. J Tersoff, DR Hamann. Phys Rev Lett 50:1988, 1983.
6b. ND Lang. Phys Rev Lett 55:230, 1985.
7. J Tersoff, DR Hamann. Phys Rev B 31:805, 1985.
8. G Burns. Solid State Physics. San Francisco, Academic Press, 1985.
9. GM McCelland, R Erlandsson, S Chiang. In: Review of Progress in Quantitative Non-Destructive Evaluation. Vol. 6B. New York, Plenum Press, 1987, p 1307.
10a. FF Abraham, IP Batra, S Ciraci. Phys Rev Lett 60:1314, 1988.
10b. SAC Gould, K Burke, PK Hansma. Phys Rev B 40:5363, 1989.
11. JA DeRose, RM Leblanc. Surf Sci Rep 22:73, 1995.
12. HG Hansma, JH Hoh. Annu Rev Biophys Struct 23:115, 1994.
13. CJ Chen. Introduction to Scanning Tunneling Microscopy. New York, Oxford University Press, 1993.
14. WM Heckl. Thin Solid Films 210/211:640, 1992.
15a. PK Hansma, J Tersoff. J Appl Phys 61:R1, 1987.
15b. D Rugar, PK Hansma. Phys Today 43:23, 1990.
16. JA DeRose, T Thundat, LA Nagahara, SM Lindsay. Surf Sci 256:102, 1991.
17. JA DeRose, DB Lampner, SM Lindsay, NJ Tao. J Vac Sci Technol A 11:776, 1993.
18. CR Clemmer, TP Beebe, Jr. Science 251:640, 1991.
19. WM Heckl, G Binnig. Ultramicroscopy 42–44:1073, 1992.
20. YL Lyubchenko, BL Jacobs, SM Lindsay. Nucleic Acids Res 20:3983, 1992.
21. T Thundat, DP Allison, RJ Warmack. Scanning Microscopy 6:911, 1992.
22. N Tao. Phys Rev Lett 76:4066, 1996.
23a. A Tazi, S Boussaad, JA DeRose, RM Leblanc. Proceedings of the 8th International Conference on STM/STS and Related Techniques. Snowmass Village, Colorado, July, 1995.
23b. A Tazi, S Boussaad, JA DeRose, RM Leblanc. J Vac Sci Technol B 14:1476, 1996.
24. L Shao, JA DeRose, B Mainsbridge, V Konka, RM Leblanc. To be published.
25. EJ Boekmar, B Hankamer, D Bald, J Krup, J Nield, AF Boonstra, J Barber, M Roogner. Proc Natl Acad Sci USA 92:175, 1995.
26. A Holzenberg, MC Bewley, FH Wilson, WV Nicholson, RC Ford. Nature 363:470, 1993.
27. MK Lyon, KM Marr, PS Furcinitti. J Struct Biol 110:133, 1993.
28. J Garnaes, T Bjørnholm, M Jørgensen, JA Zasadzinski. J Vac Sci Technol B 12:1936, 1994.
29. S Bousaad, JA DeRose, RM Leblanc. Chem Phys Lett 246:107, 1995.
30. M Kunitake, N Batina, K Itaya. Langmuir 11:2337, 1995.

31. M Radmacher, M Fritz, HG Hansma, PK Hansma. Science 265:1577, 1994.
32. FA Schabert, C Henn, A Engel. Science 268:92, 1995.
33. S Karrasch, R Hegerl, JH Hoh, W Baumeister, A Engel. Proc Natl Acad Sci USA 91:836, 1994.
34. H Yamada, Y Hirata, M Hara, J Miyake. Thin Solid Films 243:455, 1994.
35. I Lee, JW Lee, RJ Warmack, DP Allison, E Greenbaum. Proc Natl Acad Sci USA 92:1965, 1995.
36. DC Cullen, CR Lowe. J Colloid Interface Sci 166:102, 1994.
37. J Yang, J Mou, Z Shao. FEBS Lett 338:89, 1994.
38. P Wagner, P Kernen, M Hegner, E Ungewickell, G Semenza. FEBS Lett 356:267, 1994.

# 23
# Self-Assembled Amphiphiles on Surface: "Surface Rheology"

**Yihan Liu and D. Fennell Evans**
*University of Minnesota, Minneapolis, Minnesota*

When surfactant molecules adsorb from solution onto a solid surface, they spontaneously form well-ordered assemblies such as monolayers, bilayers, or other structured clusters. These "self-assembled" thin films constitute nano-scale structures which find increasingly wide use in many technologies involving optical and electronic devices, sensors and transducers, surface modifiers for improving wettability and biocompatibility, protective and lubricate layers, and patternable materials (1,2). In many of these applications, nano-scale mechanical properties play an essential role. Recent advances in instrumentation, such as atomic force microscopy (AFM) (3), provide us with the ability to characterize mechanical properties at a molecular level (4–7).

A modified version of AFM called a lateral force microscope (LFM, Figure 1) incorporates a quadrant optical detector that makes it possible not only to map surface topography but also to measure the forces between an AFM tip and the sample simultaneously in both normal and lateral directions (4a,4b). Using the LFM, we have measured shear properties of a variety of monolayers of double-chained cationic surfactants adsorbed on mica (Figure 1). These surfactants differ in chain composition, chain length, terminal functional groups, and degree of unsaturation. They form monolayer via electrostatic bonding of the cationic headgroups to the negative charge sites on mica surface (ca. 48 $Å^2$/charge). The monolayers were close-packed and molecularly smooth over several microns to tens of microns, and were robust against scanning with the AFM tip. The measurements were made in air at room temperature, with the microscope placed in

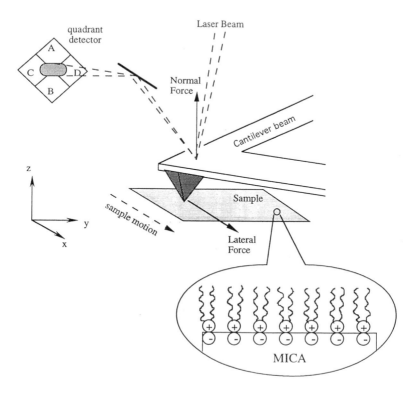

**Figure 1** Schematic illustrating the lateral force microscope (LFM) and the surfactant monolayer. The sample is scanned beneath the sharp tip (radius of curvature ≈ 20 nm), and the deflection of the cantilever registered in the optical quadrant detector permits topography mapping as well as simultaneous measurement of the normal and lateral force. The sample was made from adsorption of double-chained cationic surfactants onto a negatively charged, atomically smooth mica surface in solution, forming a close-packed monolayer with surfactant tails pointing toward the air.

a closed chamber equipped with an inlet for nitrogen to control and maintain a constant humidity during data collection. We found that the measured shear force can vary significantly for different molecules and that in most cases, it shows nonlinear dependence on velocity. In addition, hydrophilic and hydrophobic surfaces respond differently to changes in humidity. Analysis of our data provides a way to relate surface rheological properties to the surfactants' molecular structure.

On the bare substrate mica, lateral force was found to increase linearly with the logarithm of velocity; however, on surfaces coated with a surfactant monolayer, we observed two different behaviors. Figure 2A shows that on each of the three monolayer surfaces, ditetradecyldimethylammonium (2C14), dieicosyldimethylammonium (2C20), and didocosyldimethylammonium (2C22), lateral force reaches a plateau at high velocities. This behavior suggests a viscoelastic response in the monolayers analogous to the shear thinning observed from shear stress versus shear rate measurements in a bulk system. The magnitude of the lateral force also differs by more than one order of magnitude between the 2C14 monolayer and the 2C20 or 2C22 monolayers. In general, we found that the magnitude of lateral force increases with increasing chain length in a manner that correlates with the chain melting temperature $T_m$. The lateral force is small for melted chains and large for solid, rigid chains, a result we attribute to differences in the compressibility of the chains.

When $T_m$ lies just above room temperature, at which all measurements were made, the lateral force curves display a maximum, as illustrated in Figure 2B. The existence of such a peak implies a maximum of work done by the lateral force (and hence a maximum of energy transfer to the monolayer assembly), suggesting a shear-induced phase transition, which occurs prior to chain melting temperature. Similar premelting phase transitions have been observed previously on other mono- and multilayer amphiphilic thin film systems using a number of different techniques (8-15). At present, the exact nature of this transformation is not understood; interested readers are referred to discussion presented elsewhere (16).

All of our observations on the 11 surfactant monolayers we studied fall into the two general patterns displayed in Figures 2A and B. Six surfactants with $T_m$ either below or 40°C above room temperature behave similarly to those shown in Figure 2A and five surfactants with $T_m$ 10-30°C above room temperature display a maximum in their lateral force versus velocity curves, as shown in Figure 2B.

To determine how the LFM, which uses a nanometer-sharp tip with a radius of curvature around 20 nm, compares with more traditional macroscopic mechanical measurements, we modified the AFM probe by attaching a spherical glass bead, radius = 7.5 $\mu$m, to the end of a cantilever (19). Measurements on the dihexadecyldimethylammonium (2C16) monolayer with the glass bead give (Figure 3), at the low load of 40 nN, a force versus velocity curve similar to that observed with the regular AFM tip. At the higher load of 108 nN and velocities less than about 10 $\mu$m/s, we observed the same type of behavior, but above this velocity (identified by the arrow in the figure), the lateral force increased significantly. AFM imaging of the surface showed that the monolayer had been disrupted at the large load and high velocities. Imaging the glass probe by a scanning electron microscope

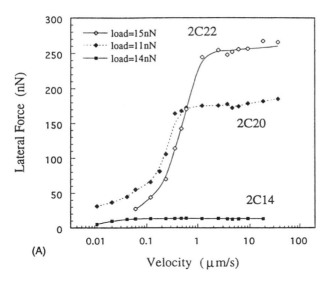

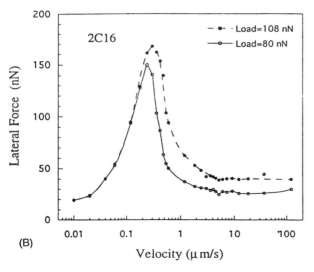

**Figure 2** Lateral force versus velocity for (A) ditetradecyldimethylammonium $(2C_{14}N^+2C_1, T_m = 4°C)$, dieicosyldimethylammonium $(2C_{20}N^+2C_1, T_m = 62°C)$ (17), and didocosyldimethylammonium $(2C_{22}N^+2C_1, T_m = 70°C)$, and (B) dihexadecyldimethylammonium $(2C_{16}N^+2C_1, T_m = 34°C)$ monolayer coated surfaces. All measurements were made at room temperature. Other surfactant monolayers we have studied include 16,16'-dihydroxydihexadecyldimethylammonium $(2(HOC_{16})N^+2C_1, T_m = 69°C)$, 16-hydroxydihexadecyldimethylammonium $(HOC_{16}C_{16}N^+2C_1, T_m = 71°C)$, and N-(α-trimethylammonioacetyl)-O,O'-bis(1H,1H,2H,2H-perfluorodecyl)-L-glutamate $(2C_8^FC_2\text{-L-Glu-}C_1N^+3C_1, T_m = 94°C)$, which behave like those in (A); and dioctadecyldimethylammonium $(2C_{18}N^+2C_1, T_m = 48°C)$ (18), dodecyloctadecyldimethylammonium $(C_{12}C_{18}N^+2C_1, T_m = 29°C)$, dimethyldioleylammonium $(2(C_9 = C_9)N^+2C_1, T_m = 36°C)$, and dimethyloctadecyloleylammonium $(C_9 = C_9C_{18}N^+2C_1, T_m = 46°C)$, which behave like that in (B).

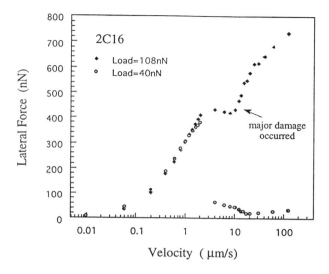

**Figure 3** Lateral force versus velocity for dihexadecyldimethylammonium monolayer measured by a modified AFM probe with a glass bead (radius = 7.5μm) attached. Measurement at the higher load and high velocity resulted in monolayer disruption, starting where the arrow indicates. SEM showed the surface of the glass bead to be porous at a submicron scale.

revealed that the glass surface was rough. We attribute the difference between the result obtained by the regular AFM tip and that obtained by the glass sphere to the roughness of the glass surface which caused the monolayer disruption by abrasive scraping.

Using LFM, we can also measure changes in surface interactions induced by environmental changes such as humidity. Figure 4 shows how an increase in humidity leads to an increase in the lateral force with hydrophobic monolayers made from surfactants such as dihexadecyldimethylammonium, but a *decrease* with hydrophilic monolayers from surfactants having ω terminal methyl groups replaced by hydroxides. We can understand this difference by analyzing the data using a generalized Amonton's Law (20):

$$F = \mu(N_{load} + N_{adh})$$

where $\mu$ is the frictional coefficient, $N_{load}$ is the applied load, and $N_{adh}$ is the adhesion force between the tip and the monolayer. With both types of monolayers, the adhesion force, $N_{adh}$, increases with humidity because at high humidity, capillary condensation of water forms a meniscus that bridges the tip and the monolayer, resulting in an increase in adhesion and

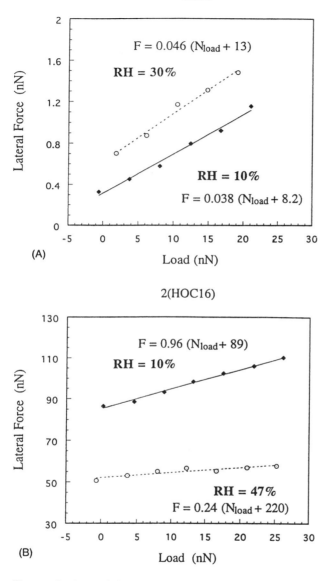

**Figure 4** Lateral force versus applied load for (A) a hydrophobic surface made from dihexadecyldimethylammonium monolayer (contact angle = 62°) and (B) a hydrophilic surface made from 16,16'-dihydroxydihexadecyldimethylammonium monolayer (contact angle = 7 ± 1°). Sliding velocity was kept fixed at 6.12 $\mu$m/s in these measurements. Linear fitting is according to a generalized Amonton's Law with the proportional constant being the frictional coefficient and the second term inside the parenthesis the adhesion force between the tip and the sample surface.

hence the lateral force. However, the opposite effect can occur if the condensed water molecules form a continuous layer, as in the case of a wettable surface (21), which then acts as a lubricant. Figure 4B demonstrates how as humidity increased, the frictional coefficient, which reflects the lubricity on the monolayer, reduced significantly on the hydrophilic monolayer surface, causing an overall decrease in the lateral force. On the other hand, for a hydrophobic surface at high humidity, isolated water clusters (which lack lubricity) form instead of a continuous film (21).

Our results show how we can directly link the nano- and micromechanical properties of surfactant monolayers determined from LFM measurements to the molecular structure of the constituent molecules and to compressibility as well as the phase properties of the monolayers. Such information provides useful insight into self-assembled thin films.

## ACKNOWLEDGMENT

This work is financially supported by the Center for Interfacial Engineering, a National Science Foundation engineering research center, and a grant from the 3M Company.

## REFERENCES

1. A Ulman. An Introduction to Ultrathin Organic Films. New York, Academic Press, 1991.
2. JD Swalen, DL Allara, JD Andrade, EA Chandross, S Garoff, J Israelachvili, TJ McCarthy, R Murray, RF Pease, JF Rabolt, JK Wynne, H Yu. Langmuir 3:932-50, 1987.
3. G Binning, CF Quate, C Gerber. Phys Rev Lett 56:930, 1986.
4a. O Marti, J Colchero, J Mlynek. Nanotechnology 1:141, 1990
4b. O Marti, J Colchero, J Mlynek. In: Nanosources and Manipulation of Atoms Under High Fields and Temperatures; Applications. Binh, VT, Et, AL, eds. Kluwer Academic Publishers, 1993, p 253.
5a. RM Overney, E Meyer, J Frommer, D Brodbeck, R Luethi, L Howald, MJ Guentherodt, M Fujihira, M Takano, Y Gotoh. Nature (London) 359:133-5, 1992.
5b. E Meyer, R Overney, D Brodbeck, L Howald, R. Luethi, J Frommer, HJ Guentherodt. Phys Rev Lett 69:1777-80, 1992.
5c. E Meyer, R Overney, R Luethi, D Brodbeck, L Howald, J Frommer, HJ Guentherodt, O Wolter, M Fujihira, H Takano, Y Gotoh. Thin Solid Films 220:132-7, 1992.
6. Y Liu, T Wu, DF Evans. Langmuir 10:2241-5, 1994.
7. G Haugstad, WL Gladfelter, EB Weberg, RT Weberg, RR Jones. Langmuir 11:3473-82, 1995.

8. H Yoshizawa, YL Chen, J Israelachvili. J Phys Chem 97:4128–40, 1993.
9. C Naselli, JF Rabolt, JD Swalen. J Chem Phys 82:2136–40, 1985.
10. C Naselli, JP Rabe, JF Rabolt, JD Swalen. Thin Solid Films 134:173–8, 1985.
11. K Kobayashi, K Takaoka, S Ochiai. Thin Solid Films 178:453–8, 1989.
12. C Bohm, R Steitz, H Riegler. Thin Solid Films 178:511, 1989.
13. Y Sasanuma, Y Kitano, A Ishitani, H Nakahara, K Fukuda. Thin Solid Films 199:359–65, 1991.
14. YI Rabinovich, DA Guzonas, RH Yoon. Langmuir 9:1168–70, 1993.
15. RL Garrell, JE Chadwick. Colloids Surf A 93:59–72, 1994.
16. Y Liu, DF Evans, Q Song, DW Grainger. Langmuir submitted.
17. YH Tsao, SX Yang, DF Evans, H Wennerstroem. Langmuir 7:3154–9, 1991.
18. RG Laughlin, RL Munyon, YC Fu, AJ Fehl. J Phys Chem 94:2546–52, 1990.
19. WA Ducker, TJ Senden, RM Pashley. Nature 353:239–41, 1991.
20. BV Derjaguin. Wear 128:19, 1988.
21. A Delville. J Phys Chem 99:2033–7, 1995.

# 24
## Suprabiomolecular Architectures at Functionalized Surfaces

**Wolfgang Knoll**
The Institute of Physical and Chemical Research (RIKEN),
Wako, Saitama, Japan,
and Max-Planck-Institute for Polymer Research, Mainz, Germany

**Masahiko Hara and Kaoru Tamada***
The Institute of Physical and Chemical Research (RIKEN),
Wako, Saitama, Japan

## I. INTRODUCTION

The functionalization of solid surfaces by molecularly controlled architectures is currently an extremely active field of research in interfacial science. In particular, it is hoped that new strategies can be developed that will allow for a purposeful design of the interface between artificial systems with technical substrates and the living world of biomolecules, cells, or even tissues. Eventually, the goal is not only to achieve a passive biocompatibilization of system components required in (bio-) sensor applications or in medical technologies, but to generate an interactive "interphase" that allows for an active communication between information processing units, e.g., neuronal cells, and microelectronic device components (1).

Three different approaches are currently taken: The first one tries to couple bimolecular lipid membranes through spacer molecules to solid supports (2). These tethers can be either oligopeptides or proteins, carbohydrates, or polyelectrolytes. The key question to be solved is, To what extent must these model membranes mimic the properties of biological membranes, e.g., in terms of fluidity, lateral organization, or incorporation of other functional units such as integral proteins. The challenge of this

---
*Current affiliation:* National Institute of Materials and Chemical Research, Tsukuba, Ibaraki, Japan.

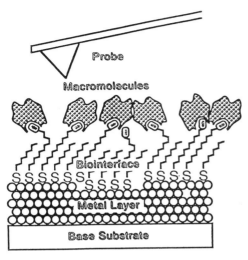

**Figure 1** Interfacial architecture built by a metal substrate/thiol SAM/protein layer.

membrane mimetic approach is to provide a buffer layer system that decouples the membrane sufficiently from its strong interaction with the substrate yet provides the physical and chemical stability for the whole system required for technical applications.

The second route in biocompatibility is based on the observation that many natural and artificial contacts between cells and substrates involve adhesion proteins, e.g., laminin or integrins. It is known, however, that only certain domains of these proteins, in some cases only a few amino acids long, are the active sites. The basis of this peptide mimetic approach is, therefore, to chemically functionalize technical substrates by amino acid sequences that promote the interaction to cells and tissues (3).

Finally, the last approach to be mentioned is based on the so-called self-assembly process observed for certain rod-shaped molecules with specific headgroup functionalities. Their strong enthalpic interaction with the substrate, e.g., thiols and disulfides with nobel metal surfaces, silanes with oxide substrates like $SiO_2$ and $TiO_2$, or $-COOH$ with $Al_2O_3$, leads to the formation of well-ordered monomolecular assemblies with tunable surface properties. The potential to chemically modify their endgroup functionality has attracted an immense interest to use these layers also for technical applications in lubrication, corrosion protection, and particularly for biocompatibilization and the build-up of complex supramolecular architectures (4).

# Suprabiomolecular Architectures

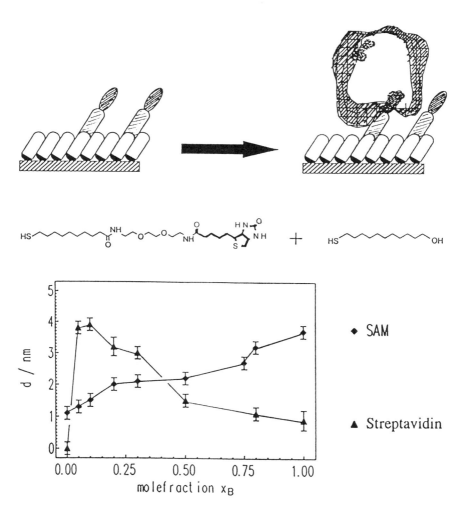

**Figure 2** An example of a binary mixed-thiol monolayer and its binding capacity.

In what follows we will concentrate on this latter technique and summarize some important aspects of the formation process of these SAMs and their use as a binding matrix for biorecognition reaction between a surface-bound ligand and a receptor protein in solution (5). A schematic illustration of such an interfacial architecture is given in Figure 1.

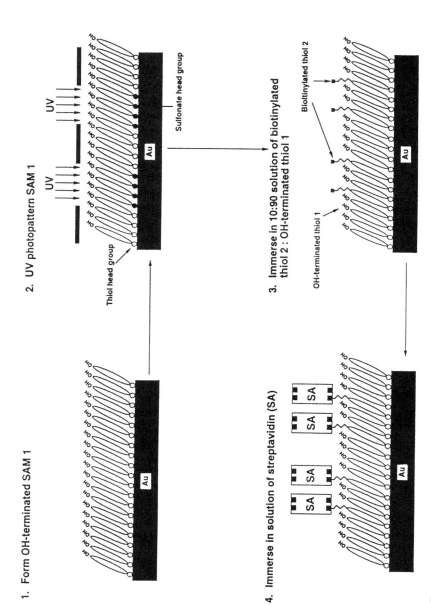

Figure 3  Photopatterning of thiol SAMs.

## II. OPTIMIZING THE BINDING CAPACITY OF A BIOTIN-DERIVATIZED THIOL MONOLAYER INVESTIGATED BY SURFACE PLASMON SPECTROSCOPY

The first aspect to be discussed concerns the optimization of the binding reaction between the ligand biotin, coupled to a self-assembled thiolmonolayer on a Au substrate, and the protein streptavidin approaching this functionalized surface from the aqueous phase. Both the formation of the thiol monolayer and the protein binding were observed on-line by surface plasmon spectroscopy. This way, kinetic data during the adsorption process (6) as well as layer thicknesses could be derived with a sensitivity in the angstrom range (7).

First, single-component thiol layers were investigated. It turned out, however, that an all-biotinylated monolayer has a binding capacity far below full coverage, because all binding sites are located too close to each other and hence block each other. The individual ligand can not be recognized by the protein. On the other hand, it was found that an OH-functionalized thiol system offers another most important surface property: it totally passivates the Au surface for nonspecific adsorption from solution.

From these studies a strategy was derived that employs the mixed assembling of a monolayer from a binary thiol solution. An example of such a binary mixture is given in Figure 2. The drawing summarizes the two important features required for maximum binding: (1) the lateral dilution of the single biotin-sites, and (2) their separation from the monolayer surface by a suitable spacer unit.

The structure formula of two molecules that were investigated is given in the inset of Figure 2. The monolayer thickness measured as a function of the mole fraction of the biotinylated compound in the solution shows a monotonous increase as the amount of the longer molecule is increased. However, the thickness increase obtained after injection of the protein solution shows a pronounced maximum at ca. 5–10 mol % of the biotinylated species. Higher biotin contents obviously lead to the above-mentioned sterical hindrance for the protein binding.

It could be shown that these ideas of a molecular "Lego" or "tinker-toy" approach can be applied to even more complex multilayer architectures built up by molecular recognition at such solid–solution interfaces. Other biotinylated systems can be bound, e.g., antibodies, lectins, or oligonucleotides (8).

Evidence that these architectures are indeed molecularly controlled comes from studies with other ligands, such as desthiobiotin, that have a

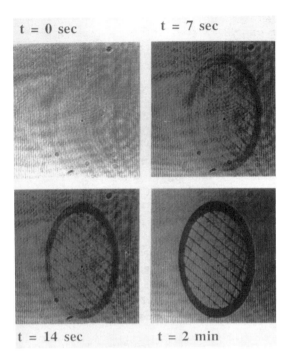

**Figure 4** Streptavidin binding to a photo-patterned SAM.

lower binding affinity to streptavidin. This allows not only for the buildup of multilayer assemblies but also for a controlled disassembly with a repetitive regeneration of a binding matrix by competitive replacement of bound streptavidin by an excess of free biotin.

## III. LATERALLY PATTERNED SAMS OBSERVED BY SURFACE PLASMON MICROSCOPY

An important aspect of surface functionalization concerns the lateral variation of surface properties by patterned SAMs. One attractive protocol for lateral structuring is based on the UV-photooxidation of thiolate groups to sulfonates that are only physisorbed to Au surfaces. This principle is sketched in Figure 3. If the (partially) irradiated SAM is immersed into another thiol solution, the exposed areas are refunctionalized by the new thiol molecules (9). In our example patches of mixed biotin-/OH-

## Suprabiomolecular Architectures

functionalized areas were generated in a pure OH matrix. Only after "decoration" with streptavidin, the illumination mask pattern was visible by surface plasmon microscopy (10). The diffusion-controlled time evolution of this binding process is shown in Figure 4. Prior to the protein injection the latent image could not be seen, because the thickness difference between the two areas is only about 1 Å, too small to be resolved by surface plasmon microscopy.

### IV. THIOL MONOLAYER FORMATION OBSERVED BY AFM AND STM

So far, we have discussed the properties of SAMs that were assembled for time intervals long enough to form a layer completely covering the substrate. However, it was found that some of these properties changed with time. Moreover, from a fundamental point of view it was important to also gather information about the assembling process itself, i.e., about the early stages of a SAM at submonolayer coverage.

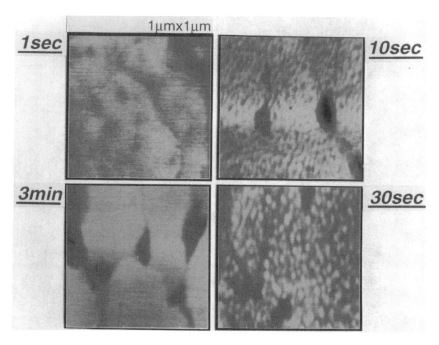

**Figure 5** Series of AFM images taken after various immersion times as indicated.

In order to address this problem, we performed experiments with highly diluted thiol solutions that allowed us to also control assembling times just enough for the nucleation of a few monolayer domains at very dilute coverages (11). Figure 5 shows a series of AFM pictures taken from SAMs at increasing immersion ($\hat{=}$ assembling) times. For these experiments, alkyl thiols ($n$-$C_{12}H_{25}SH$) at a concentration of $10^{-2}$ M in ethanol were chosen. It can be seen that within a few seconds the first domains comprising a few tens of molecules appear, grow in size, and rapidly merge to finally cover the whole gold substrate.

Another most remarkable aspect of this nucleation process is revealed by STM images of early stages of a mercaptopyridine monolayer self-assembled on Au. As is clearly seen in Figure 6 the nucleation does not

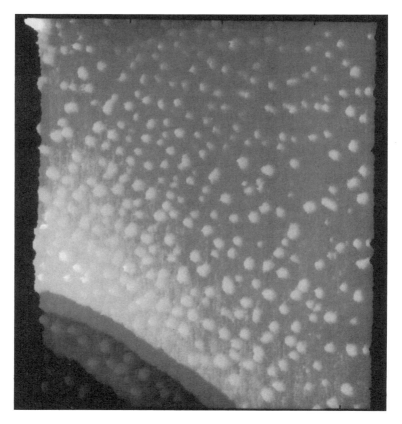

**Figure 6** STM image of 4-mercaptopyridine nucleation on Au(111) surface.

# Suprabiomolecular Architectures

occur in a random fashion but rather leads to a periodic arrangement of monolayer domains (12). Such an ordered nucleation had been seen before in STM images of the nucleation behavior of Fe on Au and had been identified to be due to the herringbone relaxation of the 22 × √3 surface reconstruction of a Au(111) surface. A particularly interesting finding is the observation that the adsorption of thiol molecules from solution seems to induce the herringbone ordering of the Au surface reconstruction.

## V. CONCLUSIONS

In conclusion, we propose a model for the self-assembling process that is similar to the well-investigated formation of monomolecular layers by vapor deposition. Some basic steps are summarized in Figure 7. For these flexible, rod-like molecules, additional steps that lead to monolayers with highly stretched well-oriented (and in some cases even crystalline) chains have to be considered in order to fully describe not only the structural details of SAMs but also the time dependence of their surface modifying properties.

However, if these processes are understood, monolayers with tailor-made, well-controlled functionalities can be built and employed for various applications.

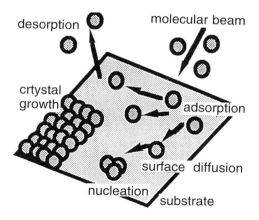

**Figure 7** Growth mechanism of thiol SAMs on Au.

## ACKNOWLEDGMENTS

This work is based on many colleagues' contributions. We are particularly grateful to L. Angermaier, G. Batz, M. Liley, D. Piscevic, P. Sluka, J. Spinke, and M. Tarlov.

## REFERENCES

1. A Offenhäusser, W Knoll. Kobunshi 44:18–22, 1995.
2. J Spinke, J Yang, H Wolf, M Liley, H Ringsdorf, W Knoll. Biophys J 63:1667, 1992.
3. M Matsuzawa, P Liesi, W Knoll. J Neurosci Meth 69:189–196, 1996.
4. J Spinke, M Liley, H-J Guder, L Angermaier, W Knoll. Langmuir 9:1821, 1993.
5. J Spinke, M Liley, F-J Schmidt, H-J Guder, L Angermaier, W Knoll. J Chem Phys 99:7012, 1993.
6. W Knoll, M Liley, D Piscevic, J Spike, MJ Tarlov. Adv Biophys 34:237–250, 1997.
7. EF Aust, S Ito, M Sawodny, W Knoll. TRIP 2:313–323, 1997.
8. D Piscevic, R Lawall, M Veith, M Liley, Y Okahata, W Knoll. Appl Surf Sci 90:425, 1995.
9. D Piscevic, MJ Tarlov, W Knoll. Supramol Sci 2:99, 1995.
10. EF Aust, M Sawodny, S Ito, W Knoll. Scanning 16:353–361, 1994.
11. K Tamada, M Hara, H Sasabe, W Knoll. Langmuir 13:1558–1566, 1997.
12. M Hara, H Sasabe, W Knoll. Thin Solid Films, 273:66, 1996.

# 25
# Langmuir–Blodgett Films of Condensation Polymers

**Masa-aki Kakimoto and Toshio Imai**
*Tokyo Institute of Technology, Tokyo, Japan*

## I. INTRODUCTION

The most fascinating point of Langmuir–Blodgett (LB) films is that they have an oriented layer structure at the molecular level. The molecules used in the LB method are amphiphilic molecules, and possess both hydrophilic and hydrophobic moieties at the same time. The typical case is fatty acids, where a long alkyl chain and a carboxylic functional group correspond to the hydrophobic and hydrophilic moieties, respectively. Although the LB method affords a nicely oriented layer structure, shortcomings of this method are that the resulting LB films are thermally and mechanically weak.

To overcome these shortcomings, recent attention has been paid to polymeric LB films. Polymers for LB films usually have a comb-like structure, where the hydrophobic long alkyl chains grow from the hydrophilic polymer backbone. Although thermal stability of the hydrophilic part can be improved by using polymers, the resulting LB films still possess thermally unstable long alkyl chains.

It was planed to remove the long alkyl chains from the polymeric LB films. Furthermore, if the polymers consist of an aromatic system, the LB films are expected to have thermal stability as well as the conjugated electron system. However, a number of aromatic polymers are insoluble in organic solvents and nonamphiphilic, so direct fabrication of such LB films is quite difficult. The precursor method was invented to solve this complicated problem. Some aromatic polymers such as polyimides, polybenzothiazoles, and poly(*p*-phenylene vinylene) can be prepared from their soluble precursors.

The strategy of the "precursor method" is as follows (Figure 1): First, the soluble precursor polymers are introduced into the amphiphilic structure. Second, the precursor LB films are prepared from the amphiphilic polymers. Then, long alkyl chains in the precursor LB films are removed away by appropriate methods. We have prepared LB films consisting of aromatic polyimides by the precursor method.

## II. PREPARATION OF POLYIMIDE MONO- AND MULTILAYER FILMS

Wholly aromatic polyimides are thermally stable engineering plastics, and have been widely used as reliable insulating materials in microelectronics. Recent developments in this field toward higher integration of devices have required ultrathin films of polyimides.

Since polyimides $\underline{5}$ (Equation 1) are essentially infusible and insoluble in organic solvents, they are processed into films at the stage of poly(amic acid)s $\underline{3}$, which are readily synthesized from tetracarboxylic dianhydrides $\underline{1}$ and diamines $\underline{2}$. Thermal treatment of poly(amic acid) films to 300°C affords polyimide films through cyclodehydration. Alternatively, chemical treatment of poly(amic acid) films with a mixture of acetic anhydride and pyridine is also effective to obtain polyimide films (4).

The preparation of polyimide LB films consisted of three steps, as illustrated in Equation 1 (2–4). In the first step, monolayer films of poly(amic acid) long alkyl amine salts $\underline{4}$ at the air–water interface were prepared.

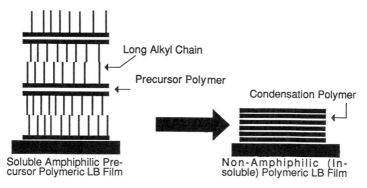

**Figure 1** Principal concept of "the precursor method."

# LB Films of Condensation Polymers

In spite of our expectation, the poly(amic acid) 3 itself, which possess hydrophilic carboxyl functional group in the polymer backbone, did not afford a stable monolayer at the air–water interface. Introduction of a hydrophobic long alkyl chain into 3 was performed by mixing poly(amic acid)s and long alkyl amines. Poly(amic acid) salts 4, thus obtained, afforded very stable monolayer films at the air–water interface. In the second step, the polyamic acid salt monolayer films were deposited on appropriate plates such as glass, quartz, or silicon wafer. Finally, polyimide multilayer films were obtained by treatment of poly(amic acid) salt multilayer films on the plates with a mixture of acetic anhydride and pyridine.

The structure and the code of various poly(amic acid)s 3 are shown in Table 1. A solution of 3 in a mixture of $N,N$-dimethylacetamide (DMAc) and benzene (1:1) prepared to a concentration of 1 mmol/l and a solution of long alkyl amines such as dimethylhexadecylamine (DMC16) in the same mixed solvent system with the same concentration were combined to produce a solution of poly(amic acid) salts 4. The solutions were then spread onto deionized water. Figure 2 shows $\pi-A$ curves of poly(amic acid) salts

**Table 1** Extrapolated Surface Area of Poly(amic acid) Salts 4

| Code | Poly(amic acid) | Surface Area (nm²/unit) |
|------|-----------------|-------------------------|
| a    |                 | 1.38                    |
| b    |                 | 1.65                    |
| c    |                 | 1.39                    |
| d    |                 | 1.60                    |
| e    |                 | 1.30                    |
| f    |                 | 1.50                    |
| g    |                 | 1.10                    |

4a. Figure 3 illustrates the molecular model of the repeat unit of poly(amic acid) 3a in which the aromatic ring coming from the pyromellitic acid lies flat on the water surface. The area circumscribed by the line around the model was 1.28 nm², which was in good agreement with a surface area of 1.38 nm² obtained by extrapolation of the steep rise of $\pi$-$A$ curve to zero pressure, considering the area additionally occupied by amine function. This fact strongly supports the interpretation that the spread film at the air-water interface has the monolayer structure.

A variety of poly(amic acid)s 3b-3f were used for the preparation of

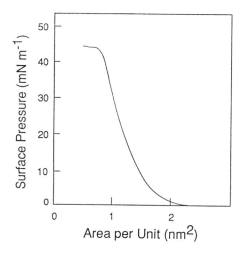

**Figure 2** $\pi$–$A$ curves of poly(amic acid) salts 4a measured with compression rate of 7 mm/min at 20°C.

monolayers to examine the behavior of monolayers at the air–water interface in more detail. Two equivalents of alkyl amine DMC16 were used to produce polyamic acid salts 4b–4f, which were spread onto the pure water under the same conditions as described above. In all cases, steeply rising $\pi$–$A$ curves were obtained. The values of each extrapolated surface area are summarized in Table 1. The difference between the area of 4a and 4b is almost equivalent to the difference between 4c and 4d, corresponding to the surface area of benzoyl function which was calculated to be about 20 nm²

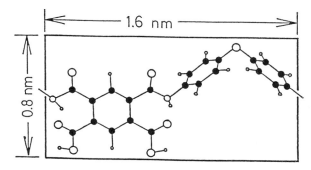

**Figure 3** Molecular model of poly(amic acid) 3a.

using the molecular model. Similarly, the observed area of phenoxy group from the difference between 4a and 4g, 4e and 4f as well as that of phenyl group from the difference between 4a and 4f were in good agreement with the corresponding calculated values.

Deposition of poly(amic acid) salt 4a was carried out at a surface pressure of 25 mN/m onto an appropriate substrate by drawing down and up through the air–water interface at a rate of 3–5 mm/min at 20°C (vertical dipping method). In the transmission FT-IR spectrum of the 4a film (200 layers) deposited on a silicon wafer, typical absorptions were observed at around 2920 $cm^{-1}$ due to the long chain hydrocarbon and carbonyl groups.

A multilayer film of 4a obtained as above was immersed overnight in a mixture of acetic anhydride, pyridine, and benzene (1 : 1 : 3) to afford a polyimide film of 5a. A linear relationship was again obtained between the number of layers and the absorbance at 284 nm of 5a films, which indicates that the layers did not come off during the imidization step.

In the IR spectrum of polyimide films of 5a, the absorption peak due to hydrocarbon group of the 4a film disappeared, and new characteristic absorptions corresponding to the imide carbonyl groups appeared at 1780 and 1720 $cm^{-1}$. This suggested that the cyclization of the poly(amic acid) salt 4a to polyimide 5a proceeded almost completely with the removal of the long chain alkyl amine. The chemical resistance of the polyimide multilayer film of 5a was determined by monitoring the change in the UV spectrum before and after immersing it in various solvents for 1 h. Although the film readily decomposed in alkaline solution, it was quite stable in strong acid and organic solvents. The X-ray interference pattern of the multilayer film of 5a (100 layers) suggested that the film has uniform thickness of around 40 nm. Furthermore, the thickness of the same film was measured directly to be 42 nm using Talystep. According to the results of the ellipsometry, the monolayer thickness is 0.45 nm.

## III. DEFECTS IN POLYIMIDE LANGMUIR–BLODGETT FILMS EVALUATED BY ELECTROCHEMICAL METHOD

The defects in LB films are one of their big problems, especially in practical uses. Although some methods such as electron microscopy and metal decoration technique have been applied to detect the defects (5), we adopted an electrochemical redox reaction that is carried out on the surface of the electrodes, and readily monitored by the cyclic voltammetry (6). The degree of defects in the LB films that cover the electrode surface should be readily evaluated, because the redox reaction proceeds only when the electrolyte

# LB Films of Condensation Polymers

ions come in contact with the electrode surface through the defects. The electrochemical evaluation of defects in the polyimide LB films by redox reaction in aqueous solution was examined.

The monolayer films of poly(amic acid) alkyl amine salts 4 were transferred onto a glassy carbon disk electrode (5 mm diameter, 19.6 mm$^2$ surface area), which was inclined in 45° against the normal direction. The LB films of 4 were converted to the LB films of polyimides 5 by the treatment with the mixture of acetic anhydride and pyridine. At the same time, the Y-type LB film of cadmium arachidate on the same electrode was prepared. Electrochemical measurements were carried out under nitrogen atmosphere at 25°C using a potentio-galvanostat. A glassy carbon rod and a saturated calomel electrode were employed as the counter and the reference electrodes, respectively.

The redox reaction between potassium ferri- and ferrocyanide was employed for the present work. The dotted line in Figure 4 shows the cyclic voltammogram of potassium ferrocyanide recorded using the uncoated electrode. The cyclic voltammograms recorded with the electrode coated with 3 layers of cadmium arachidate (7.5 nm thick) and 11 layers of polyimide 5a (5 nm thick) showed apparent decrease of the redox peak currents and increase of the redox peak to peak separation, which indicated that contact of the ferrocyanide ion with the surface of the electrode was restricted. In the electrodes deposited with two different types of LB films,

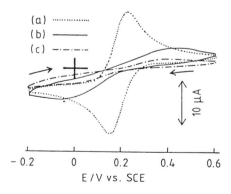

**Figure 4** Cyclic voltammograms of K$_4$[Fe(CN)$_6$] aqueous solution. Electrodes used were (a) uncoated glassy carbon electrode (area 19.6 mm$^2$), (b) coated with cadmium arachidate LB film (3 layers, 7.5 nm thick), and (c) coated with polyimide LB film of 5a (11 layers, 5 nm thick). K$_4$[Fe(CN)$_6$] (1 mM) and KCl (100 mM) aqueous solution was used as electrolyte. Scan was 100 mV/s.

the polyimide LB film had a lower degree of defects than the cadmium arachidate. The better coating ability of the polyimide LB film may be due to its amorphous nature.

In the next stage, the same measurement was carried out with polyimide 5h, which consisted of aliphatic tetracarboxylic dianhydride. The cyclic voltammograms observed with changing the number of layers are shown in Figure 5. A fair degree of defects was observed with the LB film with low number of layers, which may be explained by the direct influence of the roughness of the electrode surface. Decrease in the redox peak current and increase in the peak-to-peak separation were observed with increasing the number of layers. The ten-layer LB film of 5h was found to be sufficient for covering the surface of the electrode with no detectable defects. The

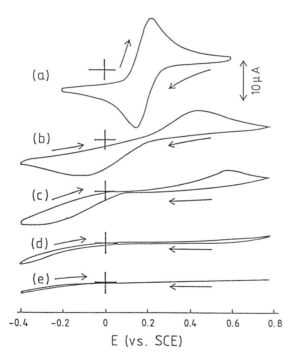

**Figure 5** Cyclic voltammograms of $K_4[Fe(CN)_6]$ aqueous solution. Electrodes used were (a) uncoated glassy carbon electrode (area 19.6 mm$^2$), (b-d) coated with polyimide LB film of 5b: (b) 1 layer, (c) 3 layers, (d) 6 layers, and (e) 10 layers. $K_4[Fe(CN)_6]$ (1 mN) and KCl (100 mM) aqueous solution was used as electrolyte. Scan rate was 100 mV/s.

excellent coating ability may be attributed to the flexible aliphatic structure of tetracarboxylic acid unit.

Thus, the cyclic voltammetry is one of the convenient methods for the evaluation of defects in LB films. The results suggested that the polyimide LB films possessed a low degree of defects compared with the LB films of fatty acids, and the degree depended markedly on the chemical structure of the polyimides employed.

## IV. ORIENTATION OF POLYMER CHAIN IN POLYIMIDE LB FILMS

Orientation of molecules in L and LB films is one of the most important problems in basic physics, as well as fabrication of "super molecular machine." Sugi and coworkers have proposed a flow orientation model to discuss the in-plane molecular orientation (7).

Recently, a unique moving wall type trough was invented, and it might be possible that organic molecules behave in a different way compared with the case in an ordinary trough. Here, an azobenzene unit, which possesses strong anisotropic absorption at 380 nm, was newly introduced into the polyamic acid backbone to measure molecular orientation of the polymer, and the molecular orientation of L and LB films of poly(amic acid) amine salts 4 was examined in detail (8).

Two kinds of poly(amic acid)s, 3i and 3j, possessing an azobenzene unit were prepared from 4,4'-diaminoazobenzene and the corresponding dianhydrides by conventional condensation (1). The preparation of L and LB films was performed using Nippon Laser and Electronics model NL-LB-240 LB film equipment. Although the trough was made for the moving-wall-type equipment, it was also used as an ordinary-type trough after removal of the moving tape. Figure 6 shows diagrams of the troughs used in this study. The remarkable difference between these troughs is the movable Teflon tape. In the moving-wall-type trough, the water surface for L film preparation is surrounded by the pressing bar, two Teflon tapes, and the substrate plate. The width of the pressing bar and the plate is the same, and the Teflon tape moves with moving the pressing bar at the same speed. In the ordinary trough the wall is fixed, and the width of the plate is usually smaller than that of the water surface. Thus, the moving wall system can minimize the stress between the L film and the wall of the trough, and prevent the complicated flow on deposition of the L film (9,10).

Absorption spectra of L films were measured using an Otsuka Electronics MCPD-1000 spectrometer. Light source and detector connected with the main systems by glass fiber cables were set up at 5 mm upper

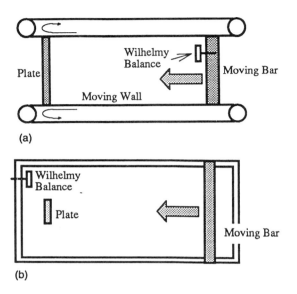

**Figure 6** Diagram of troughs used. (a) Moving wall type LB trough; (b) ordinary LB trough.

position of the L film, where the light was irradiated from the normal direction of the film, and the incident light was reflected by a mirror set at the bottom of the trough. The reflected light coming through the L film was detected at the same position as the light source.

Typical absorption spectra of the L films of 4i are shown in Figure 7. Stronger absorption was observed by irradiation of polarized light, which was parallel to the pressing bar. This phenomenon indicated that the polymer backbone is oriented parallel to the pressing bar. Here, the absorptions measured by the parallel and perpendicular polarized light to the pressing bar are noted as Apara and Aperp, respectively. From the relationship between the dichroic ratio ($D$ = Apara/Aperp) and the surface pressure, higher orientation of the L film was observed at higher surface pressure. Orientation of the molecules was also accelerated by introduction of straightforward molecular structure judging from the fact that 4j showed higher molecular orientation.

As shown in Figure 8, the moving wall trough achieved higher orientation of the L films compared with an ordinary trough. The difference in these two troughs should be frictional stress between the L film and the wall of the trough. During the course of the compression, the friction decreases the orientation of at least the edge position of L films in an ordinary trough.

# LB Films of Condensation Polymers

Next, the L films of precursor polymers 4i were deposited onto a quartz plate using a usual vertical dipping technique. When an ordinary trough was used, the absorption of the LB film for the dipping (parallel) direction was apparently larger than that for the perpendicular direction, as shown in Figure 9. This fact strongly supported observation that, considering the orientation of the polymer chain at the air-water interface, the polymer molecules turned 90°, when they were transferred from an air-water interface to the plate. The dichroic ratios ($D$) of the LB films are indicated by Apara/Aperp, where Apara and Aperp are the absorptions parallel and perpendicular to the transfer direction, respectively.

Table 2 shows $D$ of the LB films deposited using the different types of troughs. It is remarkable that $D$ is less than 1.0 for the moving wall trough. This was evidence that the molecular orientation of the polymer chain was perpendicular to the dipping direction when the moving wall trough was used. That is, the polymer chain did not turn during the deposition process, and the influence of chemical structure of the L films on the molecule orientation was also retained in the LB films in this type of trough.

Changing the molecular orientation during the deposition process seems to be one of the uncontrollable factors for preparation of LB films

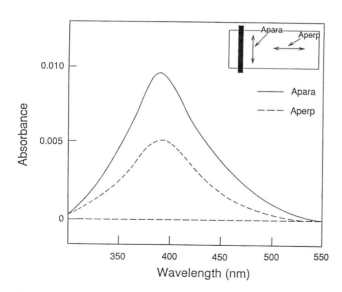

**Figure 7** Polarized absorption spectra of L film of 4a measured at a surface pressure of 30 mN/m.

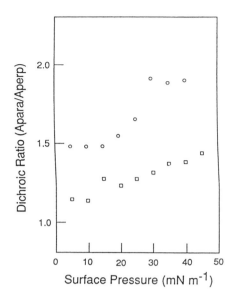

**Figure 8** Relationship between dichroic ratio and surface pressure in L films, of 4a at compression rate of 15 mm/min. Circles, moving wall trough; squares, ordinary trough.

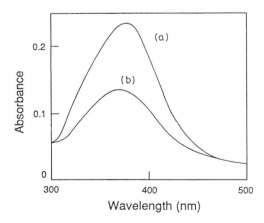

**Figure 9** Polarized absorption spectra of LB films of 4a prepared by ordinary trough. (a) Polarization is parallel to dipping direction (Apara), (b) polarization is perpendicular to dipping direction (Aperp).

**Table 2** Dichroic Ratio ($D$) of LB Films of Precursor Polymer 4

|    | Ordinary trough | Moving wall trough |
|----|-----------------|--------------------|
| 4i | 1.11            | 0.83               |
| 4j | 1.17            | 0.58               |

*Note:* In both troughs, L films were transferred onto a glass plate at a surface pressure of 20 mN/m and a dipping speed of 5 mm/min.

having well-defined molecular orientation. Thus, if one can make sufficiently oriented L films, the moving wall trough can fabricate the LB films with high molecular orientation.

## V. POLYBENZOTHIAZOLES LB FILMS

We have reported a novel synthetic method of polybenzothiazoles starting from sulfur-protected 2,5-dithio-1,4-phenylenediamines, where the protecting groups were isopropyl, 2-(methoxycarbonyl)ethyl, and 2-cyanoethyl (11a–11c). The precursor polyamides are usually soluble in organic solvents because of the substituents on the thiol group. Polybenzothiazoles are obtained by curing the precursor polymers at 300°C with subsequent elimination of the protecting groups and water. Another series of polybenzothiazoles and polyamides having β-propionic acid moiety 6 were synthesized (12). The basic structure of polymer 6, possessing two carboxylic acids in each repeating unit, was similar to polyamic acid 3.

As shown in Equation 2, polyamide 6 was mixed with tertiary amine having three long alkyl chains to afford alkyl amine salts 7. The obtained precursor polymeric salts 7 were dissolved in an equal volume mixture of $N,N$-dimethylacetamide and benzene and spread onto the air–water interface. The surface-pressure–area ($\pi$–$A$) curve of 7 is shown in Figure 10. With decreasing surface area, after the small plateau, the curve steeply rose. This indicated two-dimensional crystallization of the long alkyl side chain. The fact that the limiting surface areas of the precursor polymers possessing different chemical structures in 6 were always 1.5 nm$^2$, suggested that these areas did not indicate the areas of the repeating unit of polyamic acid, but seemed to show that of two molecules of trialkyl amine. Deposition of the monolayers of 6 was readily carried out, resulting in a Y-type deposition onto hydrophilic substrates such as glass plates or silicon wafers.

The multilayers of 7 were heated at 280°C to convert them into poly-

(2)

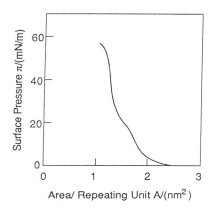

**Figure 10** $\pi$-A curves of $\underline{7}$.

benzothiazole LB films. Figure 11 shows the IR spectra of $\underline{6}$ before and after the heat treatment, where the LB film (25 layers) was prepared on a calcium difluoride plate. Alkyl chain absorptions at 2915 and 2850 cm$^{-1}$ as well as an ester carbonyl at 1730 cm$^{-1}$ disappeared after the heat treatment. On the other hand, typical absorptions of the benzothiazole ring were ob-

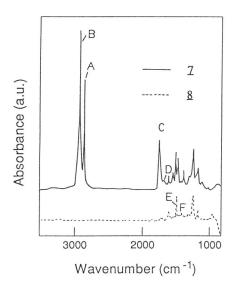

**Figure 11** FT-IR spectra of LB films of $\underline{7}$ and $\underline{8}$.

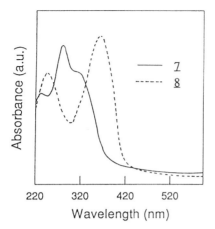

**Figure 12** Absorption spectra of LB films of 7 and 8.

served at 1490 and 1400 cm$^{-1}$ in the final LB film. These IR spectra indicated that the long alkyl amine and acrylic acid were eliminated to form a benzothiazole ring. As shown in Figure 12, the UV-visible absorption spectra changed after the heat treatment because the polybenzothiazoles have a long conjugating π-electron, which causes third-order nonlinear optical effects.

The X-ray diffraction measurement of 7 LB film showed a periodic pattern corresponding to 5.9 nm. Because the LB film was prepared by Y-type deposition, the thickness obtained from the X-ray diffraction should be that of the double layers. Monolayer thickness of the precursor 7 and polybenzothiazole 8 LB films were determined to be 2.8 and 0.34 nm, respectively, by using ellipsometry.

An anisotropic third-order nonlinear optical effect was observed in the LB film of polybenzothiazole 8. The nonlinear susceptibility $\chi^{(3)}$ parallel to the dipping direction was $3.8 \times 10^{-11}$ ESU, whereas that of the perpendicular direction was about one-fifth that of the parallel direction. Thus, the polymer chain of 8 had some ordering to the dipping direction.

## VI. POLY(P-PHENYLENE VINYLENE) LB FILM

Electrically conducting LB films are one of the most fascinating targets. Conducting polymers based on π-conjugation structure have attracted con-

# LB Films of Condensation Polymers

siderable interests for their various application potentials. Recently, poly(*p*-phenylene vinylene) (PPV) has been used as a light-emitting layer in an electroluminescent device, which is suitable for a large-area display application (13). Despite several attempts, conducting LB films obtained so far have been anisotropic, owing to the presence of insulating alkyl chains between the layers. PPV is a well-known conducting polymer that has been synthesized by thermal conversion of the precursor poly(sulfonium salts) 10. The "precursor method" was applied to preparation of PPV LB films shown in Equation 3. An LB film of poly(sulfonium salt) 10b with long-chain perfluoroalkyl substituents was chosen as the precursor (14a,14b).

An aqueous solution of poly(sulfonium salt) 10a ($X$ = Cl), which is a precursor polymer for PPV 12, was readily obtained from $\alpha,\alpha'$-bis(diethylsulfonium) *p*-xylene dichloride 9 in aqueous alkaline media. The concentration of 10a was adjusted from ca. 20 to 2 mM by diluting with ethanol and then to 1 mM with 1,1-trichloroethane. The resulting solution of 10a was mixed with the same volume of 1 mM solution of sodium perfluorononanoate in ethanol-1,1-trichloroethane (1 : 1 by volume). This means that an equimolar amount of perfluorononanoate and sulfonium groups is present in the solution. The mixed solution was applied to the LB method. The LB film of precursor polysulfonium salt 10b was successfully prepared by usual LB technique with Y-type deposition. It was well known that the precursor polysulfonium salt of PPV could be converted to PPV by thermal elimination of the sulfide and counter anion. The conversion of the LB film of the precursor polymer to PPV was carried out at 250°C/torr for various thermal treatment times. The difference of heating time should be reflected in the variation of the structure during the conversion process.

The dependence of the formation of the conjugated PPV structure on the thermal annealing time was examined by UV spectroscopy. The UV-visible absorption spectra of the PPV precursor LB films (40 layers) treated at 250°C/torr for various thermal annealing times are shown in Figure 13. It was obvious that the PPV precursor (before thermal treatment) had almost no absorption in the visible region. On the other hand, under the thermal treatment, a strong and broad absorption peak in the visible region appeared. This appearance of a strong absorption in the visible region demonstrated the formation of the conjugated PPV structure in the LB film. Furthermore, it could be observed that the intensity of the absorption

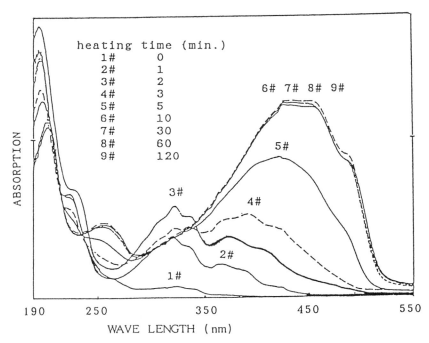

**Figure 13** UV visible absorption spectra of PPV precursor LB films (40 layers) treated at 250°C/torr for various heating times. Curve 1, without heating; curve 2, 1 min; curve 3, 2 min; curve 4, 3 min; curve 5, 5 min; curve 6, 10 min; curve 7, 30 min; curve 8, 60 min; curve 9, 120 min.

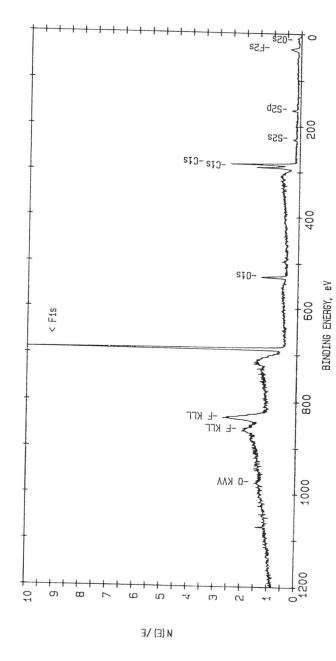

**Figure 14** XPS survey spectrum of PPV precursor LB films (30 layers) with take-off angle of 45°.

in the visible region increased, and its wavelength showed red shift as the thermal treatment time prolonged. Finally, the absorption maximum reached 450 nm with the edge longer than 500 nm. These results indicated that $\pi$-conjugated PPV structure grew up during the thermal treatment. Also, the absorption peak wavelength of the PPV LB film (450 nm) was about 30 nm longer than that of the conventional casting film, meaning that the $\pi$-conjugation could be more extended in the LB film. The time dependence of the absorbance at wavelength of 450 nm, which corresponded to the absorbance of PPV, exhibited that the $\pi$-conjugation of PPV formed under thermal treatment within 10 min.

Both the survey and C(1s) high-resolution X-ray photoelectron spectroscopy (XPS) spectra of the LB films (30 layers) were obtained at a takeoff angle of 45°. The survey spectrum of the LB film of PPV precursor (before thermal treatment) displayed the peaks characterized by carbon C(1s) at 284.2 eV, fluorine F(1s) at 688.5 eV, F(2s) at 30.6 eV, sulfur S(2s) at 229.1 eV, S(2p) at 165.6 eV, and oxygen O(1s) at 531.6 eV, as shown in Figure 14. Comparing the calculated atomic concentration, i.e., C(52.00), F(40.53), S(1.30), O(6.17), with the theoretical composition, i.e., C(51.22), F(41.46), S(2.44), O(4.88), we found that the atomic concentration agreed

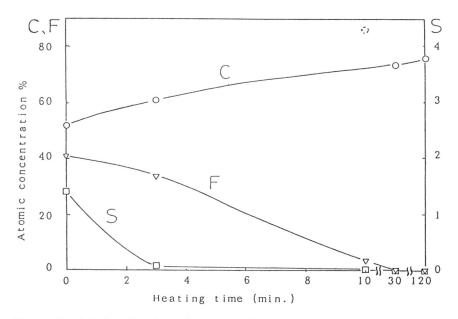

**Figure 15** Relationship of atomic concentration and heating time.

# LB Films of Condensation Polymers

well with theoretical composition of the PPV precursor except the high value of oxygen. This should be attributed to the oxygen absorbed in the surface of the LB film, which could not be removed even though the sample was handled under inert atmosphere before measurement. The variation of atomic concentration with thermal treatment times indicated a quick decrease of fluorine and sulfur atoms (as shown in Figure 15), which was proof of the removal of perfluorononanoic acid and the sulfide during thermal treatment. On the other hand, the concentration of carbon after removal of sulfur and fluorine became almost constant. We found that by 10 min of thermal treatment, sulfur loss was 94.6% and fluorine loss was 91.4% of their original concentrations. These results indicated that thermal

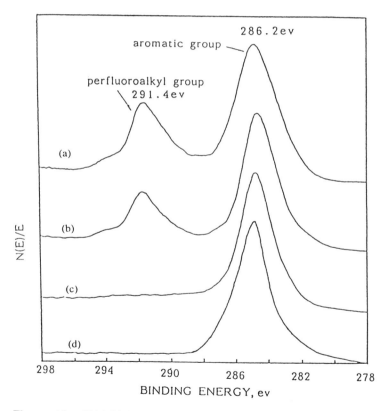

**Figure 16** C(1s) high-resolution spectra of PPV precursor LB films (30 layers) treated at 250°C/torr for various heating times: (a) without heating; (b) 3 min; (c) 10 min; (d) 120 min.

elimination was almost completed within 10 min by thermal treatment, resulting in the PPV LB film.

In the C(1s) high-resolution spectra (Figure 16), two peaks were observed for the PPV precursor 10b, the one was at 284.2 eV assigned to the carbon associated to aromatic group, and the other was at 291.4 eV assigned to the carbon associated to perfluoro alkyl group. As the thermal treatment time proceeded, the relative intensity of the peak at 291.4 eV decreased, and finally it disappeared within 10 min. These results also implied that perfluorononanoic acid was almost completely removed after 10 min of thermal treatment to produce the PPV LB film.

## VII. CONCLUSION

The polymeric LB films prepared by the, "precursor method" have successfully applied to devices such as an aligning layer for a liquid crystalline cell (15), solar battery cells (16), photomemory devices working in the photon mode (17), and EL devices (18). The characteristic feature that they do not have long alkyl chains has brought a new field in LB films.

## REFERENCES

1. CE Sroog. In: Macromolecular Synthesis. Vol. 1. Moore, JA, ed. New York, John Wiley & Sons, 1977, p 295.
2. M Suzuki, M Kakimoto, T Konishi, Y Imai, M Iwamoto, T Hino. Chem Lett 395, 1986.
3. M Kakimoto, M Suzuki, T Konishi, Y Imai, M Iwamoto, T Hino. Chem Lett 823, 1986.
4. M Kakimoto, M Suzuki, Y Imai, M Iwamoto, T Hino. In Polymers for High Technology, Electronics and Photonics. Bowden, MJ, Turner, SR, eds. ACS Symposium Series 346. Washington, DC, American Chemical Society, 1987, p 484.
5. M Vandevyver. Thin Solid Films 159:243, 1988.
6. Y Nishikata, M Kakimoto, A Morikawa, I Kobayashi, Y Imai, Y Hirata, K Nishiyama, M Fujihira. Chem Lett 861, 1989.
7. N Minari, K Ikegami, S Kuroda, K Saito, M Saito, M Sugi. J Phys Soc Jpn 58:222, 1989.
8. Y Nishikata, K Komatsu, M Kakimoto, Y Imai. Thin Solid Films 210/211:29, 1992.
9. H Kumehara, S Tasaka, S Miyata. Nippon Kagaku Kaishi 12:2330, 1987.
10. H Kumehara, T Kasuga, T Watanabe, S Miyata. Thin Solid Films 178:175, 1989.
11a. T Hattori, K Kagawa, M Kakimoto, Y Imai. Macromolecules 26:4089, 1993.

11b. T Hattori, H Akita, M Kakimoto, Y Imai. Macromolecules 25:3351, 1992.
11c. T Hattori, H Akita, M Kakimoto, Y Imai. J Polym Sci A Polym Chem 30: 197, 1992.
12. T Hattori, H Akita, M Kakimoto, Y Imai. Polym J 26:930, 1994.
13. JH Burroughes, DC Bradley, AR Brown, RN Marks, K Mackay, RH Friend, PL Burns, AB Holmes. Nature 347:539, 1990.
14a. Y Nishikata, M Katimoto, Y Imai. Thin Solid Films 179:191, 1989.
14b. A Wu, S Yokoyama, S Watanabe, M Katimoto, Y Imai. Thin Solid Films 244: 750, 1994.
15. Y Nishikata, A Morikawa, Y Takiguchi, A Kanemoto, M Kakimoto, Y Imai. Jpn J Appl Phys 27:L1163, 1988.
16. Y Nishikata, A Morikawa, M Kakimoto, Y Imai, Y Hirata, K Nishiyama, M Fujihira. J Chem Soc Chem Commun 1772, 1989.
17. S Yokoyama, M Kakimoto, Y Imai. Langmuir 10:4594, 1994.
18. A Wu, M Jikei, M Kakimoto, Y Imai, S Ukishima, Y Takahashi. Chem Lett 2319, 1994.

# 26
# Langmuir–Blodgett Film as Alignment Layers for Nematic Liquid Crystal Displays

**A. Albarici, J. A. Mann, Jr., and J. B. Lando**
*Case Western Reserve University, Cleveland, Ohio*

**J. Chen, H. Vithana and D. Johnson**
*Kent State University, Kent, Ohio*

**Masa-aki Kakimoto**
*Tokyo Institute of Technology, Tokyo, Japan*

## I. INTRODUCTION

Most of the liquid crystal displays (LCD) produced today are either twisted nematic (TN) or supertwisted nematic (STN) electrooptical effects. The first TN displays made their appearance in products about 20 years ago in watches and calculators. Later, with continued improvement in liquid crystal materials (wider temperature range and low viscosities) and device technologies (polarizers, alignment layers), TN displays expanded into other areas including automotive instrument panels, consumer electronics, and test equipment. By 1987, making use of conventional TN materials and fabrication processes, STN displays were quickly commercialized making extensive inroads into laptop computers and word processors. The basic advantages of STN over TN are better contrast ratio between the dark and white mode and better resolution (1,2). The construction and basic operation of TN displays are illustrated in Figure 1.

It is well known that different bulk liquid crystal alignments can be induced by several kinds of surface treatments. Homogenous alignment of liquid crystal molecules can be achieved by modifying substrates by oblique

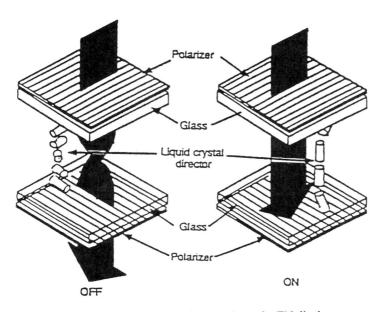

**Figure 1** Construction and basic operation of a TN display.

evaporation of inorganic materials (3) or the use of stretched polymers (4), photolithographic gratings (5), rubbed polymer films (6-10) or, more recently, Langmuir-Blodgett (LB) films (11-14) as the orienting medium. Currently, surface alignment is achieved most often by the use of polymer layers, usually polyimides and polyvinylalcohol, spin-coated onto a substrate such as conductive ITO (indium-tin-oxide) coated glass and subsequently rubbed mechanically to induce alignment. Normally, rubbing is performed by moving a rotating cylinder, covered with rubbing cloth, over a polymer-coated substrate with a constant velocity. This method is of particular interest because of its feasibility for large area treatment and mass production. However, this approach has several shortcomings, including the following: (1) rubbing can cause damage to the underlying active elements, in particular in thin film transistor driven liquid crystal displays (15); (2) the effect of rubbing is neither well understood nor easily controlled but is somewhat an art that works surprisingly well in less critical applications; and (3) rubbing can generate dust or static electricity that can have an adverse influence on the display characteristics. In general, this static electricity tends to destroy switching elements, causes poor memory capability, and largely reduces the productivity of displays. In addition, the rubbing process also has difficulty in assuring uniformity of a large area (16).

Ordered LB films have recently been shown to have enormous promise as LCD alignment layers (11-14). As many authors have reported (16,17), LB films have a fair degree of molecular orientation along the film transfer direction, which make these films able to align the molecular long axis of nematic liquid crystals along the same film transfer direction without rubbing the LB film. In this way, the use of LB films as alignment layers avoids the major problems of the rubbing technique discussed above. The first report about alignment of LC on LB films was given by Hiltrop and Stegemeyer (18), who used lecithin films, but a number of other molecules have been used as LB materials (19,20). Recently, Japanese scientists have aligned LCs on polyimide films prepared using the LB technique without the rubbing process (12,16,21).

The aim of this work is to study the alignment properties of polyimide LB films on nematic LC in order to evaluate the potential of LB film for uses in real applications and to understand the mechanisms responsible for the LC alignment. This work consists of two parts, one concerning the preparation and characterization of the polyimide Langmuir and LB films and the other concerning the characterization of the alignment properties of these LB films and comparing this to conventional rubbed films. The LB material chosen for the present study was poly[$N,N'$-bis(phenoxyphenyl)-pyromelltimidel (PMDA-ODA) because of its good thermal stability over 400°C and because of its excellent mechanical and dielectric properties (22-24).

## II. EXPERIMENTAL

### A. Preparation and Characterization of LB Films

In general, polyimides are quite insoluble in organic solvents, and the only hydrophilic groups are the carbonyl groups in the polymer backbone. Because of this they do not have a good balance between hydrophilic and hydrophobic groups in the molecules and do not possess the necessary amphiphilic nature to form a stable monolayer at the air-water interface. For these reasons it is not possible to prepare polyimide films directly using the LB technique. Although polyamic acids (the polyimide's precursor) are soluble in some organic solvents, they do not form stable monolayers either. To overcome this instability problem, the polyamic acid is mixed with a long carbon chain amine in order to get a polyamic acid alkyl amine salt (PAAS). These species, due to the fact that long carbon chains are attached to the polymer backbone, form stable Langmuir monolayers and can be deposited onto solid substrates and then converted to polyimide by chemical or thermal treatment. This "precursor method" of preparation of polyimide LB films was first proposed by Kakimoto and colleagues (21), and is shown in Scheme 1.

Isotherms, often known as pressure-area curves, provide a characteristic description of an insoluble monolayer at the air-water interface. They measure the change in surface tension, and thus surface pressure due to the change in area of a film forming molecule on a liquid surface.

All isotherm experiments were carried out on a commercially available film balance (Lauda) 70 × 15 × 0.7 cm. This trough was equipped with a moving barrier operated by an IBM PC computer. Surface pressure is the most commonly measured property of an insoluble monolayer. It was measured using a floating barrier pressure sensor with a precision of ±0.1 mN/m. The interior trough surfaces and barrier were made of Teflon.

Spreading solutions of PAAS in a mixture of $N,N'$-dimethylacetamide (DMAC) and benzene (1 : 1) at a concentration of 0.5 mM (0.47 mg/ml) were prepared from poly(pyromellitamic) acid (PA) and $N,N'$-dimethylhexadecylamine in the molar ratio 1 : 2, that is, an equimolar

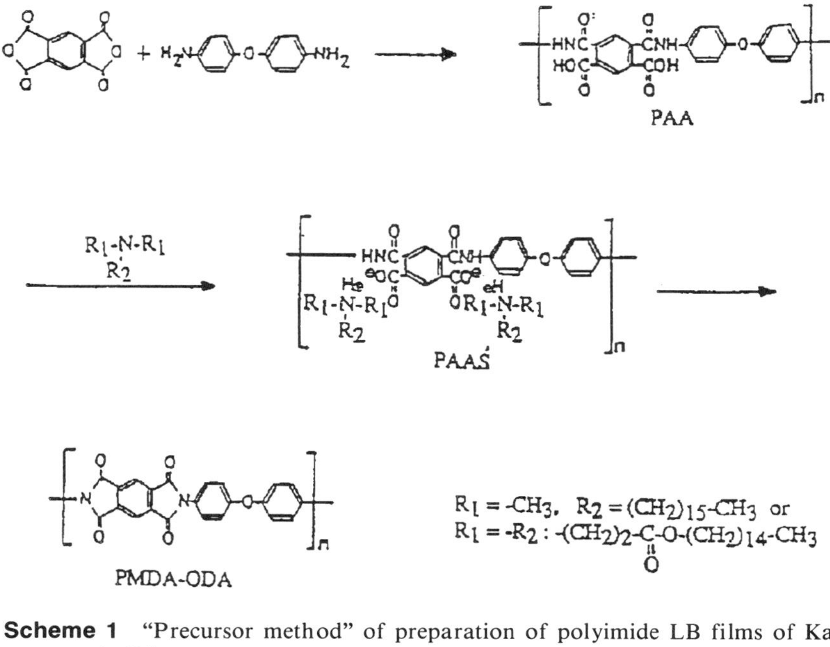

**Scheme 1** "Precursor method" of preparation of polyimide LB films of Kakimoto et al. (21).

amount of carboxylic function and amine. Then 100 μl of PAAS was spread on highly purified dionized water (Millipore™, >18 MΩ) using a microliter Hamilton syringe. After spreading, benzene was allowed to evaporate from the surface for about 10 min. The resulting monolayer was then compressed at a constant rate of 1.8 cm/min by displacing the barrier. All measurements were done at 220°C. Two kinds of long carbon chain amines were used: $N,N'$-dimethylhexadecylamine and $O,O',O''$-trihexadecanoyltriethanolamine. The use of benzene as a second solvent for preparation of the spreading solution was necessary since DMAC is soluble in water, which would make the spreading procedure quite difficult. Besides the use of the mixture of solvents, the spreading must be carefully done to avoid sinking the material to the bottom of the trough. Poly(pyromellitamic) acid (PA) and $O,O',O''$-trihexadecanoyltriethanolamine were provided by Professor Kakimoto (Tokyo Institute of Technology). PAAS solutions were prepared from a 10% stock solution of PA in DMAC (w/v). DMAC (absolute) and benzene (spectroscopic grade) were purchased from Aldrich Chemical Company and used as received. $N,N'$-dimethylhexadecylamine (technical grade) was purchased from Fukuda and distilled twice at reduced pressure. The film balance and computers are located in the Polymer Microdevice Laboratory, a large Class 100 clean room at Case Western Reserve University.

Isobaric stability is defined here as the change in average surface area with time at a constant applied surface pressure. These experiments allowed the study of the PAAS monolayer stability for both types of long carbon chain amine: $N,N'$-dimethylhexadecylamine and $O,O',O''$-trihexadecanoyltriethanolamine. Solutions of known monomer concentrations of PAAS were spread on the highly purified dionized water. After spreading, benzene was allowed to evaporate from the surface. The resulting monolayer was compressed at 1.8 cm/min until the desired surface pressure was reached. The surface pressure was kept constant during the experiment by displacing the barrier. The change in molecular area with time, after constant surface pressure was reached, was recorded.

## B. UV Spectroscopy

Because PAAS and polyimide have intense absorption in the UV range, UV spectroscopy was used to monitor the multilayer depositions by determining the relationship between the UV absorbance and the number of layers of the LB films. For these experiments, PAAS multilayer films were deposited on quartz plates using a deposition velocity of 10 mm/min and 25 mN/m of surface pressure. Quartz plates, 25 × 25 × 1.5 mm, were cleaned with "piranha" solution, that is, a mixture of concentrate sulfuric acid and hydrogen peroxide (3:1), for 30 min at 90°C, thoroughly washed with deionized water and dried with nitrogen. UV spectra were taken before and after imidization at 900 nm/min using a Varian Cary 1/3E spectrophotometer. The relationship between absorbance and number of layers was then determined.

### C. Fourier Transform Infrared Spectroscopy (FT-IR)

In order to evaluate the in-plane anisotropy of the LB films prior to and after the chemical treatment for imidization, FT-IR dichroic spectra, transmission mode, were taken and the dichroic ratios, $I_\parallel / I_\perp$, were measured. $I_\parallel$ is the absorbance measured by polarized light which is parallel to deposition direction and $I_\perp$ is that which is perpendicular to the deposition direction. The dichroic ratios were determined as a function of the deposition velocity and the surface pressure used to prepare the LB films. Substrates for transmission measurements were 13 × 25 × 2 mm ZnSe plates that had been cleaned by successive ultrasonication in chloroform, methanol, and water for 10 min each, and finally subjected to an argon plasma cleaning for 20 min. For these experiments 11 layers of the LB films were deposited on both sides of the ZnSe substrates. The infrared spectra were recorded in a Bio-Rad FTS-40 FT-IR spectrophotometer equipped with a Perkin-Elmer polarizer and an MCT detector with a resolution of 8 $cm^{-1}$. The number of scans necessary to get a good signal-to-noise ratio was 2048.

### D. FT-IR Transmission and Reflection–Absorption Spectroscopy

The molecular orientation in the LB films prior to and after imidization was also qualitatively evaluated by comparing the relative peak intensities of the FT-IR transmission spectrum with those of the reflection–absorption (RA) spectrum for the same film. When comparing the spectra, one has to be aware that in transmission only vibrations with a component of the change in the dipole moment parallel to the surface are observed, whereas in reflection only vibrations with a component normal to the surface are present in the spectrum.

Substrates for RA measurements were silicon wafers that had been cleaned with concentrated sulfuric acid and then vacuum-evaporated with a 100-nm-thick gold film. Twelve-layer LB films deposited on these gold surfaces were used for the RA measurements. The infrared spectra were recorded on a Bio-Rad FTS-40 FT-IR spectrophotometer equipped with a reflection attachment (Harrick Scientific Corp.), a polarizer, and an MCT detector with a resolution of 8 $cm^{-1}$. The incident angle used was 85° and the number of scans was 2480. For transmission measurements the same conditions and substrates for determination of dichroic ratios were used.

The spectra were obtained for samples prepared using different deposition velocities in the range of 3 to 15 mm/min and at 20, 25, and 30 mN/m of surface pressures.

## E. Optical Phase Retardation

Besides the infrared dichroism studies, optical phase retardation was also measured to evaluate the molecular anisotropy of the LB films as a function of the deposition velocity in the range of 1.8 to 20 mm/min. In our setup the optical system essentially worked as an automatic ellipsometer in which an acoustic optical modulator (AOM) was operated at 1 KHz. The polarizer and the analyzer are arranged to be in the crossed position, with the polarizer making an angle of 45° with the deposition direction of the LB film. The modulated laser light, He–Ne ($\lambda$ = 632.8 nm) with below 0.2 mW of power to avoid heating of the sample, is linearly polarized by passing through a polarizer. After transmission through the birefringent sample, the light becomes elliptically polarized, which is subsequently linearly polarized again by means of a quarter-wave plate. This linearly polarized light then reaches the analyzer that rotates to minimize the signal. It is the rotation angle of the analyzer, in degrees, that determines the optical phase retardation. Accuracy is better than 0.01°.

## F. Atomic Force Microscopy

The capabilities of atomic force microscopy (AFM) are focused on the characterization of the micrometer and molecular scale as well as defects of the surface of the LB films. In this kind of experiment, the pure force interaction between the sample surface and a cantilever tip provides the direct observation of the sample surface structure. The AFM measurements were performed with a Nanoscope III AFM system, using a pyramidal $Si_3N_4$ cantilever with a spring constant of 0.86 N/m. The scanning rate was selected to be as low 3 Hz for large scale images and 110 Hz for molecular scale scans. Typical forces were less than 10-8 N. For these experiments five-layer PAAS and polyimide LB films deposited on silicon wafers, ITO coated glass, and mica (Ted Pella, Inc.) were used.

By using AFM, besides the information regarding the surface structures of the LB films, determination of monolayer thickness was also done by analyzing stepped multilayer samples that were prepared depositing monolayers with different immersions. Typical samples were prepared using 10 mm/min of deposition velocity and 25 mN/m of surface pressure. The size of the substrates was 10 × 10 mm.

### G. Transmission Electron Microscopy and Electron Diffraction

PAAS LB films prior to and after imidization reaction were examined by transmission electron microscopy (TEM) (JEOL-JEM-100SX) and electron diffraction (ED) in order to investigate the morphology of these films. Two different ways were used to prepare the samples. In the first one, copper or gold carbon-coated grids (200 mesh; Ted Pella, Inc.) were stuck on hydrophobic glass substrates by using Formvar solution (Ted Pella, Inc.). After deposition of the multilayer LB films on these substrates and imidization reaction, the grids coated with LB films were carefully removed from the glass slides and analyzed. In the second way of sample preparation, the LB films were deposited on hydrophilic glass slides and chemically treated to yield the polyimide. Immediately after removing the plates from the imidization solution, they were put in highly purified dionized water. Due to the fact that the LB films were swollen by the imidization medium, they were released from the glass slides, floated on water, and picked up on 400 mesh TEM copper grids. For preparation of the hydrophobic substrates, glass slides previously cleaned with "piranha" solution, washed thoroughly with water, and dried with nitrogen were then kept in an atmosphere of hexamethyldisilazane (HMDS) for 24 h. The resulting plates were strongly hydrophobic due to the coating of their surfaces with hydrocarbon groups.

Calibration of the ED spacings was carried out using gold. The instrument was operated in low-dosage conditions by keeping the bias control at 4 to avoid degradation of the films.

### H. X-ray Reflectivity

The morphology and molecular orientation of the polyimide LB films were also investigated by in-plane and out-of-plane X-ray diffraction and X-ray reflectivity experiments. In-plane (gracing incidence X-ray scattering—GEKS) and out-of plane measurements were performed using synchrotron radiation on the beamline X23B at the National Synchrotron Light Source (NSLS). A Huber goniometer and Super 123 software were used for the measurements. X-ray energies of 7.76 keV and wavelength of 1.62 Å were selected. Specimens were prepared by depositing polyimide LB layers onto silicon wafer (2 × 2) cm and mica (2.5 × 2.5 cm). Samples with 11 and 23 layers were examined. Typical scans were performed from $2\theta = 1.5°$ to $45°$ and at $2\theta = 0.5°$ per minute at room temperature.

The in-plane experiments were performed with the sample in the nearly vertical position and with the deposition direction of the LB films nearly parallel to the beam direction. The incident angle for these in-plane experiments was around 0.2°. Before running the experiments, the cross-sectional area of the beam and its footprint on the sample plate were checked by using burn paper.

The thickness $D$ of the LB films was calculated based on the X-ray reflectivity curves using the following equation (25):

$$D = \pi/\Delta k_{z,o}$$

where $k_{z,o}$ is the $z$ component of the wavevector (normal to a surface) and is given by

$$k_{z,o} = (2\pi/\lambda) \sin \theta$$

$\Delta_{kz,o}$ is the separation distance between two minima in the reflectivity curve.

## I. Alignment Properties of LB Films

The liquid crystal cells were made using two identical ITO-coated glass substrates (25 × 30 × 1 mm) coated with polyimide (or PAAS) LB films and assembled with the deposition direction antiparallel to each other. Thick glass (7 mm) substrates were used to assemble cells for pretilt angle and polar anchoring energy measurements to avoid distortions in the LC cells. Polyester cells were determined by optical interference. The cells were then vacuum-filled with a nematic liquid crystal, pentyl-cyano-biphenyl (5CB), heated to 60°C and then slowly cooled down to get rid of the flow-induced alignment effect.

The orientation behavior of the liquid crystal in the assemble cells was observed under a polarizing microscope.

5CB, for which the nematic-isotropic transition is 35.3°C, was purchased from BDH Chemicals and used without further purification.

## J. Measurements of Polar Anchoring Energy and Pretilt Angle

A polarizing microscopy was used to check the LC alignment provided by the assembled cells. The crystal rotation method was used to determine the pretilt angle. Measurements were performed at 25°C for cells with 3, 5, 7, 9 and 11 layers of polyimide. Samples prepared using five-layer rubbed polyimide LB film were also analyzed.

Polar anchoring energy was measured using Yokoyama and Van Sprang's high field method. The setup used for simultaneous measurements of retardation and capacitance of the cells as a function of the applied voltage and temperature is shown in Figure 2. The range of temperature studied was from 28°C to 36°C. The sinusoidal voltage (1300 Hz) applied across the cell was varied from 0 to 35 V. The temperature-control ability was 10 mK. The rate of voltage scan for determination of Vth was set ranged from 1 mV/min near Vth to 5 mV/min near the maximum voltage.

## III. RESULTS AND DISCUSSION

### A. Isotherms

Isotherms for PAAS at 22°C are shown in Figure 2. Curve A was obtained using $N,N'$-dimethylhexadecylamine while curve B was obtained using $O,O',O''$-trihexadecanoyltriethanolamine. In both cases several cycles of compression and expansion of the monolayer at the air-water interface showed good reproducibility within 0.5 nm$^2$/molecule, implying good stability of the monolayers. The limiting area for isotherm A was found to be 1.30 ± 0.03 nm$^2$/molecule and 1.27 ± 0.02 nm$^2$/molecule for isotherm B. The similarity between these values and 1.28 nm$^2$/molecule calculated from molecular model (17) suggests that, in both experiments, the imide rings are

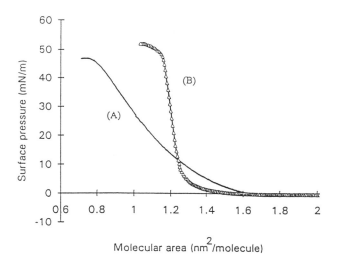

**Figure 2** Surface pressure-area isotherms of PAAS monolayers at 22°: (A) with $N,N'$-dimethylhexadecylamine and (B) with $O,O',O''$-trihexadecanoyltriethanolamine.

lying flat on the air–water interface. The isotherm A is characterized by a slow raising of the surface pressure because of the poor molecular packing of the polymer side carbon chains. This is due to the difference between the cross section area of the alkyl chains and the area per monomer (1.28 nm$^2$), which seems to be too large for the single-chain tert-amines to align vertically at the air–water interface. Isotherm B is characterized by a steep rise in surface pressure up to 47 mN/m. In this case, the packing of the side chains is more favorable since there are six side chain alkyl groups per monomer unit, which provides a better ratio between areas of the cross section of the alkyl groups and the monomer. Although no difference in the collapse pressure was observed between the two materials, both being 47 mN/m, the surface modulus (26) calculated from the slope of the linear portion of the pressure-area isotherm was 93.7 × 106 (N/m)/m$^2$ for isotherm A and 441.7 × 106 (N/m)/m$^2$ for isotherm B, which reveals higher rigidity of the material B due to the better molecular packing of the side chains. However, the surface pressure started raising at 1.6 nm$^2$ in case A and at 1.25 nm$^2$ in case B. This is probably due to the fact that the side chains in material A are highly tilted (see FT-IR results) forcing the side chains to occupy a larger area than the side chains in material B.

Atomic force microscopy analysis showed that the monolayer thickness was 0.85 nm and 2.02 nm for material A and B, respectively. When these results are compared with the calculated lengths of the side chains, i.e., around 1.90 nm and 2.20 nm for material A and B, respectively, it can be said that for material B the calculated value is much closer to the observed value than for material A, which confirms that the side chains in material B tend to be perpendicular to the water surface.

### B. Isobaric Stability

Figure 3 shows the isobaric tests for PAAS at 25 mN/m of surface pressure and 22°C. Curve A was obtained using $N,N''$-dimethylhexadecylamine and curve B using $O,O',O''$-trihexadecanoyltriethanolamine. Both monolayers showed good stability: a contraction in molecular area of less than 13% for material A and less than 3% for material B for a period of 6 h. Since the side chain groups in both materials are chemically very similar, the better stability of material B is because of its better packing.

### C. Imidization Reaction

The effectiveness of the chemical treatment for imidization of the PAAS LB films was confirmed by infrared spectroscopy. IR spectra of a 160-layer PAAS ($N,N'$-dimethylhexadecylamine) LB film in the transmission mode

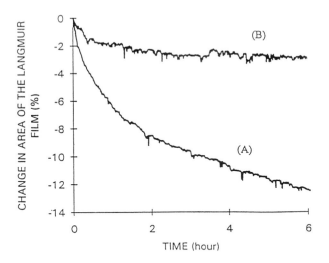

**Figure 3** Change in molecular area versus time for PAAS monolayers at 25 mN/m: (A) with $N,N'$-dimethylhexadecylamine and (B) with $O,O',O''$-trihexadecanoyl-triethanolamine.

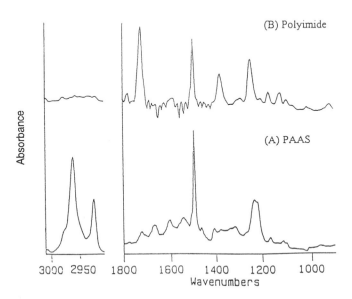

**Figure 4** Infrared transmission spectra of a 160-layer LB film on ZnSe: (A) before chemical treatment and (B) after chemical treatment.

before and after chemical treatment is shown in Figure 4. The structure of the PAAS LB film prior to the chemical treatment is characterized basically by absorption at 2924 and 2853 cm$^{-1}$ due to the aliphatic C—H stretching of the *n*-hexadecyl chains, 1670 cm$^{-1}$ due to the antisymmetric stretching vibrations of the carbonyl groups, 1500 cm$^{-1}$ due to the aromatic ring breathing mode of the phenyl rings, and 1227 cm$^{-1}$ due to antisymmetric stretching vibrations of the C—O—C groups (27). After treatment, bands at 2924 cm$^{-1}$, 2853 cm$^{-1}$, and 1670 cm$^{-1}$ disappeared, giving rise to absorption bands at 1780 and 1725 cm$^{-1}$ due to the asymmetric and symmetric stretching vibrations of the imide carbonyl, respectively, and 1378 cm$^{-1}$ due to symmetric stretching vibrations of the C—N groups (27,28). The intensities of these absorption bands did not change with time after 12 h of chemical treatment. These results strongly suggest that the cyclodehydration reaction to afford the polyimide preceded with the removal of the *N,N'*-dimethylhexadecylamine.

## D. UV Spectroscopy

Figure 5 shows the UV spectra for PAAS before and after chemical treatment. The PAAS UV spectrum before imidization is characterized by two absorption bands at 258 nm and at 300 nm. Following imidization, an absorption band is observed at 290 nm, possibly due to the transition dipole of the biphenyl groups (29). The difference in absorbance observed below 250 nm is probably due to the conversion of the amide groups to polyimide (30). The absorption maxima at 258 nm and 290 nm for the PAAS LB films prior to and after imidization, respectively, followed Beer's law at multilayer thickness in the range studied of 10 to 50 layers. Figures 6 and 7 show the plot of UV absorbance versus number of LB layers for both cases. Linear plots indicate that the deposition process is reproducible and they confirm that one monolayer was deposited on each side of the quartz substrate during each emersion or immersion. Besides that, these linear plots confirm that the multilayers were not peeled off after imidization.

For these samples as well as for all others mentioned in this work, the transfer of the monolayers took place on each downward and upward dip (Y-type), except for the first layer (Z-type) in case of hydrophilic substrates. The deposition ratio was around 90% for downward dip and around 100% for upward dip up to 10 strokes using the same spreading, when the deposition speed was 10 mm/min. However the deposition ratio degraded after the 10th layer, indicating the onset of irregular deposition. This is due to the development of inhomogeneous film surface as the number of strokes increases. The procedure then adopted was to use the same spreading for depositing no more than 10 layers.

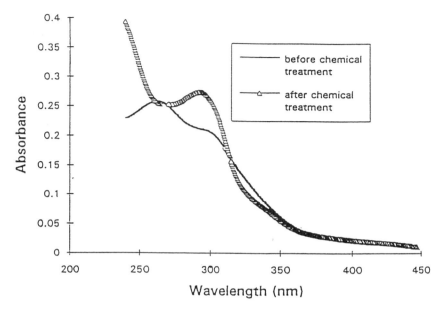

**Figure 5** UV spectra of a 50-layer LB film on quartz before and after chemical treatment.

## E. Infrared Dichroism

Figure 8 shows the IR dichroic spectra for an 11-layer PAAS LB film ($N,N'$-dimethylhexadecylamine) before the imidization reaction. The dichroic ratio ($D$) $I_\parallel/I_\perp$ calculated for the 1500 cm$^{-1}$ absorption band (aromatic ring breathing mode of the phenyl rings) was equal to 1.2. Because the structure of the PAA as well as the PMDA-ODA has a zig-zag conformation with bends at the C—O—C bond making an angle of 1200 (17,31,32), it can be said that the transition moment at 1500 cm$^{-1}$ has a tendency to be parallel to the chain axis. Thus, based on the calculated dichroic ratio we can conclude that the polymer chains are aligned along the deposition direction. On the other hand, the local molecular order in these films seems to be low since the other absorption bands relative to the backbone of the polymer chain were similar in intensity. This is possibly due to the presence of the tilted carbon side chains and to the fact that the PAA chain comprises a mixture of meta and para isomers distributed randomly along the chain that can originate structural irregularity (32). However, the overall molecular orientation along the deposition direction

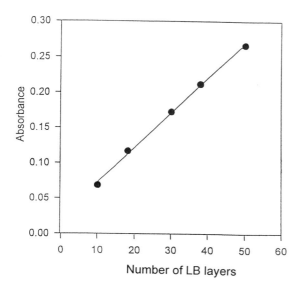

**Figure 6** UV absorption at 258 nm versus number of PAAS LB layers on quartz. The correlation coefficient is 0.996.

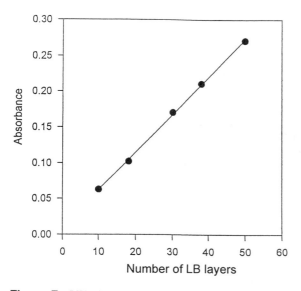

**Figure 7** UV absorption at 290 nm versus number of polyimide LB layers. The correlation coefficient is 0.999.

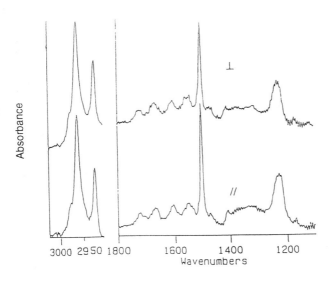

**Figure 8** Infrared dichroic spectra of an 11-layer PAAS LB film deposited on both sides of a ZnSe substrate. The parallel direction ($\parallel$) coincides with the deposition direction. The deposition speed is 10 mm/min at 25 mN/m.

in the polyimide's precursor LB films was confirmed by their capability to align nematic liquid crystals.

Figure 9 shows the IR dichroic spectra for an 11-layer polyimide LB film prepared using 10 mm/min of deposition velocity. Calculated values of $D$ for the 1725, 1500, and 1380 cm$^{-1}$ absorption bands as a function of deposition velocity are shown in Table 1. As we expected, the dichroic ratio for the imide carbonyl (1725 cm$^{-1}$) behaves in a different way than those of the phenyl rings at 1500 cm$^{-1}$ and C—N bonds at 1380 cm$^{-1}$, since the transition moment at 1725 cm$^{-1}$ tends to be perpendicular to the chain axis. These results show that the polymer chains in the polyimide LB films are clearly aligned along the deposition direction and that there are no significant differences of $D$ for the several absorption bands when different deposition speeds are compared. Thus, $D$ is not dependent upon the deposition speed in the range studied.

It was also verified by IR that samples prepared spreading some drops of PAAS solution on ZnSe, and subsequently dried and imidized, did not show any anisotropy. It indicates that the deposition process is responsible for the molecular orientation observed in the LB films.

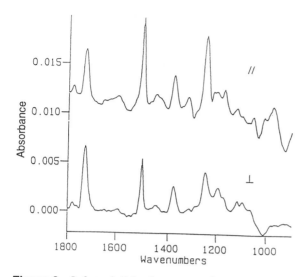

**Figure 9** Infrared dichroic spectra of an 11-layer polyimide LB film deposited on both sides of a ZnSe substrate. The parallel direction coincides with the deposition direction. The deposition speed is 10 mm/min at 25 mN/m.

## F. Molecular Anisotropy by FT-IR Transmission and RA Spectroscopy

Figure 10 shows the IR transmission and RA spectra of PAAS LB films before imidization. The antisymmetric and symmetric $CH_2$ stretching bands at 2924 and 2853 $cm^{-1}$, respectively, have significant intensities in both spectra, which means that the side carbon chains are tilted relative to the plane of the substrate. The phenyl ring breathing mode at 1500 $cm^{-1}$ gave a relatively stronger intensity in the transmission spectra than in the RA spectra, which indicates that the phenyl rings tend to be parallel to the

**Table 1** IR Dichroic Ratios of 11-Layer Polyimide LB Films

| Deposition speed (mm/min) | Imide-carbonyl (1725 $cm^{-1}$) | Phenyl rings (1500 $cm^{-1}$) | C-N bonds (1380 $cm^{-1}$) |
|---|---|---|---|
| 3 | 0.80 | 1.77 | 1.23 |
| 10 | 0.77 | 1.50 | 1.16 |
| 15 | 0.83 | 1.56 | 1.09 |

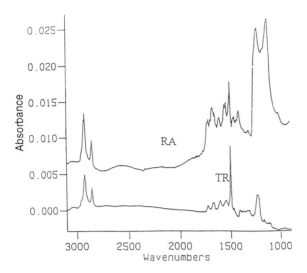

**Figure 10** Infrared transmission (TR) and reflection-absorption (RA) spectra of PAAS $N,N'$-dimethylhexadecylamine LB films. Substrates for transmission and RA measurements are ZnSe and Au, respectively.

sample surface. This observation agrees with molecular models reported in literature (17).

Figure 11 shows the IR transmission and RA spectra for polyimide LB films. The phenyl ring breathing mode at 1500 cm$^{-1}$ and the symmetric stretching vibrations of the C—N groups gave relatively stronger absorption in the transmission spectra than in the RA spectra, which strongly suggests that the transition moments relative to these bonds are parallel to the sample surface. The parallel position of the aromatic rings relative to the sample surface was also indicated by the absence of the absorption peak at 1595 cm$^{-1}$ (aromatic C=C stretch) in the RA spectra that is present in the transmission spectra. On the other hand, the asymmetric and symmetric stretching vibrations of the imide carbonyl 1780 and 1725 cm$^{-1}$, respectively, gave significant intensities in both spectra, which suggests that the imide rings, which are coplanar with the carbonyl groups, are tilted relative to the sample plane.

Since the intensities of the peaks relative to the carbonyl groups of the PAAS were also significant in the transmission and RA spectra (Figure 10), it is suggested that the imidization reaction occurred without requiring a substantial rearrangement of the spatial configuration of the PAAS chains. No reasonable explanation was found for the broad and intense peaks centered at 1200 cm$^{-1}$ observed in the RA spectra of the PAAS and polyimide.

Figure 12 shows the IR transmission and RA spectra for PAAS ($O,O',O''$-trihexadecanoyltriethanolamine) before chemical treatment for imidization. Two important features were observed in these two spectra. First, the antisymmetric and symmetric $CH_2$ stretching modes at 2924 and 2853 cm$^{-1}$, respectively, are dominant in the transmission spectra, whereas in the RA spectra these bands are as weak as the asymmetric and symmetric stretching modes of the C—H bonds of the terminal methyl groups at 2964 and 2875 cm$^{-1}$, even though the concentration of $CH_2$ groups is much higher than the $CH_3$ groups. This indicates that the carbon side chains are perpendicular to the plane of the sample. Second, progression bands are in the region 1350 to 1180 cm$^{-1}$ in the RA spectra, which indicate a high packing order in the carbon side chains.

No significant differences in these results were observed when samples prepared at 20 and 30 mN/m of surface pressure were examined.

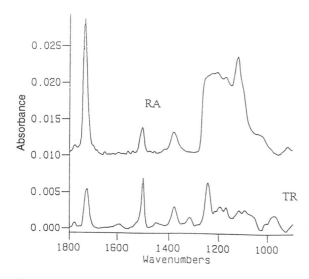

**Figure 11** Infrared transmission (TR) and reflection–adsorption (RA) spectra of polyimide LB films. Substrates for transmission and RA measurements are ZnSe and Au, respectively.

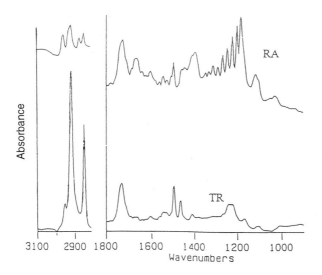

**Figure 12** Infrared transmission (TR) and reflection–absorption (RA) spectra of PAAS O,O′,O″-trihexadecanoyltriethanolamine LB films. Substrates for transmission and RA measurements are ZnSe and Au, respectively.

## G. Optical Phase Retardation

The optical phase retardation of a 10-layer polyimide LB film as a function of the deposition speed is shown in Figure 13. From these results, it can be said that the molecules are oriented and the molecular orientation is along the deposition direction. These results also agree with the ones obtained by IR dichroism, since the retardation did not depend upon the deposition velocity in the range studied. It must be pointed out that the main advantage of using this technique is the fact that the samples for retardation measurements are deposited on glass, which are the substrates used in real applications. This eliminates the possibility of errors due to interactions between substrate and film.

## H. Atomic Force Microscopy

From polyimide LB films deposited on ITO-coated glass, no molecular structures were observed due to the roughness of the substrates. However, molecularly resolved images were obtained when five-layer polyimide LB films deposited on mica were used (Figure 14). The distance between chain centers were determined to be 0.63 nm and the thickness of a monolayer

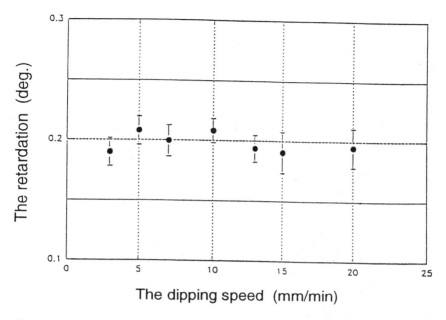

**Figure 13** Phase optical retardation versus deposition speed for 5-layer polyimide LB films deposited on both sides of glass substrates at 25 mN/m.

0.45 nm. Due to similarity of these values to the ones given in the literature for the unit cell of this polyimide (31), it can be suggested that this picture actually shows the molecular structure of the PI film. The image was stable, and rotating the scan direction did not affect the configuration of the polyimide chains, ruling out the possibility that the scanning influenced the orientation of the chains or caused some similar imaging artifact. The chains showed a zig-zag structure and the direction of alignment was always coincident with the deposition direction, which strongly suggests that the orientation of the polyimide chains is produced by the vertical deposition.

## I. Transmission Electron Microscopy and Electron Diffraction

As described in the experimental section, PAAS LB films were deposited on carbon-coated copper grids and then chemically imidized. It was found, however, that copper was oxidized by the acetic anhydride during the chemical treatment forming crystals of copper acetate. *d*-Spacings calculated from electron diffraction patterns obtained from these crystals confirmed

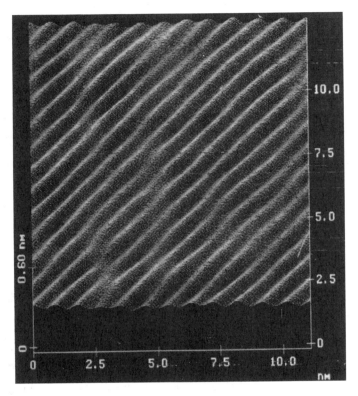

**Figure 14** AFM image of a 5-layer polyimide LB film surface deposited on mica. The deposition direction is from the bottom left to the top right.

this observation. Figure 15 shows a transmission electron micrograph of the copper acetate crystals. It was noticed that the longest dimension of these crystals was always aligned along the deposition direction. Based on this observation it again was confirmed that the LB films had the molecules aligned along the deposition direction and, as a result, they could act as substrates for epitaxial crystallization of the copper acetate.

In order to prevent formation of any extraneous byproduct in the imidization medium, the copper grids were substituted by carbon-coated gold grids. Samples prepared using a striping technique were also examined. No discrete reflections were observed in the electron diffraction analysis other than diffuse rings, which suggests that the polyimide LB film is essentially amorphous. The same result was obtained with samples of PAAS ($N,N'$-dimethylhexadecylamine) LB film. However, 30-layer ($-600$ Å)

# Nematic Liquid Crystal Displays

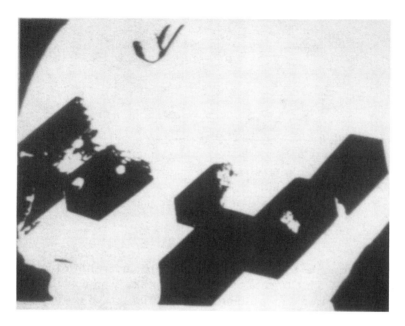

**Figure 15** Transmission electron micrograph of copper acetate crystals on polyimide LB film. The deposition direction of the LB layers is from the bottom left to the top right.

PAAS ($O,O',O''$-trihexadecanoyltriethanolamine) LB films gave $d$-spacings for the observed rings of 4.0, 2.4, and 2.1 Å. The most intense line, 4.0 Å, is probably related to the distance between the well-ordered amine carbon chains that are packing perpendicular to the polymer backbone. These results agree with the X-ray diffraction data reported by Kakimoto and coworkers (33).

## J. X-ray Reflectivity

A typical X-ray reflectivity curve for a 23-layer polyimide LB film is shown in Figure 16. No diffraction peaks were observed in the range of $2\theta = 2°$ to $60°$. These results imply that the films are probably essentially amorphous or the crystallinity is low with very small crystals, which agrees with the results from ED.

The in-plane analysis did not show any diffraction peak either, which could imply the absence of in-plane molecular order in the polyimide LB films. However, this result does not agree with the IR, AFM, or optical

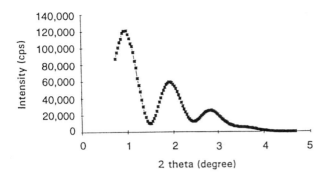

**Figure 16** X-ray reflectivity curve of a 23-layer polyimide LB film deposited on mica.

phase retardation results or with the alignment capabilities showed by the polyimide LB films. This absence of diffraction peaks is probably due to difficulties that involved the operation of the instruments to get in-plane data. For instance, for the range of incident angles used, 0.11 to 0.3°, it would be necessary to have a perfect initial leveling of the sample, which was not done since no instruments were available for such a procedure. Besides that, the very thinness of the sample used (around 100 Å) probably made the data collection more difficult, even though synchrotron radiation was used.

Although a 23-layer film was used, only three Fresnel fringes were observed in Figure 16. The reason may be the flatness of the LB layers due to the absence of lateral groups, which prevent the sharpness between interfaces when they are stacked one on another. From the fringes, the monolayer thickness was determined to be 0.46 nm, which is in good agreement with the value determined by AFM.

## K. Alignment Properties of the LB Films

### 1. LC Alignment on LB Films

When observed under the polarizing microscope, antiparallel nematic liquid crystal cells having polyimide LB films as alignment layers showed a completely dark picture when the deposition direction was parallel to the polarization axis of the polarizer (or the analyzer), whereas a bright picture was observed when the cell was rotated an angle lower than 90°. Based on these observations, it was concluded that polyimide LB films were able to align the liquid crystals homogeneously with liquid crystal director oriented along the deposition direction. It was verified that at least three layers of polyimide LB films were necessary to obtain a uniform alignment, although even one layer was able to align very large domains of liquid crystals. The reason may be that three layers are needed to totally screen the interaction between the liquid crystal molecules and the isotropic ITO-coated glass substrate and also to smooth the roughness and fully cover the ITO surface. All LC cells with the number of polyimide LB layers varying from 3 to 15 gave very uniform homogeneous alignment.

Similar results were observed when PAAS LB films were used as alignment layers, which indicates that the LB films of the precursor are already oriented. On the other hand, when an antiparallel LC cell was assembled using two identical ITO-coated glass plates on which six layers of polyimide LB films were deposited in such a way that the deposition direction of the sixth layers made an angle of 45° with the deposition direction of the previous five layers, it was found that the alignment direction was along the deposition direction of the uppermost LB layer. This means that the interaction distance between the alignment layer and the LC molecules that is responsible for the LC alignment is rather short, that is, approximately one LB layer (0.46 nm).

## 2. LC Alignment Mechanism

At the present time there are two models described in the literature to explain the LC alignment mechanisms on rubbed films. One is the groove model, which says that the LC alignment is induced by the grooves generated on the film surface by the rubbing process (34). The other is referred to as orientational epitaxy (OE). The meaning that is ascribed to OE is simply that the LC director will prefer to align uniformly along a direction established by the molecular orientation of the alignment films, as a result of the interactions between the LC molecules and the oriented films at all length scales down to the molecular level (13,16). In order to better understand the LC alignment mechanism, polyimide LB layers were deposited on spin-coated polyimide rubbed films with the deposition direction perpendicular to the rubbing direction. Antiparallel LC cells were then assembled using these plates as alignment layers and, in order to probe the direction of the LC alignment, the cells were filled with a mixture of 5CB and a blue dye (guest-host cell). In this way, the LC alignment direction was easily determined by observing the transmitted light through the system cell-polarizer. A bright blue color indicated that the LC alignment direction was coincident with the polarization axis of the polarizer. Figure 17 shows a picture of the guest-host LC cells on a polarizer sheet where different num-

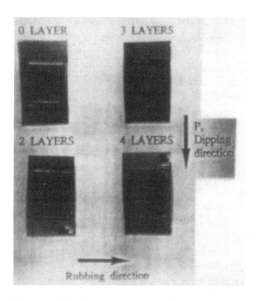

**Figure 17** Guest-host nematic LC cells. The number of polyimide LB layers deposited on the spin-coated rubbed polyimide films and the direction of polarization (P) are indicated.

bers of LB layers were deposited on the rubbed polyimide films. By examining this picture it was concluded that just three LB layers ($=13$ Å) are necessary to screen out the interaction between the LC molecules and switch the LC alignment direction. Figure 18 shows the topography of the polyimide rubbed films before the deposition of the LB layers. The depth of the grooves was found to be around 80 A. However, the groove structure still exists after deposition of the LB layers, as shown in Figure 19. These results suggest that the grooves have no effect on the LC alignment and that the mechanism of alignment for both polyimide LB layers and spin-coated rubbed polyimide films is essentially the same: the interaction between the LC and the oriented polymers align the LC in a preferential direction.

## 3. Pretilt Angle Measurements

The pretilt angles measured using LC cells with different number layers of polyimide LB film as alignment layers are displayed in Figure 20. These results show that the pretilt angle is very small or zero and does not depend upon the number of LB layers in the range studied. When an electric field of 6 V was applied, tilt domains were observed under the polarizing micro-

# Nematic Liquid Crystal Displays

**Figure 18** AFM image of a spin-coated rubbed polyimide film. The direction of the grooves formed due to the rubbing process correspond to the rubbing direction.

scope in all of these cells (Figure 21), which confirmed the pretilt angle measurements.

In order to evaluate the rubbing effect on the pretilt angle given by polyimide LB films, pretilt angle measurements were performed on an LC cell in which rubbed five-layer polyimide LB films were used as alignment layers. The results are shown in Figure 22. Clearly, the symmetric point ($\Psi_x$) was shifted from zero after rubbing, giving a pretilt angle of 2°. However, no grooves were observed in these rubbed LB films, which would be difficult given the thickness of the LB layers. This observation confirms what was suggested previously, that is, the grooves have no effect on the LC alignment mechanism.

It has been claimed that the generation of pretilt angles on polyimide alignment films can be controlled by the degree of branching of the polymer chains and that the pretilt may be attributed to steric interaction of the side chains with the LC molecules (22,35,36). However, it has been shown in the literature (37) that the topography of the rubbed layer can play an important role as far as the generation of pretilt angle is concerned, which seems to explain pretilt angles given by rubbed polyimide films with flat molecular structures like PMDA-ODA. Based on this, we suggested that the polyimide

**Figure 19** AFM image of a three-layer polyimide LB film deposited on spin-coated rubbed polyimide perpendicular to the rubbing direction. Note that the grooves are still present.

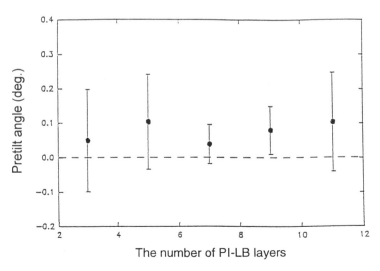

**Figure 20** Pretilt angle versus number of polyimide LB layers used for LC alignment.

# Nematic Liquid Crystal Displays

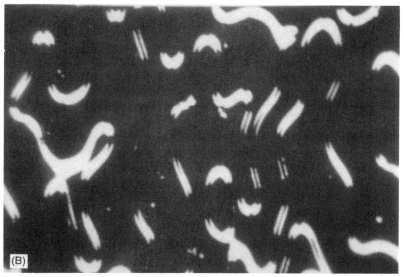

**Figure 21** (A,B) Micrographs of an LC cell under the polarizing microscope. Tilt domains appear when an electric field is applied (B). Five-layer polyimide LB films are used for LC alignment.

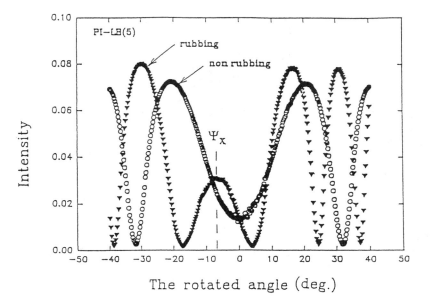

**Figure 22** Angular dependence of transmitted intensity for nematic LC cells with rubbed and nonrubbed polyimide LB films as alignment layers. The symmetric point $\Psi_x$ is shifted from zero after rubbing, giving a pretilt angle of 2°.

LB films were unable too provide pretilt angles due to the fact that the polymer backbones tend to be parallel to the surface substrate, as was shown by the IR and AFM results.

### L. Polar Anchoring Energy Measurements

Figure 23 shows an example of the plot $R/R_o$ versus $1/CV$ on a cell with five-layer polyimide LB film as alignment layers at two different temperatures. From these plots the polar anchoring energies and the extrapolation lengths were determined.

Figure 24 shows the temperature dependence of the polar anchoring energy and extrapolation length for 5CB aligned on five-layer polyimide LB film. The polar anchoring energies were shown to decrease with temperature and rapidly reach very low values as the *NI* (nematic–isotropic) transition point is approached. Opposite behavior was observed with extrapolation lengths. These results indicate that as the temperature increases, the anisotropy of the LC layer decreases, and consequently the interaction between the alignment films and the LC layers decreases.

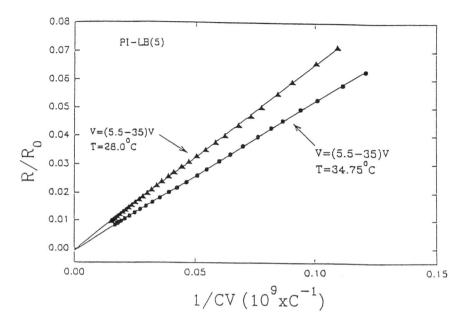

**Figure 23** A typical plot of $R/R_o$ versus $1/CV$ for an LC cell with five-layer polyimide LB film as alignment layers at two temperatures.

Polar anchoring energy and extrapolation length measurements were also performed in cells assembled using five-layer rubbed polyimide LB film as alignment layers. The results are shown in Figure 25, and they were found to be very similar to those from Figure 24. Although the rubbing strength used was very weak to avoid removing the film from the substrate, it was enough to generate a pretilt angle, but is seems that the rubbing process did not enhance the order of the LB film. This suggests that the LB films were already well oriented.

## M. Obtaining Pretilt Angle with Polyimide LB Films

When the angle of evaporation of $SiO_x$ was 60°, a sawtooth structure formed due to the shadowing effect, and the direction of alignment of the LC molecules was along (parallel to) the sawtooth structure. After depositing the LB layers across the sawtooth structure, it was found that the LC molecules still align homogeneously, and the direction of alignment was switched to the LB deposition direction. Three layers of polyimide LB

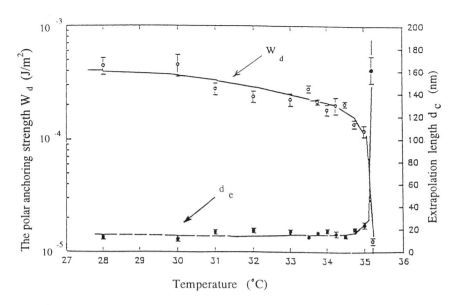

**Figure 24** Temperature dependence of the polar anchoring energy and extrapolation length for an LC cell having five-layer polyimide LB films as alignment layers.

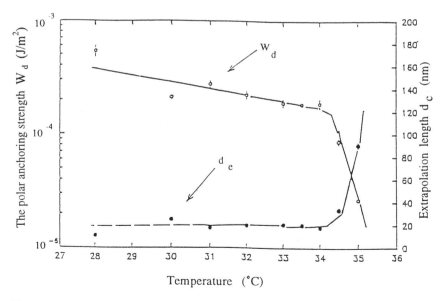

**Figure 25** Temperature dependence of the polar anchoring energy and extrapolation length for an LC cell having rubbed five-layer polyimide LB films as alignment layers.

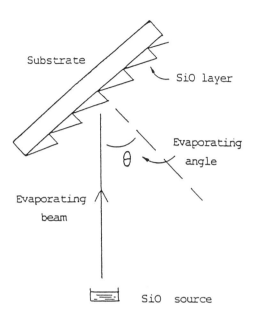

**Figure 26** Diagram of the system used for deposition of $SiO_x$. $\Theta$ is the angle between the plate normal and the evaporation beam.

films were enough to switch the alignment direction. In the case of an 85° evaporation angle, it is claimed that a needle-like structure pointing toward the $SiO_x$ source forms and the LC molecules align homogeneously along the direction of the needles with a large pretilt angle. In this case the LB deposition direction was along the direction of the needles (deposition direction facing the needles). The diagram shown in Figure 26 illustrates the preparation of the $SiO_x$ layers. Table 2 shows the variation of the pretilt angle as a function of the evaporation angle and the number of LB layers.

Based on these results, a model of the LC alignment is shown in

**Table 2** Pretilt Angle as a Function of the Evaporation Angle and Number of Polyimide LB Layers

| Number of LB layers | Evaporation angle 60° | Evaporation angle 85° |
|---|---|---|
| 3 | 0.61° | 3.1° |
| 5 | 0° | 0.7° |

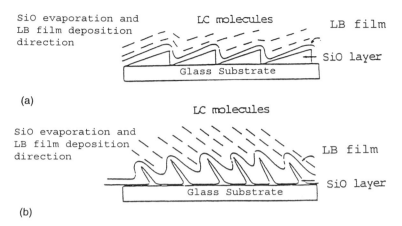

**Figure 27** Model of alignment of LC molecules on polyimide LB films deposited on SiO$_x$ layers: (a) for 60° evaporation angle, (b) for 85° evaporation angle.

Figure 27. In both cases, three layers of polyimide LB films provided a high enough pretilt angle to prevent the formation of tilt domains when an electric field was applied and to align the LC molecules along the deposition direction. However, in the 600 case (Figure 27), when the number of the LB layers increased to five, tilt domains started appearing, suggesting that the sawtooth structure was smearing out. In the 850 case, the fact that the length of the needles may be larger than the molecular length of the LC molecules could be responsible for the high pretilt angles. After depositing the LB layers, the surface could be smoothed but still keep the needle structure to a lesser extent, even with LB layers (Figure 27).

These results, allied to the fact that the pretilt angle was zero when the LB layers were deposited parallel to the sawtooth structure in the case of a 60° evaporation angle, suggest that the topography of the alignment surface plays a very important role as far as the generation of pretilt angle is concerned.

## IV. CONCLUSIONS

It was shown that the LB technique is suitable for preparation of oriented polyimide ultrathin films that are capable of homogeneously aligning nematic liquid crystals (LC) with no need for rubbing.

The PAAS Langmuir monolayers were found to be essentially homogeneous, and the stability and surface modulus was higher for PAAS when a multichain tertamine was used mainly due to the better packing of the methylene side chains. A limiting area of 130 nm$^2$/molecule for both PAAS studied indicated that the imide rings are lying flat on the water surface.

The polyimide chains in the LB films were found to be oriented along the deposition direction with the imide rings tilted relative to the plane of the sample. However, the degree of anisotropy did not depend upon the deposition velocity or surface pressure within the range studied. The local molecular orientation in the PAAS LB films appeared to be lower than in the polyimide LB films. The polyimide LB films were found to be essentially amorphous.

The LC alignment direction always followed the direction of the vertical deposition of the LB films. It was found that at least three layers of polyimide LB film were necessary to obtain a uniform alignment. However, the LC alignment direction was solely determined by the deposition direction of the uppermost polyimide LB layer, which showed that the interaction distance between the alignment film and the LC molecules responsible for the LC alignment is rather short, that is, approximately one LB layer (0.46 nm). It was found, however, that polyimide LB films were not able to generate a pretilt angle on untreated surfaces.

Three layers of polyimide LB films deposited on spin-coated rubbed films perpendicular to the rubbing direction were enough to switch the LC alignment to the deposition direction, even though the grooves were still present. This result led to the conclusion that the LC alignment mechanism is due to the interaction between the LC molecules and the oriented polymer and the grooves have no effect on the LC alignment.

The polar anchoring energy provided by polyimide LB films was found to decrease with temperature. Similar results were obtained with rubbed polyimide LB films.

Measurements of pretilt angles generated by polyimide LB films deposited on silicon oxide showed that the topography of the alignment layer plays an important role as far as the generation of pretilt angles are concerned.

## REFERENCES

1. B Bahadur. Liquid Crystal—Application and Uses. Vol. 1. Singapore: World Scientific Publishing Co., 1990.
2. E Kaneko. Liquid Crystal TV Displays—Principles and Applications of Liquid Crystal Displays. Tokyo:KTK Scientific Publishers, 1987.
3. W Urbach, M Boix, E Guyon. Appl Phys Lett 25:479, 1974.
4. H Aoyma, Y Yamazahi, M Matsuura, S Kobayashi. Mol Cryst Liq Cryst Lett 72:127, 1981.

5. ES Lee, P Vetter. Jpn J Appl Phys V32:L1436-L1438, 1993.
6. M Barmentlo, NAJM Van Aerle, RWJ Hollering, JPM Damen. J Appl Phys 71, 1992.
7. MY Han, P Vetter, T Uchida. Jpn J Appl Phys V32:L1242-L1244, 1993.
8. M Barmentlo, RWJ Hollering, J Van Aerle. NAJM Liquid Crystals 14:475-481, 1993.
9. MB Feller, W Chen, YR Shen. Phys Rev A, 43:6778-6792, 1990.
10. NAJM Van Aerle, M Barmentlo, RWJ Hollering. J Appl Phys 74, 1993.
11. T Seki, M Sakuragi, Y Kawanishi, T Suzuki, T Tamaki. Thin Solid Films 210/211:836-838, 1992.
12. S Baka, A Seki, J Seto. Thin Solid Films 180:263-270, 1989.
13. JY Fang, ZH Lu, AW Ming, ZM Ai, Y Wei. Phys Rev A 46:16, 1992.
14. M Murate, M Uekita, Y Nakajima, K Saitoh. Jpn J Appl Phys 3 (pt 2, no. 5A), 1993.
15. H Ikeno, A Oh-Saki, M Nitta, K Nakaya. Sid Digest 45-48, 1988.
16. M Murata, H Anaji, M Isurugi. Jpn J Appl Phys 31:L189-L192, 1992.
17. I Fujiara, C Ishimoto, J Seto. J Vac Sci Technol 89, 1991.
18. K Hiltrop, H Stegemeyer. Mole Cryst Liq Cryst 49:61, 1978.
19. FC Saunders, J Staronlynsca, GW Smith. Mol Cryst Liq Crystal 122:297, 1985.
20. H Idunose, M Seizuld, T Goto. Mol Cryst Liq Cryst 203:25, 1991.
21. M Suzuki, M Kakimoto, T Konishi. Chem Lett 14:395-398, 1986.
22. K Ogawa, N Mino, K Nakajima. Jpn J Appl Phys 29:L1689-L1692, 1990.
23. E Gattiglia, TP Russel. J Polymer Sci B Polymer Phys 27:2131-2144, 1989.
24. TP Russel. J Polymer Sci Polymer Phys 22:1105-1117, 1984.
25. TP Russel. Mater Sci Rep Rev J, 171-271, 1990.
26. K O'Brien, CE Rogers, JB Lando. Thin Solid Films 102:131-140, 1983.
27. H Ishida, ST Wellinghoff, E Baer, JL Koenig. Macromolecules 13:826-834, 1980.
28. AIC Saini, CM Carlin, HH Patterson. J Polymer Sci A Polymer Chem 30:419-427, 1992.
29. FIL Schoch, WA Sei, MG Burke. Langmuir 9:278-283, 1993.
30. S Ito, K Kanno, S Ohmori. Macromolecules 24:659-665, 1991.
31. TP Russel, MF Toney. Macromolecules 26:2847-2859, 1993.
32. N Takahasshi, R Yoon, W Parrish. Macromolecules 17:2583-2588, 1984.
33. K Hirano, M Sato, H Fukuda, M Kakimoto, Y Imai. Langmuir 8, 1992.
34. D Berreman. Phys Rev Lett 28:1683, 1972.
35. H Fukuro, S Kobayashi. Mol Cryst Liq Cryst 163:157-162, 1988.
36. M Murata, M Vekita, Y Nakajima, K Saito. Mol Cryst Liq Cryst 237:111, 1993.
37. H Hatoh, T Yamamoto; Y Morizumi. Appl Phys Lett 64, 1994.

# 27
# Dye-Sensitized Solar Cells Based on Redox Active Monolayers Adsorbed on Nanocrystalline Oxide Semiconductor Films

**K. Kalyanasundaram and M. Grätzel**
*Swiss Federal Institute of Technology, Lausanne, Switzerland*

## I. INTRODUCTION

Photochemical conversion and storage of solar energy in the form of useful chemical fuels or electricity has drawn the attention of a large number of chemists in the last 3 decades. For many years the research work at Lausanne focused on the photochemistry in microheterogeneous media and their applications to energy conversion and storage (1–5). Early work using organized assemblies of micelles, vesicles, etc., was followed by studies of photoprocesses on semiconductors in the form of electrodes, particulates, and finely divided colloids (6–9). In recent years focus has been on nanocrystalline semiconducting oxide films, particularly on the development of a photovoltaic solar cell for the direct conversion of sunlight to electricity (10–13). The cell is based on the concept of sensitization of monolayers of dyes adsorbed onto nanocrystalline semiconductor films. Figure 1 shows the main components of this solar cell. Efficient solar cells with sunlight to electrical conversion efficiency around 10% have been developed (14,15).

Success in this area has been largely due to significant developments in the last 2 decades in a number of areas that form the main theme of this book: colloid, surface and photochemistry applied to microheterogeneous systems composed of micelles, microemulsions, finely divided semiconductor colloids, and particulates. Since an assessment of quarter century of

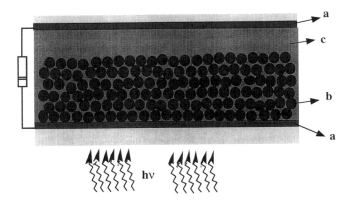

**Figure 1** Dye-sensitized photoelectrochemical cell based on nanocrystalline films of $TiO_2$. (a) Conducting glasselectrode, (b) $TiO_2$ with the adsorbed dye, (c) redox electrolyte ($I^-/I_3^-$).

progress is one of the goals of this book, in this paper we first review some of the earlier work that laid the foundations of the dye-sensitized solar cells, and this is followed by a description of the current state of the art in the design of the dye-sensitized solar cell.

## II. PHOTOVOLTAIC SOLAR CELLS

Various approaches have been proposed for light energy conversion over the years. Herein we focus on solar cells, systems designed for the direct conversion of sunlight to electricity. Earliest form of solar cells developed largely by the physicist community are solid state solar cells such as those based on Si. A conventional photovoltaic solar cell consists of two layers of a semiconducting material. One is chemically treated to have an excess of electrons (n-type) and the other carries an excess of positively charged holes (p-type). When the two layers are brought to contact, electrons flow from the n to the p side, producing an electric field at the interface. When photons of light energy from the sun strike the semiconductor, they excite the electrons to higher energy levels, leaving the "holes" behind. These electron-hole pairs recombine rapidly unless the electrons are carried away quickly to create useful electric current. When the electron-hole pairs are created near the p-n junction, the built-in electric field forces the positively charged holes to the p side and the negatively charged electrons to the n

# Dye-Sensitized Solar Cells

side. This movement of free charges causes a current to flow between the positive p region and the negative n region. Together with the voltage difference, electric power is generated. This in essence constitutes the principles of commercially available solar cells based on Si.

## III. PHOTOGALVANIC CELLS

Chemists have been interested in photoelectrochemical cells based on light-induced electron transfer reactions. Electron transfer reactions play an important role in chemical and biological systems and are extensively studied for this reason. Their importance is recognized with awards of the Noble Prize to Henry Taube and Rudy Marcus. In the last 2 decades there have been a number of studies devoted to photoredox reactions as well. In photoelectrochemical cells, visible light absorption leads to generation of charge-separated products, and these redox products are subsequently processed (reoxidized/reduced) at suitable electrodes with the generation of electricity. The earliest form of photoelectrochemical cells studied by chemists are based on photogalvanic effects, first proposed by Rabinowitch in the 1940s (16,17).

Figure 2 shows schematically the operation of a photogalvanic cell. Electronic excitation of an organic or inorganic dye in the presence of an

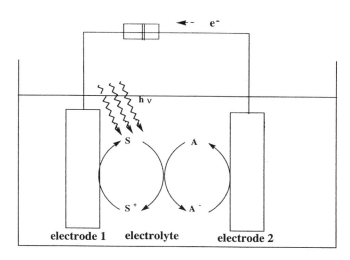

**Figure 2** Principles of operation of photogalvanic cell.

electron acceptor (or donor) in a homogeneous solution leads to electron transfer:

$$S + A \leftrightarrows S^+ + A^- \tag{1}$$

The photogenerated redox products are subsequently oxidized/reduced at suitable metal/catalytic electrodes:

$$S^+ + e^- \xrightarrow{\text{electrode 1}} S \tag{2}$$

$$A^- \xrightarrow{\text{electrode 2}} A + e^- \tag{3}$$

Thus, regeneration of the starting material at the electrodes is accompanied by conversion of light to electricity. Conceptually these are the simplest form of solar cells. Performance of these solar cells has been very poor for a number of reasons. Cells of this kind never gave sunlight to electrical conversion efficiency ($\eta$) even 0.5%.

It is useful to review some of the reasons for the poor performance of photogalvanic cells, since these factors do play an important role in the dye-sensitized photoelectrochemical cells we will discuss later. Firstly, kinetics of electron transfer reactions at the electrodes are much slower as compared to forward and back electron transfer steps of photoinduced processes. Hence, reasonable photocurrents can be obtained if and only if large differences in the forward and back electron transfer steps exist. If the electron transfer products do not live long, then only the light absorbed in the vicinity of the electrode is processed and most of the light absorbed in the bulk of the solution is lost. If the back electron transfer is 100 times slower than in the forward direction, then the equilibrium under steady state conditions will shift toward products by a small percentage—conditions under which even sluggish electrodes can intervene. For charge collection at high efficiencies the system must have several orders of magnitude difference between the forward and back electron transfer rates. Most often, in energy-storing reactions of interest, the back electron transfer is much faster than that in the forward direction.

Second, efficient performance of the photogalvanic cell requires high selectivity on the part of the electrodes (to carry out only the reactions as indicated in Reactions 2 and 3 above and not respond to other redox products present). If the electrodes are not selective, they only function as additional recombination centers for photogenerated products. Electrochemists addressed this problem using modified electrodes but only with very limited success. Attempts to increase performance by thick coating of the dye on the electrode yielded poor results. Thicker layers of the dye led to efficient intermolecular quenching of the excited states (process known as "concentration quenching") in competition with the desired electron transfer steps and "inner filter" effects.

## IV. PHOTOCHEMISTRY IN MICROHETEROGENEOUS MEDIA

Overall the photogalvanic cell system clearly showed the need to look for efficient ways of processing photoredox reactions. Key steps such as charge separation, rates of forward and back electron transfer, and charge transport to the collector electrodes must be controlled microscopically. Plant photosynthesis is the supreme example of how nature controls photoredox reactions using organized structures. Hence chemists resorted to model system studies on various organized and microheterogeneous media (such as micelles, vesicles, monolayers, cavities of zeolites, cyclodextrins, etc.). Significant advances have been made in a number of areas in the field of photochemistry in microheterogeneous systems. A number of monographs describing these results are available and hence we will confine ourselves to stating some of the key advances relevant to the topic under discussion (5, 18,19).

By appropriate design of the host system, it is possible to control the distribution of solutes at the microscopic level and also introduce constraints on their internal mobility. Also it is possible to manipulate the presence of charged interfaces at these microheterogeneous hosts to selectively adsorb or desorb reactants to the surface. These two effects provide means of controlling the efficiency of bimolecular processes between reactants and of their products. Simply by varying the ratio of micelle to quencher or choosing appropriately charged (anionic/cationic) micelles, it is possible to nearly eliminate collisional encounters of excited probe and the quenchers. It has been possible to nearly eliminate fluorescence quenching of solubilized dyes (including excimer formation) and enhance phosphorescence yields to levels such that it is possible to observe phosphorescence even at room temperature. Particularly striking are the effects on photoinduced processes. Cage escape yields of redox products and rate constants of photoinduced reactions have been found to vary by orders of magnitude by variations in the nature of the reactant distribution in the micropores/cavities of organized hosts.

The use of biomimetic systems to control charge separation process is modeled on the natural photosynthesis and can be illustrated with the following example. Consider an artificial system consisting of a sensitizer (chlorophyll *a*) and a quencher (duroquinone) cosolubilized in an anionic micelles of sodium lauryl sulfate. Photoexcitation of chlorophyll (Chl) leads to electron transfer to the acceptor quinone:

$$Chl^* + DQ \rightarrow Chl^+ + DQ^{-\cdot} \tag{4}$$

A radical ion pair is produced within the anionic micelle. The duroquinone anion ($DQ^{-\cdot}$) is clearly destabilized with respect to the aqueous bulk solution and will therefore be ejected into the water. Conversely $Chl^{+\cdot}$ cation is electrostatically stabilized by the micelle and remains associated with it (cf. Scheme 1). Once $Chl^{+\cdot}$ and $DQ^{-\cdot}$ are separated, their diffusional re-encounter will be obstructed by the ultrathin barrier of the micellar double layer. The efficiency of the charge separation process will critically depend on the relative rates of $DQ^{-\cdot}$ ejection and intramicellar back electron transfer from $DQ^{-\cdot}$ to $Chl^{+\cdot}$. Monitoring of the transient absorptions at wavelength corresponding to these redox species (465 and 685 nm) following a short laser-pulse excitation of the chlorophyll showed that the charge separation is indeed efficient.

## V. REGENERATIVE SOLAR CELLS BASED ON SEMICONDUCTOR ELECTRODES

Concurrent to the studies of photochemistry in microheterogeneous systems, attention was paid in the last 2 decades to photoelectrochemical cells based on semiconductor electrodes (20–23). The principles of operation of such cells are outlined in Figure 3 for an n-type semiconductor. Immersion of a semiconductor electrode in a redox electrolyte leads to an equilibration of the energy levels of the electrons in these materials (known as "Fermi level") with that of the electrolyte. This in turn leads to development of a depletion layer that allows separation of charge carriers. Irradiation of such semiconductor electrodes with light (of energy equal or greater than the bandgap energy) leads to promotion of an electron to upper energy level

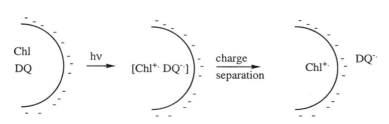

**Scheme 1** Photoinduced charge separation in an anionic micelle.

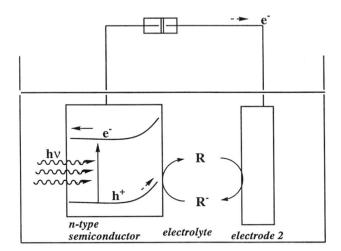

**Figure 3** Methods of preparation of nanocrytalline films of TiO$_2$.

(conduction band), leaving a positive hole behind in the valence band. The presence of the space charge region at the semiconductor–electrolyte interface provides an internal means of efficient charge separation. In n-type semiconductors the direction of the field is such that holes migrate toward the surface where they undergo chemical reaction (oxidation of the redox mediator). The electrons flow through the semiconductor layer to arrive at the back contact, flow through the external circuit and carry out the inverse reaction, namely reduction of the redox mediator. Thus the overall effect of bandgap irradiation is to drive electrons through the external circuit or conversion of light to electricity. In view of the key role played by the liquid junction interface, these cells are also known as *liquid junction solar cells*.

The efficiency of these regenerative solar cells for conversion of visible light to electricity depend on several factors: bandgap energy, efficiency of charge separation in the bulk of the semiconductor, charge transport through the semiconductor, and the efficiency of electrochemical redox processes at the illuminated semiconductor and the counterelectrode. Photogenerated holes are strong oxidants and if they are not processed rapidly, they cause undesirable oxidations, including that of the semiconductor itself (photocorrosion). Nonoxide semiconductors have low bandgap ($\leq 3$ eV) but they are prone to severe photocorrosion. Oxide semiconductors, in general, are more stable. But unfortunately their bandgaps are rather large ($\geq 3.2$ eV) to be of any utility for visible light irradiation. A number of regenerative solar cells with sunlight to electrical conversion efficiency in excess of 15% have been identified. Charge separation and transport processes work optimally with single crystal electrodes, and hence high efficiencies are obtained only using cells that use single crystal electrodes.

## VI. DYE SENSITIZATION OF SEMICONDUCTOR ELECTRODES

The photosensitized electron transfer across the semiconductor–solution interface plays a vital role in silver halide photography and in electroreprography (24,25). In the silver halide photography, organic dyes are used to extend the spectral range. The optimized efficiency of spectral sensitization is known to be very high. Sensitization of semiconductor electrodes using adsorbed monolayers of dyes has been examined as a simple model system by a number of scientists including Hauffe, Tributsch, Gerischer, Memming, Honda, and Tsubomura (26-28). The observed efficiency of sensitization was quite low in these model systems, due to two principal reasons: concentration quenching of the excited state of thicker layers of the dye on semiconductor layers (as has been observed earlier in photogalvanic cells) and enhanced nonradiative decay of excited states of the dye on semiconductor surfaces by excitonic processes. Thicker layers of the dye are also insulating in many cases. In recent years Spitler and Parkinson (29a,29b) adsorbed a number of organic dyes on two-dimensional layer-type dichalcogenides such as $MoSe_2$, $WSe_2$, and $SnS_2$ and obtained high photocurrent yields per photon absorbed. Light absorption by a monolayer of the dye is extremely low ($\ll 1\%$) and hence overall white light efficiency is rather small.

## VII. PHOTOINDUCED PROCESSES IN COLLOIDAL SEMICONDUCTORS

The confidence gained in the control of photoinduced processes on organized assemblies such as micelles and in regenerative solar cells led to investigation of similar processes on finely divided particulate dispersions and colloidal form of semiconductors (7-10). In colloidal and particulate semiconductor systems both the electrons and holes reach the surface and hence both oxidation and reduction processes are induced by light absorption by the semiconductor (cf. Scheme 2). Most often a sacrificial electron donor/

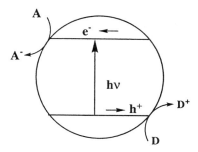

**Scheme 2** Photoinduced generation of electrons and holes in particulate semiconductors.

acceptor is used in large excess to rapidly scavenge one of the charge carriers (holes or electrons) so that the other is used to carry out some useful net chemistry.

Colloidal and particulate semiconductors exhibit several unique properties. The size of the colloidal particles is generally small enough to render their solutions optically translucent for convenient analysis of elementary steps involved in charge generation and subsequent trapping steps. The size and distribution of colloidal particles can be controlled by varying the experimental parameters such as concentration of the reactants, reaction temperature, and nature of the stabilizer and the medium. In colloidal particles, the charge separation process occurs largely by diffusion, and this diffusion of charge carriers from the interior to the particle surface occurs more rapidly than the recombination. Thus, it is feasible to obtain quantum yields for photoredox processes approaching unity. Whether such high efficiencies can really be observed depends on the rapid removal of at least one of the charge carrier upon their arrival at the interface. The translucent nature of the colloidal semiconductor solutions allows facile fast kinetic (laser pulse photolysis) and luminescence measurements outside the bandgap absorption region (6,9).

Luminescence observed from colloidal semiconductors serves as an important tool to study the dynamics of charge carrier recombination. Photogenerated charge carriers ($e^-/h^+$) move around within the semiconductor where they undergo recombination. This process is accompanied by an emission with energy close to the bandgap energy. If there are defect/chemical sites one of the charge carriers can be trapped, and results in recombination through these defect sites. The emission in this case is much lower in energy. Studies of quantum yields and lifetimes of luminescence provide a direct measure of the dynamics of charge recombination and trapping. Temperature dependence and sensitivity to quenching by adsorbed species allow probing the kinetic aspects of charge migration within the semiconductor. There been a number of studies on the above lines (mostly on dichalcogenides) and they provided rich information on charge carrier dynamics in colloidal and particulate semiconductor systems.

## VIII. TiO$_2$ As A Semiconductor

For most of our discussion hereafter, unless stated otherwise, we will use TiO$_2$ as the light-absorbing semiconductor. TiO$_2$ is one of the most extensively studied oxide materials with a bandgap energy of $\approx 3.2$ eV. Two crystalline forms readily accessible are rutile and anatase (the third form, brookite, is very difficult to synthesize). In addition to use as a photoelectrode, diverse applications of thin films are being developed, in areas such as capacitor dielectrics, heat-reflecting and UV-absorbing layers, coatings to improve chemical and mechanical stability of glass, etc. Since the reports of Fujishima and Honda (30) on the possible use of TiO$_2$ as a photoelectrode to decompose water to oxygen, there have been a large number of photoelectrochemical studies on this and related oxides. Basic studies of TiO$_2$ employed mainly single crystal (rutile) electrodes, but polycrystalline (anatase, rutile) electrodes prepared by variety of methods have also been investigated. Some of the most popular methods are aerosol pyrolysis of Ti-alkoxides on Ti support, anodic or thermal oxidation of Ti, RF-sputtering, and sintering of TiO$_2$ powders and colloids. Chemical vapor deposition methods employ titanium tetra(ethoxide), titanium tetra(isopropoxide), titanium acetylacetonate, and TiCl$_4$, which are mixed with additional gaseous reactants (O$_2$, H$_2$O, dopants).

## IX. PHOTOREDOX PROCESSES INVOLVING PARTICULATE SEMICONDUCTORS

A form of photocatalysis that is gaining increased importance for practical applications is usage of particulate semiconductors (micron-size) to carry out light-induced oxidation/reduction processes. Irradiation of semiconductor colloids and particulates leads to generation of electrons and holes. Photogenerated hole of TiO$_2$, for example, is a strong oxidant ($E_{ox} > +2$ eV) capable of oxidizing a large number of organic molecules. When photolysis is carried out in aerated/oxygenated solutions, the conduction band electrons can be rapidly removed, allowing the holes to carry out chemical oxidations of interest—degradation of toxic wastes and pollutants for example. Thus, polychlorinated toxic/carcinogenic compounds such as trichloroethylene can be efficiently degraded to $CO_2$ and HCl (to concentrations as low as 10 ppm). This form of application is come to be known as "heterogeneous photocatalysis" especially with widely abundant, cheap, nontoxic materials such as $TiO_2$ (4,7-9,31,32). One commercially available, widely used form of $TiO_2$, Degussa P-25, is prepared by flame/aerosol pyrolysis of $TiCl_4$. It is a complex oxide with coexisting amorphous, anatase, and rutile phase, and this unusual coexistence of several phases and moderately high surface area could account for its unusually high photocatalytic activity. The number of groups investigating applications in the area of environmental protection and fight against pollution is so many that we now have an international conference devoted to this subject.

## X. ADSORPTION/SURFACE CHELATION OF DYES

It is important to realize that the overall charge of the amphoteric oxide colloids in aqueous solution depends on the pH. The surface of oxides such as $TiO_2$ is filled with hydroxyl group and the degree of ionization of these groups is determined by the operational pH with respect to point of zero zeta potential (PZZP). At pHs below PZZP, the surface is positively charged and it becomes negatively charged at higher pHs (cf. Scheme 3).

Dyes can be either adsorbed or chemically derivatized to the oxide surface of the semiconductor. The adsorption of dyes is controlled largely by electrostatic forces. On $TiO_2$, for example, cationic dyes readily adsorp at pH ≥ 6 and anionic dyes strongly adsorb at pH ≤ 6. Depending on the method of preparation, the isoelectric point (IP) or the PZZP of $TiO_2$ is situated in the range of 4-6. Studies of aryl carboxylic acids adsorption on

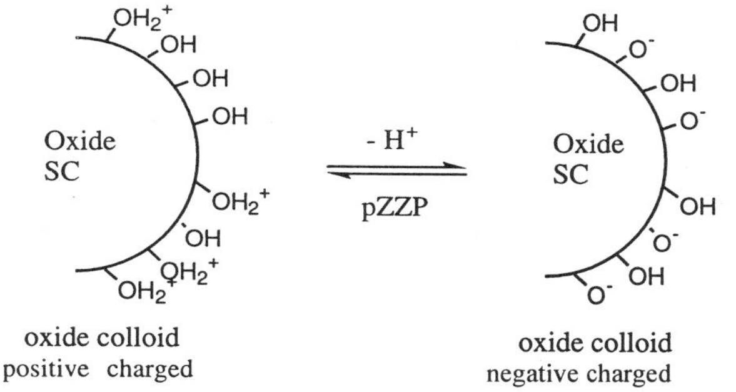

**Scheme 3** Point of zero zeta potential for oxide colloids in aqueous solutions.

the surface of colloidal $TiO_2$ showed that surface complexation of these molecules play an important role in promoting interfacial electron transfer. Laser photolysis studies on methyl viologen showed that at monolayer coverage $m$-phthalate increased the rate by 1700 times. Trapping of electrons by Ti(V) surface states and the removal of such traps by complexation of aryl derivatives is evoked to rationalize these observations. The positive role of chelated complexes in promoting the electron transfer has been exploited in the dye-sensitized solar cell by using photosensitizers that have one or more carboxyl groups, e.g., 4,4'-dicarboxy-2,2'-bypyridine complexes of Ru.

## XI. STUDIES OF DYE-SEMICONDUCTOR COLLOID MIXTURES

A logical extension to the studies of electron hole-transfer processes following bandgap excitation was to study the photophysics/photochemistry of dyes adsorbed onto the semiconductor colloids. Studies of luminescence and photoinduced electron processes using time-resolved techniques showed that, in suitable cases, efficient quenching of the excited state of the dye by electron transfer to the conduction band of the $TiO_2$ can be observed (6,9):

$$S^* \rightarrow S^+ + e^-_{cb} \tag{5}$$

In the absence of other reagents, the oxidized dye is reduced back to the ground state by the conduction band electrons. If we assume that the decreased excited state for the adsorbed dye is entirely due to the charge injection process and all other (radiative, nonradiative decay) occur at the same rate as in neat solvents, the observed lifetime can be correlated to a specific rate constant $k_{et}$ for electron by the expression:

$$1/\tau_{ads} = 1/\tau + k_{et} \tag{6}$$

where $\tau$ and $\tau_{ads}$ are the lifetimes of the excited state in neat solvent and adsorbed on the surface of $TiO_2$, respectively. Measured rate constants $k_{et}$ are typically in the range of $(4-30) \times 10^8 \, s^{-1}$.

Laser photolysis studies allow monitoring of the electron transfer process in the forward and reverse direction. Monitoring of the singlet and triplet excited state properties as a function of pH showed that the quenching of the excited state is intimately linked to the state of adsorption. Efficient quenching of excited state(s) takes place if and only if the dye is adsorbed onto the semiconductor. Studies with coumarin and Ru–bpy complexes have shown that the forward electron transfer (injection of electrons into the conduction band by the electronically excited state of the dye) is extremely rapid (less than a few picoseconds). The dye cation radical is

## Dye-Sensitized Solar Cells

formed with a yield of nearly 100%. The subsequent recombination reactions occurs over a much longer timescale (several microseconds or longer). Thus the conditions are ideal for light-induced electron transfer to occur with maximum quantum efficiency.

### XII. NANOCRYSTALLINE SEMICONDUCTOR ELECTRODES

Nanocrystalline semiconducting films or thin layers are being investigated in many laboratories. In addition to their use in solar cells, research focuses also on possible applications in photoconductors, electrochromic displays, optical and electrical switches, biosensors, and intercalation batteries to mention a few. Porous nanocrystalline films represent the current trend wherein there is growing interest in studying lower dimensional materials: three-dimensional homogeneous solid → multilayers with quasi-two-dimensional structure (e.g., A/B/A/B/A/B layers) → quasi-one-dimensional structures (nanowires in an insulating matrix) → nanoclusters with quasi zero-dimensional structure (clusters immersed in a solid or liquid suspension matrix) → porous nanocrystalline films.

The nanocrystalline semiconductor electrodes distinguish themselves by their porosity and high surface-to-volume ratio. Nanometer-sized porous films avoid one of the serious problems of dye distribution/limited light absorption encountered when dyes are coated onto flat semiconductor electrodes. Large effective surface area (1000-fold higher) allow efficient distribution of the dye as a monolayer retaining high light absorption properties. An intriguing feature of nanocrystalline $TiO_2$ films is that the charge transport of the photoinjected electrons passing through all the particles and grain boundaries is highly efficient, the quantum yields being nearly unity.

### XIII. PREPARATION OF NANOCRYSTALLINE $TiO_2$ FILMS

The preparation of $TiO_2$ films have been described in several publications (33a,33b). Figure 4 outlines some of the key steps involved in these methods. Nanocrystalline semiconductor films are often made two steps: prepare a colloidal solution of monodispersed nanosized particles of the semiconductor and then use the colloid to form a few micron-thick films with good electrical conduction properties on substrates of interest. Alternatively nanocrystalline particles are formed directly onto the substrate by an electrochemical or chemical deposition process.

In our laboratories, the electrodes are prepared by spreading a paste

| (a) | (b) | (c) |
|---|---|---|
| Ti(OR)$_4$<br>↓ hydroysis<br>Ti(OH)$_x$ + ROH<br>↓ peptisation<br>colloid<br>↓ autoclave<br>nanoparticles<br>↓ application<br>nanoporous films | TiCl$_4$ + O$_2$<br>↓ flame oxygenolysis<br>TiO$_2$ powder (P-25)<br>↓ Tx-100<br>dispersion<br>↓ application<br>nanoporous films | TiCl$_4$ + BCl$_3$<br>↓ Δ /ITO<br>CVD<br>TiO$_2$ + B$_2$O$_3$<br>↓ - HBO$_2$<br>nanoporous films |

**Figure 4** Schematic representation of a regenerative solar cell based on semiconductor electrodes. (a) In situ hydrolysis of Ti-alkoxides (sol-gel methods); (b) flame hydrolysis of TlCl$_4$; (c) chemical vapor deposition (CVD).

of nanosized colloidal TiO$_2$ particles on a conducting glass support (ITO). Sintering at 350–450°C produces electronic contact not only between the particles and the support but also between all the particles constituting the film. Thus a sponge-like structure is obtained and the colloidal TiO$_2$ film is porous (typically a porosity of 50% is achieved) from the outer layers to the ITO contact. The pores between the colloidal particles are interconnected and can be filled with an electrolyte, i.e., the semiconductor–electrolyte interface is accessible throughout the whole colloidal membrane. A roughness factor, defined as the ratio between the real and the projected surface of these films, of about 1000 has been estimated for a 10-μm-thick TiO$_2$ film. It should be pointed out that the porosity of the film can vary with the sintering temperature, dependent on the material. For example, ZnO nanocrystalline films can be made with a high porosity at sintering temperatures of ≈200°C, whereas a more compact film is obtained after sintering at 400°C. Care should therefore be taken in interpreting colloidal semiconductor films with respect to their porosity.

The optical properties of the TiO$_2$ films can be drastically changed due to the autoclaving step with implements for different devices. The purpose of the autoclaving is to increase the particle size and hence better diffusion properties in the electrolyte. Autoclaving a colloidal solution at 200°C yields a transparent nanocrystalline TiO$_2$ (100% anatase) with a relatively uniform particle size distribution of 15 nm. Increasing the autoclaving temperature to 250°C gives a film showing a high degree of light scattering. In this white film aggregates of small particles and usually some larger particles (>100 μm) are found and there is some content of rutile. To produce film thickness above 5 μm and for better ion diffusion in the electrolyte, it is an advantage if aggregates and/or larger aggregates are included in the colloidal paste. Therefore, the photovoltaic cell is based on

# Dye-Sensitized Solar Cells

a 200–250°C autoclaved colloidal solution with a film thickness of 10 μm, whereas the electrochromic electrode is made from a solution autoclaved at 200°C giving a transparent 3–4-μm-thick film.

## XIV. CHARGE SEPARATION PROCESSES IN NANOCRYSTALLINE TIO₂ FILMS

Figure 5 presents schematically the charge separation processes that occur in nanocrystalline oxide films following excitation of the semiconductor with light of energy equal or greater than the bandgap energy. There have been a number of studies devoted to the mechanism of charge separation in nanosized semiconductor films and their results have been reviewed elsewhere (34a,34b). Here we cite only some key points. In these nanocrystalline porous films, the electrolyte penetrates the whole colloidal film up to the surface of the back contact and a semiconductor–electrolyte junction occurs thus at each nanocrystal, much like a normal colloidal system. Dur-

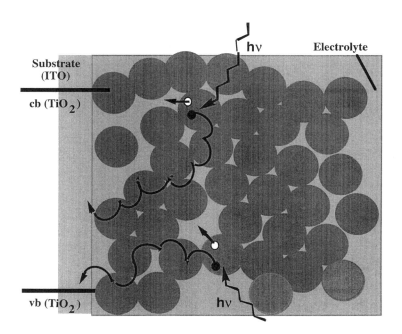

**Figure 5** Charge separation, transport processes in nanocrystalline semiconductor oxide films.

ing illumination, light absorption in any individual colloidal particle will generate an electron-hole pair. Assuming that the kinetics of charge transfer to the electrolyte is much faster for one of the charges (holes for $TiO_2$) than the recombination processes, the other charge (electrons) can create a gradient in the electrochemical potential between the particle and the back contact. In this gradient, the electrons (for $TiO_2$) can be transported through the interconnected colloidal particles to the back contact, where they are withdrawn as a current. The charge separation in a nanocrystalline semiconductor does not depend on a built-in electric field, i.e., a Schottsky barrier, but is mainly determined by kinetics at the semiconductor–electrolyte interface. The creation of light-induced electrochemical potential for the electrons in $TiO_2$ also explains the building up of a photovoltage.

There will be an increased probability of recombination with increased film thickness, as the electron has, on average, to be transported across an increasing number of colloidal particles and grain boundaries. This indeed has been observed experimentally. Thus there exists an optimal thickness to obtain maximum photocurrent. Another loss mechanism due to increasing film thickness is a resistance loss leading to a decrease in photovoltage and fill factor.

## XV. DYE-SENSITIZED PHOTOVOLTAIC CELL BASED ON NANOCRYSTALLINE FILMS

We have now discussed some of the key problems to be addressed when one attempts to utilize photoredox reactions for light energy conversion, particularly direct conversion to electricity. Studies in organized media and semiconductor dispersions and colloids allow manipulation of these. Efficient electron transfer quenching coupled with huge differences observed in the rate constants of forward and back electron transfer observed in dye–semiconductor colloid mixtures strongly suggested that a highly organized form of this may provide an efficient way of light energy harvesting. Results of electron transfer processes in thin nanosized porous films also appeared promising. The above factors in essence laid the foundations to our efforts to pursue dye-sensitized photoelectrochemical cells based on nanocrystalline films (10–13).

Figure 6 presents schematically the principles of operation of such a cell. Optical excitation of the dye molecule S leads to formation of its electronically excited state, which injects electrons into the conduction band of the semiconductor. The oxidized dye ($S^+$) is subsequently reduced back to its ground state S by an electron donor. The injected electron, after hopping through several semiconductor particles, arrives at the back con-

# Dye-Sensitized Solar Cells

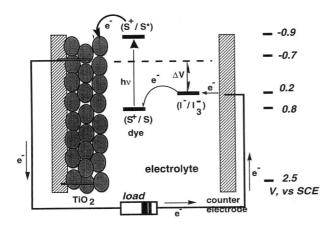

**Figure 6** Schematic presentation of the dye-sensitized photoelectrochemical cell.

tact. The electron then flows through the external load to the counter electrode where the reverse reaction (reduction of the oxidized donor/mediator) takes place. The overall effect of light irradiation is to drive an electron through the external circuit, that is, conversion of light to electricity. This form of the solar cell can be considered as a hybrid version of the two cells described earlier, viz., photogalvanic cells based on dye excitation in homogeneous solvents and regenerative cells involving direct/bandgap excitation of semiconductor electrode immersed in an redox electrolyte.

Indicated on the right of the schematic representation of the cell are the electron energy levels of the dye and the semiconductor at various stages. The data refer to $Ru(dcbpy)_2(SCN)_2$ as the photosensitizer adsorbed on to nanocrystalline $TiO_2$ films and $(I^-/I_3^-)$ as the redox electrolyte. The value of the redox potential (0.2V versus SCE) is the formal potential in the solution, whereas the oxidation potential of the dye is the standard potential. The absorption threshold of the Ru-complex adsorbed on $TiO_2$ is about 800 nm, corresponding to a potential of 1.6 V, which added to the oxidation potential of the dye gives a reduction potential of $-0.8$ V versus SCE. From the potential of the conduction bandedge, $V_{cb}$ and the bandgap energy $E_{bg}$ of 3.2 eV for $TiO_2$ (anatase), the potential of the valence band is obtained. The cell voltage observed under illumination corresponds to the difference, $\Delta V$, between the quasi-Fermi level of $TiO_2$ and the electrochemical potential of the electrolyte. S stands for the sensitizer, S* for its electronically excited form, and $S^+$ stands for the oxidized form of the dye.

## XVI. PERFORMANCE CHARACTERISTICS OF THE SOLAR CELL

The performance of the solar cell can be described using several criteria. The efficiency with which monochromatic photons are converted into electrons (photocurrent) is given by the *incident photon-to-current conversion efficiency* IPCE), defined as

$$\text{IPCE} = \frac{\text{number of electrons flowing through the external circuit}}{\text{number of photons incident}}$$

The IPCE values are determined by three factors: (1) light harvesting efficiency (depends on the spectral and photophysical properties of the dye), (2) the charge injection yield (depends on the excited state redox potential and the lifetime), and (3) the charge collection efficiency (depends on the structure and morphology of the $TiO_2$ layer). The dependence of the IPCE values on the excitation wavelength is given by the photocurrent action spectrum. The sunlight to electrical conversion efficiency ($\eta$) is determined by integrating the IPCE values over the solar spectrum.

As far as choice of sensitizers is concerned, our emphasis has been on polypyridyl complexes of Ru. These complexes exhibit strong visible light absorption arising from charge transfer transitions from filled d orbitals ($t_{2g}$) of the the central metal ion (Ru) to the empty $\pi^*$ orbitals bipyridine ligand. The MLCT excited states are fairly long-lived in fluid solutions and undergo efficient electron-transfer reactions. In the last 2 decades a vast amount of literature has grown on the photophysics and photochemical properties of these complexes (34a–34c). Quantitative analysis of the spectral, electrochemical, and photophysical properties of several hundred complexes have led to clear understanding of the CT transitions, that it is now feasible to tailor-make complexes with desired properties. Probably it is no exaggeration to state that most of our current understanding of excited state properties of metal complexes comes from the studies of polypyridine complexes. In order to expand the limited spectral response of mononuclear complexes, polynuclear complexes consisting of several chromophoric units are also being examined (35). The latter also serve as model systems for light-harvesting antenna.

In our studies, to promote efficient charge injection, one or more of the bipyridine ligands carried a carboxyl group. Among hundreds of complexes screened as potential sensitizers, two complexes have been identified to give maximum white-light conversion efficiency: [Ru(dcbpy)$_2$-(SCN)$_2$)] and [(CN)(bpy)Ru-CN-Ru(dcbpy)$_2$-CN-Ru(bpy)$_2$(CN)] (Scheme 4).

Figure 7 shows the photocurrent action spectrum obtained for $TiO_2$ films coated with some mononuclear Ru-complexes in a thin layer cell

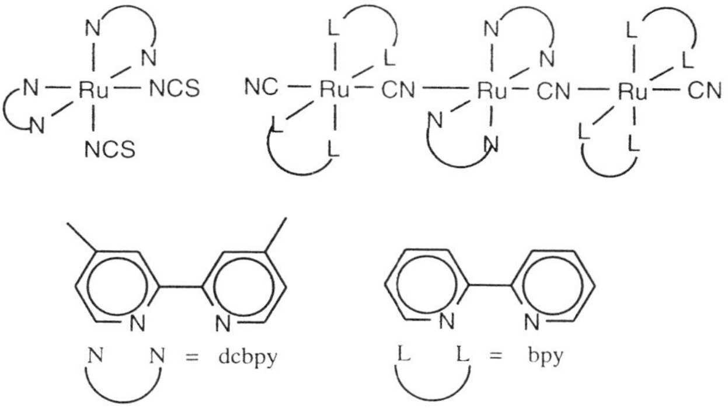

# Dye-Sensitized Solar Cells

**Scheme 4** Structures of polypyridyl complexes of Ru used as photosensitizers.

containing 0.03 M $I_2$ and 0.3 M LiI in acetonitrile as the redox electrolyte. The photocurrent was measured at short circuit where the $TiO_2$ film is poised at a potential around 0.1 V measured against SCE. The IPCE values exceeded 80% in the wavelength region between 480 and 600 nm, attaining plateau values of 85–90% in the 510–570 nm region. The yields were uncorrected for the absorption and scattering (ca. 10%) by the conducting glass substrate. Taking this into account, the conversion of photon flux into electrical current is nearly quantitative in this wavelength domain.

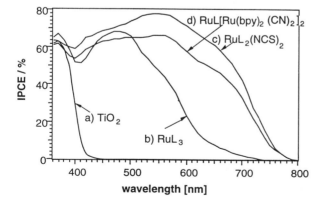

**Figure 7** Dependence of incident photon-to-current-conversion efficiency (IPCE) values with the excitation wavelength for various Ru-complexes and naked $TiO_2$ film in the solar cell.

**Table 1** Performance Characteristics of Solar Cells Based on Nanocrystalline $TiO_2$ Films Sensitized by [Ru(dcbpy)$_2$(SCN)$_2$)] Complex and [I$^-$/I$_3^-$] as the Redox Electrolyte

| Light intensity (mW/cm$^2$) | $i_{sc}$ (mA/cm$^2$) | $V_{oc}$ (mV) | ff | $\eta$(%) |
|---|---|---|---|---|
| 24.1 | 5.0 | 640 | 0.76 | 10.4 |
| 38.2 | 7.9 | 660 | 0.76 | 10.4 |
| 53.6 | 11.5 | 670 | 0.74 | 10.3 |
| 96.0 | 18.2 | 720 | 0.73 | 10.0 |

*Source:* Ref. 15.

Table 1 presents data on the performance characteristics of the solar cell based on the photosensitizer, [Ru(dcbpy)$_2$(SCN)$_2$], dcbpy = 4,4'-dicarboxy-2,2'-bipyridine. It may be recalled that the solar energy input at tropical midday is about 100 mW/cm$^2$. The overall light-to-electrical conversion efficiency is in the range of 10%. The short circuit current increases linearly with light intensity up to about 1 sun. This implies that the photocurrent is not limited by diffusion of the triiodide ions within the nanocrystalline film up to current densities of at least 20 mA/cm$^2$.

## XVII. OPTIMIZATION OF THE SOLAR CELL OUTPUT

The performance of the solar cell has been optimized using two important procedures. The first one concerns the reduction of redox mediator at the counterelectrode. Conducting glass (ITO) electrode, by itself, is a poor electrode material for the (I$^-$/I$_3^-$) redox couple. Its performance can be improved considerably by sputtering a shiny mirror/islands of Pt. A shiny mirror also reflects the light, ensuring maximum light absorption by the dye-coated layer. A light-reflecting counter-electrode increases the photocurrent yields in the long wavelength region (600–800 nm) where the dye absorption is rather poor.

The second treatment concerns optimization of the power output features of the cell. Exposure of the dye-coated electrode to a solution of an amine (donor) such as 4-*tert*-butylpyridine was found to improve dramatically the fill factor (ff) and the open-circuit voltage ($V_{oc}$) of the device without affecting the short-circuit photocurrent ($i_{sc}$) in a significant fashion. For example, the untreated electrode gave $i_{sc}$ = 17.8 mA/cm$^2$, $V_{oc}$ = 0.38 V and ff = 0.48, corresponding to an overall conversion efficiency ($\eta$) of

3.7%. After the electrode is dipped in 4-*tert*-butylpyridine, $V_{oc}$ increases to 0.66, ff to 0.63, and $\eta$ to 8.5%. The increase in the open circuit voltage and the fill factor is due to the suppression of the dark current at the semiconductor–electrolyte junction. The effect of the substituted pyridine can be rationalized in terms of its adsorption at the $TiO_2$ surface, blocking the surface states that are active intermediates in the heterogeneous charge transfer.

## XVIII. STABILITY

For practical applications, the stability or the long-term performance of the cell is an important parameter to consider. A weak point of the solar cells based on nanocrystalline $TiO_2$ films is their photosensitivity to the UV component of the sunlight. Membrane films produced from colloids contain ca. 30% rutile and the rest anatase. Rutile form is more sensitive ($E_{bg} \approx 3.0$ eV) as compared to anatase ($E_{bg} \approx 3.2$ eV), and a substantial amount of charge carriers can be generated by sunlight. As mentioned earlier, photogenerated holes are strong oxidants, capable of oxidizing the organic dye and the solvent. A polycarbonate film that cuts off all the light below 390 nm placed on the cell the stability is greatly enhanced. Long-term stability measurements have been, in general, very encouraging. In view of the organic components, the sensitizer can be considered as the weakest point. Based on the photocurrent obtained (charge passed through the cell) in laboratory experiments, it has been found that the sensitizer can sustain $10^7$–$10^8$ turnovers without degradation—turnovers corresponding to ca. 20 years' operation in sunlight.

## XIX. KINETICS OF ELECTRON TRANSFER

A quantitative picture of the light harvesting process on the solar cell is beginning to emerge. Figure 8 shows some of the important electron transfer steps involved in the operation. On the basis of different measurements it is possible to indicate the orders of magnitude for the rate constants of electron transfer steps involved. The primary *light absorption* step by the sensitizer giving rise to the formation of charge transfer excited state (step 1) is known to be extremely fast ($\approx$ femtoseconds). Following the light absorption of the Ru-complex, the *electron injection into the conduction band* (step 2) takes place in the picosecond range. Fast emission decay measurements indicate that the electron injection from Ru-dcbpy based

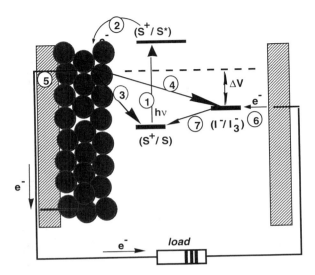

**Figure 8** Principal electron transfer steps involved in the operation of the dye-sensitized photoelectrochemical cell.

dyes adsorbed on $TiO_2$ to be ultrafast ($\leq$ ps) with a quantum yield of $\approx 100\%$. The rate constant for the back electron transfer (step 3) however is much smaller for several reasons, typically several microseconds. By transient absorption measurements it is possible to directly follow this back electron transfer process. Another important recombination process is *reduction of $I_3^-$* (step 4) in the electrolyte by conduction band electrons. The exchange current density, $j_o$, of the reverse saturation current of this process has been measured recently. Values between $10^{-11}$–$10^{-9}$ A/cm$_2$ were obtained depending on the electrolyte. Surface treatment of the electrode can alter these values drastically.

The *electron movement in the nanocrystalline $TiO_2$* electrode to the back contact (step 5) is significantly slower than in single crystal $TiO_2$. Measurements in our laboratory have shown that the photocurrent transients, following UV excitation of $TiO_2$ particles from a nanosecond pulsed laser, decay in the millisecond to second range. The exchange current density for the reduction of triiodide at the counter electrode ITO (step 6) coated with a catalytic amount of Pt has been measured to be $(0.01-2) \times 10^{-1}$ A/cm$^2$. Finally, the reduction of the oxidized dye by iodide (step 7) occurs on a timescale of $10^{-8}$ s.

## XX. APPLICATIONS OF NANOCRYSTALLINE $TiO_2$ FILMS

Mention was made earlier on some of the possible applications one can envisage for nanocrystalline films. In Lausanne we have been pursuing some of these avenues as well. The anodes of lithium batteries employ oxides such as $LiMO_2$ (M = Co, Ni, or Mn) and sulfides (e.g., $TiS_2$) in the form of micron-sized particles compressed pellets mixed with carbon and a polymeric binder. The morphology of the anode is such that large pores/channels present therein allow reversible insertion of Li and extraction into and from the lattice. Studies on nanocrystalline $TiO_2$ electrodes (36a,36b) showed that efficient, reversible, and rapid intercalation of lithium occurs and the materials can serve as electrochromic switches.

Viologens form a group of redox indicators that undergo drastic color changes upon oxidation/reduction. The reduced form of methyl viologen, for example, is deep blue while the oxidized form is colorless. Efficient reduction of adsorbed viologen compounds by conduction band electrons of $TiO_2$ can be used as a basis of electrochromic display. Upon electroreduction, transparent nanocrystalline film of $TiO_2$ containing viologen develop strong color, and the film can be decolorized by reversing the potential. Varying the chemical structure and redox potentials of the viologens, it is possible to tune the color and hence build a series of electrochromic display devices. Early results in this area appear very promising.

## XXI. PROGRESS AND NEW HORIZONS

In the first half of this chapter, progress in some of the key areas pertinent to the dye-sensitized solar cell were reviewed. Our understanding of the structure and dynamics of organized microstructures (be it a self-assembled one such as micelles or vesicles or colloidal/particulate dispersions of the semiconductor) has improved considerably in the last 25 years, so that it is now possible to design tailored experiments wherein the distribution and local structure of the reactants can be controlled. Such tailor-made systems allow exploring of novel applications. Nanocrystalline semiconductor films technology is a growing field. Nanosized films are finding increasing applications in a number of electronic/high-tech areas and some of these were cited in the article. Microengineering, a long-sought dream of chemists, is fast becoming a reality. Organized microstructures are no longer scientific curiosities of a few individuals. As we approach the 21st century, it is very likely that they form the backbone of a number of devices that will be part of our day-to-day life.

As mentioned in the introductory paragraphs, it is difficult to visualize evolution of dye-sensitized solar cell that we currently examine in the absence of basic colloid, surface, and photochemical studies on microheterogeneous systems of various kinds. It is a typical example of the situation where technological applications grow out of solid foundations laid in the form of fundamental studies.

## REFERENCES

1. M. Grätzel, ed. Energy Resources through Photochemistry and Catalysis. New York, Academic Press, 1983.
2. J Kiwi, K Kalyanasundaram, M Grätzel. Struc Bonding 49:37, 1982.
3. K Kalyanasundaram, M Grätzel, E Pelizzetti. Coord Chem Rev 69:57, 1986.
4. K Kalyanasundaram. In: Photoelectrochemistry, Photocatalysis and Photoreactors. NATO Adv Study Inst Series, Vol C146. Dordrecht, Reidel Publishers, 1985, p 219.
5. K Kalyanasundaram. Photochemistry in Microheterogeneous Systems. New York, Academic Press, 1986.
6. PV Kamat. In: Kinetics and Catalysis in Microheterogeneous Systems. Kalyanasundaram, K, Grätzel, M, eds. Surfactant Science Series, Vol. 38, New York, Marcel Dekker, 1992, p 375.
7. N Serpone, E Pelizzetti, eds. Photocatalysis: Fundamentals, Applications. New York, Wiley, 1992.
8a. A Henglein. Chem Rev 89:1861, 1989.
8b. A Henglein. Top Curr Chem 143:113, 1988.
9. PV Kamat. Progr React Kinet 19:277, 1994.
10. J DeSilvestro, M Grätzel, L Kavan, J Augustynski. J Am Chem Soc 107:2988, 1985.
11. N Vlachopoulos, P Liska, J Augustynski, M Grätzel. J Am Chem Soc 110: 1216, 1988.
12. P Liska, N Vlachopoulos, MK Nazeeruddin, P Comte, M Grätzel. J Am Chem Soc 110:3686, 1988.
13. MK Nazeeruddin, P Liska, J Moser, N Vlachopoulos, M Grätzel. Helv Chim Acta 73:1788, 1990.
14. B O'Regan, M Grätzel. Nature 353:737, 1991.
15. MK Nazeeruddin, A Kay, I Rodicio, R Humphry-Baker, E Müller, P Liska, N Vlachopoulos, M Grätzel. J Am Chem Soc 115:6382, 1993.
16. E Rabinowitch. J Chem Phys 8:551,560, 1940.
17. WJ Albery. Acc Chem Res 15:142, 1982.
18. V Ramamurthy, ed. Photochemistry in Organized and Constrained Media. New York, VCH Publishers, 1993.
19. R Zana, ed. Surfactant Solutions: New Methods of Investigation. Surfactant Science Series, Vol. 22, New York, Marcel Dekker, 1987.
20. K Kalyanasundaram. Solar Cells 15:93, 1985.

21. HO Finklea. Semiconductor Electrodes. Amsterdam, Elsevier, 1988.
22. YV Pleskov, YY Guervich. Semiconductor Photoelectrochemistry (translation from Russian). New York, Consultants Bureau, 1986.
23a. R Memming. In: Electroanalytical Chemistry, Bard, AJ, ed. New York, Marcel Dekker, 1979, p. 1.
23b. AJ Nozik. Ann Rev Phys Chem 29:189, 1978.
23c. AJ Bard. J Phys Chem 86:172, 1982.
23d. MS Wrighton. Acc Chem Res 12:303, 1979.
23e. A Heller. Acc Chem Res 14:154, 1981.
23f. AJ Bard, MA Fox. Acc Chem Res 28:141, 1995.
24. TH James, ed. Theory of Photographic Processes. 4th edition. New York, Macmillan, 1977.
25. JW Weigl. Angew Chem Intennat Edn 16:374, 1977.
26. F Willig, H Gerischer. Top Curr Chem 61:31, 1976.
27. K Hauffe. Photograph Sci Eng 20:124, 1976.
28. M Grätzel, K Kalyanasundaram. In: Photosensitization and Photocatalysis Using Inorganic and Organometallic Compounds, Kalyanasundaram, K, Grätzel, M, eds. Dordrecht, Kluwer Academic Publishers, 1993, p 247.
29a. MT Spitler, BA Parkinson. Langmuir 2:549, 1986.
29b. MT Spitler. J Electroanal Chem 228:69, 1987.
30. A Fujishima, K Honda. Nature (London) 238:37, 1972.
31. MA Fox, M Chanon, eds. Photoinduced Electron Transfer. Part D, Inorganic Substrates, Applications). New York, Elsevier, 1991.
32. MA Fox. Acc Chem Res 16:314, 1983.
33a. L Kavan, M Grätzel. Electrochim Acta 40:643, 1995, and refs. cited therein.
33b. L Kavan, A Kay, B O'Regan, M Grätzel. J Electroanal Chem 346:291, 1993.
34a. K Kalyanasundaram. Photochemistry of Polypyridine and Porphyrin Complexes. New York, Academic Press, 1993.
34b. A Juris, F Barigelletti, V Balzani, S Campagna, P Belser, A von Zelewsky. Coord Chem Rev 84:85, 1988.
34c. K Kalyanasundaram. Coord Chem Rev 46:159, 1982.
35. V Balzani, F Scandola. Supramolecular Photochemistry. Chicester, U.K., Ellis Horwood, 1991.
36a. A Hagfeldt, N Vlachopoulos, M Grätzel. J Electrochem Soc 141:L82, 1994.
36b. S-Y Huang, L Kavan, A Kay, M Grätzel. I Exnar. Active Passive Elec Comp 19:23–30, 1995.

# Index

Abnormal water, 59
Acoustic optical modulator (AOM), 549
Adsorption, 589
Aggregation, 463
Aggregation number, 10
Alkali-surfactant-polymer, 249
Alkane-carbon-number (ACN), 235
Amphiphile, 459
Amphiphilic molecule, 519
Amphiphilic squaraines, 463
Anisotropic, 527
Anomalous phase, 73, 94
Anomalous swelling, 85
Anti conformation, 57
Antiferromagnetic, 457
Aromatic polymers, 519
Artificial systems, 509
Atomic force microscope (AFM), 481, 501, 549, 562
Azeotropic discontinuity, 88
Azeotropic point, 88

Bicontinuous, 168, 194
Bicontinuous sponge phase, 107
Bilayer lipid membrane, 290
Bimolecular lipid membranes, 509
Biocompatibility, 501, 510
Biomimetic system, 583
Biorecognition reaction, 511
Biotin, 513
Biotinylated, 513
Birefringence, 145
Bis-squaraines, 472
Bjerrum transition, 158
Black liquid phase, 388
Blue I phase, 74
*Bordetella pertussis*, 495
Brewster angle microscopy (BAM), 449
Brownian fluctuations, 193

Capillary forces, 250
Capillary number, 250, 280
Chain length compatibility
 foams, 34, 11, 34
 microemulsions, 34

**605**

Colloidal microcrystaline cellulose, 345
Composite, 327
Contaminated soils, 221
Copolymer, 327
Coreflood assembly, 267
Cosolvent, 305, 307, 308
Coulombic repulsion, 16
Critical micelle concentration (CMC), 290, 298
Critical temperature, 21
Cross-linked micelles, 103
Cumulative oil recovery, 267, 269
Curvature, 169
Cyclic voltammetry, 524

Darcy velocity, 250
Deinococcus radiodurans, 449
Depletion phenomenon, 134
Derjaguin-Landau equation, 135
Desthiobiotin, 513
Destructible surfactant, 69
Dichroic ratio, 548
Dielectric constant, 188
Diffusion order, 351
Dilong-chain surfactant, 85
Dipole moment, 548
Dipole–dipole coupling, 472
Double emulsion, 359
Double relaxation, 62
Droplet clustering, 103
Droplet size, 7
Dry spots, 39
Dye-semiconductor, 590
Dye sensitization, 586
Dynamic interfacial tension, 6
Dynamic surface tension, 3, 7, 433

Effective diffusion coefficient, 341
Effective pair interaction, 142
Effective surfactant layer thickness, 377
Electrochemical measurements, 525

Electroluminescent device, 535
Electron conversion efficiency, 596
Electron diffraction (ED), 550, 563
Electron paramagnetic resonance (EPR), 457
Electrooptical effects, 543
Emulsification, 6, 7, 12
Emulsions, 333
   porous media, 21
   pressure drop in emulsion flow, 22
Enhanced oil recovery (EOR), 15, 175, 250
Enzymatic reaction, foam, 30, 33
Enzymatic synthesis, 30
Equilibrium surface density, 429
*Escherichia coli*, 491
Ethoxylogs, 77
Extracellular surface, 491
Extractions, 226

Facial amphiphile, 67
Fermi level ($E_F$), 482
Fingering, 252
First order transition, 442
Flooding surfactant polymer, 17
Fluid nanoscopic photoreaction vessels, 64
Fluorescein, 42
Foam, 33, 39
Foam enzymatic reaction, 33
Foaming ability, 3, 4
Foam surfactant concentration, 4
Frictional coefficient, 505
Functionalization, 509

Ganglia, 309
Garti's modification, 349
Gauche confirmation, 57
Gibbs triangle, 216
Gracing incidence x-ray scattering (GEKS), 550

# Index

Gauche confirmation, 57
Gibbs triangle, 216
Gracing incidence x-ray scattering (GEKS), 550
Groove model, 567

H aggregate, 463
Hard sphere, 138
H dimerization, 474
Hexagonal close-packed, 492
Hexagonally packed intermediate (HPI), 491
Hoffman prescription, 158
Hydrodynamic effect, 48
Hydrophilic–lipophilic balance (HLB), 93, 172, 236, 334
Hydrophobes, 233
Hydrophobic effect, 151
Hydrophobic–hydrophilic interface, 463
Hydrophobicity, 308
Hydrophobic spacer, 354
Hyperextended tails, 69

Imidization reaction, 553
Immunoglobulin, 495
Infrared dichroism, 553
Initial dead time, 433
Initial delay time, 433
Interaggregate interaction, 121
Interfacial area, 31
Interfacial concentration, 58
Interfacial tension, 16, 20
    oil–water interface, 16
    ultra, 16
Interfacial viscosity, 16
Interferogram, 128
Interparticle forces, 363
Interparticle spacing, 291
Interphase, 509
Inverse micelles, 343
Isobaric stability, 547, 553

Isoplethal method, 91
Isoplethal phase studies method, 96
Isotherm, 546
J aggregate, 463

Lamellar phases, 165
Langmuir–Blodgett (LB) films, 447, 463, 484, 519, 544
Langmuir monolayers, 437, 449, 545
Laplace pressure, 415
Laser photolysis, 590
Lateral force, 502
Lattice spacing, 443
Lattice structures, 437
LC alignment, 566
Lecithins, 27
Levitational effect, 48
Light-emitting layer, 535
Light-harvesting antenna, 596
Lipid monolayer, 39
Liquid crystal displays (LCD), 543
Liquid crystalline phases, 152
Liquid dispersion, 364
Liquid junction solar cells, 585
Local density of states (LDOS), 482
Loss modulus, 149
Lower phase, 306, 307, 308
Luminescence, 587
Lysozyme, 489

Magnetic nanoparticles, 289, 305, 306, 579
Magnetic phase transitions, 458
Mean molecular area (MMA), 424
Medium effect, 64
Meissner effect, 22
Meissner shieldings, 22
Mercaptopyridine, 516
Micellar aggregates, 1
Micellar custering, 103

[Micelles]
   intermicellar distance, 9
   mixed, 11
   mixed oppositely charged surfactant, 11
   organic solvents, 66
   reactions, 64, 65
   relaxation time (T1, T2), 1, 62
   solubilization capacity, 61
Microdomains, 319
Microemulsion globule, 316
Microemulsions, 12, 161, 193, 215, 216, 233
   enhanced oil recovery, 15
   formation of nanoparticles, 19, 24
   phase transitions, 14
   reactions, 43
Microheterogeneous media, 463, 579
Microlatexes, 325
Microlatex particles, 48
Middle phase, 307, 308
Minireactors, 182
Miscibility gap, 81, 83, 84
Mixed monolayers, 32
Mixed polymer solutions, 184
Mixed surfactant, 36
Molecular anisotropy, 559
Monochromator crystal, 438
Monolayer, 26, 27, 291
   enzymatic reactions, 30
   intermolecular distance, 29
   phase transition, 32
   retardation of evaporation, 32
   thickness, 48
Monomers, 1
Monte Carlo, grand canonical ensemble (GCEMC), 138
Morphology, 550
Multilayer assemblies, 514
Mustard, 201

Myelinic texture, 85, 86

Nanocrystalline semiconductor films, 591
Nanoparticles, 19
Nanosize particles, 291, 293, 299
National Synchrotron Light Source (NSLS), 550
Natural interfacial curvature, 367
Nematic liquid crystals, 558
Nematic-isotropic transition, 551
$N$-hydroxysuccinimide, 497
Non-DLVO surface forces, 127
Nuclear magnetic resonance (NMR), 167

Octadecyltrichlorosilane (OTS), 450
Oil gangila, 177
Optical interferences, 551
Optical phase retardation, 549
Optimal salinity, 234, 309, 310
Organic synthesis, 182
Orientation, 527
Orientational epitaxy (OE), 567
Original oil in place, 250
Ornstein–Zernike theory, 135

Packing density, 487
Packing parameter concept, 171
Paraffin balls, 60
Particulate semiconductors, 588
Peritectic line, 83
Phase boundries, 442
Phase transfer reagents, 184
Phospholipids, 491
Photochemical conversions, 579
Photochemistry, 579
Photocurrent, 589
Photoinduced process, 586
Photoreaction units (PRU), 492
Photovoltaic cells, 581
Pivotal structure, 91

# Index

Plasmon peak, 294
Plastic crystalline, 365
Point of zero zeta potential (PZZP), 596
Polar anchoring energy, 551, 572
Polarized light, 528
Polybenzothiazoles, 531
Polycyclic aromatic hydrocarbon (PAH), 215
Polydispersity, 291, 293
Polymer incompatibility, 112
Polymerization, 193, 200
Porous media, 17, 21
Potentio-galvanostat, 525
Precipitation, 20
Precursors method, 520, 546
Pressure-area curves, 546
Pressure-area isotherm, 467
Pretilt angle measurements, 568
Primary micelles, 103
Protein complex photo system, 485
Protein immobilization, 188
Protein rejecting surfaces, 189
Pseudoemulsion, 397
Pseudophase model, 63

Quantum dots, 289
Quasi-elastic light scattering (QELS), 201
Quencher, 583

Radial distribution function, 139
Reactant/micelle association constant, 63
Real diffusion coefficient, 341
Redox reaction, 525
Regioselectivity, 187
Relative hydrophilicity, 74
Relative lipophilicity, 74
Relaxation time, 1
    surfactant concentration, 2
Remanence value, 302
Residual monomers, 203

Reverse micellar mechanism, 339
Reverse, micelles, 290, 299, 348, 463
Rheological properties, 147
*Rhodopseudomonas viridis*, 492
Rodlike micelles, 60
Rubbing process, 567

Salinity, 18, 19, 22, 306
    optimal, 23
Saturated fatty acid, 437
Scanning electron microscope (SEM), 23
Scanning force microscope (SFM), 481
Scanning probe microscope (SPM), 481
Scanning tunneling microscope (STM), 481
Secondary micelles, 103
Self-assembled, 501
Semiconductors, 579
Shadowing effect, 573
Shear-induced phase transitions, 503
Shear viscosity, 145
Skeletonization, 447
Slow relaxation time, 62
Soil remediation, 215
Solid micelles, 153
Solid-solution interface, 513
Solubilization parameter, 255
Soluble oil, 309
Soluble surfactant behavior, 93
Solution properties, 8
Specular reflectivity, 438
Spherical micelles, 60
Sponge phases, 74, 94
Spontaneous curvature, 121
Spreading coefficient, 395
Squeeze-out hypothesis, 420
Stern region, 66
Sticky collisions, 319
Storage modulus, 149
Streptavidin, 514
String-of-beads, 103

Superconductors, 27
Super molecular machine, 527
Superstructure phases, 390
Supertwisted nematic (STN), 543
Supramolecular architectures, 510
Surface alignment, 544
Surface bound ligand, 511
Surface chelation, 589
    sensitizer, 583
Surface potential, 29
Surface pressure, 28, 29, 528
Surface refining hypothesis, 420
Surface rheology, 502
Surfactant enhanced alkaline
    flooding, 258
Surfactant monomers, 1
Surfactant phase, 73, 74, 77, 307
Surfactant–polymer flooding, 17
Swollen, 305
Synchrotron radiation, 437, 447

Tear film, 39, 44
    thinning, 44
Thermal disturbance, 37
Thermal gelation, 115
Total acid number (TAN), 257, 262
Transferred films, 447
Transmission electron microscope
    (TEM), 21, 147, 481, 550, 563
Tubules, 153

Tunable surface, 510
Tunneling current, 482
Tunneling spectroscopy (TS), 492
Twisted nematic (TN), 543
Two-phase mixtures, 73

Ultralow interfacial tension, 13, 233
Undulation, 384
Unit aggregate, 465
Upper phase, 306, 308

Vertical dipping technique, 529
Vesicles, 60, 153, 463
Virial coefficient, 134
Viscoelastic, 502
Viscoelasiticity, 151
Viscous forces, 250

Wettability, 501
    reversal, 280
Wetting, 5
    surfactant concentration, 5
Winsor phases, 147
Worm-like micelles, 162

X-ray diffraction, 437, 534, 550
X-ray photoelectron spectroscopy
    (XPS), 450
X-ray reflectivity, 550, 565